Computer Processing of Remotely-Sensed Images

Computer Processing of Remotely-Sensed Images

An Introduction

Third Edition

Paul M. Mather

The University of Nottingham

CD-ROM exercises contributed by Magaly Koch, Boston University.

John Wiley & Sons, Ltd

Telephone (+44) 1243 779777

Email (for orders and customer service enquiries): cs-books@wiley.co.uk
Visit our Home Page on www.wileyeurope.com or www.wiley.com

Reprinted July 2005

Other Wiley Editorial Offices

John Wiley & Sons Inc., 111 River Street, Hoboken, NJ 07030, USA

Jossey-Bass, 989 Market Street, San Francisco, CA 94103-1741, USA

Wiley-VCH Verlag GmbH, Boschstr. 12, D-69469 Weinheim, Germany

John Wiley & Sons Australia Ltd, 33 Park Road, Milton, Queensland 4064, Australia

John Wiley & Sons (Asia) Pte Ltd, 2 Clementi Loop #02-01, Jin Xing Distripark, Singapore 129809

John Wiley & Sons Canada Ltd, 22 Worcester Road, Etobicoke, Ontario, Canada M9W 1L1

Wiley also publishes its books in a variety of electronic formats. Some content that appears in print may not be available in electronic books.

Library of Congress Cataloging-in-Publication Data

Mather, Paul M.
Computer processing of remotely-sensed images: an introduction / Paul M. Mather.–3rd ed.
p. cm.
Includes bibliographical references and index.
ISBN 0-470-84918-5 (cloth : alk. paper) – ISBN 0-470-84919-3 (pbk.: alk. paper)
1. Remote sensing–Data processing. I. Title.

G70.4.M38 2004
621.36′78–dc22 2004005079

British Library Cataloguing in Publication Data

A catalogue record for this book is available from the British Library

ISBN 10: 0-470-84919-3 (P/B)
ISBN 13: 978-0-470-84919-4 (P/B)
ISBN 10: 0-470-84918-5 (H/B)
ISBN 13: 978-0-470-84918-7 (H /B)

Typeset in 9/11 pt Times by TechBooks, New Delhi, India
Printed and bound in Great Britain by Antony Rowe Ltd, Chippenham, Wiltshire
This book is printed on acid-free paper responsibly manufactured from sustainable forestry in which at least two trees are planted for each one used for paper production.

'I hope that posterity will judge me kindly, not only as to the things which I have explained but also as to those which I have intentionally omitted so as to leave to others the pleasure of discovery.'

René Descartes

'I am none the wiser, but I am much better informed.'

Queen Victoria

Contents

CONTENTS OF CD-ROM

Preface to the First Edition

Environmental remote sensing is the measurement, from a distance, of the spectral features of the Earth's surface and atmosphere. These measurements are normally made by instruments carried by satellites or aircraft, and are used to infer the nature and characteristics of the land or sea surface, or of the atmosphere, at the time of observation. The successful application of remote sensing techniques to particular problems, whether they be geographical, geological, oceanographic or cartographic, requires knowledge and skills drawn from several areas of science. An understanding of the way in which remotely sensed data are acquired by a sensor mounted onboard an aircraft or satellite needs a basic knowledge of the physics involved, in particular environmental physics and optics. The use of remotely sensed data, which are inherently digital, demands a degree of mathematical and statistical skill plus some familiarity with digital computers and their operation. A high level of competence in the field in which the remotely sensed data are to be used is essential if full use of the information contained in those data is to be made. The term "remote sensing specialist" is thus, apparently, a contradiction in terms, for a remote-sensing scientist must possess a broad range of expertise across a variety of disciplines. While it is, of course, possible to specialise in some particular aspect of remote sensing, it is difficult to cut oneself off from the essential multidisciplinary nature of the subject.

This book is concerned with one specialised area of remote sensing, that of digital image processing of remotely-sensed data but, as we have seen, this topic cannot be treated in isolation, and so Chapter 1 covers in an introductory fashion the physical principles of remote sensing. Satellite platforms currently or recently in use, as well as those proposed for the near future, are described in Chapter 2, which also contains a description of the nature and sources of remotely-sensed data. The characteristics of digital computers as they relate to the processing of remotely sensed image data is the subject of chapter 3. The remaining five chapters cover particular topics within the general field of the processing of remotely-sensed data in the form of digital images, and their application to a range of problems drawn from the Earth and environmental sciences. Chapters 1 to 3 can be considered to form the introduction to the material treated in later chapters.

The audience for this book is perceived as consisting of undergraduates taking advanced options in remote sensing in universities and colleges as part of a first degree course in geography, geology, botany, environmental science, civil engineering or agricultural science, together with postgraduate students following taught Masters courses in remote sensing. In addition, postgraduate research students and other research workers whose studies involve the use of remotely sensed images can use this book as an introduction to the digital processing of such data. Readers whose main scientific interests lie elsewhere might find here a general survey of this relatively new and rapidly developing area of science and technology. The nature of the intended audience requires that the formal presentation is kept to a level that is intelligible to those who do not have the benefit of a degree in mathematics, physics, computer science or engineering. This is not a research monograph, complete in every detail and pushing out to the frontiers of knowledge. Rather, it is a relatively gentle introduction to a subject which can, at first sight, appear to be overwhelming to those lacking mathematical sophistication, statistical cunning or computational genius. As such it relies to some extent on verbal rather than numerical expression of ideas and concepts. The author's intention is to provide the foundations upon which readers may build their knowledge of the more complex and detailed aspects of the use of remote sensing techniques in their own subject rather than add to the already extensive literature which caters for a mathematically-orientated readership. Because of the multidisciplinary nature of the intended audience, and since the book is primarily concerned with techniques, the examples have been kept simple, and do not assume any specialist knowledge of geology, ecology, oceanography, or other branch of Earth science. It is expected that the reader is capable of working out potential applications in his or her own field, or of following up the references given here.

It is assumed that most readers will have access to a digital image processing system, either within their own department or institution, or at a regional or national remote-sensing centre. Such processors normally have a built-in software package containing programs to carry out most, if not all, of the operations described in this book. It is hoped that the material presented here will provide such readers with the background necessary to make sensible use of the facilities at their disposal. Enough detail is provided, however, to allow interested readers to develop computer programs to implement their own ideas or to modify the operation of a standard package. Such ventures are to be encouraged, for software skills are an important part of the remote sensing scientist's training. Furthermore, the development and testing of individual ideas and conjectures provides the opportunity for experiment and innovation, which is to be preferred to the routine use of available software. It is my contention that solutions to problems are not always to be found in the user's manual of a standard software package.

I owe a great deal to many people who have helped or encouraged me in the writing of this book. Michael Coombes of John Wiley and Sons took the risk of asking me to embark upon the venture, and has proved a reliable and sympathetic source of guidance as well as a model of patience. The Geography Department, University of Nottingham, kindly allowed me to use the facilities of the Remote Sensing Unit. I am grateful also to Jim Cahill for many helpful comments on early drafts, to Michael Steven for reading part of chapter 1 and for providing advice on some diagrams, to Sally Ashford for giving a student's view and to George Korybut-Daszkiewicz for his assistance with some of the photography. An anonymous referee made many useful suggestions. My children deserve a mention (my evenings on the word processor robbed them of the chance to play their favourite computer games) as does my wife for tolerating me. The contribution of the University of Nottingham and the Shell Exploration and Development Company to the replacement of an ageing PDP11 computer by a VAX 11/730-based image processing system allowed the continuation of remote sensing activities in the university and, consequently, the successful completion of this book. Many of the ideas presented here are the result of the development of the image processing software system now in use at the University of Nottingham and the teaching of advanced undergraduate and Masters degree courses. I am also grateful to Mr. J. Winn, Chief Technician, Mr. C. Lewis and Miss E. Watts, Cartographic Unit, and Mr. M.A. Evans, Photographic Unit, Geography Department, University of Nottingham, for their invaluable and always friendly assistance in the production of the photographs and diagrams. None of those mentioned can be held responsible for any errors or misrepresentations that might be present in this book; it is the author's prerogative to accept liability for these.

Paul M. Mather *March 1987*

Remote Sensing Unit
Department of Geography
The University
Nottingham NG7 2RD

Preface to the Second Edition

Many things have changed since the first edition of this book was written, more than ten years ago. The increasing emphasis on scientific rigour in remote sensing (or Earth observation by remote sensing, as it is now known), the rise of interest in global monitoring and large-scale climate modelling, the increasing number of satellite-borne sensors in orbit, the development of Geographical Information Systems (GIS) technology, and the expansion in the number of taught Masters courses in GIS and remote sensing are all noteworthy developments. Perhaps the most significant single change in the world of remote sensing over the past decade has been the rapid increase in and the significantly reduced cost of computing power and software available to students and researchers alike, which allows them to deal with growing volumes of data and more sophisticated and demanding processing tools. In 1987 the level of computing power available to researchers was minute in comparison with that which is readily available today. I wrote the first edition of this book using a BBC Model B computer, which had 32 Kb of memory, 100 Kb diskettes and a processor which would barely run a modern refrigerator. Now I am using a 266 Mz Pentium II with 64 Mb of memory and a 2.1 Gb disc. It has a word processor that corrects my spelling mistakes (though its grammar checking can be infuriating). I can connect from my home to the University of Nottingham computers by optic fibre cable and run advanced software packages. The cost of this computer is about one percent of that of the VAX 11/730 that is mentioned in the preface to the first edition of this book.

Although the basic structure of the book remains largely unaltered, I have taken the opportunity to revise all of the chapters to bring them up to date, as well as to add some new material, to delete obsolescent and uninteresting paragraphs, and to revise some infelicitous and unintelligible passages. For example, Chapter 4 now contains new sections covering sensor calibration, plus radiometric and topographic correction. The use of artificial neural networks in image classification has grown considerably in the years since 1987, and a new section on this topic is added to chapter 8, which also covers other recent developments in pattern recognition and methods of estimating Earth surface properties. Chapter 3, which provides a survey of computer hardware and software, has been almost completely re-written. In Chapter 2 I have tried to give a brief overview of a range of present and past sensor systems but have not attempted to give a full summary of every sensor, because details of new developments are now readily available via the World Wide Web. I doubt whether anyone would read this book simply because of its coverage of details of individual sensors.

Other chapters are less significantly affected by recent research as they are concerned with the basics of image processing (filtering, enhancement, and image transforms), details of which have not changed much since 1987, though I have added new references and attempted to improve the presentation. I have, however, resisted the temptation to write a new chapter on GIS, largely because there are several good books on this topic that are widely accessible (for example, Bonham-Carter (1994) and McGuire *et al* (1991)), but also because I feel that this book is primarily about image processing. The addition of a chapter on GIS would neither do justice to that subject nor enhance the reader's understanding of digital processing techniques. However, I have referred to GIS and spatial databases at a number of appropriate points in the text. My omission of a survey of GIS techniques does not imply that I consider digital image processing to be a 'stand-alone' topic. Clearly, there are significant benefits to be derived from the use of spatial data of all kinds within an integrated environment, and this point is emphasised at a number of places in this book. I have added a significant number of new references to each of the chapters, in the hope that readers might be encouraged to enjoy the comforts of his or her local library.

I have added a number of 'self-assessment' questions at the end of each chapter. These questions are not intended to constitute a sample examination paper, nor do they provide a checklist of 'important' topics (the implication being that the other topics covered in the book are unimportant). They are simply a random set of questions – if you can answer them then you probably understand the contents of the chapter. Readers should use the MIPS software described in Appendices A and B to try out the methods mentioned in these questions. Datasets are also available on the accompanying CD, and are described in Appendix C.

Perhaps the most significant innovation that this book offers is the provision of a CD containing software and images. I am not a mathematician, and so I learn by trying out ideas rather than exclusively by reading or listening. I learn new methods by writing computer programs and applying them to various data sets. I am including a small selection of the various programs that I have produced over the past 30 years, in the hope that others may find them useful. These programs are described in Appendix B. I have been teaching a course on remote sensing for the last 14 years. When this course began there were no software packages available, so I wrote my own (my students will remember NIPS, the Nottingham Image Processing System, with varying degrees of hostility). I have completely rewritten and extended NIPS so that it now runs under Microsoft Windows 95. I have renamed it as Mather's Image Processing System (MIPS), which is rather an unimaginative name, but is nevertheless pithy. It is described in Appendix A. Many of the procedures described in this book are implemented in MIPS, and I encourage readers to try out the methods discussed in each chapter. It is only by experimenting with these methods, using a range of images, that you will learn how they work in practice. MIPS was developed on an old 486-based machine with 12 Mb of RAM and a 200 Mb disc, so it should run on most PCs available in today's impoverished universities and colleges. MIPS is not a commercial system, and should be used only for familiarisation before the reader moves on to the software behemoths that are so readily available for both PCs and UNIX workstations. Comments and suggestions for improving MIPS are welcome (preferably by e-mail) though I warn readers that I cannot offer an advisory service nor assist in research planning!

Appendix C contains a number of Landsat, SPOT, AVHRR and RADARSAT images, mainly extracts of size 512×512 pixels. I am grateful to the copyright owners for permission to use these data sets. The images can be used by the reader to gain practical knowledge and experience of image processing operations. Many university libraries contain map collections, and I have given sufficient details of each image to allow the reader to locate appropriate maps and other back-up material that will help in the interpretation of the features shown on the images.

The audience for this book is seen to be advanced undergraduate and Masters students, as was the case in 1987. It is very easy to forget that today's student of remote sensing and image processing is starting from the same level of background knowledge as his or her predecessors in the 1980s. Consequently, I have tried to restrain myself from including details of every technique that is mentioned in the literature. This is not a research monograph or a literature survey, nor is it primarily an exercise in self-indulgence and so some restriction on the level and scope of the coverage provided is essential if the reader is not to be overwhelmed with detail and thus discouraged from investigating further. Nevertheless, I have tried to provide references on more advanced subjects for the interested reader to follow up. The volume of published material in the field of remote sensing is now very considerable, and a full survey of the literature of the last 20 years or so would be both unrewarding and tedious. In any case, online searches of library catalogues and databases are now available from networked computers. Readers should, however, note that this book provides them only with a background introduction – successful project work will require a few visits to the library to peruse recent publications, as well as practical experience of image processing.

I am most grateful for comments from readers, a number of whom have written to me, mainly to offer useful suggestions. The new edition has, I hope, benefited from these ideas. Over the past years, I have been fortunate enough to act as supervisor to a number of postgraduate research students from various countries around the world. Their enthusiasm and commitment to research have always been a factor in maintaining my own level of interest, and I take this opportunity to express my gratitude to all of them. My friends and colleagues in the Remote Sensing Society, especially Jim Young, Robin Vaughan, Arthur Cracknell, Don Hardy and Karen Korzeniewski, have always been helpful and supportive. Discussions with many people, including Mike Barnsley, Giles Foody and Robert Gurney, have added to my knowledge and awareness of key issues in remote sensing. I also acknowledge with gratitude the help given by Dr. Magaly Koch, Remote Sensing Center, Boston University, who has tested several of the procedures reported in this book and included on the CD. Her careful and thoughtful advice, support, and encouragement have kept me from straying too far from reality on many occasions. My colleagues in the School of Geography in the University of Nottingham continue to provide a friendly and productive working environment, and have been known occasionally to laugh at some of my jokes. Thanks especially to Chris Lewis and Elaine Watts for helping to sort out the diagrams for the new edition, and to Dee Omar for his patient assistance and support. Michael McCullagh has been very helpful, and has provided a lot of invaluable assistance. The staff of John Wiley and Sons has been extremely supportive, as always. Finally, my wife

Rosalind deserves considerable credit for the production of this book, as she has quietly undertaken many of the tasks that, in fairness, I should have carried out during the many evenings and weekends that I have spent in front of the computer. Moreover, she has never complained about the chaotic state of our dining room, nor about the intrusive sound of Wagner's music dramas. There are many people, in many places, who have helped or assisted me; it is impossible to name all of them, but I am nevertheless grateful. Naturally, I take full responsibility for all errors and omissions.

Paul M. Mather *June 1998*
Nottingham

Preface to the Third Edition

In the summer of 2001, I was asked by Lyn Roberts of John Wiley and Sons to prepare a new edition of this book. Only minor updates would be needed, I was told, so I agreed. A few weeks later was presented with the results of a survey of the opinions of the 'great and the good' as to what should be included in and what should be excluded from the new edition. You are holding the result in your hands. The 'minor updates' turned into two new chapters (a short one on computer basics, replacing the old Chapter 3, and a lengthier one on the advanced topics of interferometry, imaging spectroscopy and lidar, making a new Chapter 9) plus substantial revisions of the other chapters. In addition, I felt that development of the MIPS software would be valuable to readers who did not have access to commercial remote sensing systems. Again, I responded to requests from postgraduate students to include various modules that they considered essential, and the result is a Windows-based package of 90 000+ lines of code.

Despite these updates and extensions both to the text of the book and the accompanying software, my target audience is still the advanced undergraduate taking a course in environmental remote sensing. I have tried to introduce each topic at a level that is accessible to the reader who is just becoming aware of the delights of image processing while, at the same time, making the reasonable assumption that my readers are, typically, enthusiastic, aware and intelligent, and wish to go beyond the basics. In order to accommodate this desire to read widely, I have included an extensive reference list. I am aware, too, that this book is used widely by students taking Masters level courses. Some of the more advanced material, for example in chapters 6, 8 and 9, is meant for them; for example, the new material on wavelets and developments in principal components analysis may stimulate Masters students to explore these new methods in their dissertation work. The first three chapters should provide a basic introduction to the background of remote sensing and image processing; Chapters 4 to 8 introduce essential ideas (noting the remark above concerning parts of Chapters 6 and 8), while chapter 9 is really for the postgraduate or the specialist undergraduate.

I am a firm believer in learning by doing. Reading is not a complete substitute for practical experience of the use of image-processing techniques applied to real data that relates to real problems. For most people, interest lies in the meaning of the results of an operation in terms of the information that is conveyed about a problem rather than in probing the more arcane details of particular methods, though for others it is the techniques themselves that fascinate. The level of mathematical explanation has therefore been kept to a minimum and I have attempted to use an approach involving examples, metaphors and verbal explanation. In particular, I have introduced a number of examples, separate from the main text, which should help the reader to interpret image-processing techniques in terms of real-world applications.

Many of these examples make use of the MIPS software that is provided on the CD that accompanies this book. MIPS has grown somewhat since 1999, when the second edition of this book was published. It has a new user interface, and is able to handle images of any size in 8-, 16- or 32-bit representation. A number of new features have been added, and it is now capable of providing access to many of the techniques discussed in this book. I would appreciate reports from readers about any difficulties they experience with MIPS, and I will maintain a web site from which updates and corrections can be downloaded. The URL of this web site, and my e-mail address, can be found in the file *contactme.txt* which is located in the root directory of the CD.

Many of the ideas in this book have come from my postgraduate students. Over the past few years, I have supervised a number of outstanding research students, whose work has kept me up to date with new developments. In particular, I would like to thank Carlos Vieira, Brandt Tso, Taskin Kavzoglu, Premelatha Balan, Mahesh Pal, Juazir Hamid, Halmi Kamarrudin and Helmi Shafri for their tolerance and good nature. Students attending my Masters classes in digital image processing have also provided frank and valuable feedback. I would also like to acknowledge the valuable assistance provided by Rosemary Hoole and Karen Laughton of the School of Geography, University of Nottingham. The help of Dr Koch of Boston University, who made many useful comments on the manuscript and the MIPS software as they have progressed, is also gratefully acknowledged, as is the kindness of Professor J. Gumuzzio and his group at the Autonomous University of Madrid for allowing me access to DAIS images of their La Mancha study site. Dr. Koch has also provided a set of four advanced exercises, which can be found in the Examples folder of the accompanying CD. I am very grateful to her for this contribution, which I am sure significantly enhances the value of this book. Magaly Koch as also provided a set of four advanced exercises, which can be found in the *Examples* folder of the CD. I am very grateful to her for this contribution which – I am sure – will significantly enhance the value of the book. My wife continues to tolerate what she quietly considers to be my over-ambitious literary activities, as well as my predilection for the very loudest bits of Mahler, Berlioz, Wagner and others. Colleagues and students of the School of Geography, University of Nottingham, have helped in many ways, not least by humouring me. Finally, I would like to thank Lyn Roberts, Keily Larkins, and the staff of John Wiley who have helped to make this third edition a reality, and showed infinite patience and tolerance.

A book without errors is either trivial or guided by a divine hand. I cannot believe that this book is in the latter category, and it is possible that it is not in the former. I hope that the errors that you do find, for which I take full responsibility, are not too serious and that you will report them to me.

Paul M. Mather *August 2003*
Nottingham

paul.mather@nottingham.ac.uk

List of Examples

1

Remote Sensing: Basic Principles

'Electromagnetic radiation is just basically mysterious.'

B.K. Ridley (*Time, Space and Things*, (second edition). Cambridge: Cambridge University Press, 1984)

1.1 INTRODUCTION

The science of remote sensing comprises the analysis and interpretation of measurements of electromagnetic radiation that is reflected from or emitted by a target and observed or recorded from a vantage point by an observer or instrument that is not in contact with the target. Earth observation (EO) by remote sensing is the interpretation and understanding of measurements made by airborne or satellite-borne instruments of electromagnetic radiation that is reflected from or emitted by objects on the Earth's land, ocean, or ice surfaces or within the atmosphere, and the establishment of relationships between these measurements and the nature and distribution of phenomena on the Earth's surface or within the atmosphere. An important principle underlying the use of remotely-sensed data is that different objects on the Earth's surface and in the atmosphere reflect, absorb, transmit or emit electromagnetic energy in different proportions, and that such differences allow these components to be identified. Sensors mounted on aircraft or satellite platforms record the magnitude of the energy flux reflected from or emitted by objects on the Earth's surface. These measurements are made at a large number of points distributed either along a one-dimensional profile on the ground below the platform or over a two-dimensional area below or to one side of the ground track of the platform. Figure 1.1 shows an image being collected by a nadir-looking sensor.

Data in the form of one-dimensional profiles are not considered in this book, which is concerned with the processing of two-dimensional (spatial) data collected by imaging sensors. Imaging sensors are either nadir (vertical) or side looking. In the former case, the ground area to either side of the point immediately below the satellite or aircraft platform is imaged, while in the latter case an area of the Earth's surface lying to one side of the satellite or aircraft track is imaged. The most familiar kinds of images, such as those collected by the nadir-looking Thematic Mapper (TM) and Enhanced Thematic Mapper Plus (ETM+) instruments carried by US Landsat series of Earth satellites, or by the HRV instrument (which can be side-looking or nadir-pointing) onboard the SPOT satellites, are scanned line by line (from side to side) as the platform moves forward along its track. This forward (or along track) motion of the satellite or aircraft is used to build up an image of the Earth's surface by the collection of successive scan lines (Figure 1.1).

Two kinds of scanners are used to collect the electromagnetic radiation that is reflected or emitted by the ground surface along each scan line. Electro-mechanical scanners have a small number of detectors, and they use a mirror that moves back and forth to collect electromagnetic energy across the width of the scan line (AB in Figure 1.2(a)). The electromagnetic energy reflected by or emitted from the portion of the Earth's surface, which is viewed at a given instant in time, is directed by the mirror onto these detectors (Figure 1.2(a)). The second type of scanner, the push-broom scanner, uses an array of solid-state charge-coupled devices (CCDs), each one of which 'sees' a single point on the scan line (Figure 1.2(b)). Thus, at any given moment, each detector in the CCD array is observing a small area of the Earth's surface along the scan line. This ground area is called a *pixel.* A remotely-sensed image is made up of a rectangular matrix of measurements of the flux or flow of electromagnetic radiation (EMR) emanating from individual pixels, so that each pixel value represents the magnitude of upwelling electromagnetic radiation for a small ground area. This upwelling radiation contains information about (i) the nature of the Earth surface material present in the pixel area, (ii) the topographic position of the pixel area (i.e., whether it is horizontal, on a sunlit slope or on a shaded slope) and (iii) the state of the atmosphere through which the EMR has to pass. This account of image acquisition is a very simplified one, and more detail is provided in Chapter 2. The nature of Earth surface materials and their interaction with EMR are covered in section 1.3. Topographic and atmospheric interactions are described in sections 4.7 and 4.4, respectively.

Computer Processing of Remotely-Sensed Images: An Introduction, Third Edition. Paul M. Mather.
 ISBNs: 0-470-84918-5 (HB); 0-470-84919-3 (PB)

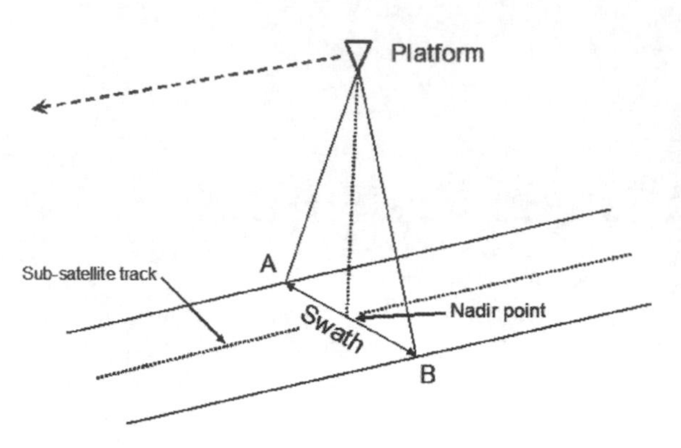

Figure 1.1 A sensor carried onboard a platform, such as an Earth-orbiting satellite, builds up an image of the Earth's surface by taking repeated measurements across the swath AB. As the satellite moves forward, successive lines of data are collected and a two-dimensional image is generated. The distance AB is the swath width. The point immediately below the platform is the nadir point, and the imaginary line traced on the Earth's surface by the nadir point is the sub-satellite track.

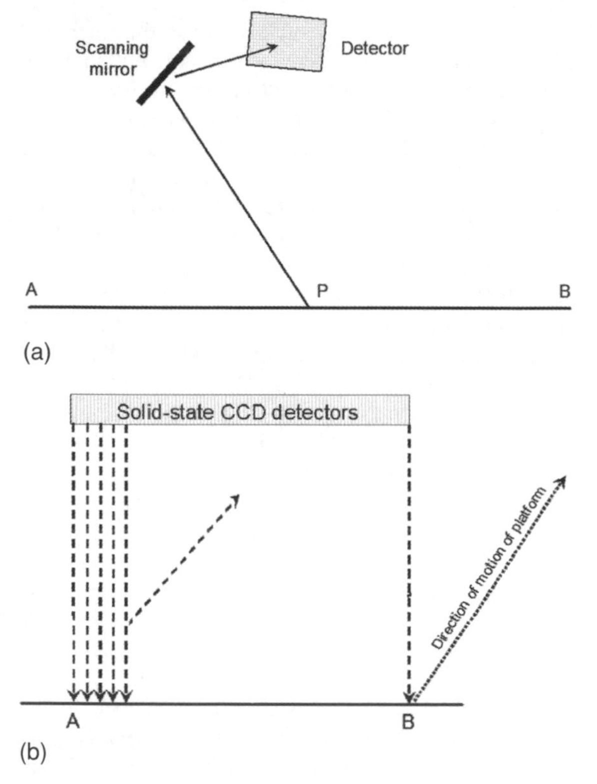

Figure 1.2 (a) Upwelling energy from point P is deflected by a scanning mirror onto the detector. The mirror scans across a swath between points A and B on the Earth's surface. (b) An array of solid-state (CCD) detectors images the swath AB. The image is built up by the forward movement of the platform.

The magnitude of the radiance reflected or emitted by the small ground area represented by a pixel is represented numerically by a number, usually a nonzero integer (a whole number) lying within a specified range, such as 0–255. Remotely sensed images thus consist of rectangular arrays of numbers, and because they are numerical in nature, so computers are used to display, enhance and manipulate them. The main part of this book deals with techniques used in these types of processing. Spatial patterns evident in remotely-sensed images can be interpreted in terms of geographical variations in the nature of the material forming the surface of the Earth. These Earth surface materials may be vegetation, exposed soil and rock, or water surfaces. Note that the characteristics of these materials are not detected directly by remote sensing. Their nature is inferred from the properties of the electromagnetic radiation that is reflected, scattered or emitted by these materials and recorded by the sensor. Another characteristic of digital image data is that they can be calibrated so as to provide estimates of physical measurements of properties of the target such as radiance, reflection or albedo. These values are used, for example, in models of climate or crop growth. Examples of the uses of remotely-sensed image data in Earth science and environmental management can be found in Calder (1991). Kaufman *et al.* (1998) demonstrate the wide variety of applications of remote sensing data collected by the instruments on the American Terra satellite. A number of World Wide Web sites provide access to image libraries for those with computers connected to the Internet. A good starting point is my own page of links at

http://www.geog.nottingham.ac.uk/~ mather/
useful_links.html

a copy of which is included on the accompanying CD. The NASA tutorial by Nicholas Short at

http://code935.gsfc.nasa.gov/Tutorial/TofC/
Coverpage.html

is also recommended.

Aerial photography is a familiar form of Earth observation by remote sensing. Past generations of air photographs differ from digital images in that they are analogue in nature. The term 'analogue' means: using some alternative physical representation to display some property of interest. For example, a photographic film represents a range of light levels, in terms of the differential response of silver halide particles in the film emulsion. Analogue images cannot be processed by computer unless they are converted to digital form, using a scanning device. Computer scanners operate much in the same way as those carried by satellites in that they view a small area of the photograph, record the proportion of incident light that is reflected back by that small area, and convert that proportion to a number, usually in the range 0 (no reflection, or black) to 255 (100% reflection, or white). The numbers between 0 and 255 represent increasingly lighter shades of grey.

Nowadays, digital cameras are increasingly being used in aerial photography. Images acquired by such cameras are similar in nature to those produced by the push-broom type of sensor mentioned above. Instead of a film, a digital camera has a two-dimensional array of CCDs (rather than a one-dimensional CCD array, as used by the SPOT satellite's HRV instrument, mentioned above). The amount of light from the scene that impinges on an individual CCD is recorded as a number in the range 0 (no light) to 255 (detector saturated). A two-dimensional set of CCD measurements produces a greyscale image. Three sets of CCDs are used to produce a colour image, just as three layers of film emulsion are used to generate an analogue colour photograph. The three sets of CCDs measure the amounts of red, green and blue light that reach the camera. Nowadays, digital imagery is relatively easily available from digital cameras, from scanned analogue photographs, as well as from sensors carried by aircraft and satellites.

The nature and properties of electromagnetic radiation are considered in section 1.2, and are those which concern its interaction with the atmosphere, through which the electromagnetic radiation passes on its route from the Sun (or from another source such as a microwave radar) to the Earth's surface and back to the sensor mounted onboard an aircraft or satellite. The interactions between electromagnetic radiation and Earth surface materials are summarised in section 1.3. It is by studying these interactions that the nature and properties of the material forming the Earth's surface are inferred.

1.2 ELECTROMAGNETIC RADIATION AND ITS PROPERTIES

1.2.1 Terminology

The terminology used in remote sensing is sometimes understood only imprecisely, and is therefore occasionally used loosely. A brief guide is therefore given in this section. It is neither complete nor comprehensive, and is meant only to introduce some basic ideas. The subject is dealt with more thoroughly by Bird (1991a, b), Chapman (1995), Elachi (1987), Rees (1990), Slater (1980) and Schowengerdt (1997). Note that the term *electromagnetic radiation* is abbreviated to *EMR* in the following sections.

Electromagnetic radiation (EMR) transmits energy. As the name implies, EMR has two components. One is the electric field; the other is the magnetic field. These two

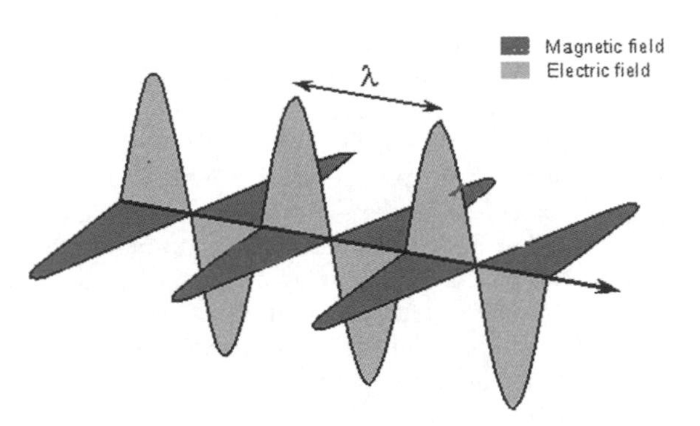

Figure 1.3 Electromagnetic energy behaves as if it is composed of two component waves, representing magnetic and electric fields, which are at right angles to each other. If the electric field is vertical (i.e., moving in the Y–Z plane) and the magnetic field is horizontal (i.e., moving in the X–Z plane) then the energy is vertically polarised. If the positions are reversed, with the electric field in the X–Z plane and the magnetic field in the Y–Z plane, then the signal is horizontally polarised. The energy is propagating in the Z direction, with the transmitting antenna located at the origin of the coordinate axes. Y is the Earth normal vector, which is a line from the centre of the Earth to the antenna position. The symbol λ indicates the wavelength of the electromagnetic energy. Adapted with permission from a figure by Nick Strobel at http://www.astronomynotes.com/light/s2.htm.

fields are mutually perpendicular, and are also perpendicular to the direction of travel (Figure 1.3). There is no 'right way up' – EMR can be transmitted with a horizontal electric field and a vertical magnetic field, or vice versa. The disposition of the two fields is described by the polarisation state of the EMR, which can be either horizontal or vertical. Polarisation state is used in microwave remote sensing (section 2.4).

Energy is the capacity to do work. It is expressed in *joules* (J), a unit that is named after James Prescott Joule, an English brewer whose hobby was physics. *Radiant energy* is the energy associated with EMR. The rate of transfer of energy from one place to another (for example, from the Sun to the Earth) is termed the *flux* of energy, the term derived from the Latin word meaning 'flow'. It is measured in *watts* (W), after James Watt (1736–1819), the Scottish inventor who was instrumental in designing an efficient steam engine while he was working as a technician at Glasgow University (he is also credited with developing the first rev counter). The interaction between electromagnetic radiation and surfaces such as that of the Earth can be understood more clearly if the concept of *radiant flux density* is introduced. Radiant flux is the rate of transfer of radiant (electromagnetic) energy. Density implies variability over the two-dimensional surface on which the radiant energy falls, hence radiant flux density is the magnitude of the radiant flux that is incident upon or, conversely, is emitted by a surface of unit area (measured in watts per square metre or W m^{-2}). The topic of emission of EMR by the Earth's surface in the form of heat is considered at a later stage. If radiant energy falls (is incident) upon a surface then the term *irradiance* is used in place of radiant flux density. If the energy flow is away from the surface, as in the case of thermal energy emitted by the Earth or solar energy that is reflected by the Earth, then the term *radiant exitance* or *radiant emittance* (measured in units of W m^{-2}) is appropriate.

The term *radiance* is used to mean the radiant flux density transmitted from a unit area on the Earth's surface as viewed through a unit solid (three dimensional) angle (just as if you were looking through a hole at the narrow end of an ice-cream cone). This solid angle is measured in *steradians*, the three-dimensional equivalent of the familiar radian (defined as the angle subtended at the centre of a circle by a sector which cuts out a section of the circumference that is equal in length to the radius of the circle). If, for the moment, we consider that the irradiance reaching the surface is backscattered in all upward directions (Figure 1.4(b)), then a proportion of the radiant flux would be measured per unit solid viewing angle. This proportion is the radiance (Figure 1.5). It is measured in watts per square metre per steradian (W m^{-2} sr^{-1}). The concepts of the radian and steradian are illustrated in Figure 1.6.

Reflectance, ρ, is the dimensionless ratio of the irradiance and the radiant emittance of an object. The reflectance

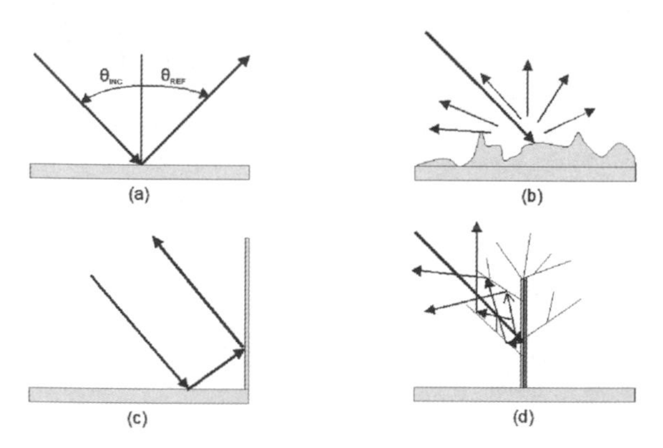

Figure 1.4 Types of scattering of electromagnetic radiation. (a) Specular, in which incident radiation is reflected in the forward direction, (b) Lambertian, in which incident radiation is equally scattered in all upward directions, (c) corner reflector, which acts like a vertical mirror, especially at microwave wavelengths and (d) volume scattering, in which (in this example) branches and leaves produce primary and secondary scattering.

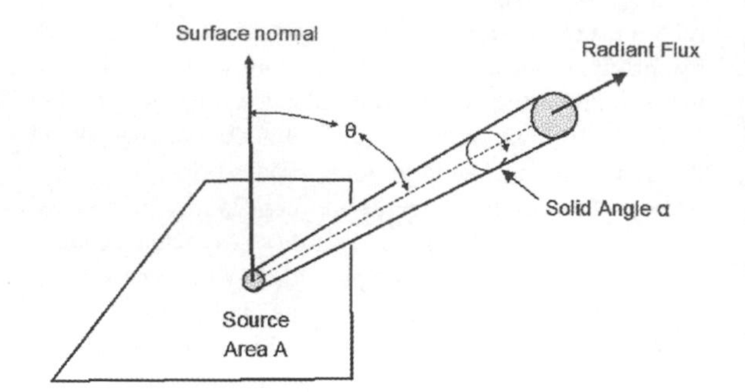

Figure 1.5 Radiance is the flux of electromagnetic energy leaving a source area A in direction θ per solid angle α. It is measured in watts per square metre per steradian (Wm^{-2} sr^{-2}).

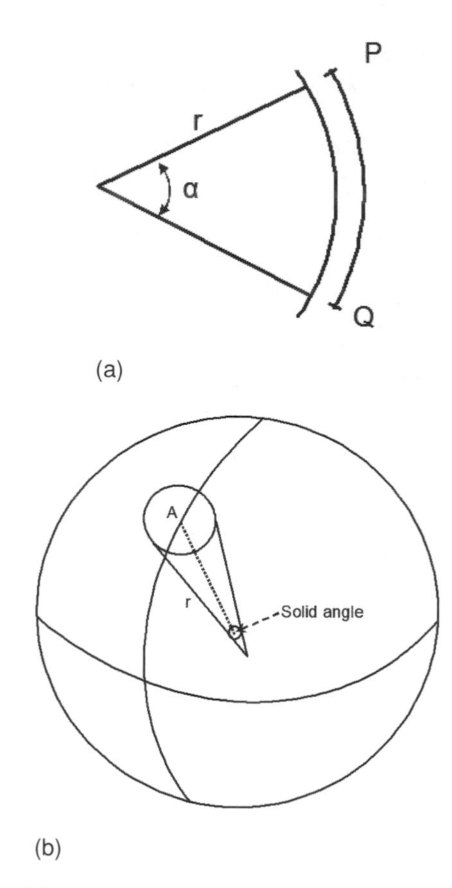

Figure 1.6 (a) The angle α formed when the length of the arc PQ is equal to the radius of the circle r is equal to one radian or approximately 57°. Thus, angle $\alpha = \text{PQ} \div r$ radians. There are 2π radians in a circle (360°). (b) A steradian is a solid three-dimensional angle formed when the area A delimited on the surface of a sphere is equal to the square of the radius r of the sphere. A need not refer to a uniform shape. The solid angle shown is equal to $A \div r^2$ steradians (sr). There are 4π steradians in a sphere.

of a given object is independent of irradiance, as it is a ratio. When remotely-sensed images collected over a time-period are to be compared, it is common practice to convert the radiance values recorded by the sensor into reflectance factors in order to eliminate the effects of variable irradiance over the seasons of the year. This topic is considered further in section 4.6.

The quantities described above can be used to refer to particular wavebands rather than to the whole electromagnetic spectrum (section 1.2.3). The terms are then preceded by the adjective *spectral*; for example, the spectral radiance for a given waveband is the radiant flux density in that waveband (i.e. spectral radiant flux density) per unit solid angle. Terms such as *spectral irradiance*, *spectral reflectance* and *spectral exitance* are defined in a similar fashion.

1.2.2 Nature of electromagnetic radiation

An important point of controversy in physics over the last 250 years has concerned the nature of EMR. Newton, while not explicitly rejecting the idea that light is a wave-like form of energy (the wave theory), inclined to the view that it is formed of a stream of particles (the corpuscular theory). The wave-corpuscle dichotomy was not to be resolved until the early years of the twentieth century with the work of Planck and Einstein. The importance to remote sensing of the nature of EMR is fundamental, for we need to consider radiation both as a waveform and as a stream of particles. The wave-like characteristics of EMR allow the distinction to be made between different manifestations of such radiation (for example, microwave and infrared radiation) while, in order to understand the interactions between EMR and the Earth's atmosphere and surface, the idea that EMR consists of a stream of particles is most easily used. Building on the work of Planck, Einstein proposed in 1905 that light consists of particles called *photons*, which, in most respects, were similar to other sub-atomic particles such as protons and neutrons. It was found that, at the sub-atomic level, both wave-like and particle-like properties were exhibited, and that phenomena at this level appear to be both waves and particles. Erwin Schrödinger (1867–1961) wrote as follows in *Science, Theory and Man* (New York: Dover, 1957):

> 'In the new setting of ideas, the distinction [between particles and waves] has vanished, because it was discovered that all particles have also wave properties, and vice-versa. Neither of the concepts must be discarded, they must be amalgamated. Which aspect obtrudes itself depends not on the physical object but on the experimental device set up to examine it.'

Thus, from the point of view of quantum mechanics, EMR is both a wave and a stream of particles. Whichever view is taken will depend on the requirements of the particular situation. In section 1.2.5, the particle theory is best suited to explain the manner in which incident EMR interacts with the atoms, molecules and other particles which form the Earth's atmosphere. Readers who, like myself, were resistant in their formative years to any kind of formal training in basic physics will find Gribben (1984) to be readable as well as instructive, while Feynman (1985) is a clear and well-illustrated account of the surprising ways that light can behave.

1.2.3 The electromagnetic spectrum

The Sun's light is the form of EMR that is most familiar to human beings. Sunlight that is reflected by physical objects travels in most situations in a straight line to the observer's eye. On reaching the retina, it generates electrical signals that are transmitted to the brain by the optic nerve. The brain uses these signals to construct an image of the viewer's surroundings. This is the process of vision, which is closely analogous to the process of remote sensing; indeed, vision is a form – perhaps the basic form – of remote sensing (Greenfield, 1997). A discussion of the human visual process can be found in section 5.2. Note that the process of human vision involves image acquisition (essentially a physiological process) and image understanding (a psychological process), just as Earth observation by remote sensing does. Image interpretation and understanding in remote sensing might therefore be considered to be an attempt to simulate or emulate the brain's image understanding functions.

Visible light is called so because the eye detects it, whereas other forms of EMR are invisible to the unaided eye. Sir Isaac Newton (1643–1727) investigated the nature of white light, and in 1664 concluded that it is made up of differently coloured components, which he saw by passing white light through a prism to form a rainbow-like spectrum. Newton saw the *visible spectrum*, which ranges from red through orange, yellow, and green to blue, indigo and violet. Later, the astronomer Friedrich Wilhelm (Sir William) Herschel (1728–1822) demonstrated the existence of EMR with wavelengths beyond those of the visible spectrum; these he called *infrared*, meaning *beyond the red*. It was subsequently found that EMR also exists beyond the violet end of the visible spectrum, and this form of radiation was given the name *ultraviolet*. Herschel, incidentally, started his career as a band-boy with the Hanoverian Guards and later came to live in England.

Other forms of EMR, such as X-rays and radio waves, were later discovered, and it was eventually realised that all were manifestations of the same kind of radiation which travels at the speed of light in a wave-like form, and which can propagate through empty space. The speed of light (c_0) is 299,792,458 m s^{-1} (approximately 3×10^8 m s^{-1}) in a vacuum, but is reduced by a factor called the index of refraction if the light travels through media such as the atmosphere or water. EMR reaching the Earth comes mainly from the Sun and is produced by thermonuclear reactions in the Sun's core. The set of all electromagnetic waves is called the *electromagnetic spectrum*, which includes the range from the long radio waves, through the microwave and infrared wavelengths to visible light waves and beyond to the ultraviolet and to the short-wave X- and gamma rays (Figure 1.7).

Symmetric waves can be described in terms of their *frequency* (f), which is the number of waveforms passing a fixed point in unit time. This quantity used to be known as *cycles per second* (cps) but nowadays the preferred term is *Hz* (Hertz, after Heinrich Hertz (1857–1894), who discovered radio waves in 1888). Alternatively, the concept of *wavelength* can be used (Figure 1.8). The wavelength is the distance between successive peaks (or successive troughs) of a waveform, and is normally measured in metres or fractions of a metre (Table 1.1). Both frequency and wavelength convey the same information and are often

Table 1.1 Terms and symbols used in measurement

Factor	Prefix	Symbol	Factor	Prefix	Symbol
10^{-18}	Atto	a			
10^{-15}	Femto	f			
10^{-12}	Pico	p	10^{12}	Tera	T
10^{-9}	Nano	n	10^{9}	Giga	G
10^{-6}	Micro	μ	10^{6}	Mega	M
10^{-3}	Milli	m	10^{3}	Kilo	K

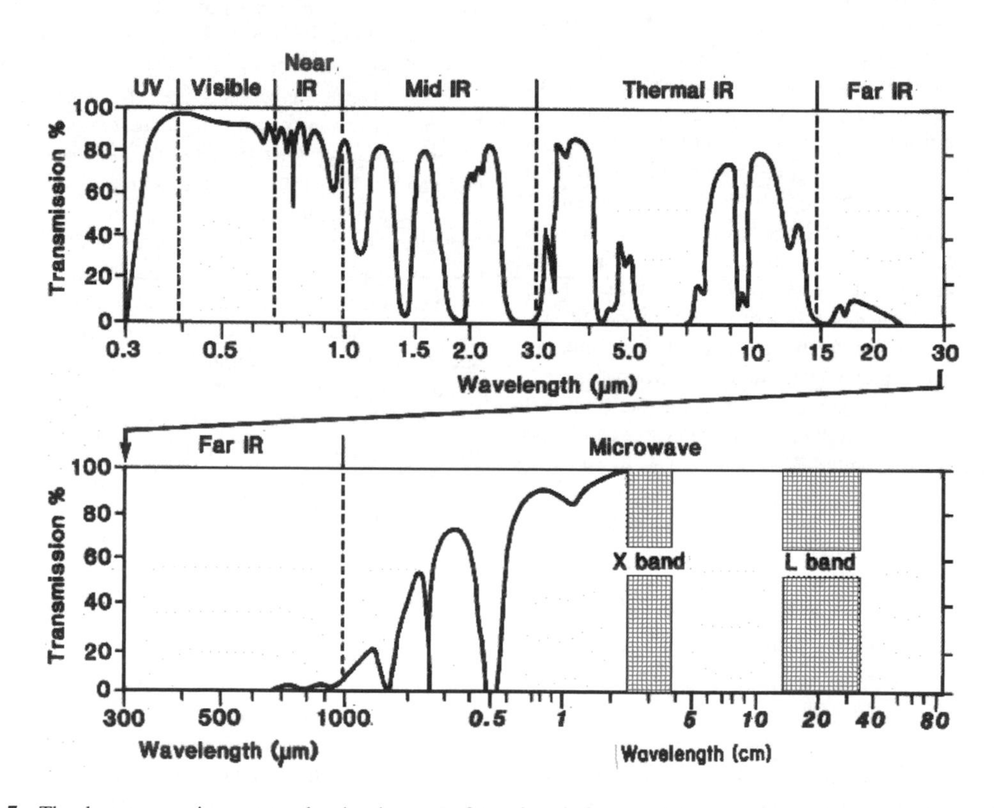

Figure 1.7 The electromagnetic spectrum showing the range of wavelengths between 0.3 μm and 80 cm. The vertical dashed lines show the boundaries of wavebands such as ultraviolet (UV) and near-infrared (near-IR). The shaded areas between 2 and 35 cm wavelength indicate two microwave wavebands (X band and L band) that are used by imaging radars. The curve shows atmospheric transmission. Areas of the electromagnetic spectrum with a high transmittance are known as atmospheric windows. Areas of low transmittance are opaque and cannot be used to remotely sense the Earth's surface. Reprinted with permission from A.F.H Goetz and L.C. Rowan, 1981, Geologic remote sensing, Science, **211**, 781–791 (figure 1). © American Association for the Advancement of Science.

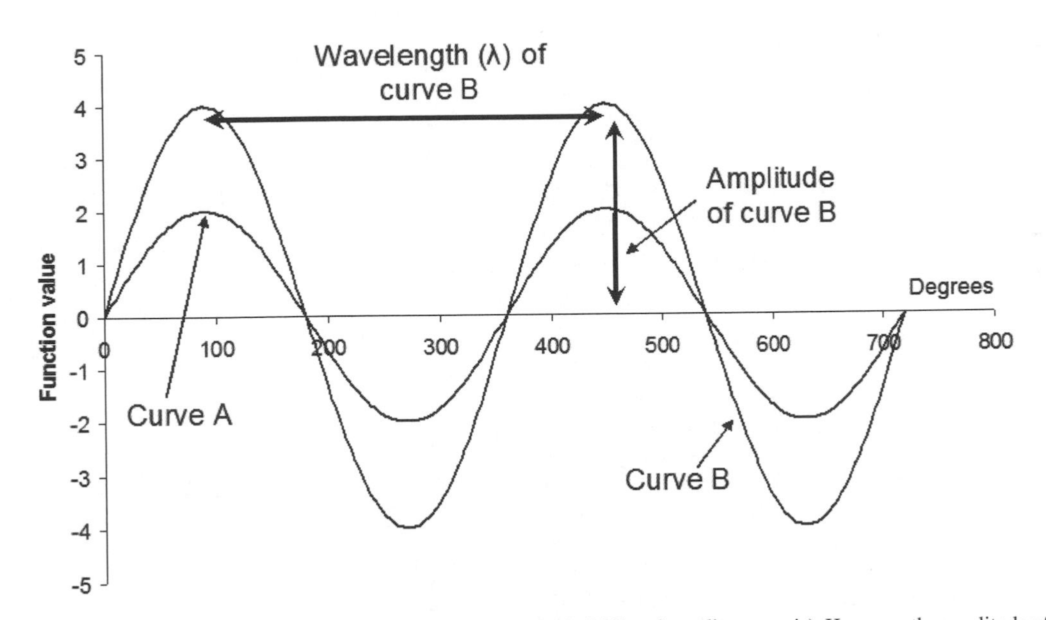

Figure 1.8 Two curves (waveforms) A and B have the same wavelength (360° or 2π radians, x-axis). However, the amplitude of curve A is two units (y-axis) while curve B has an amplitude of four units. If we imagine that the two curves repeat to infinity and are moving to the right, like traces on an oscilloscope, then the frequency is the number of waveforms ($0 - 2\pi$) that pass a fixed point in unit time (usually measured in cycles per second or Hertz, Hz). The period of the waveform is the time taken for one full waveform to pass a fixed point. These two waveforms have the same wavelength, frequency and period, and differ only in terms of their amplitude.

used interchangeably. Another measure of the nature of a waveform is its period (T). This is the time, in seconds, needed for one full wave to pass a fixed point. The relationships between wavelength, frequency and period are given by:

$$f = c/\lambda$$
$$\lambda = c/f$$
$$T = 1/f = \lambda/c$$

In these expressions, c is the speed of light. The velocity of propagation (v) is the product of wave frequency and wavelength, i.e.

$$v = \lambda f$$

The *amplitude* (A) of a wave is the maximum distance attained by the wave from its mean position (Figure 1.8). The amount of energy, or intensity, of the waveform is proportional to the square of the amplitude. Using the relationships specified earlier we can compute the frequency given the wavelength, and vice versa. If, for example, wavelength λ is 0.6 μm or 6×10^{-7} m, then, since velocity v equals the product of wavelength and frequency f, it follows that:

$$v = 6 \times 10^{-7} f$$

so that:

$$f = \frac{c_0}{v} = \frac{3 \times 10^8}{6 \times 10^{-7}} \text{ Hz}$$

i.e.

$$f = 0.5 \times 10^{15} \text{ Hz} = 0.5 \text{ PHz}$$

1 PHz (Petaherz) equals 10^{15} Hz (Table 1.1). Thus, electromagnetic radiation with a wavelength of 0.6 μm has a frequency of 0.5×10^{15} Hz. The *period* is the reciprocal of the frequency, so one wave of this frequency will pass a fixed point in 2×10^{-15} seconds. The amount of energy carried by the waveform, or the squared amplitude of the wave, is defined for a single photon by the relationship

$$E = hf$$

where E is energy, h is a constant known as Planck's constant (6.625×10^{-34} J s) and f is frequency. Energy thus increases with frequency, so that high frequency, short-wavelength electromagnetic radiation such as X-rays carries more energy than visible light or radio waves.

While electromagnetic radiation with particular temporal and spatial properties is used in remote sensing to convey information about a target, it is interesting to note that both

Table 1.2 Wavebands corresponding to perceived colours of visible light

Colour	Waveband (μm)	Colour	Waveband (μm)
Red	0.780–0.622	Green	0.588–0.492
Orange	0.622–0.597	Blue	0.492–0.455
Yellow	0.597–0.577	Violet	0.455–0.390

time and space are defined in terms of specific characteristics of electromagnetic radiation. A second is the duration of 9 192 631 770 periods of the caesium radiation (in other words, that number of wavelengths or cycles are emitted by caesium radiation in one second; its frequency is approximately 9 GHz or a wavelength of around 0.03 m). A metre is defined as 1 650 764.73 vacuum wavelengths of the orange-red light emitted by krypton-86.

Visible light is defined as electromagnetic radiation with wavelengths between (approximately) 0.4 μm and 0.7 μm. We call the shorter wavelength end (0.4 μm) of the visible spectrum 'blue' and the longer wavelength end (0.7 μm) 'red' (Table 1.2). The eye is not uniformly sensitive to light within this range, and has its peak sensitivity at around 0.55 μm, which lies in the green part of the visible spectrum (Figures 1.7 and 1.9). This peak in the response function of the human eye corresponds closely to the peak in the Sun's radiation emittence distribution (section 1.2.4).

The process of atmospheric scattering, discussed in section 1.2.5 below, deflects light rays from a straight path and thus causes blurring or haziness. It affects the blue end of the visible spectrum more than the red end, and consequently the blue waveband is not used in many remote-sensing systems.

Figure 1.10(a)–(c) shows three greyscale images collected in the blue/green, green and red wavebands, respectively, by a sensor called the *Thematic Mapper* (TM) that is carried by the American Landsat-5 satellite (Chapter 2). The different land cover types reflect energy in the visible spectrum in a differential manner, although the clouds and cloud shadows in the upper centre of the image are clearly visible in all three images. Various crops in the fields round the village of Littleport (north of Cambridge in eastern England) can be discriminated, and the River Ouse can also be seen as it flows northwards in the right-hand side of the image area. (It is dangerous to rely on visual interpretation of images such as these. Since they are digital in nature they can be manipulated by computer so that interpretation is more objective – that is what this book is about!)

Electromagnetic radiation with wavelengths shorter than those of visible light (less than 0.4 μm) is divided into three spectral regions, called gamma rays, X-rays and ultraviolet

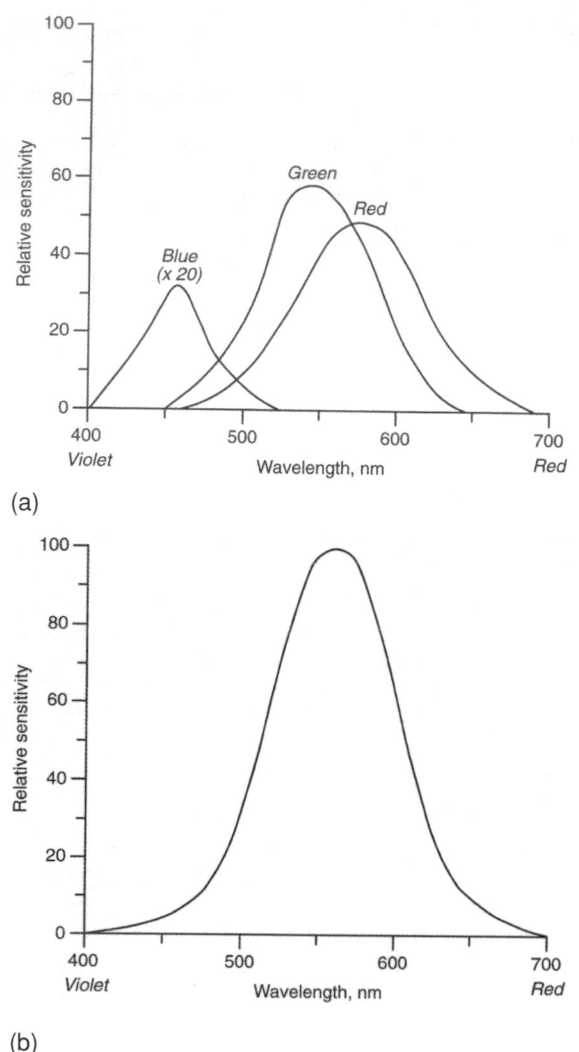

Figure 1.9 (a) Response function of the red-, green- and blue-sensitive cones on the retina of the human eye. (b) Overall response function of the human eye. Peak sensitivity occurs near 550 nm (0.55 μm).

radiation. Because of the effects of atmospheric scattering and absorption (section 4.4; Figure 4.11), none of these wavebands is used in satellite remote sensing, though low-flying aircraft can detect gamma-ray emissions from radioactive materials in the Earth's crust. Radiation in these wavebands is dangerous to life, so the fact that it is mostly absorbed or scattered by the atmosphere allows life to exist on Earth. In terms of the discussion of the wave/particle duality in section 1.2.2, it should be noted that gamma radiation has the highest energy levels and is the most

(a)

(b)

(c)

Figure 1.10 Images collected in (a) band 1 (blue-green), (b) band 2 (green) and (c) band 3 (red) wavebands of the optical spectrum by the Thematic Mapper sensor carried by the Landsat satellite. The area shown is near the town of Littleport in Cambridgeshire, eastern England. Original data © ESA 1994, distributed by Eurimage.

'particle-like' of all electromagnetic radiation, whereas radio frequency radiation is most 'wave-like' and has the lowest energy levels.

Wavelengths that are longer than the visible red are subdivided into the infrared (IR), microwave, and radio frequency wavebands. The *infrared* waveband, extending from 0.7 μm to 1 mm, is not a uniform region. Short-wavelength (SWIR) or near-infrared (NIR) energy, with wavelengths between 0.7 and 0.9 μm, behaves like visible light and can be detected by special photographic film. Infrared radiation with a wavelength of up to 3.0 μm is primarily of solar origin and, like visible light, is reflected by the surface of the Earth. Hence, these wavebands are often known as the *optical* bands. Figure 1.11(a) shows an image of the area shown in Figure 1.10 collected by the Landsat Thematic Mapper sensor in the near-infrared region of the spectrum (0.75–0.90 μm). This image is considerably clearer than the visible spectrum images shown in Figure 1.10. We will see in section 1.3.2 that the differences in reflection between vegetation, water, and soil are probably greatest in this near-infrared band. An image of surface reflection in the Landsat Thematic Mapper mid-infrared waveband (2.08–2.35 μm) of the same area is shown in Figure 1.11(b).

In wavelengths longer than around 3 μm, IR radiation emitted by the Earth's surface can be sensed in the form of heat. The amount and wavelength of this radiation depends on the temperature of the source (section 1.2.4). Because these longer IR wavebands are sensed as heat, they are called the *thermal infrared* (TIR) wavebands. Much of the TIR radiation emitted by the Earth is absorbed by, and consequently heats, the atmosphere thus making life possible on the Earth (Figure 1.7). There is, however, a 'window' between 8 and 14 μm which allows a satellite sensor above the atmosphere to detect thermal radiation emitted by the Earth, which has its peak wavelength at 9.7 μm. Note, though, that the presence of ozone in the atmosphere creates a narrow absorption band within this window, centred

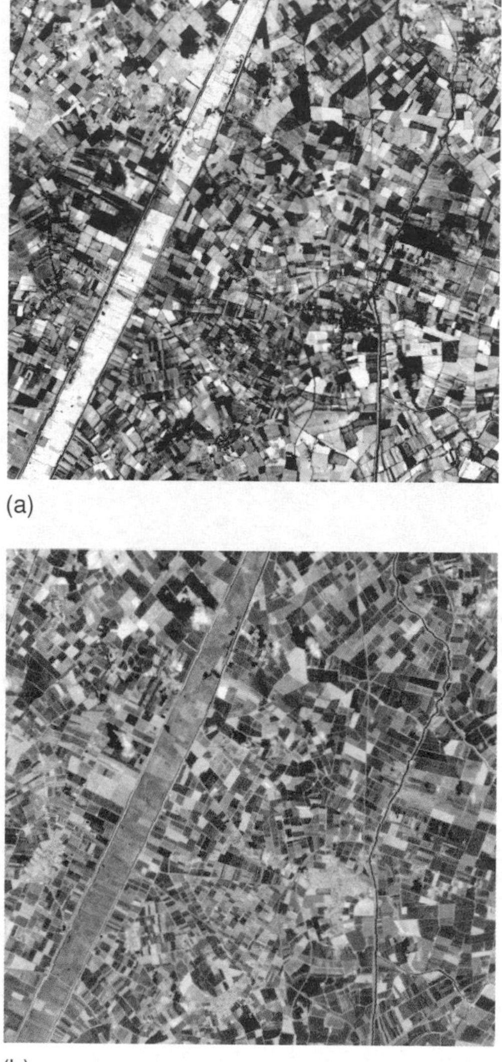

Figure 1.11 Image of ground surface reflectance in the 0.75–0.90 μm band (near infrared) and (b) middle infrared (2.08–2.35 μm) image of the same area as that shown in Figure 1.10. These images were collected by the Landsat-5 Thematic Mapper (bands 4 and 7). Original data © ESA 1994, distributed by Eurimage.

at 9.5 μm. Absorption of longer-wave radiation by the atmosphere has the effect of warming the atmosphere. This is called the natural greenhouse effect. Water vapour (H_2O) and carbon dioxide (CO_2) are the main absorbing agents, together with ozone (O_3). The increase in the carbon dioxide content of the atmosphere over the last century, due to the burning of fossil fuels, is thought to enhance the greenhouse effect and to raise the temperature of the atmosphere above its natural level. This could have long-term climatic consequences. An image of part of W. Europe acquired by the Advanced Very High Resolution Radiometer (AVHRR) carried by the US NOAA-14 satellite is shown in Figure 1.12. The different greyscales show different levels of emitted thermal radiation in the 11.5–12.5 μm waveband. Before these grey levels can be interpreted in terms of temperatures, the effects of the atmosphere as well as the nature of the sensor calibration must be considered. Both these topics are covered in Chapter 4. For comparison, a visible band image of Europe and North Africa produced by the Meteosat-6 satellite is shown in Figure 1.13. Both images were collected by the UK NERC-funded satellite receiving station at Dundee University, Scotland.

That region of the spectrum composed of electromagnetic radiation with wavelengths between 1 mm and 300 cm is called the *microwave band*. Most satellite-borne sensors that operate in the microwave region use microwave radiation with wavelengths between 3 and 25 cm. Radiation at these wavelengths can penetrate cloud, and the microwave band is thus a valuable region for remote sensing in temperate and tropical areas where cloud cover restricts the collection of optical and thermal infrared images. Some microwave sensors can detect the small amounts of radiation at these wavelengths that are emitted by the Earth. Such sensors are called *passive* because they detect EMR that is generated externally, for example, by emittance by or reflectance from a target. Passive microwave radiometers such as the Scanning Multichannel Microwave Radiometer (SMMR) produce imagery with a low spatial resolution (section 2.2.1) that is used to provide measurements of sea-surface temperature and wind speed, and also to detect sea ice.

Because the level of microwave energy emitted by the Earth is very low, a high-resolution imaging microwave sensor generates its own electromagnetic radiation at centimetre wavelengths, transmits this energy towards the ground, and then detects the strength of the return signal that is scattered by the target in the direction of the sensor. Devices that generate their own electromagnetic energy are called *active* sensors to distinguish them from the *passive* sensors that are used to detect and record radiation of solar or terrestrial origin in the visible, infrared, and microwave wavebands. Thus, active microwave instruments are not dependent on an external source of radiation such as the Sun or, in the case of thermal emittance, the Earth. It follows that active microwave sensors can operate independently by day or by night. An analogy that is often used is that of a camera. In normal daylight, reflected

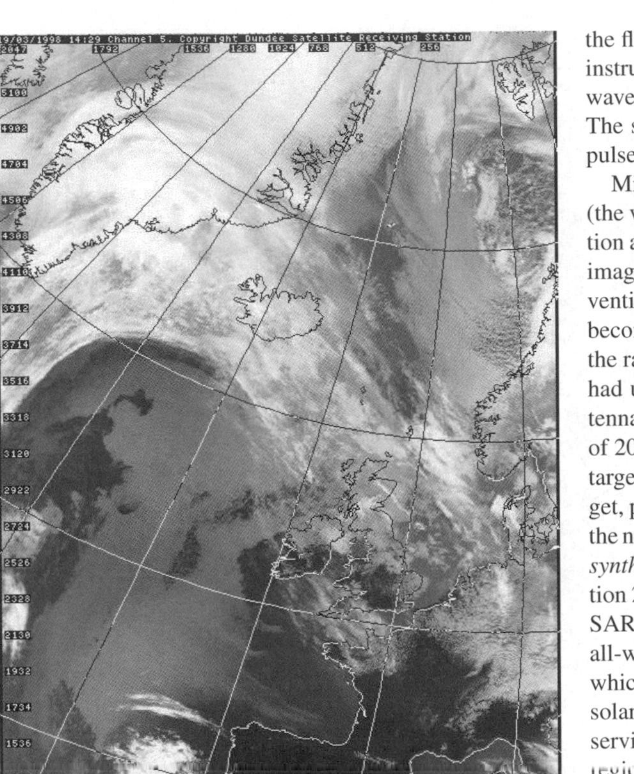

Figure 1.12 NOAA AVHRR band 5 image (thermal infrared, 11.5–12.5 μm) of western Europe and NW Africa collected at 1420 on 19 March 1998. The image was downloaded by the NERC Satellite Receiving Station at Dundee University, Scotland, where the image was geometrically rectified (Chapter 4) and the latitude/longitude grid and digital coastline were added. Dark areas indicate greater thermal emissions. The position of a high-pressure area (anticyclone) can be inferred from cloud patterns. Cloud tops are cold and therefore appear white. The NOAA satellite took just over 15 minutes to travel from the south to the north of the area shown on this image. © Dundee Satellite Receiving Station, Dundee University.

radiation from the target enters the camera lens and exposes the film. Where illumination conditions are poor, the photographer employs a flashgun that generates radiation in visible wavebands, and the film is exposed by light from the flashgun that is reflected by the target. The microwave instrument produces pulses of energy, usually at centimetre wavelengths, that are transmitted by an antenna or aerial. The same antenna picks up the reflection of these energy pulses as they return from the target.

Microwave imaging sensors are called *imaging radars* (the word *radar* is an acronym, derived from Radio Detection and Ranging). The spatial resolution (section 2.2.1) of imaging radars is a function of their antenna length. If a conventional ('brute force') radar is used, then antenna lengths become considerable. Schrier (1993b, p. 107) notes that if the radar carried by the Seasat satellite (launched in 1981) had used a 'brute force' approach then its 10 m long antenna would have generated images with a spatial resolution of 20 km. A different approach, using several views of the target as the satellite approaches, reaches and passes the target, provides a means of achieving high resolution without the need for excessive antenna sizes. This approach uses the *synthetic aperture radar* (SAR) principle, described in section 2.4, and all satellite-borne radar systems have used the SAR principle. The main advantage of radar is that it is an all-weather, day-night, high spatial resolution instrument, which can operate independently of weather conditions or solar illumination. This makes it an ideal instrument for observing areas of the world such as the temperate and tropical regions, which are often covered by clouds and therefore inaccessible to optical and infrared imaging sensors.

A radar signal does not detect either colour information (which is gained from analysis of optical wavelength sensors) or temperature information (derived from data collected by thermal infrared sensors). It can detect both surface roughness and electrical conductivity information (which is related to soil moisture conditions). Because radar is an active rather than a passive instrument, the characteristics of the transmitted signal can be controlled. In particular, the wavelength, depression angle, and polarisation of the signal are important properties of the radiation source used in remote sensing. Radar wavelength (Table 1.3)

Table 1.3 Radar wavebands and nomenclature.

Band designation	Frequency (MHz)	Wavelength (cm)
P	300–1000	30–100
L	1000–2000	15–30
S	2000–4000	7.5–15
C	4000–8000	3.75–7.5
X	8000–12000	2.5–3.75
K_u	12000–18000	1.667–2.5
K	18000–27000	1.111–1.667
K_a	27000–40000	0.75–1.111

Figure 1.13 Portion of a Meteosat-6 visible channel image of Europe and N. Africa taken at 1800 on 17 March 1998, when the lights were going on across Europe. Image received by Dundee University, Scotland. © EUMETSAT/Dundee Satellite Receiving Station, Dundee University.

determines the observed roughness of the surface, in that a surface that has a roughness with a frequency less than that of the microwave radiation used by the radar is seen as smooth. An X-band (c. 3 cm wavelength) image of the area around the city of Perpignan in S.W. France is shown in Figure 1.14. Radar sensors are described in more detail in section 2.4.

Beyond the microwave region is the *radio* band. Radio wavelengths are used in remote sensing, but not to detect Earth surface phenomena. Commands sent to a satellite utilise radio wavelengths. Image data is transmitted to ground receiving stations using wavelengths in the microwave region of the spectrum; these data are recorded on the ground by high-speed tape-recorders while the satellite is in direct line of sight of a ground receiving station. Image data for regions of the world that are not within range of ground receiving stations are recorded by onboard tape recorders or solid-state memory and these recorded data are subsequently transmitted together with currently scanned data when the satellite is within the reception range of a ground receiving station. The first three Landsat satellites (section 2.3.6) used onboard tape recorders to supplement data that were directly transmitted to the ground. The latest Landsat (number 7) relies on the Tracking and Data Relay Satellite (TDRS) system, which allows direct broadcast of data from Earth resources satellites to one of a set of communications satellites located above the Equator in geostationary orbit (meaning that the satellite's orbital velocity is just sufficient to keep pace with the rotation of the Earth). The signal is relayed by the TDRS system to a ground receiving station at White Sands, New Mexico. European satellites use a similar system called Artemis, which became operational in 2003.

1.2.4 Sources of electromagnetic radiation

All objects whose temperature is greater than absolute zero, which is approximately –273°C or 0 K (Kelvin), emit radiation. However, the distribution of the amount of radiation at each wavelength across the spectrum is not uniform. Radiation is emitted by the stars and planets; chief of these, as far as the human race is concerned, is the Sun, which provides the heat and light radiation needed to sustain life on Earth. The Sun is an almost-spherical body with a diameter of 1.39×10^6 km and a mean distance from Earth of 150×10^6 km. Its chief constituents are hydrogen and helium. The conversion of hydrogen to helium in the Sun's core provides the energy that is radiated from the outer layers. At the edge of the Earth's atmosphere the power received from the Sun, measured over the surface area of the Earth, is approximately 3.9×10^{22} MW which, if it were distributed evenly over the Earth, would give an incident

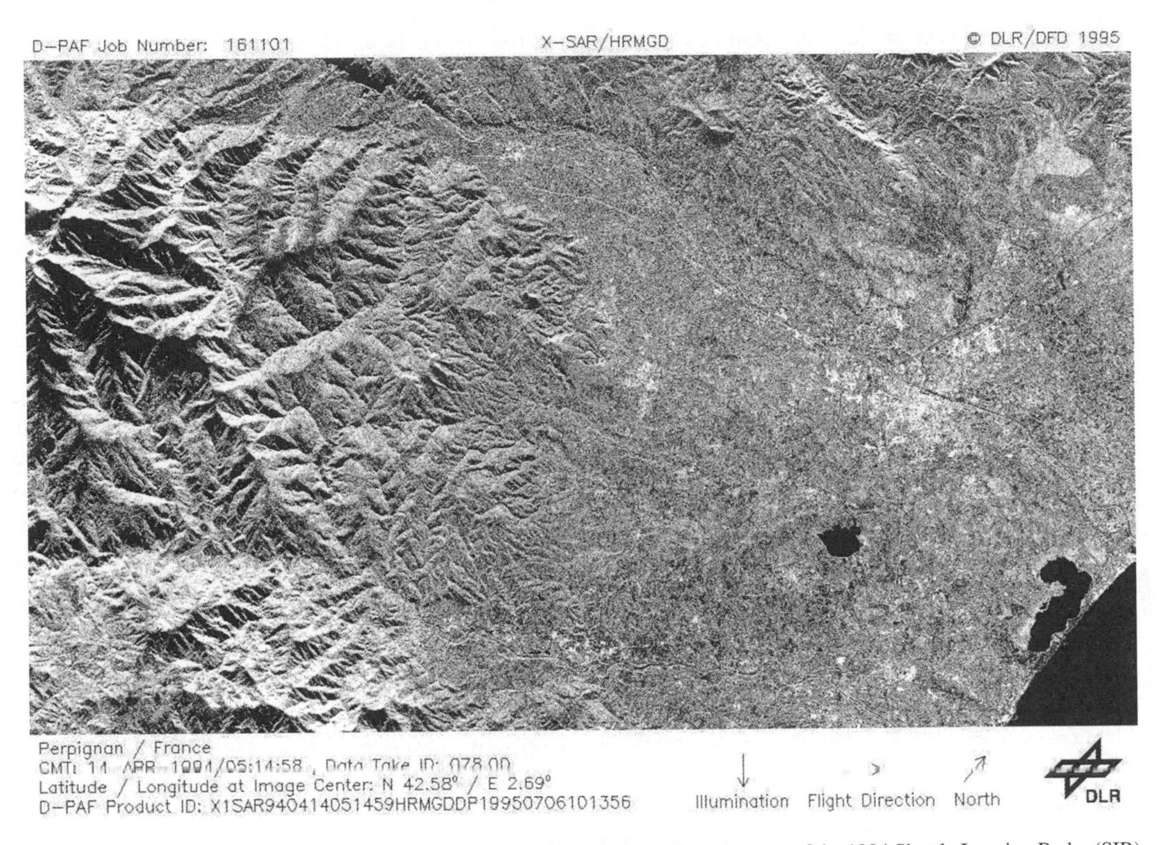

Figure 1.14 X-band radar image of part of SW France around Perpignan collected as part of the 1994 Shuttle Imaging Radar (SIR) – C/X – SAR experiment. © 1997 Deutsches Zentrum fûr Luft- und Raumfahrt e.V.

radiant flux density of 1367 W m^{-2}. This value is known as the *solar constant*, even though it varies throughout the year by about ±3.5%, depending on the distance of the Earth from the Sun (and this variation is taken into account in the radiometric correction of remotely-sensed images; see section 4.6). Bonhomme (1993) provides a useful summary of a number of aspects relating to solar radiation. On average, 35% of the incident radiant flux is reflected from the Earth (including clouds and atmosphere), the atmosphere absorbs 17%, and 47% is absorbed by the materials forming the Earth's surface. From the Stefan–Bolzmann Law (below) it can be shown that the Sun's temperature is 5777 K if the solar constant is 1367 W m^{-2}. Other estimates of the Sun's temperature range from 5500 K to 6200 K. The importance of establishing the surface temperature of the Sun lies in the fact that the distribution of energy emitted in the different regions of the electromagnetic spectrum depends upon the temperature of the source.

If the Sun were a perfect emitter, it would be an example of a theoretical ideal, called a *blackbody*. A blackbody transforms heat energy into radiant energy at the maximum rate that is consistent with the laws of thermodynamics (Suits, 1983). *Planck's Law* describes the spectral exitance (i.e., the distribution of radiant flux density with wavelength, section 1.2.1) of a blackbody as:

$$M_\lambda = \frac{c_1}{\lambda^5(\exp[c_2/\lambda T] - 1)}$$

where

$$\begin{aligned} c_1 &= 3.742 \times 10^{-16}\ \text{W m}^{-2} \\ c_2 &= 1.4388 \times 10^{-2}\ \text{m K} \\ \lambda &= \text{wavelength (m)} \\ T &= \text{temperature (Kelvin)} \\ M\lambda &= \text{spectral exitance per unit wavelength.} \end{aligned}$$

Curves showing the spectral exitance of blackbodies at temperatures of 1000 K, 1600 K and 2000 K are shown in Figure 1.15. The total radiant energy emitted by a blackbody is dependent on its temperature, and as temperature increases the wavelength at which the maximum spectral exitance is achieved is reduced. The dotted line in Figure 1.15 joins the peaks of the spectral exitance curves. It is described by *Wien's Displacement Law*, which gives the wavelength of maximum spectral exitance (λ_m) in terms of temperature:

$$\lambda_m = \frac{c_3}{T}$$

and

$$c_3 = 2.898 \times 10^{-3} \text{ m K}$$

The total spectral exitance of a blackbody at temperature T is given by the Stefan–Boltzmann Law as:

$$M = \sigma T^4$$

In this equation, $\sigma = 5.6697 \times 10^{-8}$ W m^{-2} K^{-4}.

The distribution of the spectral exitance for a blackbody at 5900 K closely approximates the Sun's spectral exitance curve, while the Earth can be considered to act like a blackbody with a temperature of 290 K (Figure 1.16). The solar radiation maximum occurs in the visible spectrum, with maximum irradiance at 0.47 μm. About 46% of the total energy transmitted by the Sun falls into the visible waveband (0.4 to 0.76 μm).

Wavelength-dependent mechanisms of atmospheric absorption alter the actual amounts of solar irradiance that reach the surface of the Earth. Figure 1.17 shows the spectral irradiance from the Sun at the edge of the atmosphere (solid curve) and at the Earth's surface (broken line). Further discussion of absorption and scattering can be found in section 1.2.5. The spectral distribution of radiant energy emitted by the Earth (Figure 1.16) peaks in the thermal infrared wavebands at 9.7 μm. The amount of terrestrial emission is low in comparison to solar irradiance. However, the solar radiation absorbed by the atmosphere is balanced by terrestrial emission in the thermal infrared, keeping the temperature of the atmosphere approximately constant. Furthermore, terrestrial thermal infrared emission provides sufficient energy for remote sensing from orbital altitudes to be a practical proposition. The characteristics of the radiation sources used in remote sensing impose some limitations on the range of wavebands available for

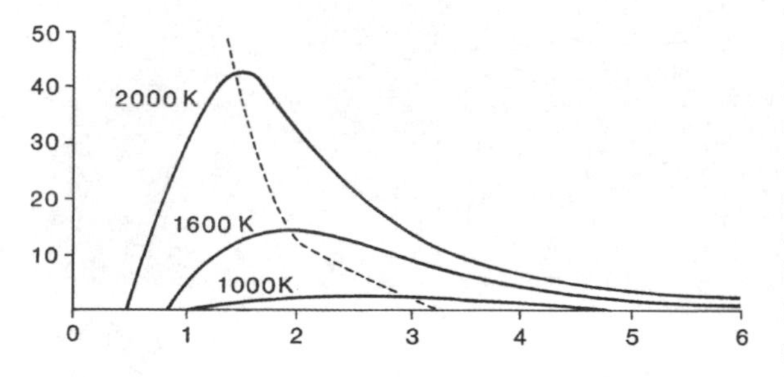

Figure 1.15 Spectral exitance curves for blackbodies at temperatures of 1000, 1600 and 2000 K. The dotted line joins the emittance peaks of the curves and is described by Wien's displacement law (see text).

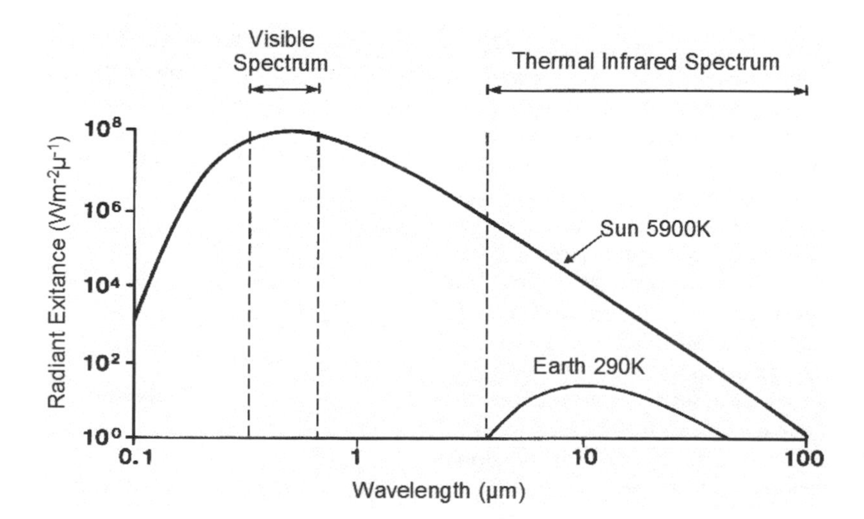

Figure 1.16 Spectral exitance curves for blackbodies at 290 and 5900 K, the approximate temperatures of the Earth and the Sun.

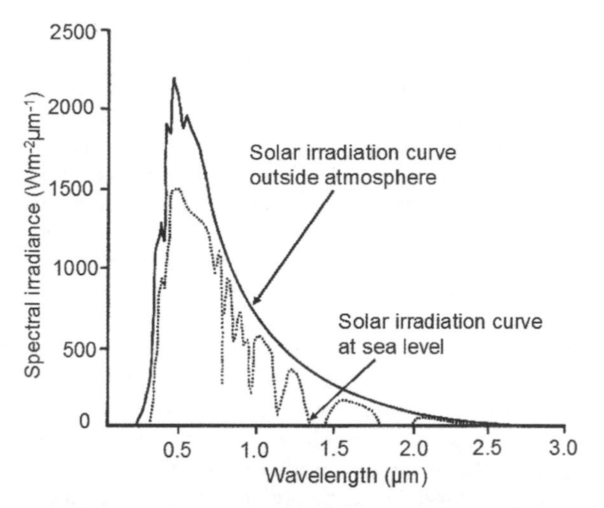

Figure 1.17 Solar irradiance at the top of the atmosphere (solid line) and at sea level (dotted line). Differences are due to atmospheric effects as discussed in the text. See also figure 1.5. Based on Manual of Remote Sensing, 2nd edition, ed. R.N. Colwell, 1983, figure 1.5. Reproduced by permission of the American Society for Photogrammetry and Remote Sensing.

use. In general, remote sensing instruments that measure the spectral reflectance of solar radiation from the Earth's surface are restricted to the wavelengths shorter than 2.5 μm. Instruments to detect terrestrial radiant exitance operate in the spectral region between 3 and 14 μm. Because of atmospheric absorption by carbon dioxide, ozone and water vapour, only the 3–5 μm and 8–14 μm regions of the thermal infrared band are useful in remote sensing. An absorption band is also present in the 9–10 μm region. As noted earlier, the Earth's emittance peak occurs at 9.7 μm, so satellite-borne thermal sensors normally operate in the 10.5–12.5 μm spectral region. The 3–5 μm spectral window can be used to detect local targets that are hotter than their surroundings, for example, forest fires. Since the 3–5 μm region also contains some reflected solar radiation it can only be used for temperature sensing at night.

Wien's Displacement Law (Figure 1.15) shows that the radiant power peak moves to shorter wavelengths as temperature increases, so that a forest fire will have a radiant energy peak at a wavelength shorter than 9.7 μm. Since targets such as forest fires are sporadic in nature and require high-resolution imagery the 3–5 μm spectral region is used by aircraft-mounted thermal detectors. This is a difficult region for remote sensing because it contains a mixture of reflected and emitted radiation, the effects of which are not easy to separate.

The selection of wavebands for use in remote sensing is therefore seen to be limited by several factors, primarily (i) the characteristics of the radiation source, as discussed in this section, (ii) the effects of atmospheric absorption and scattering (section 1.2.5) and (iii) the nature of the target. This last point is considered in section 1.3.

1.2.5 Interactions with the Earth's atmosphere

In later chapters, we consider measurements of radiance from the Earth's surface made by instruments carried by satellites such as Landsat and SPOT that operate in the optical wavebands, that is, those parts of the electromagnetic spectrum with properties similar to those of visible light. It was noted at the beginning of this chapter that one aim of remote sensing is to identify the nature, and possibly the properties, of Earth-surface materials from the spectral distribution of EMR that is reflected from, or emitted by, the target and recorded by the sensor. The existence of the atmosphere causes problems, because EMR from the Sun that is reflected by the Earth (the amount reflected depending on the reflectivity or albedo of the surface) and detected by the satellite or aircraft-borne sensor must pass through the atmosphere twice, once on its journey from the Sun to the Earth and once after being reflected by the surface of the Earth back to the sensor. During its passage through the atmosphere, EMR interacts with particulate matter suspended in the atmosphere and with the molecules of the constituent gases. This interaction is usually described in terms of two processes. One, called *scattering*, deflects the radiation from its path while the second process, *absorption*, converts the energy present in electromagnetic radiation into the internal energy of the absorbing molecule. Both absorption and scattering vary in their effect from one part of the spectrum to another. Remote sensing of the Earth's surface is impossible in those parts of the spectrum that are seriously affected by scattering and/or absorption, for these mechanisms effectively render the atmosphere opaque to incoming or outgoing radiation. As far as remote sensing of the Earth's surface is concerned, the atmosphere

> '. . . appears no other thing to me but a foul and pestilential congregation of vapours'
>
> *Hamlet*, Act 2, Scene 2

Atmospheric absorption properties can, however, be useful. Remote sensing of the atmosphere uses these properties. A good example is the discovery and monitoring of the Antarctic ozone hole.

Regions of the spectrum that are relatively (but not completely) free from the effects of scattering and absorption are called *atmospheric windows*; electromagnetic radiation in these regions passes through the atmosphere with less modification than does radiation at other wavelengths (Figure 1.7). This effect can be compared to the way in which the bony tissues of the human body are opaque to X-rays, whereas the soft muscle tissue and blood are transparent. Similarly, glass is opaque to ultraviolet radiation but is transparent at the visible wavelengths. Figure 1.17 shows a plot of wavelength against the percentage of incoming radiation transmitted through the atmosphere; the window regions are those with a high transmittance. The same information is shown in a different way in Figure 1.7.

The effect of the processes of scattering and absorption is to add a degree of haze to the image, that is, to reduce the contrast of the image, and to reduce the amount of radiation returning to the sensor from the Earth's surface. A certain amount of radiation that is reflected from the neighbourhood of each target may also be recorded by the sensor as originating from the target. This is because scattering deflects the path taken by electromagnetic radiation as it travels through the atmosphere, while absorption involves the interception of photons or particles of radiation. Our eyes operate in the visible part of the spectrum by observing the light reflected by an object. The position of the object is deduced from the assumption that this light has travelled in a straight line between the object and our eyes. If some of the light reflected towards our eyes from the object is diverted from a straight path then the object will appear less bright. If light from other objects has been deflected so that it is apparently coming to our eyes from the direction of the first object then that first object will become blurred. Taken further, this scattering process will make it appear to our eyes that light is travelling from all target objects in a random fashion, and no objects will be distinguishable. Absorption reduces the amount of light that reaches our eyes, making a scene relatively duller. Both scattering and absorption, therefore, limit the usefulness of some portions of the electromagnetic spectrum for remote sensing purposes. They are known collectively as *attenuation* or *extinction*.

Scattering is the result of interactions between electromagnetic radiation and particles or gas molecules that are present in the atmosphere. These particles and molecules range in size from microscopic (with radii approximately equal to the wavelength of the electromagnetic radiation) to raindrop size (100 μm and larger). The effect of scattering is to redirect the incident radiation, or to deflect it from its path. The atmospheric gases that primarily cause scattering include oxygen, nitrogen, and ozone. Their molecules have radii of less than 1 μm and affect electromagnetic radiation with wavelengths of 1 μm or less. Other types of particles reach the atmosphere both by natural causes (such as salt particles from oceanic evaporation or dust entrained by aeolian processes) or because of human activities (for instance, dust from soil erosion caused by poor land management practices, and smoke particles from industrial and domestic pollution). Some particles are generated by photochemical reactions involving trace gases such as sulphur dioxide or hydrogen sulphide. The former may reach the atmosphere from car exhausts or from the combustion of fossil fuels. Another type of particle is the raindrop, which tends to be larger than the other kinds of particles mentioned previously (10–100 μm compared to 0.1–10 μm radius). The concentration of particulate matter varies both in time and over space. Human activities, particularly agriculture and industry, are not evenly spread throughout the world, nor are natural processes such as wind erosion or volcanic activity. Meteorological factors cause variations in atmospheric turbidity over time, as well as over space. Thus, the effects of scattering are spatially uneven (the degree of variation depending on weather conditions) and vary from time to time. Remotely sensed images of a particular area will thus be subjected to different degrees of atmospheric scattering on each occasion that they are produced. Differences in atmospheric conditions over time are the cause of considerable difficulty in the quantitative analysis of time-sequences of remotely-sensed images.

The mechanisms of scattering are complex, and are beyond the scope of this book. However, it is possible to make a simple distinction between selective and non-selective scattering. Selective scattering affects specific wavelengths of electromagnetic radiation, while non-selective scattering is wavelength-independent. Very small particles and molecules, with radii far less than the wavelength of the electromagnetic radiation of interest, are responsible for *Rayleigh scattering*. The effect of this type of scattering is inversely proportional to the fourth power of the wavelength, which implies that shorter wavelengths are much more seriously affected than longer wavelengths. Blue light (wavelength 0.4–0.5 μm) is thus more powerfully scattered than red light (0.6–0.7 μm). This is why the sky seems blue, for incoming blue light is so scattered by the atmosphere that it seems to reach our eyes from all directions, whereas at the red end of the visible spectrum scattering is much less significant so that red light maintains its directional properties. The sky appears to be much darker blue when seen from a high altitude, such as from the top of a mountain or from an aeroplane, because the degree of scattering is reduced due to the reduction in the length of the path traversed through the atmosphere by the incoming solar radiation. Scattered light reaching the Earth's surface is termed *diffuse* (as opposed to *direct*) irradiance or, more simply, *skylight*. Radiation that has been scattered within the atmosphere and which reaches the sensor without having made contact with the Earth's surface is called the *atmospheric path radiance*.

Mie scattering is caused by particles that have radii between 0.1 and 10 μm, that is, approximately the same magnitude as the wavelengths of electromagnetic radiation in the visible, near infrared and thermal infrared regions of the spectrum. Particles of smoke, dust and salt have radii of these dimensions. The intensity of Mie scattering is inversely proportional to wavelength, as in the case of Rayleigh scattering. However, the exponent ranges in value from −0.7 to −2 rather than the −4 of Rayleigh scattering. Mie scattering affects shorter wavelengths more than longer wavelengths, but the disparity is not as great as in the case of Rayleigh scattering.

Non-selective scattering is wavelength-independent. It is produced by particles whose radii exceed 10 μm. Such particles include water droplets and ice fragments present in clouds. All visible wavelengths are scattered by such particles. We cannot see through clouds because all visible wavelengths are non-selectively scattered by the water droplets of which the cloud is formed. The effect of scattering is, as mentioned earlier, to increase the haze level or reduce the contrast in an image. If contrast is defined as the ratio between the brightest and darkest areas of an image, and if brightness is measured on a scale running from 0 (darkest) to 100 (brightest), then a given image with a brightest area of 90 and a darkest area of 10 will have a contrast of 9. If scattering has the effect of adding a component of upwelling radiation of 10 units then the contrast becomes 100:20 or 5. This reduction in contrast will result in a decrease in the detectability of features present in the image. Figure 1.18 shows relative scatter as a function of wavelength for the 0.3 μm–1.0 μm region of the spectrum for a variety of levels of atmospheric haze.

Absorption is the second process by which the Earth's atmosphere interacts with incoming electromagnetic radiation. Gases such as water vapour, carbon dioxide, and ozone absorb radiation in particular regions of the electromagnetic spectrum called absorption bands. The processes involved are very complex and are related to the vibrational and rotational properties of the molecules of water vapour, carbon dioxide, or ozone, and are caused by transitions in the energy levels of the atoms. These transitions occur at characteristic wavelengths for each type of atom and at these wavelengths absorption rather than scattering is dominant. Remote sensing in these absorption bands is thus rendered impossible. Fortunately, other regions of the spectrum with low absorption (high transmission) can be used. These regions are called 'windows', and they cover the 0.3–1.3 μm (visible/near infrared), 1.5–1.8, 2.0–2.5 and 3.5–4.1 μm (middle infrared) and 7.0–15.0 μm (thermal infrared) wavebands. The utility of these regions of the electromagnetic spectrum in remote sensing is considered at a later stage.

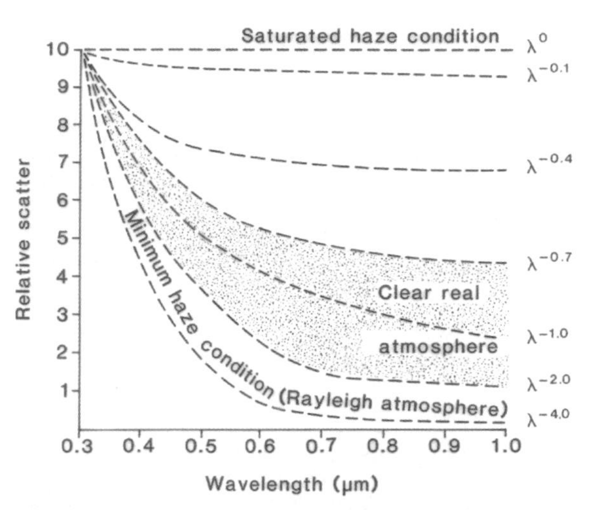

Figure 1.18 Relative scatter as a function of wavelength for a range of atmospheric haze conditions. Based on R.N. Colwell (ed.), 1983, Manual of Remote Sensing, 2nd edition, Figure 6.15. Reproduced by permission of the American Society for Photogrammetry and Remote Sensing.

1.3 INTERACTION WITH EARTH SURFACE MATERIALS

1.3.1 Introduction

Electromagnetic energy reaching the Earth's surface from the Sun is reflected, transmitted or absorbed. Reflected energy travels upwards through, and interacts with, the atmosphere; that part of it which enters the field of view of the sensor (section 2.2.1) is detected and converted into a numerical value that is transmitted to a ground receiving station on Earth. The amount and spectral distribution of the reflected energy is used in remote sensing to infer the nature of the reflecting surface. A basic assumption made in remote sensing is that specific targets (soils of different types, water with varying degrees of impurities, rocks of differing lithologies, or vegetation of various species) have an individual and characteristic manner of interacting with incident radiation that is described by the spectral response of that target. In some instances, the nature of the interaction between incident radiation and Earth surface material will vary from time to time during the year, such as might be expected in the case of vegetation as it develops from

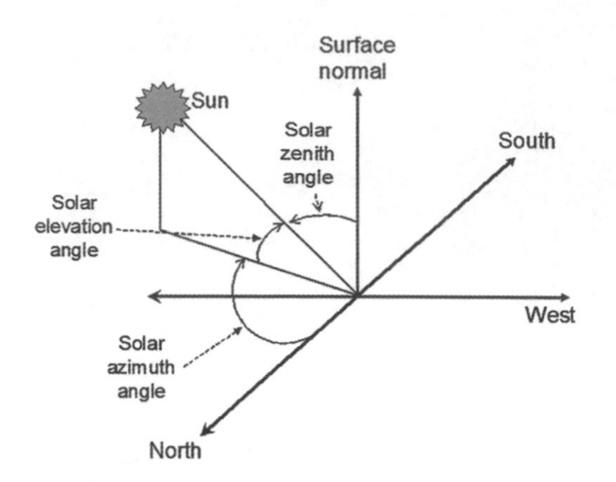

Figure 1.19 Solar elevation, zenith and azimuth angles. The elevation angle of the Sun – target line is measured upwards from the horizontal plane. The solar zenith angle is measured from the surface normal, and is equal to (90° – elevation angle). Azimuth is measured clockwise from north. In this figure, the Sun is in the south east with an azimuth of approximately 120° and a zenith angle of approximately 60°.

the leafing stage, through growth to maturity and, finally, to senescence.

The spectral response of a target also depends upon such factors as the orientation of the Sun (solar azimuth, Figure 1.19), the height of the Sun in the sky (solar elevation angle), the direction that the sensor is pointing relative to nadir (the look angle) and the state of health of vegetation, if that is the target. Nevertheless, if the assumption that specific targets are characterised by an individual spectral response were invalid then the Earth observation by remote sensing would be an impossible task. Fortunately, experimental studies in the field and in the laboratory, as well as experience with multispectral imagery, have shown that the assumption is generally a reasonable one. Indeed, the successful development of remote sensing of the environment over the last decade bears witness to its validity. Note that the term *spectral signature* is sometimes used to describe the spectral response curve for a target. In view of the dependence of spectral response on the factors mentioned above, this term is inappropriate for it gives a misleading impression of constancy.

In this section, spectral reflectance curves of vegetation, soil, rocks and water are examined in order to emphasise their characteristic features. The results summarised in the following paragraphs must not be taken to be characteristic of all varieties of materials or all observational circumstances. One of the problems met in remote sensing is that the spectral reflectance of a given Earth surface cover type is influenced by a variety of confusing factors. For example, the spectral reflectance curve of a particular agricultural crop such as wheat is not constant over time, nor is it the same for all kinds of wheat. The spectral reflectance curve is affected by factors such as soil nutrient status, the growth stage of the vegetation, the colour of the soil (which may be affected by recent weather conditions), the solar azimuth and elevation angles and the look angle of the sensor. The topographic position of the target in terms of slope orientation with respect to solar azimuth and slope angle also has an effect on the reflectance characteristics of the target, as will the state of the atmosphere. Methods for dealing with some of these difficulties are described in sections 4.5 and 4.7. Hence, the examples given in this section are idealised models rather than templates.

Before turning to the individual spectral reflectance features of Earth surface materials, a distinction must be drawn between two kinds of reflectance that occur at a surface. *Specular* reflection is that kind of reflection in which energy leaves the reflecting surface without being scattered, with the angle of incidence being equal to the angle of reflectance (Figure 1.4(a)). Surfaces that reflect *specularly* are smooth relative to the wavelength of the incident energy. *Diffuse* or Lambertian reflectance occurs when the reflecting surface is rough relative to the wavelength of the incident energy, and the incident energy is scattered in all directions (Figure 1.4(b)). A mirror reflects specularly while a piece of paper reflects diffusely. In the visible part of the spectrum, many terrestrial targets are diffuse reflectors, whereas calm water can act as a specular reflector. At microwave wavelengths, however, some terrestrial targets are specular reflectors, while volume reflectance (scattering) can occur at optical wavelengths in the atmosphere and the oceans, and at microwave wavelengths in vegetation (Figure 1.4(d)).

A satellite sensor operating in the visible and near-infrared spectral regions does not detect all the reflected energy from a ground target over an entire hemisphere. It records the reflected energy that is returned at a particular angle (see the definition of radiance in section 1.2.1). To make use of these measurements, the distribution of radiance at all possible observation and illumination angles (called the *bidirectional reflectance distribution function* or BRDF) must be taken into consideration. Details of the BRDF are given by Slater (1980) who writes:

> '. . . the reflectance of a surface depends on both the direction of the irradiating flux and the direction along which the reflected flux is detected.'

Hyman and Barnsley (1997) demonstrate that multiple images of the same area taken at different viewing angles provide enough information to allow different land cover types to be identified as a result of their differing BRDF. The Multi-Angle Imaging SpectroRadiometer (MISR) instrument, carried by the American Terra satellite, collects multi-directional observations of the same ground area over a timescale of a few minutes, at nadir and at fore and aft angles of view of 21.1°, 45.6°, 60° and 70.5° and in four spectral bands in the visible and near-infrared regions of the electromagnetic spectrum. The instrument therefore provides data for the analysis and characterisation of reflectance variation of Earth surface materials over a range of angles (Diner *et al.*, 1991). Chopping *et al.* (2003) use a BDRF model to extract information on vegetation canopy physiognomy.

It follows from the foregoing that, even if the target is a diffuse reflector such that incident radiation is scattered in all directions, the assumption that radiance is constant for any observation angle θ measured from the surface normal does not generally hold. A simplifying assumption is known as *Lambert's Cosine Law*, which states that the radiance measured at an observation angle θ is the same as that measured at an observation angle of 0° adjusted for the fact that the projection of the unit surface at a view angle of θ is proportional to $\cos\theta$ (Figure 1.20). Surfaces exhibiting this property are called 'Lambertian', and a considerable body of work in remote sensing either explicitly or implicitly assumes that Lambert's law applies. However, it is usually the case that the spectral distribution of reflected flux from a surface is more complex than the simple

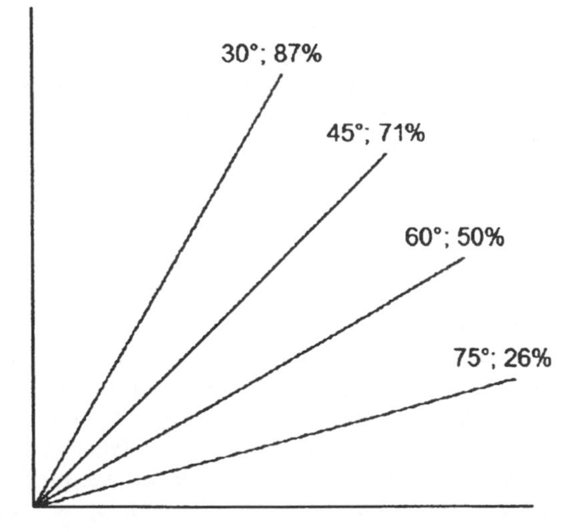

Figure 1.20 Lambert's cosine law. Assume that the illumination angle is 0°. A range of view angles is shown, together with the percentage of incoming radiance that is scattered in the direction of the view angle.

description provided by Lambert's law, for it depends on the geometrical conditions of measurement and illumination. The topic of correction of images for sun and view angle effects is considered further in Chapter 4.

1.3.2 Spectral reflectance of Earth surface materials

In this section, typical spectral reflectance curves for characteristic types of Earth surface materials are discussed. The remarks in section 1.3.1 should not be overlooked when reading the following paragraphs. The Earth surface materials that are considered in this section are vegetation, soil, bare rock and water. The short review by Verstraete and Pinty (1992) is recommended. Hobbs and Mooney (1990) provide a useful survey of remote sensing of the biosphere.

1.3.2.1 Vegetation

The reflectance spectra of three pixels selected from a Hymap imaging spectrometer data set (section 9.3) covering a forested area near the town of Thetford, eastern England (Figure 1.21) show that real-world vegetation spectra conform to the ideal pattern, though there is significant variation, especially in the near-infrared region. Both the idealised and the real curves shows relatively low values in the red and the blue regions of the visible spectrum, with a minor peak in the green region. These peaks and troughs are caused by absorption of blue and red light by chlorophyll and other pigments. Typically, 70–90% of blue and red light is absorbed to provide energy for the process of photosynthesis. The slight reflectance peak between 0.5 and 0.6 μm is the reason that actively growing vegetation appears green to the human eye. Non-photosynthetically active vegetation lacks the 'green peak'.

For photosynthetically active vegetation, the spectral reflectance curve rises sharply between about 0.65 μm and 0.76 μm, and remains high in the near-infrared region between 0.75 and 1.35 μm due to interactions between the internal leaf structure and EMR at these wavelengths. Internal leaf structure has some effect between 1.35 and 2.5 μm, but reflectance is largely controlled by leaf-tissue water content, which is the cause of the minima recorded near 1.45 μm and 1.95 μm. The status of the vegetation (in terms of photosynthetic activity) is frequently characterised by the position of a point representative of the steep rise in reflectance at around 0.7 μm. This point is called the *red edge point*, and its characterisation and uses are considered in section 9.3.2.3.

As the plant senesces, the level of reflectance in the near-infrared region (0.75–1.35 μm) declines first, with reflectance in the visible part of the spectrum not being

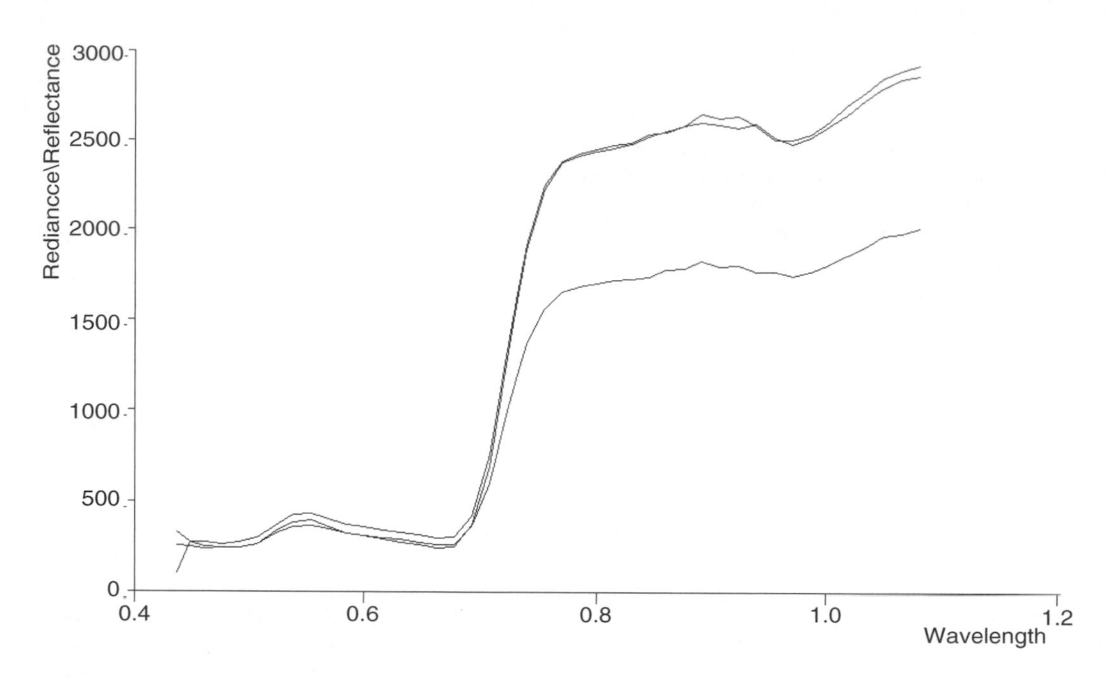

Figure 1.21 Reflectance spectra of three vegetation pixels selected from a Hymap image of the Thetford area of Norfolk, eastern England. The y-axis is graduated in units of reflectance $\times$ 1000. This plot was produced using the MIPS *Hyperspectral|Plot Hyperspectral Image Pixel* module.

affected significantly. This effect is demonstrated by the lowest of the three reflectance spectra shown in Figure 1.21. The slope of the curve from the red to the near-infrared region of the spectrum is lower, as is the reflectance in the area of the 'infrared plateau'. However, changes in reflectance in the visible region are not so apparent. As senescence continues, the relative maximum in the green part of the visible spectrum declines as pigments other than chlorophyll begin to dominate, and the leaf begins to lose its greenness and to turn yellow or reddish, depending on species. The wavelength of the red edge point also changes. Stress caused by environmental factors such as drought or by the presence or absence of particular minerals in the soil can also produce a spectral response that is similar to senescence. Areas of vegetation showing adverse effects due to the presence (or absence) of certain minerals in the soil are called geobotanical anomalies, and their distribution has been used successfully to determine the location of mineral deposits (Goetz *et al.*, 1983). Hoffer (1978) is a useful source of further information.

The shape of the spectral reflectance curve is used to distinguish vegetated and non-vegetated areas on remotely-sensed imagery. Differences between species can also be considerable, and may be sufficient to permit their discrimination, depending on the number, width, and location of the wavebands used by the sensor (section 2.2). Such discrimination may be possible on the basis of relative differences in the spectral reflectance curves of the vegetation or crop types. Absolute reflectance values (section 4.6) may be used to estimate physical properties of the vegetation, such as leaf area index (LAI) or biomass production. In agriculture, the estimation of crop yields is often a significant economic requirement. Ratios of reflectance values in two or more spectral bands are widely used to characterise vegetation (section 6.2.4). It is important to remember, however, the points made in section 1.3.1; there is no single, ideal spectral reflectance curve for any particular vegetation type, and the recorded radiance from a point on the ground will depend upon the viewing and illumination angles, as well as other variables. The geometry of the crop canopy will strongly influence the bidirectional reflectance distribution function (BRDF) (section 1.3.1), while factors such as the transmittance of the leaves, the number of leaf layers, the actual arrangement of leaves on the plant and the nature of the background (which may be soil, or leaf litter, or undergrowth) are also important. In order to distinguish between some types of vegetation, and to assess growth rates from remotely-sensed imagery, it is necessary to use *multi-temporal imagery*, that is, imagery of the same area collected at different periods in the growing season.

1.3.2.2 Geology

Geological use of remotely-sensed imagery relies, to some extent, upon knowledge of the spectral reflectance curves of vegetation, for approximately 70% of the Earth's land surface is vegetated and the underlying rocks cannot be observed directly, and differences in soil and underlying bedrock can be seen in the distribution of vegetation species, numbers of species, and vigour. Even in the absence of vegetated surfaces, weathering products generally cover the bedrock. It was noted in the preceding section that geobotanical anomalies might be used to infer the location of mineral deposits. Such anomalies include peculiar or unexpected species distribution, stunted growth or reduced ground cover, altered leaf pigmentation or yellowing (chlorosis), and alteration to the phenological cycle, such as early senescence or late leafing in the spring. It would be unwise to suggest that all such changes are due to soil geochemistry; however, the results of a number of studies indicate that the identification of anomalies in the vegetation cover of an area can be used as a guide to the presence of mineral deposits. If the relationship between soil formation and underlying lithology has been destroyed, for example by the deposition of glacial material over the local rock, then it becomes difficult to make associations between the phenological characteristics of the vegetation and lithology of the underlying rocks.

In semi-arid and arid areas, the spectral reflectance curves of rocks and minerals may be used directly in order to infer the lithology of the study area, though care should be taken because weathering crusts with spectra that are significantly different from the parent rock may develop. Laboratory studies of reflectance spectra of minerals have been carried out by Hunt and co-workers in the United States (Hunt, 1977, 1979; Hunt and Ashley, 1979; Hunt and Salisbury, 1970, 1971; Hunt *et al.*, 1971). Spectral libraries, accessible over the Internet from the Jet Propulsion Laboratory (the ASTER Spectral Library and the US Geological Survey Digital Spectral Library), contain downloadable data derived from the studies of Hunt, Salisbury, and others. These studies demonstrate that rock-forming minerals have unique spectral reflectance curves. The presence of absorption features in these curves is diagnostic of the presence of certain mineral types. Some minerals, for example quartz and feldspars, do not have strong absorption features in the visible and near-infrared regions, but can be important as dilutants for minerals with strong spectral features such as the clay minerals, sulphates and carbonates. Clay minerals have a decreasing spectral reflectance beyond 1.6 μm, while carbonate and silicate mineralogy can be inferred from the presence of absorption bands in the mid-infrared region, particularly from 2.0 to 2.5 μm. Kahle and Rowan (1980) show that multispectral thermal infrared imagery in the 8–12 μm region can be used to distinguish silicate and non-silicate rocks.

Some of the difficulties involved in the identification of rocks and minerals from the properties of spectral reflectance curves include the effects of atmospheric scattering and absorption, the solar flux levels in the spectral regions of interest (section 1.2.5) and the effects of weathering. Buckingham and Sommer (1983) indicate that the nature of the spectral reflectance of a rock is determined by the mineralogy of the upper 50 μm, and that weathering, which produces a surface layer that is different in composition from the parent rock, can significantly alter the observed spectral reflectance.

The use of multi-band imaging spectrometers mounted on aircraft and satellites can now measure the spectra of ground surface materials at a large number of closely spaced points. The interpretation of these spectra requires a detailed knowledge of the chemistry of the materials concerned. Imaging spectrometers are described in Chapter 9. Clarke (1999) gives an accessible survey of the use of imaging spectrometers in identifying surface materials.

Figure 1.22 was produced from spectral reflectance and emittance data downloaded from the NASA Jet Propulsion Laboratory's ASTER Spectral Library. Spectra for basalt and limestone are shown, for the optical and infrared regions of the spectrum. Distinctive differences are apparent. The dark basalt reflects less incoming energy than the limestone sample in the optical region of the spectrum, and shows more variability in the thermal infrared between 3 and 6 μm. The limestone sample exhibits large peaks in emissivity at around 7 and 11.5 μm. Good introductions to geological remote sensing are Drury (1993), Goetz (1989) and Gupta (1991).

1.3.2.3 Water bodies

The characteristic spectral reflectance curve for water shows a general reduction in reflectance with increasing wavelength, so that in the near infrared the reflectance of deep, clear water is virtually zero. However, the spectral reflectance of water is affected by the presence and concentration of dissolved and suspended organic and inorganic material, and by the depth of the water body. Thus, the intensity and distribution of the radiance upwelling from a water body are indicative of the nature of the dissolved and suspended matter in the water, and of the water depth. Figure 1.23 shows how the information that oceanographers and hydrologists require is only a part of the total signa received at the sensor. Solar

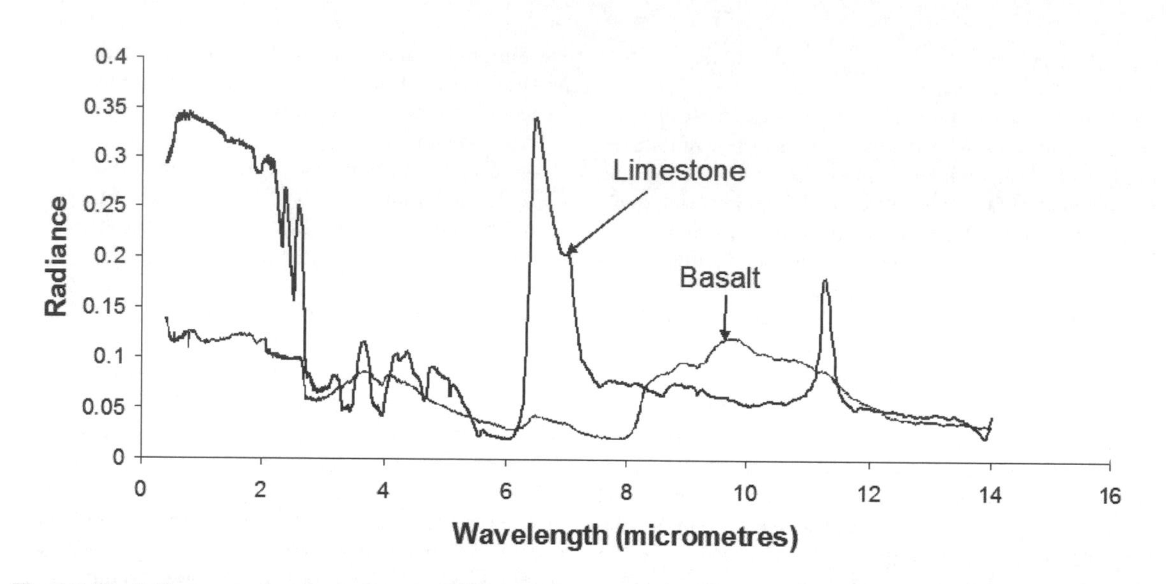

Figure 1.22 Reflectance and emittance spectra of limestone and basalt samples. Data from the ASTER Spectral Library through the courtesy of the Jet Propulsion Laboratory, California Institute of Technology, Pasadena, California. © 1999, California Institute of Technology. All Rights Reserved.

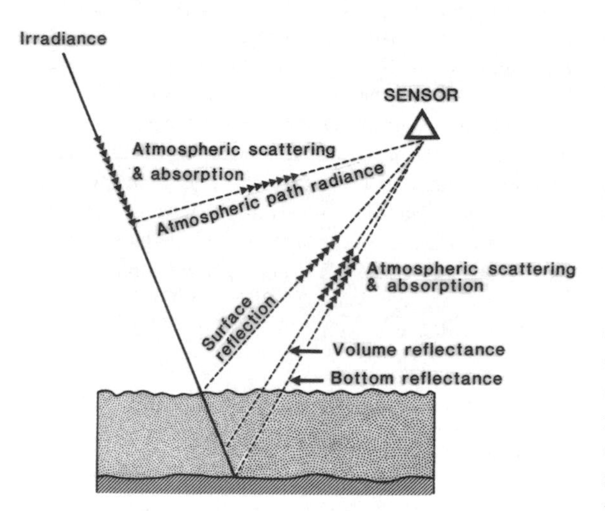

Figure 1.23 Processes acting upon solar radiant energy in the visible region of the spectrum over an area of shallow water. From T. Alföldi, 1982, Figure 1. Reproduced by permission of the Soil and Water Conservation Society of America.

irradiance is partially scattered by the atmosphere, and some of this scattered light (the path radiance) reaches the sensor. Next, part of the surviving irradiance is reflected by the surface of the water body. This reflection might be specular under calm conditions, or its distribution might be strongly influenced by surface waves and the position of the sun relative to the sensor, giving rise to *sunglint*. Once within the water body, EMR may be absorbed by the water (the degree of absorption being strongly wavelength-dependent) or selectively absorbed by dissolved substances, or backscattered by suspended particles. This latter component is termed the *volume reflectance*. At a depth of 20 m only visible light (mainly in the blue region) is present, as the near-infrared component has been completely absorbed. Particulate matter, or suspended solids, scatters the down-welling radiation, the degree of scatter being proportional to the concentration of particulates, although other factors such as the particle size distribution and the colour of the sediment are significant. Over much of the observed low to medium range of concentrations of suspended matter a positive, linear relationship between suspended matter concentration and reflectance in the visible and near-infrared bands has been observed, though the relationship becomes non-linear at increasing concentrations. Furthermore, the peak of the reflectance curve moves to progressively longer wavelengths as concentration increases, which may lead to inaccuracy in the estimation of concentration levels of suspended materials in surface waters from remotely-sensed data. Another source of error is the inhomogeneous distribution of suspended matter through the water body, which is termed *patchiness*.

The presence of chlorophyll is an indication of the trophic status of lakes and is also of importance in estimating the level of organic matter in coastal and estuarine environments. Whereas suspended matter has a generally broadband reflectance in the visible and near infrared, chlorophyll exhibits absorption bands in the region below 0.5 μm and between 0.64 and 0.69 μm. Detection of the presence of chlorophyll therefore requires an instrument with a higher spectral resolution (section 2.2.2) than would be required to detect suspended sediment. Furthermore, the level of backscatter from chlorophyll is lower than that produced by suspended sediment; consequently, greater radiometric sensitivity is also required. Ocean observing satellites (section 2.3.4) carry instruments that are 'tuned' to specific wavebands that match reflectance peaks or absorption bands in the spectra of specific oceanic materials such as chlorophyll.

Although the spectral reflectance properties of suspended matter and chlorophyll have been described separately, it is not uncommon to find both are present at one particular geographical locality. The complications in separating-out the contribution of each to the total observed reflectance are considerable. Furthermore, the suspended matter or chlorophyll may be unevenly distributed in the horizontal plane (the patchiness phenomenon noted above) and in the vertical plane. This may cause problems if the analytical technique used to determine concentration levels from recorded radiances is based on the assumption that the material is uniformly mixed at least to the depth of penetration of the radiation. In some cases, a surface layer of suspended matter may ride on top of a lower, colder, layer with a low suspended matter concentration, giving rise to considerable difficulty if standard analytical techniques are used. Reflection from the bed of the water body can have unwanted effects if the primary aim of the experiment is to determine suspended sediment or chlorophyll concentration levels, for it adds a component of reflection to that resulting from backscatter from the suspended or dissolved substances. In other instances, collection of the EMR reflected from the sea bed might be the primary focus of the exercise, in which case the presence of organic or inorganic material in the water would be a nuisance.

1.3.2.4 Soils

The spectral reflectance curves of soils are generally characterised by a rise in reflectivity as wavelength increases - the opposite, in fact, of the shape of the spectral reflectance curve for water (Figure 1.24). Reflectivity in the visible wavebands is affected by the presence of organic matter in the soil, and by the soil moisture content, while at 0.85–0.93 μm there is a ferric iron absorption band. As ferric iron also absorbs ultraviolet radiation in a broad band, the presence of iron oxide in soils is expressed visually by a reddening of the soil, the redness being due to the absorption of the wavelengths shorter (and longer) than the red. Between 1.3–1.5 μm and 1.75–1.95 μm water absorption bands occur, as mentioned in section 1.2.5. Soil reflectance in the optical part of the electromagnetic spectrum is usually greatest in the region between these two water absorption bands, and declines at wavelengths longer than 2 μm with clay minerals, if present, being identifiable by their typical,

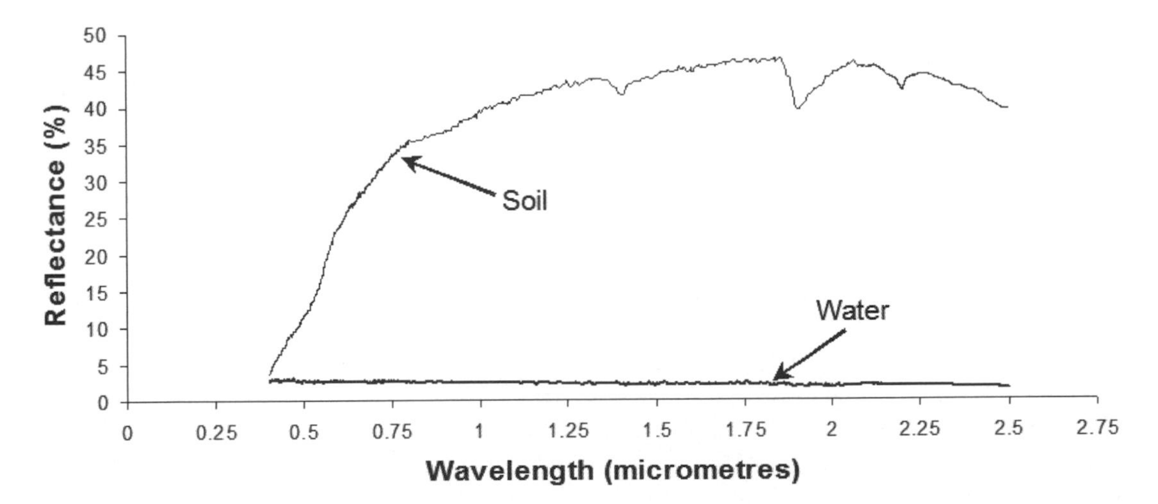

Figure 1.24 Reflectance spectra of water and soil from 0.4 μm to 2.5 μm. Data from the ASTER Spectral Library through the courtesy of the Jet Propulsion Laboratory, California Institute of Technology, Pasadena, California.

narrow-band absorption features in the 2.0–2.5 μm region. Irons *et al.* (1989) provide a comprehensive survey of factors affecting soil reflectance, while Huete (1989) gives a summary of the influence of the soil background on measurements of vegetation spectra.

1.4 SUMMARY

In the opening paragraph of section 1.3.1, a basic principle of applied remote sensing is set out. This states that individual Earth surface cover types are distinguishable in terms of their spectral reflection and emission characteristics. Changes in the spectral response of objects can also be used as an indication of changes in the properties of the object, for example the health or growth stage of a plant or the turbidity of a water body. In this chapter, the basic principles of electromagnetic radiation are reviewed briefly, and the relevant terminology is defined. An understanding of these basic principles is essential if the methods described in the remainder of this book are to be applied sensibly. Chapter 2 provides details of the characteristics of sensors that are used to measure the magnitude of electromagnetic radiation that is reflected from or emitted by the Earth surface. It is from these measurements that the spectral reflectance curves of Earth surface elements can be derived. Further details of the derivation of absolute spectral reflectance values from remotely-sensed measurements are provided in Chapter 4, while the use of data from imaging spectrometers, which can record data in tens or hundreds of bands, is described in Chapter 9. However advanced the processing techniques are, however, their results cannot be properly understood if the user does not have a good working knowledge of the material covered in Chapter 1. Readers are encouraged to consult the main references given in the text, and to familiarise themselves with the reflection spectra of natural targets, for example, by using the MIPS software that is included on the accompanying CD. MIPS contains a small spectral library which is accessed from *Plot|Plot Library Spectrum*. It includes a selection of spectra from the ASTER spectral library.

2

Remote Sensing Platforms and Sensors

2.1 INTRODUCTION

This chapter contains a description of the nature and characteristics of digital images of the Earth's surface produced by aircraft and satellite-borne sensors operating in the visible, infrared and microwave regions of the electromagnetic spectrum. The properties of specific sensors operating in each of these spectral regions are also described, and examples of typical applications are discussed. No attempt was made to provide full details of all planned, current and past platforms and sensors as that subject would require a book in itself that, furthermore, would be out of date before it was written. Examples of the most widely used satellite-borne imaging instruments, such as the HRV (carried by the SPOT satellite), TM/ETM+ (Landsat), ASTER (Terra), and SAR (RADARSAT and ERS-1/2) are used to illustrate those sections of this chapter that deal with instruments for remote sensing in the optical, near-infrared, thermal infrared and microwave wavebands. Three developing areas of remote sensing using imaging spectrometers, interferometric synthetic aperture radar, and lidar sensors are described in more detail in Chapter 9.

Remote sensing of the surface of the Earth has a long history, going back to the use cameras carried by balloons and pigeons in the eighteenth and nineteenth centuries but in its modern connotation the term *remote sensing* can be traced back to the aircraft-mounted systems that were developed during the early part of the twentieth century, initially for military purposes. Airborne film camera systems are still a very important source of remotely-sensed data (Lillesand and Keifer, 2000, Chapter 3) and spaceborne film camera systems, initially used in low Earth-orbit satellites for military purposes, have also been used for civilian remote sensing from space; for example, the NASA Large Format Camera (LFC) flown on the American Space Shuttle in October, 1984. Astronauts onboard the International Space Station can also take photographs of Earth, using any of a variety of hand-held cameras, both analogue (film) and digital.

Although analogue photographic imagery has its many and varied uses, this book is concerned with the processing of image data collected by scanning systems that ultimately generate digital image products. Analogue cameras and non-imaging (profiling) instruments such as radar altimeters are thereby excluded from direct consideration, although hard copy products from these systems can be converted into digital form by scanning and the techniques described in later chapters can be applied.

The general characteristics of imaging remote sensing instruments operating in the optical wavebands (with wavelengths less than about 3 μm) – namely, their spatial, spectral, temporal and radiometric resolution – are the subject of section 2.2. In section 2.3, the properties of images collected in the optical, near-infrared and thermal infrared regions of the electromagnetic spectrum are described. The properties of microwave imaging sensors are outlined in section 2.4.

Spatial, spectral, temporal and radiometric resolution are properties of remote sensing instruments. A further important property of the remote sensing system is the temporal resolution of the system, that is, the time that elapses between successive dates of imagery acquisition for a given point on the ground. This revisit time may be measured in minutes if the satellite is effectively stationary with respect to a fixed point on the Earth's surface (i.e., in geostationary orbit, but note that not all geostationary orbits produce fixed observation points; see Elachi, 1987) or in days or weeks if the orbit is such that the satellite moves relative to the Earth's surface. The Meteosat satellite is an example of a geostationary platform, from which imaging instruments view an entire hemisphere of the Earth from a fixed position above the equator (Figure 1.13). The NOAA (Figure 1.12), Landsat (Figures 1.10 and 1.11) and SPOT satellites are polar orbiters, each having a specific repeat cycle time (or temporal resolution) of the order of hours (NOAA) or days/weeks (Landsat, SPOT). Both the Terra and Aqua satellites are in an orbit similar to that of Landsat. Since Terra's equatorial crossing time is 10:30 and Aqua's is 13:30, it is possible to measure diurnal variations in oceanic and terrestrial systems. The temporal resolution of a polar orbiting satellite is determined by the choice of

Computer Processing of Remotely-Sensed Images: An Introduction, Third Edition. Paul M. Mather.
 ISBNs: 0-470-84918-5 (HB); 0-470-84919-3 (PB)

orbit parameters (such as orbital altitude, shape and inclination), which are related to the objectives of the particular mission. Satellites that are used for Earth-observing missions normally have a near-circular polar orbit, though the Space Shuttle flies in an equatorial orbit and some meteorological satellites use a geostationary orbit. Bakker (2000) and Elachi (1987, Appendix B) give details of the mathematics of orbit determination. The relationship between the orbit period T and the orbit radius r is given by Elachi's equation B-5:

$$T = 2\pi r\sqrt{\frac{r}{g_s R^2}}$$

in which g_s is the acceleration due to gravity (0.009 81 km s^{-1}), R is the Earth's radius (approximately 6380 km) and h is the orbital altitude (note that $r = R + h$). If, for example, $h = 705$ km then $T \approx 6052$ s ≈ 98.82 minutes. These calculations refer to the orbits of Landsat-4, -5 and -7 which is shown in Figure 2.1(a). Thus, by varying the altitude of a satellite in a circular orbit the time taken for a complete orbit is also altered; the greater the altitude, the longer the orbital period, as illustrated in Figure 2.1(b).

The angle between the orbital plane and the Earth's equatorial plane is termed the *inclination* of the orbit, which is usually denoted by the letter i. Changes in the orbit are due largely to precession, caused mainly by the slightly non-spherical shape of the Earth. If the orbital precession is the same as the Earth's rotation round the Sun then the relationship between the node line and the Sun is always the same, and the satellite will pass over a given point on the Earth's surface at the same Sun time each day. Landsat and SPOT have this kind of orbit, which is said to be *Sun-synchronous*. Figure 2.1(a) illustrates an example of a circular, near-polar, Sun-synchronous orbit, that of the later Landsat satellites, and the principle of Sun-synchronicity is illustrated in Figure 2.1(c). Many satellites carrying Earth-observing instruments use a near-polar, Sun-synchronous orbit because the fact that the Sun's azimuth angle is the same for each date of observation means that shadow effects are reduced (the shadows are always in approximately the same direction). However, the Sun's zenith angle (Figure 1.19) changes throughout the year, so that seasonal differences do occur. Some applications, such as geology, may benefit from a the use of an orbit that is not Sun-synchronous, because different views of the region of interest taken at different Sun azimuth positions may reveal structural features on the ground that are not visible at one particular Sun azimuth angle (Figure 1.19).

As noted earlier, not all Earth-observing platforms are in near-polar, Sun-synchronous orbits. The Space Shuttle has an equatorial orbit, which describes an S-shaped curve on the Earth's surface between the approximate latitudes of 50° N and 50° S. Thus, the orbit selected for a particular satellite determines not just the time taken to complete one orbit (which is one of the factors influencing temporal resolution) but also the nature of the relationship between the satellite and the solar illumination direction. The temporal resolution is also influenced by the swath width, which is the length on the ground equivalent to one scan line. Landsat TM/ETM+ has a swath width of 185 km whereas the AVHRR sensor carried by the NOAA satellites has a swath width of approximately 3000 km. The AVHRR can therefore provide much more frequent images of a fixed point on the Earth's surface than can the TM, though the penalty is reduced spatial resolution (1.1 km at nadir, compared with 30 m). A pointable sensor such as the SPOT HRV can, in theory, provide much more frequent temporal coverage than the orbital pattern of the SPOT satellite and the swath width of the HRV would indicate.

2.2 CHARACTERISTICS OF IMAGING REMOTE SENSING INSTRUMENTS

The characteristics of imaging remote sensing instruments operating in the visible and infrared spectral region can be summarised in terms of their spatial, spectral and radiometric resolutions. Other important features are the manner of operation of the scanning device that collects the image (electromechanical or electronic) and the geometrical properties of the images produced by the system. The interaction between the spatial resolution of the sensor and the orbital period of the platform determines the number of times that a given point on the Earth will be viewed in any particular time period. For observations of dynamic systems, such as the atmosphere and the oceans, a high temporal resolution is required but, because such systems are being observed globally, a coarse spatial resolution is appropriate.

2.2.1 Spatial resolution

The *spatial resolution* of an imaging system is not an easy concept to define. It can be measured in a number of different ways, depending on the user's purpose. In a comprehensive review of the subject, Townshend (1980) uses four separate criteria on which to base a definition of spatial resolution. These criteria are the geometrical properties of the imaging system, the ability to distinguish between point targets, the ability to measure the periodicity of repetitive targets, and the ability to measure the spectral properties of small targets. These will be considered briefly here; a fuller discussion can be found in Billingsley (1983), Forshaw *et al.* (1983), Simonett (1983) and Townshend (1980).

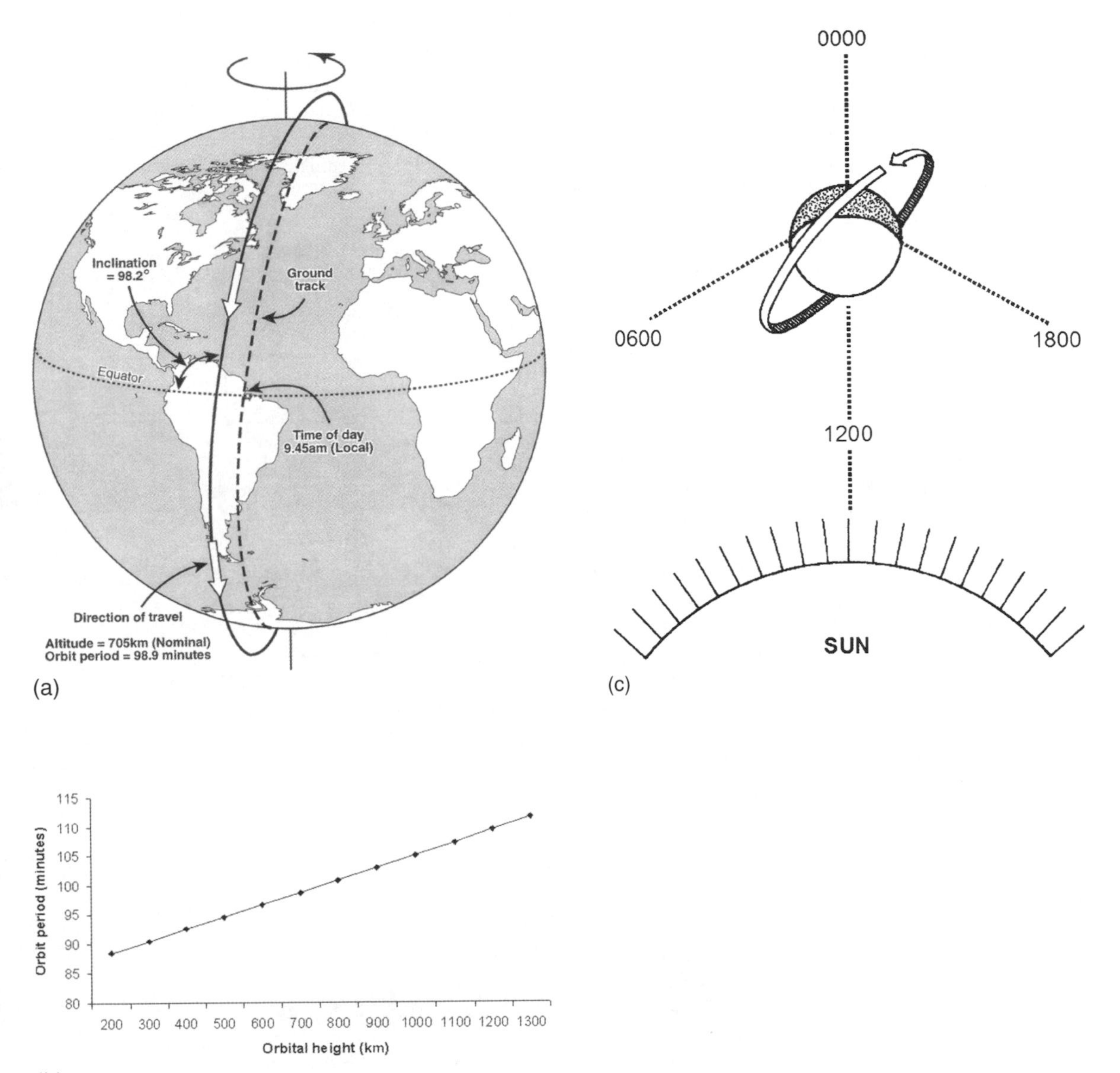

Figure 2.1 (a) Example of a Sun-synchronous orbit. This is the Landsat-4 and -5 orbit, which has an equatorial crossing time of 0945 (local Sun time) in the descending node. The satellite travels southwards over the illuminated side of the Earth. The Earth rotates through 24.7° during a full satellite orbit. The satellite completes just over 14.5 orbits in a 24-hour period. The Earth is imaged between 82°N and S latitudes over a 16-day period. Landsat-7 orbit is similar. Landsat-7 travels about 15 minutes ahead of Terra, which is in the same orbit, so that sensors onboard Terra view the Earth under almost the same conditions as does the Landsat-7 ETM+. Based on Figure 5.1, Landsat 7 Science Data Users Handbook, NASA Goddard Spaceflight Center, Greenbelt, Maryland. http://ltpwww.gsfc.nasa.gov/IAS/handbook/handbook_ toc.htm (b) Relationship between the orbit period and orbit height for a polar-orbiting satellite. (c) Illustrating the principles of a Sun-synchronous orbit. The rotation of the Earth around the Sun, the inclination of the orbit and the orbital period are such that the satellite crosses the Equator at the same local Sun time on each orbit.

The most commonly used measure, based on the geometric properties of the imaging system, is its *instantaneous field of view* (IFOV). The IFOV is defined as the area on the ground that, in theory, is viewed by the instrument from a given altitude at any given instant in time. The IFOV can be measured in one of two ways, as the angle α or as the equivalent distance X–Y on the ground in Figure 2.2. Note that Figure 2.2 is a cross section, and X–Y is, in fact, the diameter of a circle. The actual, as distinct from the nominal, IFOV depends on a number of factors. No satellite has a perfectly stable orbit; its height above the Earth will vary, often by tens of kilometres. Landsat-1 to -3 had a nominal altitude of 913 km, but the actual altitude of these satellites varied between 880 and 940 km. The IFOV becomes smaller at lower altitudes and increases as the altitude increases, although the spatial resolution of Landsat-1 to -3 Multispectral Scanner (MSS) (section 2.3.6.1) is generally specified as 79 m, the actual resolution (measured by the IFOV) varied between 76 and 81 m.

The IFOV is the most frequently cited measure of resolution, though it is not necessarily the most useful. In order to explain why this is so, we must consider the way in which radiance from a point source is expressed on an image. A highly reflective point source on the ground does not produce a single bright point on the image but is seen as a diffused circular region, due to the properties of the optics involved in imaging. A cross-section of the recorded or imaged intensity distribution of a single point source is shown in Figure 2.3, from which it can be seen that the intensity of a point source corresponds to a Gaussian-type distribution. The actual shape will depend upon the properties of the optical components of the system and the relative brightness of the point source. The distribution function shown in

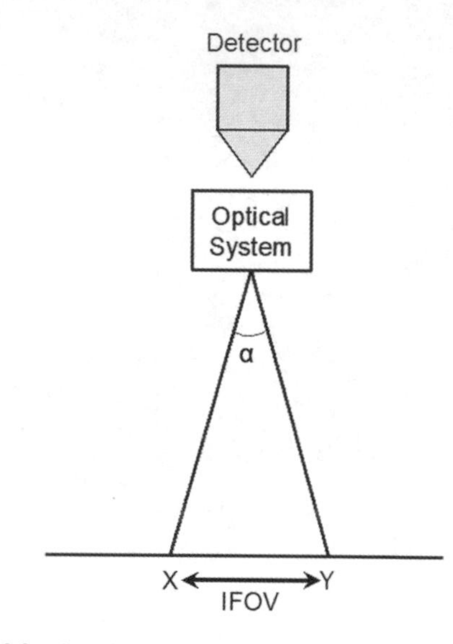

Figure 2.2 Angular instantaneous field of view (IFOV), α, showing the projection X–Y on the ground. Note that X–Y is the diameter of a circle.

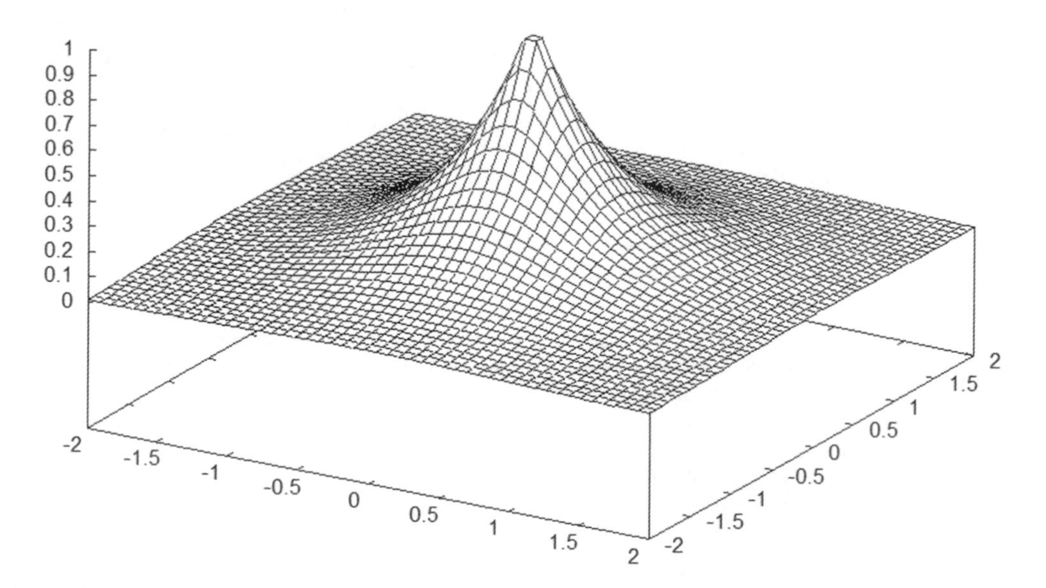

Figure 2.3 Point spread function. The area of the pixel being imaged runs from $-0.5 \leq x \leq 0.5$ and $-0.5 \leq y \leq 0.5$, i.e. centred at (0,0,) but the energy collected by the sensor is nonzero outside this range. The ideal point spread function would be a square box centred at (0,0) with a side length of 1.0.

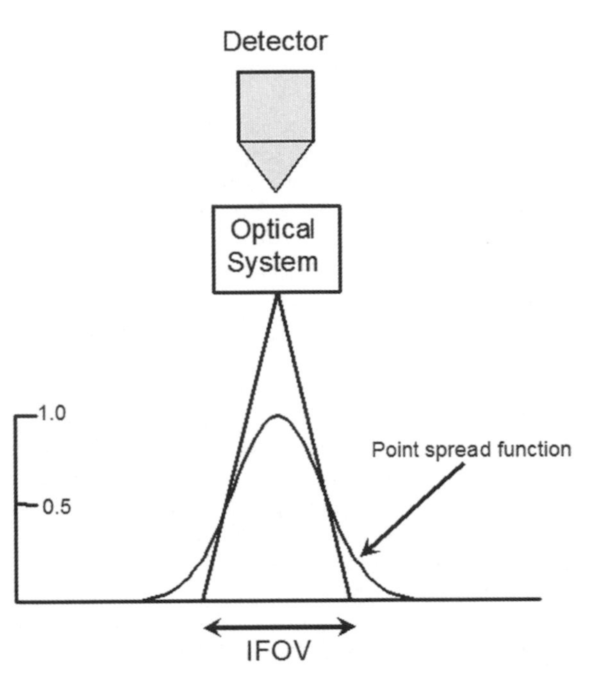

Figure 2.4 Instantaneous field of view defined by the amplitude of the point spread function.

Figure 2.3 is called the *point spread function* or PSF (Moik, 1980; Slater, 1980; Billingsley, 1983).

An alternative measure of IFOV is, in fact, based on the PSF (Figure 2.4) and the 30 m spatial resolution of the Landsat-4 and -5 Thematic Mapper (TM) (section 2.3.6.2) is based upon the PSF definition of the IFOV. The IFOV of the Landsat MSS using this same measure is 90 m rather than 79 m (Townshend, 1980, p. 9). The presence of relatively bright or dark objects within the IFOV of the sensor will increase or decrease the amplitude of the PSF to make the observed radiance either higher or lower than that of the surrounding areas. This is why high-contrast features such as narrow rivers and canals are frequently visible on Landsat ETM+ images, even though their width is less than the sensor's spatial resolution of 30 m. Conversely, targets with dimensions larger than the Landsat ETM+ IFOV of 30 m may not be discernible if they do not contrast with their surroundings. The blurring effects of the PSF can be partially compensated for by image processing involving the use of the Laplacian function (section 7.3.2). Other factors causing loss of contrast on the image include atmospheric scattering and absorption, which are discussed in Chapters 1 and 4.

The definition of spatial resolving power based on the IFOV is therefore not a completely satisfactory one. As it is a geometrical definition, it does not take into account the spectral properties of the target. If remote sensing is based upon the detection and recording of the radiance of targets, the radiance being measured at a number of discrete points, then a definition of spatial resolution that takes into account the way in which this radiance is generated might be reasonable. This is the basis of the definition of the *effective resolution element* or ERE which is defined by Colvocoresses (cited by Simonett, 1983) as

> 'the size of an area for which a single radiance value can be assigned with reasonable assurance that the response is within 5 per cent of the value representing the actual relative radiance'.

Colvocoresses estimated the ERE for the Landsat MSS system as 86 m and 35 m for the TM. These values might be more relevant than the IFOV for a user interested in classification of multispectral images (Chapter 8).

Other methods of measuring the spatial resolving power of an imaging device are based upon the ability of the device to distinguish between specific targets. There are two such measures in use, and both are perhaps more easily defined for photographic sensors. The first method uses the fact that the PSF of a point source is a bright central disc with bright and dark rings around it. The Rayleigh criterion assumes two equally bright point sources and specifies that the two sources will be distinguishable on the image if the bright central disc of one falls on the first dark ring of the PSF of the other. The minimum separation between the point sources to achieve this degree of separation on the image is a measure of the spatial resolving power of the imaging system. The second method assumes that the targets are not points but linear and parallel objects with a known separation that is related to their spatial frequency. If the objects contrast strongly with their background, then one could consider moving them closer together until the point is reached where they are no longer distinguishable. The spatial frequency of the objects such that they are just distinguishable is a measure of spatial resolving power. This spatial frequency is expressed in terms of line pairs per millimetre on the image or as cycles per millimetre. In order to calculate the spatial resolving power by this method it is common to measure the contrast of the targets and their background. The measure most often used is *modulation* (M), defined as:

$$M = \frac{E_{\max} - E_{\min}}{E_{\max} + E_{\min}},$$

where $E_{\max}$ and $E_{\min}$ are the maximum and minimum radiance values recorded over the area of the image. For a nearly homogeneous image, M would have a value close to zero while the maximum value of M is 1.0. Note that a perfectly homogeneous image would have an undefined modulation value since the calculation would involve division by zero. Returning to the idea of the parallel linear targets, we could find the ratio of the modulation measured from the image (M_I) to the modulation of the objects themselves (M_O) This gives the modulation transfer factor. A graph of this factor against spatial frequency shows the *Modulation Transfer Function* or MTF. The spatial frequency at which the MTF falls to a half of its maximum value is termed the *Effective Instantaneous Field of View* or EIFOV. The EIFOV of Landsat MSS imagery has been computed to be 66 m while the 30 m resolution of the Landsat TM is computed from the spatial frequency at the point where the MTF has 0.35 of its maximum value (Townshend, 1980, p. 12), whereas the IFOV measure gives resolutions of 79 and 45 m, respectively. The EIFOV measure is based on a theoretical target rather than on real targets, and as such gives a result that is likely to exceed the performance of the instrument in actual applications. Townshend and Harrison (1984) describe the calculation and use of the MTF in estimating the spatial resolving power of the Landsat-4 and -5 Thematic Mapper instrument (section 2.3.6.2).

The IFOV, ERE and EIFOV should not be confused with the pixel size. A digital image is an ordered set of numeric values, each value being related to the radiance from a ground area represented by a single cell or *pixel*. The pixel dimensions need not be related to the IFOV, the ERE or the EIFOV. For instance, the size of the pixel in a Landsat MSS image is specified as 56 by 79 m. The IFOV at the satellite's nominal altitude of 913 km is variously given as 79 m, 76.2 m and 73.4 m while the ERE is estimated as 87 m (Simonett, 1983, Tables 1 and 2). Furthermore, pixel values can be interpolated over the cells of the digital image to represent any desired ground spacing, using one of the resampling methods described in Chapter 4. The ground area represented by a single pixel of a Landsat MSS image is thus not necessarily identical to the spatial resolution as measured by any of the methods described above.

The discussion in the preceding paragraphs, together with the description given below of the way in which satellite-borne sensors operate, should make it clear that the individual image pixel is not 'sensed' uniquely. This would require a stop/start motion of the satellite and, in the case of the electromechanical scanner, of the scan mirror. The value recorded at the sensor corresponding to a particular pixel position on the ground is therefore not just a simple average of the radiance upwelling from that pixel. There is likely to be a contribution from areas outside the IFOV, which is termed the 'environmental radiance' (section 4.4). Disregarding atmospheric effects (which are also discussed in section 4.4), the radiance attributable to a specific pixel sensed over an area of terrestrial vegetation is in reality the sum (perhaps the weighted sum) of the contributions of the different land cover components within the pixel, plus the contribution from radiance emanating from adjacent areas of the ground. Fisher (1997) and Cracknell (1998) examine the concepts underlying the idea of the pixel in some detail, and readers should be aware of the nature of the image pixel in remote sensing before attempting to interpret patterns or features seen on images.

Spatial resolving power is an important attribute of remote sensing systems because differing resolutions are relevant to different problems; indeed, there is a hierarchy of spatial problems that can use remotely-sensed data, and there is a spatial resolution appropriate to each problem. To illustrate this point, consider the use of an image with a spatial resolution (however defined) of 1 m. Each element of the image, assuming its pixel size was 1 m × 1 m, might represent the crown of a tree, part of a grass verge by the side of a suburban road or the roof of a car. This imagery would be useful in providing the basis for small-scale mapping of urban patterns, analysis of vegetation variations over a small area or the monitoring of crops in small plots. At this scale it would be difficult to assess the boundaries of, or variation within, a larger spatial unit such as a town; a spatial resolution of 10 m might be more appropriate to this problem. A 10 m resolution would be a distinct embarrassment if the exercise was concerned with the mapping of sea-surface temperature patterns in the Pacific Ocean, for which data from an instrument with a spatial resolution of 500 m or larger could be used. For continental or global-scale problems a spatial resolution of 1 km and 5 km respectively would produce data which contained the information required (and no more) and, in addition, was present in handleable quantities. The illusion that higher spatial resolution is necessarily better is commonplace; one should always ask 'better for what?' See Atkinson and Curran (1997) for further discussion of this point.

2.2.2 Spectral resolution

The second important property of an optical imaging system is its *spectral resolution*. Microwave images collected by instruments onboard satellites such as ERS-1 and -2, J-ERS and RADARSAT (section 2.4) are generally recorded in a single waveband, as are panchromatic images collected by sensors such as IKONOS, SPOT HRV and Landsat-7 ETM+. A panchromatic (literally 'all colours')

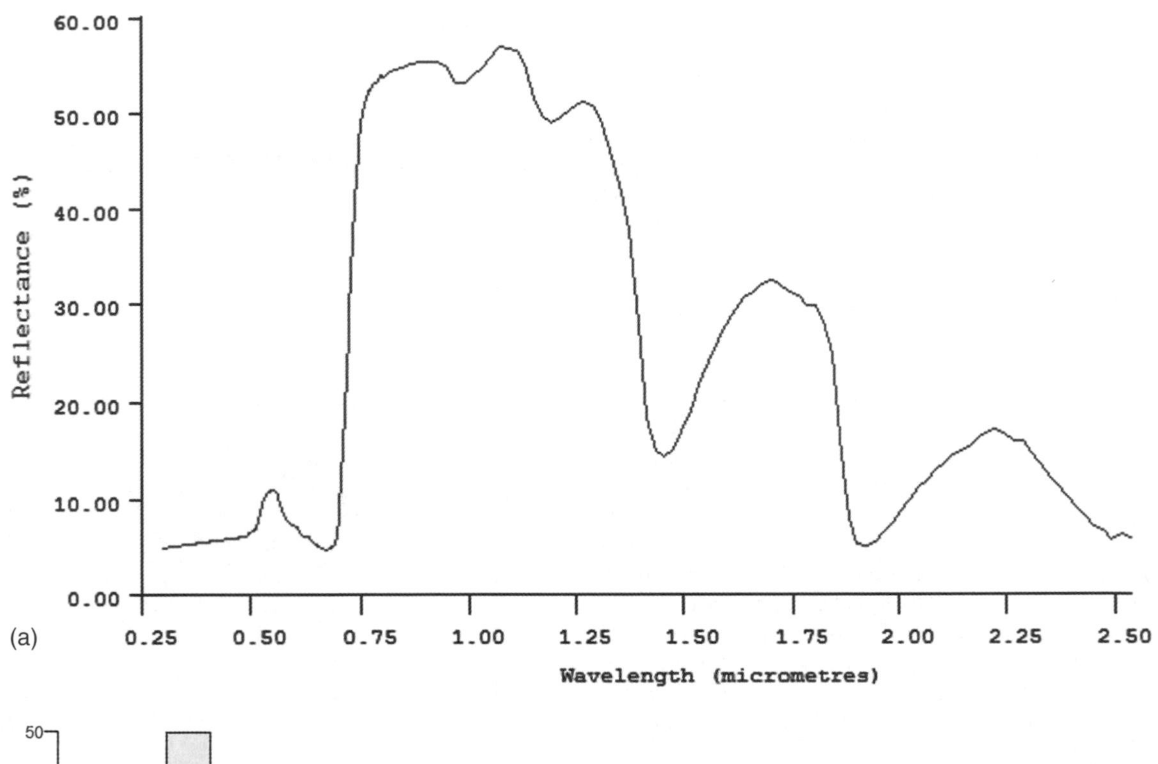

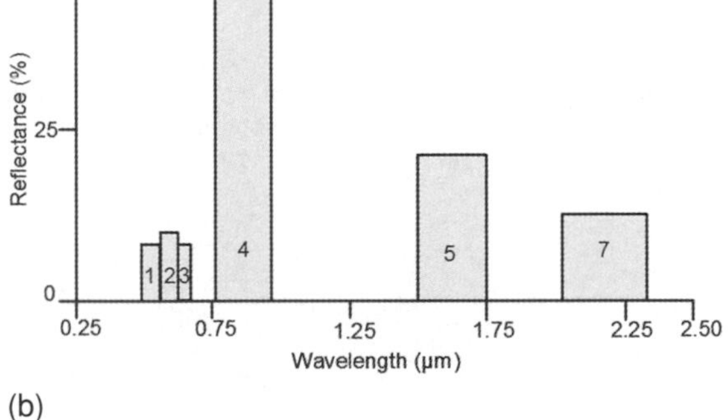

Figure 2.5 (a) Spectral reflectance curve for a leaf from a deciduous tree. Data from the ASTER Spectral Library through the courtesy of the Jet Propulsion Laboratory, California Institute of Technology, Pasadena, California. © 1999, California Institute of Technology. All Rights Reserved. (b) The reflectance spectrum shown in Figure 2.5 (a) as it would be recorded by Landsat ETM+ bands 1–5 and 7.

image is a single band image that measures upwelling radiance in the visible wavebands. Most sensors operating in the visible and infrared bands collect multispectral or multi-band images, which are sets of individual images that are separately recorded in discrete spectral bands, as shown in Figures 1.10 and 1.11. The term *spectral resolution* refers to the width of these spectral bands measured in micrometres (µm) or nanometres (nm). The following example illustrates two important points, namely, that (i) the position in the spectrum, width and number of spectral bands determines the degree to which individual targets (vegetation species, crop or rock types) can be discriminated on the multispectral image, and (ii) the use of multispectral imagery can lead to a higher degree of discriminating power than any single band taken on its own.

The reflectance spectra of vegetation, soils, bare rock and water are described in section 1.3. Differences between the reflectance spectra of various rocks, for example, might be very subtle and the rock types might therefore be separable only if the recording device was capable of detecting the spectral reflectance of the target in a narrow waveband. A wide-band instrument would simply average the differences. Figure 2.5(a) is a plot of the reflection from a leaf from a deciduous tree against wavelength. Figure 2.5(b) shows the reflectance spectra for the same target as recorded by a broad-band sensor such as the Landsat-7 ETM+. Broad-band sensors will, in general, be unable to distinguish subtle differences in reflectance spectra, perhaps resulting from disease or stress. To provide for the more reliable identification of particular targets on a remotely-sensed image the spectral resolution of the sensor must match as closely as possible the spectral reflectance curve of the intended target. This principle is demonstrated by the design of the Coastal Zone Colour Scanner (CZCS) carried by the Nimbus-7 satellite, a design that is common to most ocean-observing satellites, namely, the width and positions of the spectral bands are determined by the spectral reflectance curve of the target. Of course, other considerations must be balanced, such as frequency of coverage or temporal resolution and spatial resolution (section 2.2.1), as well as practical factors.

Only in an ideal world would it be possible to increase the spectral resolution of a sensor simply to suit the user's needs. There is a price to pay for higher resolution. All signals contain some noise or random error that is caused by electronic noise from the sensor and from effects introduced during transmission and recording. The signal-to-noise ratio (SNR) is a measure of the purity of a signal. Increasing the spectral resolution reduces the SNR of the sensor output because the magnitude of the radiance (the signal strength) in narrower spectral bands is less than that of wider bands while the inherent noise level remains constant. Smith and Curran (1996) provide details of SNR calculations, and estimate values for the AVHRR, Landsat TM and SPOT HRV as 38:1, 341:1 and 410:1, respectively.

A compromise must be sought between the twin requirements of narrow bandwidth (high spectral resolution) and a low signal-to-noise ratio. The *push-broom* type of sensor is a linear array of individual detectors with one detector per scan line element. The forward movement of the sensor forms the image one scan line at a time (Figure 1.2(b)). This arrangement provides for a longer 'look' at each scan line element, so more photons reflected from the target are collected, which results in a better signal-to-noise ratio (SNR) than does an electromechanical scanner employing a single detector, which observes each scan line element sequentially (Figure 1.2(a)). The time available to measure the energy emanating from each point along the scan line (termed the dwell time or integration time) is greater for push-broom scanners, because the individual detector 'sees' more photons coming from a given point than the detector in an electromechanical scanner, which looks at the point only for a short time. Hence, narrower bandwidths and a larger number of quantisation levels are theoretically possible without decreasing the SNR to unacceptable levels. Further discussion of different kinds of sensors is contained in section 2.3.

Justification for the use of multiple rather than single measures of the characteristics of individual Earth surface objects helps to discriminate between groups of different objects. For example, Mather (1976, p. 421) contends that:

> '... The prime justification of adopting a multivariate approach [is] that significant differences (or similarities) may well remain hidden if the variables are considered one at a time and not simultaneously'.

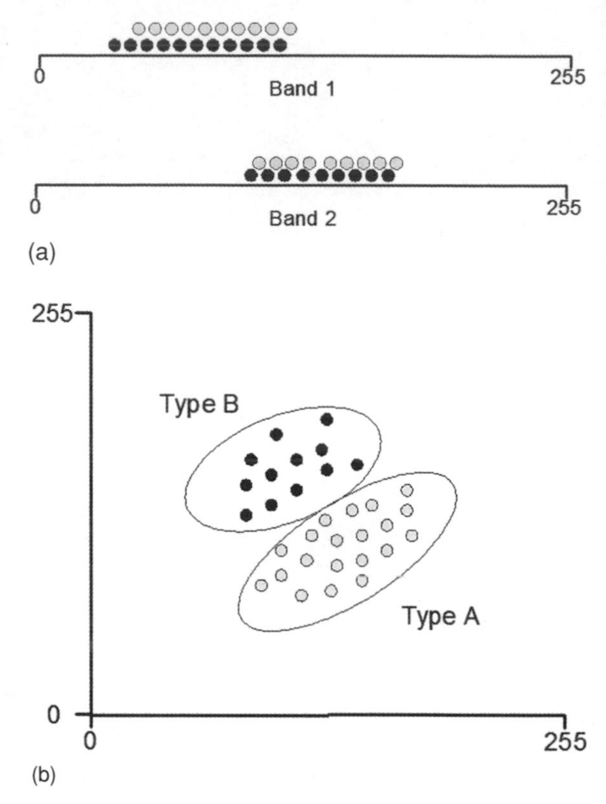

Figure 2.6 Hypothetical land cover types A and B measured on two spectral bands. (a) Shown as plots on separate bands; the two cover types cannot be distinguished. (b) Shown as a plot on two bands jointly; now the two cover types can be separated in feature space.

In Figure 2.6(a) the measurements of the spectral reflectance values for individual members of two hypothetical land-cover types are plotted separately for two spectral bands, and there is an almost complete overlap between them. If the measurements on the two spectral bands are plotted together, as in Figure 2.6(b), the difference between the types is clear, as there is no overlap. The use of well chosen and sufficiently numerous spectral bands

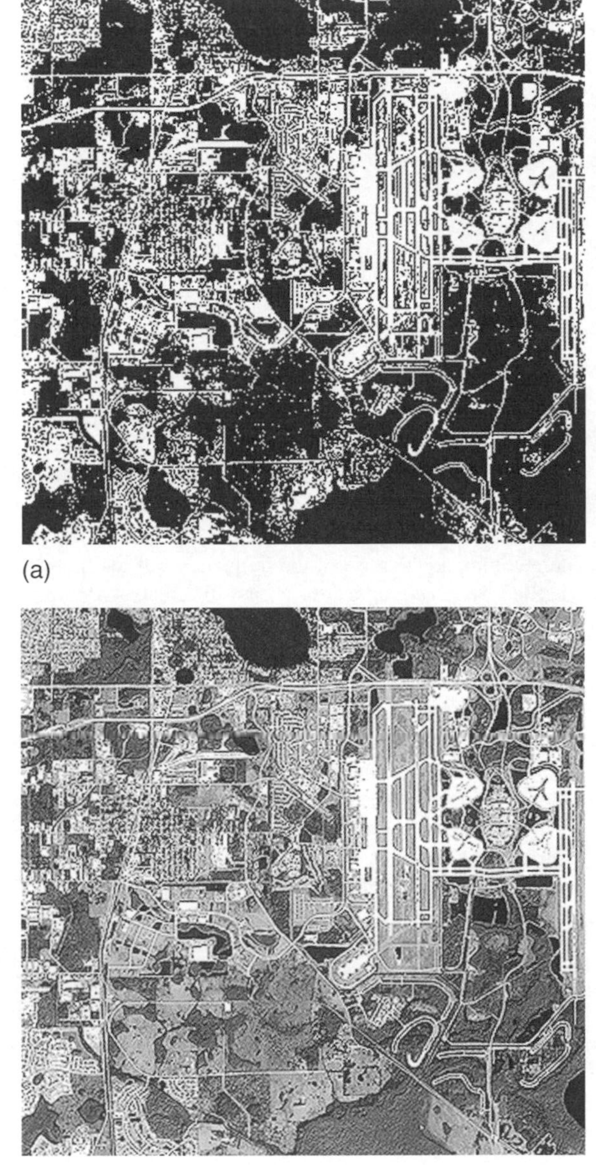

(a)

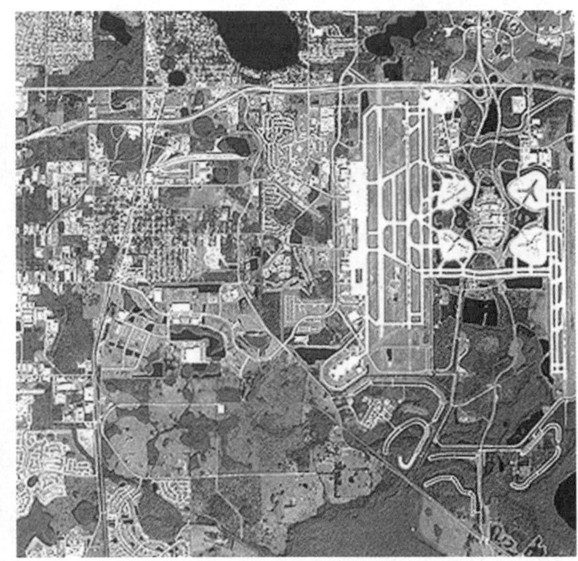

(b)

(c)

Figure 2.7 SPOT HRV panchromatic image of part of Orlando, Florida, displayed in (a) two grey levels, (b) 16 grey levels, and (c) 256 grey levels. © CNES – SPOT Image Distribution. Permission to use the data was kindly provided by SPOT Image, 5 rue des Satellites, BP 4359, F-31030 Toulouse, France (http://www.spotimage.fr).

is a requirement, therefore, if different targets are to be successfully identified on remotely-sensed images.

2.2.3 Radiometric resolution

Radiometric resolution or radiometric sensitivity refers to the number of digital quantisation levels used to express the data collected by the sensor. In general, the greater the number of quantisation levels the greater the detail in the information collected by the sensor. At one extreme one could consider a digital image composed of only two levels (Figure 2.7(a)) in which level 0 is shown as black and level 1 as white. As the number of levels increases to 16 (Figure 2.7(b)) so the amount of detail visible on the image increases. With 256 levels of grey (Figure 2.7(b)) there is no discernible difference, though readers should note that this is as much a function of printing technology as of their visual systems. Nevertheless, the eye is not as sensitive to changes in intensity as it is to variations in hue.

The number of grey levels is commonly expressed in terms of the number of binary (base two) digits (bits) needed to store the value of the maximum grey level[1]. Just two binary digits, 0 and 1, are used in a base two number rather than the 10 digits (0–9) that are used in base 10 representation. Thus, for a two level or black/white representation, the number of bits required per pixel is 1 (giving two states – 0 and 1), while for 4, 16, 64 and 256 levels the number of bits required is 2, 4, 6, and 8, respectively. Thus, '6-bit' data has 64 possible quantisation

[1] Base 2 representation of numbers is considered at greater length in Chapter 3.

levels, represented by the integer values 0 to 63 inclusive (000000 to 111111 in base 2 notation), with '0' representing black, '63' representing white, and '31' indicating mid-grey. These 'numbers' are enclosed in quotation marks to show that they are not indicating a direct measurement of ground-leaving radiance. They are the steps or quanta into which a range of physical values is divided, hence the term 'quantisation', which – in this context – means 'breaking down a continuous range into a discrete number of steps'. If, for instance, a sensor can detect radiance in the range 0.0 to 10.0 W m^{-2} sr^{-1} μm^{-1} and the signal-to-noise ratio of the instrument is such that this range can be divided into 256 levels (for reasons explained in the next paragraph) then level '0' would represent 0.0 W m^{-2} sr^{-1} μm^{-1} and level '255' would represent 10.0 W m^{-2} sr^{-1} μm^{-1}. The intermediate levels ('1' to '254') represent equal steps of $(10.0–0.0)/255.0 = 0.00392$ W m^{-2} sr^{-1} μm^{-1}.

It is needless to mention that the number of levels used to express the signal received at a sensor cannot be increased simply to suit the user's preferences. The signal-to-noise ratio (SNR) of the sensor, described above, must be taken into consideration. The step size from one level to the next cannot be less than the noise level, otherwise it would be impossible to say whether a change in level was due to a real change in the radiance of the target or to a change in the magnitude of the noise. A low-quality instrument with a high noise level would, therefore, necessarily have a lower radiometric resolution compared with a high-quality, high signal-to-noise-ratio instrument. Slater (1980) discusses this point in some detail.

Tucker (1979) investigated the relationship between radiometric resolution and the ability to distinguish between vegetation types; he found only a 2 to 3% improvement for 256 levels over 64-level imagery. Bernstein *et al.* (1984) used a measure known as entropy to compare the amount of information (in terms of bits per pixel) for 8-bit and 6-bit data for two images of the Chesapeake Bay area collected by Landsat-4's TM and MSS sensors. The entropy measure, H, is given by:

$$H = -\sum_{i=0}^{k} p(i) \log_2 p(i)$$

where k is the number of grey levels (for example, 64 or 256) and $p(i)$ is the probability of level i, which can be calculated from:

$$p(i) = \frac{F(i)}{nm}$$

In this formula, $F(i)$ is the frequency of occurrence of each grey level from 0 to $k-1$, and nm is the number of pixels in the image. Moik (1980, p. 296) suggests that the use of this estimate of $p(i)$ will not be accurate because adjacent pixel values will be correlated (the phenomenon of spatial autocorrelation, as described by Cliff and Ord, 1973) and he recommends the use of the frequencies of the first differences in pixel values; these first differences are found by subtracting the value at a given pixel position from the value at the pixel position immediately to the left, with obvious precautions for the leftmost column of the image. Bernstein *et al.* (1984) do not indicate which method of estimating entropy they use but, since the same measure was applied to both the 6-bit and 8-bit images, the results should be comparable. The 8-bit resolution image added, on average, one to one-and-a-half bits to the entropy measure compared with a 6-bit image of the same scene. These results are shown in Table 2.1. Moik (1980, Table 9.1) used both measures of entropy given above; for the first measure (based on the levels themselves) he found average entropy to be 4.4 bits for Landsat MSS (6-bit) and, for the second measure (based on first differences) he found average

Table 2.1 Entropy by band for Landsat TM and MSS sensors based on Landsat-4 image of Chesapeake Bay area, 2 November 1982 (scene E-40109-15140). See text for explanation. Based on Table 1 of Bernstein *et al.* (1984).

	Thematic Mapper		Multispectral Scanner	
Band number	Waveband (μm)	Bits per pixel	Waveband (μm)	Bits per pixel
1	0.45–0.52	4.21	0.5–0.6	2.91
2	0.52–0.60	3.78	0.6–0.7	3.57
3	0.63–0.69	4.55	0.7–0.8	4.29
4	0.76–0.90	5.19	0.8–1.1	3.63
5	1.55–1.75	5.92		
6	10.5–12.5	3.53		
7	2.08–2.35	5.11		
Average		4.61		3.60

entropy to be 4.2 bits. This value is slightly higher than that reported by Bernstein *et al.* (1984) but still less than the entropy values achieved by the 8-bit TM data. It is interesting to note that, using data compression techniques, the 6-bit data could be re-expressed in 4 or so bits without losing any information, while the 8-bit data conveyed, on average, approximately 5 bits of information. Given the enormous amounts of data making up a multi-band image (nearly 300 million pixel values for one Landsat TM image) such measures as entropy are useful both for comparative purposes and for indicating the degree of redundancy present in a given data set.

Entropy measures can also be used to compare the performance of different sensors. For example, Kaufmann *et al.* (1996) use entropy to compare the German MOMS sensor performance with that of Landsat TM and the SPOT HRV, and Masek *et al.* (2001) use the same measure to compare images of a test area acquired simultaneously by Landsat-5's TM and Landsat-7's ETM+ instruments.

The concepts of spatial, radiometric and temporal resolution are discussed concisely by George (2000).

2.3 OPTICAL, NEAR-INFRARED AND THERMAL IMAGING SENSORS

The aim of this section is to provide brief details of a number of sensor systems operating in the visible and infrared regions of the electromagnetic spectrum, together with sample images, in order to familiarise the reader with the nature of digital image products and their applications. Further details of the instruments described in this section, as well as others that are planned for the next decade, are available from the Committee on Earth Observation Satellites (CEOS) via the World Wide Web. A comprehensive survey is provided by Joseph (1996). In this chapter, attention is focused on sensors that in current use, especially the Landsat and SPOT systems, which have generated large and widely used data sets since 1972 and 1986, respectively. Sufficient detail is provided for readers to inform themselves of the main characteristics of each system. More information is provided in the references cited. Readers can also access the image files stored on the accompanying CD (which will be found in the MIPS *Images* folder if the software was properly installed) to view examples of remotely-sensed images acquired by sensors operating in the visible, near-infrared and thermal regions of the spectrum.

2.3.1 Along-Track Scanning Radiometer (ATSR)

The ATSR was developed by a consortium of British universities and research institutes led by the UK Rutherford-Appleton Laboratory (RAL). It has three thermal infrared channels centred at 3.7 μm, 10.8 μm and 12 μm, plus a near-infrared channel centred at 1.6 μm. These thermal channels correspond to bands 3, 4 and 5 of the NOAA AVHRR. The primary purpose of this instrument is to make accurate measurements of global sea surface temperatures for climate research purposes. It was flown on the ERS-1 satellite, which was launched in July 1991. ATSR-2 is a development of ATSR, and has additional channels centred at wavelengths of 0.555 μm, 0.659 μm and 0.865 μm that are intended for land applications. It is carried by the ERS-2 satellite, which was launched in April 1995. ATSR is an interesting instrument (Harries *et al.*, 1994) as it provides two views of the target from different angles – one from the nadir and one in the forward direction. Each view includes an internal calibration system; hence, the data can be corrected for atmospheric effects (which can be estimated from the two views) and calibrated more precisely. The thermal sensor has a nominal accuracy of ±0.05 K. The sensor produces images with a spatial resolution of 1 km for a 500 km swath width. See Example 2.1 for more information on ATSR.

EXAMPLE 2.1: ALONG-TRACK SCANNING RADIOMETER (ATSR)

An ATSR image of the Bering Straits on 5 February 1992 is shown in Figure 2.1. The Bering Straits separate Alaska and Siberia. The image is taken in the 1.6 μm band, which

Example 2.1 Figure 1 ATSR image of the Bering Straits taken in the 1.6 μm band.

covers a transmittance peak (atmospheric window) in the near-infrared region of the electromagnetic spectrum. ATSR has a spatial resolution of 1 km at nadir, and the region shown in the image has an areal extent of 512 × 512 km. Enhancement methods (section 5.3) are used to improve the contrast of the image, and filtering using an 'image minus Laplacian' operation is used to sharpen the detail (section 7.3.2). A tongue of cloud is the brightest feature on the right-hand side of the image. Clear water appears dark. The many ice floes cover the remaining water. Sea ice monitoring is a major application of wide field of view imagery such as this. Different types of ice can be identified using classification methods (Chapter 8).

Image source: http://www.atsr.rl.ac.uk/images/sample/atsr-1/index.shtml

An extended version of ATSR-2, called the Advanced Along Track Scanning Radiometer, is carried by the European Space Agency's ENVISAT-1 satellite, which was launched in 2002. The AATSR has seven channels (0.55, 0.67, 0.86, 1.6, 3.7, 11 and 12 μm). The spatial resolution of the visible and near-infrared channels (bands 1–4) is 500 m (with a swath width of 500 km). The infrared channels (bands 5–7) have a spatial resolution of 1 km.

2.3.2 Advanced Very High Resolution Radiometer (AVHRR)

The AVHRR, which is carried by the American NOAA (National Oceanic and Atmospheric Administration) series of satellites, was intended to be a meteorological observing system. The imagery acquired by AVHRR has, however, been widely used in land cover monitoring at global and continental scales. Two NOAA satellites are in orbit at any one time, giving morning, afternoon, evening and night-time equatorial crossing times of approximately 0730, 1400, 1930 and 0200 respectively, though it should be noted that the illumination and view geometry is not the same for all of these overpasses. The NOAA satellite orbit repeats exactly after 9 days. The orbital height is 833–870 km with an inclination of 98.7° and a period of 102 minutes. The instrument has five channels and a resolution, at nadir, of 1.1 km. Data with a 1.1 km nominal resolution are known as *Local Area Coverage* (LAC) data. A lower resolution sample of the LAC data, called *Global Area Coverage* (GAC) data is recorded for the entire 102-minute orbit. About 11 minutes of LAC data can be tape-recorded on each orbit, and so LAC data are generally downloaded to ground receiving stations such as that at the UK Natural Environment Research Council (NERC) ground station at Dundee University in Scotland, which provided the image shown in Figure 1.12. The swath width of the AVHRR is of the order of 3000 km, and so spatial resolution at the edges of the image is considerably greater than the nominal (nadir) figure of 1.1 km. Correction of the geometry of AVHRR is mentioned further in section 4.3. The effects of off-nadir viewing are not only geometrical for, as noted in Chapter 1, the radiance observed for a particular target depends on the angles of illumination and view. Thus, bidirectional reflectance factors should be taken into consideration when using AVHRR imagery.

The five channels of the AVHRR cover the following wavebands: channel 1 (green and red regions of the visible spectrum, 0.58–0.68 μm), channel 2 (near-infrared, 0.725–1.1 μm), channel 3 (thermal, 3.55–3.93 μm), channel 4 (thermal, 10.3–11.3 μm), and channel 5 (thermal, 11.5–12.5 μm). The thermal channels are used for sea surface temperature determination and cloud mapping, whereas the visible and near-infrared channels are used to monitor land surface processes, such as snow and ice melt, as well as vegetation status using vegetation indices such as the NDVI (section 6.2.4). The definitive source of information on the AVHRR sensor is Cracknell (1997).

2.3.3 MODIS (MODerate Resolution Imaging Spectrometer)

MODIS is a wide field of view instrument that is carried by both the US Terra and Aqua satellites. As noted earlier, the Terra satellite has an equatorial crossing time of 10:30 while the Aqua satellite has an equatorial crossing time of 13:30, so that the two MODIS instruments can be used to collect information relating to diurnal variations in upwelling radiance that relate to the characteristics and conditions of land surface, oceanic and atmospheric variables. It is a conventional radiometer, using a scanning mirror, and measures in 36 spectral bands in the range from 0.405 μm to 14.835 μm (Table 2.2; note that the bands are not listed contiguously in terms of wavelength). Bands 1–36 are collected during the daytime segment of the orbit, and bands 20–36 are collected during the night-time segment. The spatial resolution of bands 1 and 2 is 250 m. Bands 3–7 have a spatial resolution of 500 m, while the remaining bands (8–36) have a spatial resolution at nadir of 1 km. Estimates of land surface characteristics are derived from the higher-resolution bands (1–7) and ocean/atmosphere measurements are estimated from the coarser resolution (1 km) bands. The swath width is 2330 km, so that each point on the Earth's surface is observed at least once every two days. MODIS data are relayed to White Sands, NM, via the Tracking and Data Relay Satellite

Table 2.2 MODIS wavebands and key uses. Bands 13 and 14 operate in high low gain mode. Bands 21 and 22 have the wavelength range but band 21 saturates at about 500 K, whereas band 22 saturates at about 335 K. Derived from information obtained from http://www.sat.dundee.ac.uk/modis.html

Band number	Range (μm)	Primary use
1	0.620–0.670	Absolute land cover transformation, vegetation chlorophyll
2	0.841–0.876	Cloud amount, vegetation land cover transformation
3	0.459–0.479	Soil/Vegetation differences
4	0.545–0.565	Green vegetation
5	1.230–1.250	Leaf/Canopy differences
6	1.628–1.652	Snow/Cloud differences
7	2.105–2.155	Cloud properties, land properties
8	0.405–0.420	Chlorophyll
9	0.438–0.448	Chlorophyll
10	0.483–0.493	Chlorophyll
11	0.526–0.536	Chlorophyll
12	0.546–0.556	Sediments
13h	0.662–0.672	Atmosphere, sediments
13l	0.662–0.672	Atmosphere, sediments
14h	0.673–0.683	Chlorophyll fluorescence
14l	0.673–0.683	Chlorophyll fluorescence
15	0.743–0.753	Aerosol properties
16	0.862–0.877	Aerosol properties, atmospheric properties
17	0.890–920	Atmospheric properties, cloud properties
18	0.931–0.941	Atmospheric properties, cloud properties
19	0.915–0.965	Atmospheric properties, cloud properties
20	3.660–3.840	Sea surface temperature
21	3.929–3.989	Forest fires and volcanoes
22	3.929–3.989	Cloud temperature, surface temperature
23	4.020–4.080	Cloud temperature, surface temperature
24	4.433–4.498	Cloud fraction, troposphere temperature
25	4.482–4.549	Cloud fraction, troposphere temperature
26	1.360–1.390	Cloud fraction (thin cirrus), troposphere temperature
27	6.535–6.895	Mid troposphere humidity
28	7.175–7.475	Upper troposphere humidity
29	8.400–8.700	Surface temperature
30	9.580–9.880	Total ozone
31	10.780–11.280	Cloud temperature, forest fires & volcanoes, surface temperature
32	11.770–12.270	Cloud height, forest fires & volcanoes, surface temperature
33	13.185–13.485	Cloud fraction, cloud height
34	13.485–13.785	Cloud fraction, cloud height
35	13.785–14.085	Cloud fraction, cloud height
36	14.085–14.385	Cloud fraction, cloud height

(TDRS)., and thence to the Goddard Space Flight Centre for further processing.

MODIS data are available in standard form, as calibrated radiances. Also available are 41 further 'products' derived from MODIS data. These products include surface reflectance, land surface temperature and emissivity, land cover, leaf area index, sea ice cover, suspended solids in ocean waters and sea surface temperature, thus demonstrating the wide range of applications (relating to terrestrial, atmospheric, oceanographic and cryospheric systems). The processing steps required to generate these products are described by Masuoka *et al.* (1998). Example 2.2 shows a

EXAMPLE 2.2: MODIS

These images (next page) of North and Central America were derived from MODIS data by the process of mosaicing, which involves the fitting together of a set of adjacent images. In this case, the images were first transformed before mosaicing, using the Normalised Difference Vegetation Index, NDVI, (top) and the Enhanced Vegetation Index or EVI (bottom). The pixel values in Vegetation Index images are correlated with vegetation cover, density and vigour. Time sequences of such images can be used to monitor changes in biomass and other vegetation-related factors at a regional, continental, or global scale. Methods of transforming spectral data into vegetation indices are considered in section 6.2.4

The images above represent the state of vegetation in North America during the period 3–20 March 2000. Black areas represent water (for example, the Great Lakes and the Atlantic and Pacific Oceans). Areas with low values of NDVI and EVI are shown in white, and progressively darker colours over land indicate higher values of the index, meaning that vegetation is more vigorous, or denser, or less stressed. Where the temporal frequency of observations is sufficiently high, for example 24 hours, a single 'cloud free' image can be derived weekly from seven individual images 'cloud-free' pixel at each position from the seven available.

Vegetation extent and health are both important inputs to climate models, which are used to predict changes in the global climate.
Image source:

http://visibleearth.nasa.gov/cgi-bin/viewrecord?2091

MODIS vegetation product (showing the spatial distribution of two vegetation indices (section 6.2.4) over North and Central America.

2.3.4 Ocean-observing instruments

A number of dedicated ocean observing satellites have been placed in orbit in the past 25 years. Measurements from instruments carried by these satellites have been used to study the marine food chain and the role of ocean in biogeochemical cycles such as the carbon cycle (Esaias *et al.*, 1998). These measurements have to be very accurate, as the radiance upwelling from the upper layers of the oceans typically does not exceed 10% of the signal recorded at the sensor, the rest being contributed by the atmosphere viascattering (section 1.2.5). It is thus essential that the procedures for atmospheric correction, i.e. removal of atmospheric effects (section 4.4), from the measurements made by ocean-observing instruments should be of a high quality. The sensors that are described briefly in this section are the Coastal Zone Colour Scanner (CZCS), the Sea-viewing Wide Field-of-view Sensor (SEAWiFS), and the Moderate Resolution Imaging Spectrometer (MODIS). The last of these is described elsewhere (section 2.3.3) in more detail.

The *Coastal Zone Colour Scanner* (CZCS) instrument was carried by the Nimbus-7 satellite between 1978 and 1986, and was primarily designed to map the properties of the ocean surface, in particular chlorophyll concentration, suspended sediment distribution, gelbstoffe (yellow stuff) concentrations as a measure of salinity, and sea surface temperatures. Chlorophyll, sediment and gelbstoffe were measured by five channels in the optical region, four of which were specifically chosen to target the absorption and reflectance peaks for these materials at 0.433–0.453 μm, 0.510–0.530 μm, 0.540–0.560 μm and 0.660–0.680 μm. The 0.7–0.8 μm band was used to detect terrestrial vegetation and sea-surface temperatures were derived from the thermal infrared band (10.5–12.5 μm). The spatial resolution of CZCS was 825 m and the swath width was 1566 km. The sensor could be tilted up to 20° to either side of nadir in order to avoid sunglint, which is sunlight that is specularly reflected from the ocean surface. The Nimbus satellite had a near-polar, Sun-synchronous orbit with an altitude of 955 km. Its overpass time was 1200 local time, and the satellite orbits repeated with a period of 6 days. The CZCS experiment was finally terminated in December 1986.

A more recent ocean-observing instrument was the *Ocean Colour and Temperature Sensor* (OCTS) that was carried aboard the Japanese Advanced Earth-Observing Satellite (ADEOS). ADEOS was launched in August 1996 but lasted only until June 1997, when a catastrophic failure occurred. Next, the *Sea-viewing Wide Field-of-view Sensor* (SEAWiFS) was launched in 1997 on a commercial satellite operated by Orbital Sciences Corporation. SEAWiFS provides data in eight bands (centred at 412, 443, 490, 510, 555, 670, 705 and 865 nm). The bandwidth of channels 1–6 is 20 nm, and that of bands 7 and 8 is 40 nm. The sensor can be tilted forward or backwards by up to 20° in order to reduce sunglint. Local Area Coverage (LAC) is used to acquire data at a spatial resolution of 1.1 km at nadir, whereas Global Area Coverage (GAC) data has a spatial resolution of 4.5 km at nadir

As noted in section 2.3.3, two *Moderate Resolution Imaging Spectrometers* (MODIS) are carried by the NASA Terra and Aqua satellites. MODIS has 36 bands in the

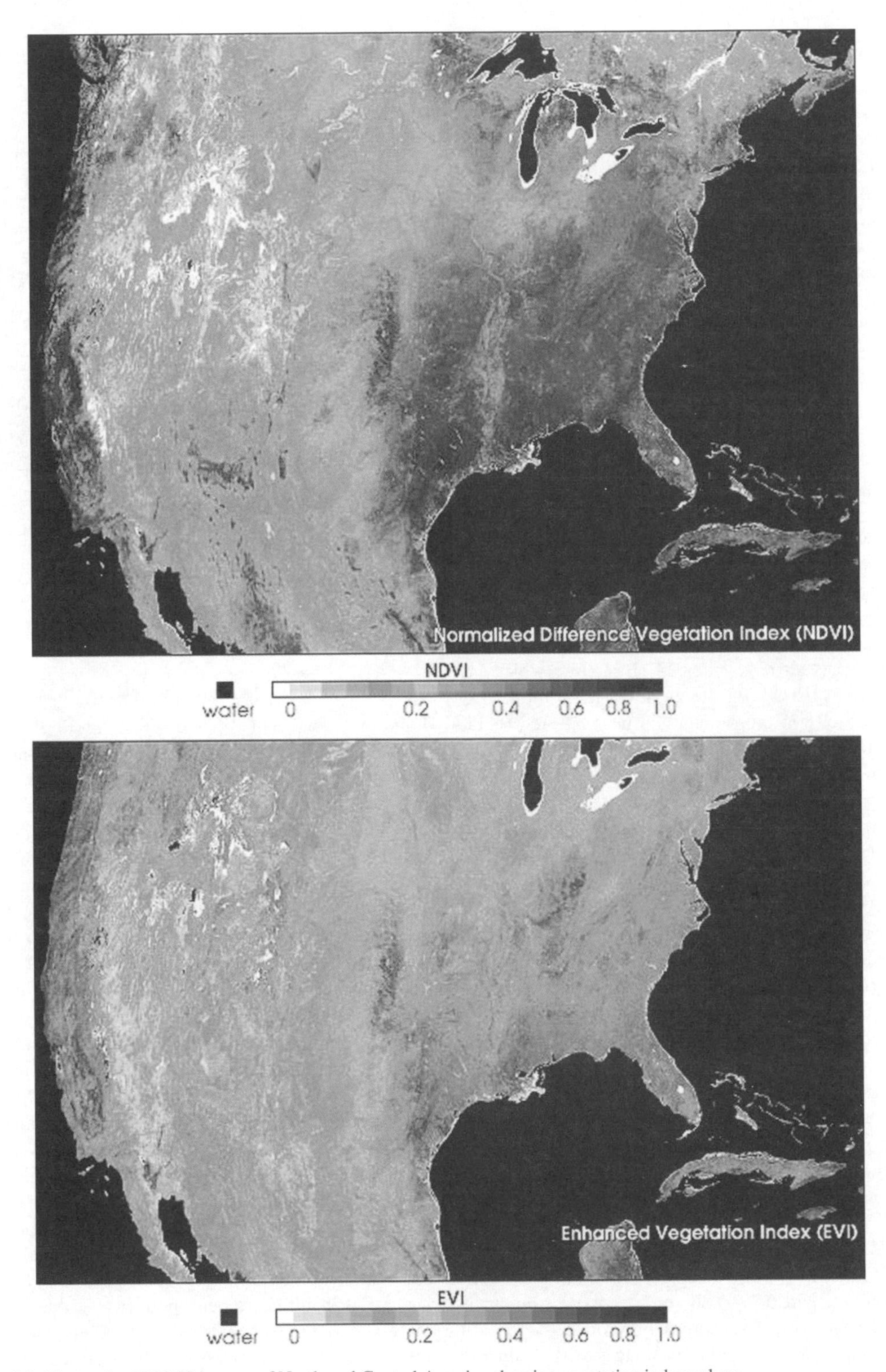

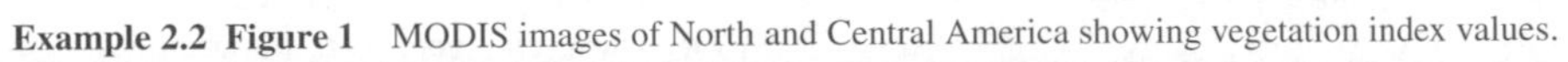

Example 2.2 Figure 1 MODIS images of North and Central America showing vegetation index values.

visible and infrared spectral regions with spatial resolution varying from 250 m for visible bands to 1 km for thermal infrared bands (Table 2.2). Bands 8–16 are similar to the spectral wavebands used by CZCS, OCTS and SEAWiFS. Esaias *et al.* (1998) provide an overview of the capabilities of MODIS for scientific observations of the oceans. They note that the MODIS sensors have improved onboard calibration facilities, and that the MODIS bands that are most similar to the SEAWiFS bands are narrower than the corresponding SEAWiFS bands, giving greater potential for better atmospheric correction (section 4.4).

A recent survey of remote sensing applications in biological oceanography by Srococz (2000) provides further details of these and other sensors that collect data with applications in oceanography.

2.3.5 IRS-1 LISS

The Indian Government has an active and successful remote-sensing programme. The first Indian Remote Sensing (IRS) satellite was launched in 1989, and carried the LISS-1 sensor, a multispectral instrument with a 76 m resolution in four wavebands. A more advanced version of LISS, LISS-2, is carried by the IRS-1B satellite, which was launched in 1991. LISS-2 senses in the same four wavebands as LISS-1 in the optical and near infrared (0.45–0.52 μm, 0.52–0.59 μm, 0.62–0.68 μm and 0.77–0.86 μm) but has a spatial resolution of 36 m. IRS-1C (launched 1995) carries an improved LISS, numbered 3, plus a 5 m panchromatic sensor. LISS-3 includes a mid-infrared band in place of the blue-green band (channel 1 of LISS-1 and LISS-2). The waveband ranges for LISS-3 are 0.52–0.59 μm, 0.62–0.68 μm, 0.77–0.86 μm and 1.55–1.70 μm. The spatial resolution improves from 36 m to 25 m in comparison with LISS-2. The panchromatic sensor (0.50–0.75 μm) provides imagery with a spatial resolution of 5 m. An example of an IRS-1C panchromatic image of Denver airport is given in Figure 2.8. IRS-1D carried a similar payload, but did not reach its correct orbit. However, some useful imagery is being obtained from its sensors. The panchromatic sensor is, like the SPOT HRV instrument, pointable so that oblique views can be obtained, and off-track viewing provides the opportunity for image acquisition every 5 days, rather than the 22 days for nadir viewing.

IRS-1C also carries the Wide Field Sensor, WIFS, which produces images with a pixel size of 180 m in two bands (0.62–0.68 μm and 0.77–0.86 μm). The swath width is 774 km. Images of a fixed point are produced every 24 days, though overlap of adjacent images will produce a view of the same point every 5 days.

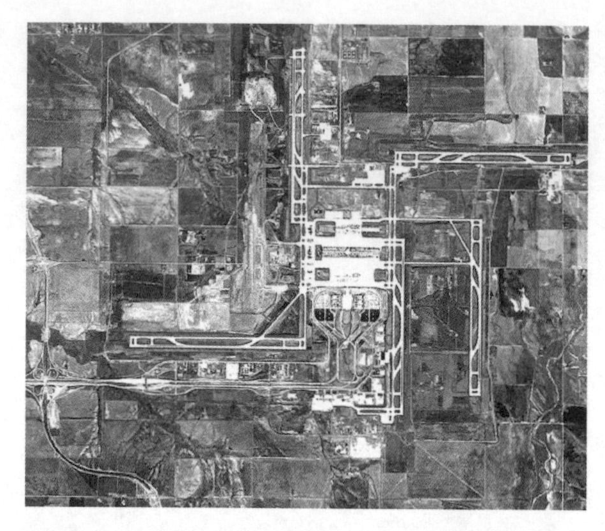

Figure 2.8 Image of Denver airport obtained from the panchromatic sensor onboard the IRS-1C satellite. The spatial resolution of the image is 5 m. Image courtesy of Space Imaging, Thornton, Colorado.

2.3.6 Landsat instruments

2.3.6.1 Landsat Multispectral Scanner (MSS)

Landsat-1, originally called ERTS-1 (Earth Resources Technology Satellite One), was the first civilian land observation satellite. It was launched into a 919 km Sun-synchronous orbit on 23 July 1972 by the US National Aeronautics and Space Administration (NASA), and operated successfully until January 1978. A second, similar, satellite (Landsat-2) was placed into orbit in January 1975. Landsats-3, -4 and -5 followed in 1978, 1982 and 1984 respectively. A sixth was lost during the launch stage. The Landsat-7 launch took place on 15 April 1999. Landsat-2 and -3 had orbits similar to Landsat-1, but the later satellites use a lower, 705 km, orbit, with a slightly different inclination of 98.2° compared with the 99.09° of Landsats 1–3. The Landsat orbit parameters are such that its instruments are capable of imaging the Earth between 82° N and 82° S latitudes. A special issue of the journal *Photogrammetric Engineering and Remote Sensing* (volume 63, number 7, 1997) is devoted to an overview of the Landsat programme.

Landsat satellites numbered 1 to 5 inclusive carried the Multispectral Scanner (MSS), which is described here for historical reasons, as is no longer operational, though a substantial archive of imagery remains available. The MSS was a four-band instrument, with two visible channels in the green and red wavebands, respectively, and two near-infrared channels (0.5–0.6 μm, 0.6–0.7 μm, 0.7–0.8 μm and 0.8–1.1 μm). These channels were numbered 4–7 in

Landsats 1–3 because the latter two satellites carried a second instrument, the Return Beam Vidicon (RBV), which generated channels 1–3. The RBV was a television-based system. It operated by producing an instantaneous image of the scene, and scanning the image, which was stored on a photosensitive tube. Landsat-4 and -5 did not carry the RBV and so the MSS channels were renumbered 1–4. Since Landsat–4 and –5 operate in a lower orbit than Landsat-1 to -3, the optics of the MSS were altered slightly to maintain the swath width of 185 km and a pixel size of 79m (along track) × 57 m (across track).

The MSS was an electromechanical scanner, using an oscillating mirror to direct reflected radiation onto a set of six detectors (one set of six for each waveband). The six detectors each record the magnitude of the radiance from the ground area being scanned, which represents six adjacent scan lines (Figure 2.9). The analogue signal from the detectors is sampled at a time interval equivalent to a distance of 57 m along-scan, and converted to digital form before being relayed to a ground receiving station. A distance of 57 m represents some over-sampling, as the instantaneous field of view of the system is equivalent to 79 m on the ground. The detectors are active only as the mirror scans in the forward direction. Satellite velocity is such that it moves forward by an amount equal to 6 × 79 m during the reverse scan cycle, so that an image is built up in each band in sets of six scan lines. The detectors deteriorate at different rates, so MSS images may show a phenomenon called six-line banding. Methods of removing this kind of 'striping' effect are discussed further in section 4.2.2.

Landsat MSS images were collected routinely for a period of over 20 years, from the launch of ERTS-1 in July 1972 to November 1997, when the Australian ground receiving station acquired its last MSS image. Although somewhat old-fashioned by today's standards, the MSS performed well and exceeded its design life. The MSS archive provides a unique historical record, and is used in studies of temporal change, for example urban growth, land degradation, and forestry, as illustrated in Example 2.3, which looks at urban change in Las Vegas, NV, using two MSS images taken 20 years apart.

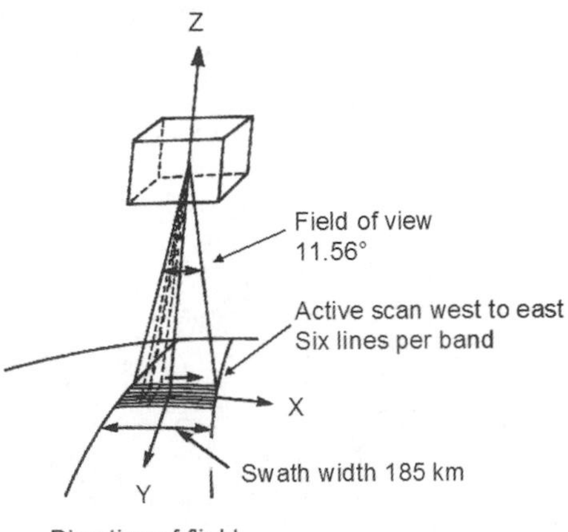

Figure 2.9 Landsat Multispectral Scanner (MSS) operation.

EXAMPLE 2.3: LANDSAT MSS

The left-hand image in Example 2.3 Figure 1 shows the city of Las Vegas, Nevada, on 13 September 1972. The image was collected by the MSS onboard Landsat-1. The right-hand image covers the same geographical area. This image was collected on 10 September 1992 by the MSS onboard Landsat-5. The growth of the city (the dark area in the centre of the image) over the 20-year period is clearly apparent. Image processing methods such as image subtraction (section 6.2.2) are used to obtain a digital representation of areas that have changed over time. The analyst must remember that quantitative measures of change include all aspects of the differences between the two images being compared. Atmospheric differences (sections 1.2.5, 4.4), instrument calibration variations (section 4.5), as well as other factors such as illumination differences caused by seasonal variations (section 4.6).When pointable sensors such as the SPOT HRV are the source of the imagery, then viewing geometry effects should also be taken into account.

The analysis of the nature, direction and drivers of change is useful in many areas of environmental management. Changes in land cover, for example, can be both a response to and a factor contributing to climate change. Variations in the termini of Alpine glaciers are an indication of local temperatures over a number of years. Change analysis is also an important aspect of urban and regional planning. The long (25-year) archive of Landsat MSS images, together with the moderate spatial resolution and wide-area coverage of the MSS, make it one of the most useful sources of data for change analysis studies.

2.3.6.2 *Landsat Thematic Mapper (TM)*

The Thematic Mapper (TM) instrument is the primary imaging sensor carried by Landsat–4 and –5. Landsat-4 (launched 16 July 1982) was switched off in August 1993 after failure of the data downlinking system. Landsat-5 continues to operate, though only a direct data downlink facility is available. Like the MSS, the TM uses a fixed set of detectors for each band and an oscillating mirror. TM has 16, rather than six, detectors per band (excluding the thermal

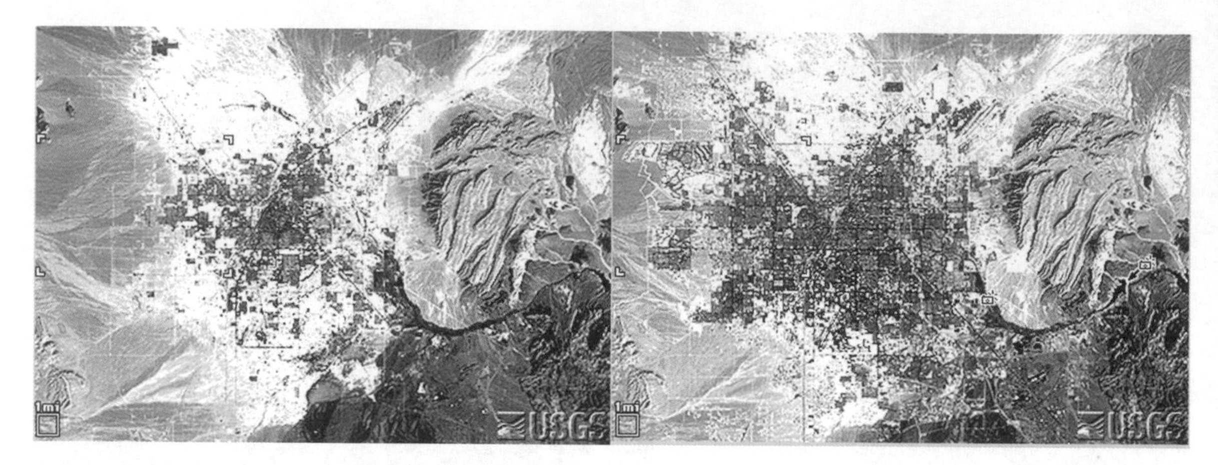

Example 2.3 Figure 1 Landsat MSS images of Las Vegas, Nevada, taken on 13 September 1972 and 10 September 1992. Image source: http://edc.usgs.gov/earthshots/slow/LasVegas/LasVegastext

infrared channel) and scans in both the forward and the reverse directions. It has seven, rather than four, channels covering the visible, near- and mid-infrared, and the thermal infrared, and has a spatial resolution of 30 m. The thermal channel uses four detectors and has a spatial resolution of 120 m. The data are quantised onto a 0–255 range. In terms of spectral and radiometric resolution, therefore, the TM design represents a considerable advance on that of the MSS.

The TM wavebands are as follows: channels 1–3 cover the visible spectrum (0.45–0.52 μm, 0.52–0.60 μm and 0.63–0.70 μm, representing visible blue-green, green and red). Channel 4 has a wavelength range of 0.75–0.90 μm in the near infrared. Channels 5 and 7 cover the mid-infrared (1.55–1.75 μm and 2.08–2.35 μm) while channel 6 is the thermal infrared channel (10.4–12.5 μm). The rather disorderly channel numbering is the result of the late addition of the 2.08–2.35 μm band.

Data from the TM instruments carried by the Landsat-4 and -5 satellites and the ETM+ carried by Landsat-7 are transmitted to a network of ground receiving stations. The European stations are located near Fucino, Italy, and Kiruna, Sweden. Data are also transmitted via the system of Tracking and Data Relay Satellites (TDRS), which are in geostationary orbits. At least one of the satellites making up the TDRS constellation are in line of sight of Landsats-4, -5 and -7. TDRS transmits the image data to a ground station at White Sands, New Mexico, from where it is relayed to the data processing facility at Norman, OK, using the US domestic communications satellite DOMSAT. TM data can also be downlinked to ground receiving stations around the world. Following a problem with the ETM+ instrument in mid-2003, Landsat-5 TM data are again being down-linked

EXAMPLE 2.4: LANDSAT ETM+

The TM and ETM+ instruments have seven spectral bands that cover the visible, near infrared and the short-wave infrared regions of the electromagnetic spectrum. Rocks, minerals and surface crusts can be discriminated in these regions (section 1.3.2.2) and so the use of TM/ETM+ imagery for geological mapping and exploration is an important application area. Techniques such as band ratioing (section 6.2.4) and colour enhancement (e.g., the decorrelation stretch, section 6.4.3) are commonly used techniques. Since each of the 30 × 30 m pixels making up the image is unlikely to cover a homogeneous area, techniques such as linear spectral unmixing (section 8.5.1) can be used to try to identify the proportions of different materials that are present within each pixel area.

Deposits of gold and silver (which are found in Tertiary volcanic complexes) have been exploited in the area shown on Example Figure 4.1 for many years, at places such as Tonopah, Rhyolite and Goldfield in Nevada. Associated with these mineral deposits are hydrothermally altered rocks, which contain iron oxide and/or hydroxyl-bearing minerals. These minerals show absorption bands in the VNIR and SWIR region of the spectrum. Both the TM and ETM+ sensors can detect radiance in these spectral regions (bands 5 and 7).

Further Reading: Abrams, M., Ashley, R., Rowan, L., Goetz, A. and Kahle, A. (1977) Mapping of hydrothermal alteration in the Cuprite Mining District, Nevada, using aircraft scanner images for the spectral region from 0.46 to 2.36 μm. *Geology*, **5**, 713–718.

Example 2.4 Figure 1 Landsat ETM+ image of the Goldfield/Cuprite area, NV. Original in colour. Image source: http://edcdaac.usgs.gov/samples/goldfield.html.

to ground receiving stations, almost 20 years after the satellite was launched. Examples of Landsat TM imagery are provided in Figures 1.10 and 1.11, and in Example 2.4.

2.3.6.3 Enhanced Thematic Mapper Plus (ETM+)

Landsat-6, which was lost on launch in October 1993, was carrying a new version of the Landsat TM called the Enhanced Thematic Mapper (ETM). Landsat-7, which was launched on 15 April 1999, and which operated successfully until mid-2003, carries an improved version of ETM, called the ETM+. This instrument measures upwelling radiance in the same seven bands as the TM, and has an additional 15 m resolution panchromatic band. The spatial resolution of the thermal infrared channel is 60 m rather than the 120 m of the TM thermal band. In addition, an onboard calibration system will allow accurate (±5%) radiometric resolution (section 2.2.3). Landsat-7 has substantially the same operational characteristics as Landsat-4 and -5, namely, a Sun-synchronous orbit with an altitude of 705 km and an inclination of 98.2°, a swath width of 185 km, and an equatorial crossing time of 10:00. The orbit is the same as that of Terra (equatorial crossing time 10:30), Aqua (equatorial crossing time 13:30) and Earth Observing-1. A sketch of the main components of the Landsat-7 satellite is shown in Figure 2.10, and the orbital characteristics are shown in Figure 2.1(a).

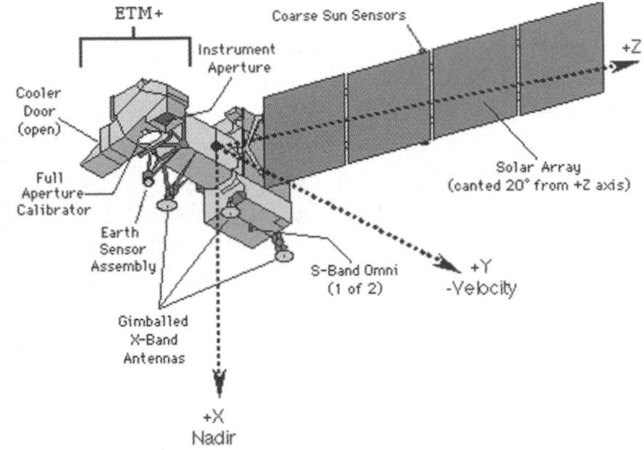

Figure 2.10 Schematic diagram of the Landsat-7 spacecraft, showing the solar array that supplies power, and the ETM+ instrument. The spacecraft travels in the $+Y$ direction, with the $+X$ direction pointing towards the Earth's centre. Attitude and altitude are controlled from the ground using thruster motors (not shown). Based on Figure 2.2, Landsat 7 Science Data Users Handbook, NASA Goddard Spaceflight Center, Greenbelt, Maryland. http://ltpwww.gsfc.nasa.gov/IAS/handbook/handbook_toc.html

2.3.6.4 Landsat follow-on programme

It is unlikely that a Landsat-8 will be built. US Government policy (as of May, 2003) is to encourage commercial companies to build and operate a satellite carrying a sensor based on the Advanced Land Imager (ALI) carried by Earth Observing-1 (EO-1), an experimental satellite that also carries the Hyperion imaging spectrometer (Chapter 9) (USGS, 2003). ALI produces imagery with comparable or improved Landsat spatial and spectral resolution with substantial mass, volume and cost savings. However, under the US Commercial Remote Sensing policy, it is unlikely that low-cost imagery with global coverage will be available following the demise of Landsat-7, which had a design life of five years (1999–2004). In fact, the ETM+ sensor developed a problem in June 2003, and Landsat-5 was brought out of retirement to collect TM imagery in order to maintain data continuity. The future of medium-resolution land remote sensing is unclear at the time of writing, as the SPOT programme also appears to be coming to an end (section 2.3.7.3).

2.3.7 SPOT sensors

2.3.7.1 SPOT High Resolution Visible (HRV)

The SPOT (Satellite Pour l'Observation de la Terre) programme is funded by the governments of France, Belgium, and Sweden and is operated by the French Space Agency, CNES (Centre National d'Etudes Spatiales), located in Toulouse, France. SPOT-1 was launched on 22 February 1986, and is still nominally operational. It carries an imaging sensor called *High Resolution Visible* (HRV) that is capable of measuring upwelling radiance in three channels (0.50–0.59 μm, 0.61–0.68 μm and 0.79–0.89 μm) at a spatial resolution of 20 m, or in a single panchromatic channel (0.51–0.73 μm) at a spatial resolution of 10 m. All channels are quantised to a 0–255 scale. HRV does not use a scanning mirror like Landsat MSS and TM. Instead, it uses a linear array of charge-coupled devices, or CCDs, so that all the pixels in an entire scan line are imaged at the same time (Figure 1.2(b)). As it has no moving parts the CCD pushbroom system might be expected to last longer than the electro-mechanical scanners carried by Landsats-1 to -7, though all of the Landsat sensors except the ETM+ have exceeded their design lives by substantial amounts. A more important consideration is the fact that, since all of the pixel data along a scan line are collected simultaneously rather than sequentially, the individual CCD detector has a longer 'look' at the pixel area on the ground, and collects more ground-leaving photons per pixel than an instrument based on an electromechanical scanner such as ETM+. This increased dwell time means that the signal is estimated more accurately and the image has a higher signal-to-noise ratio (section 2.2.2).

The SPOT orbit is near polar and Sun-synchronous at an altitude of 832 km, an inclination of 98.7°, with a period of 101.5 minutes. The field of view of the HRV sensor is 4.13° and the resulting swath width is 60 km. Two identical HRV instruments are carried, so when they are working in tandem the total swath width is 117 km (there is a 3 km overlap when both sensors point at nadir). The orbital repeat period is 26 days with and equatorial crossing time of 1030. However, the system potentially has a higher revisit capability because the HRV sensor is pointable. It can be moved in steps of 0.6° to a maximum of ±27° away from nadir. This feature allows HRV to collect image data within a strip 475 km to either side of nadir. Apart from improving the sensor's revisit capability, the collection of oblique views of a given area (from the left and right sides, respectively) provides the opportunity to generate digital elevation models (DEM) of the area, using photogrammetric methods. However, the use of non-nadir views introduces problems when physical values are to be derived from the quantised counts (section 4.6).

SPOT-1 was retired at the end of 1990, following the successful launch of SPOT-2 in January 1990. SPOT-3 followed in September 1993. All three carry an identical instrument pack. SPOT-3 was lost following a technical error, and SPOT-1 and -2 have been brought out of retirement.

Table 2.3 Spatial resolution and swath widths for the SPOT-5 instruments HRG (High Resolution Geometric), Vegetation -2, and HRS (High Resolution Stereoscopic) instruments carried by SPOT-5. Note that 2.5 m panchromatic imagery is obtained by processing the 5 m data using a technique called 'Supermode' (see text for details).

Spectral Band (μm)		Spatial Resolution (m)		
		HRG	Vegetation-2	HRS
P	0.49–0.69	2.5 or 5		10
*B*0	0.43–0.47		1000	
*B*1	0.50–0.59	10		
*B*2	0.61–0.68	10	1000	
*B*3	0.79–0.89	10	1000	
SWIR	1.58–1.75	20	1000	
Swath width (km)		60	2250	120

SPOT-4 was successfully launched on 24 March 1998. The HRV instrument on SPOT-4 is extended to provide an additional 20 m resolution channel in the mid-infrared region (1.58–1.75 μm) and the new instrument is called the HRV-IR. This sensor can be used in multispectral mode (X), or panchromatic mode (M), or in a combination of X and M modes. SPOT-4 carries an onboard tape recorder, which provides the capacity to store 20 scenes for downloading to a ground station.

The SPOT-5 satellite was launched in May 2002. It carries a new stereo instrument, an enhanced panchromatic channel, and a four-band multispectral imager. Details of these instruments are given in Table 2.3. The spatial resolution of HRG is 10 m (20 m in the short wave infrared (SWIR) band), and panchromatic images can be obtained at 5 or 2.5 m spatial resolution, compared with 10 m for SPOT-1 to -4. A new instrument, called HRS (High Resolution Stereoscopic) simultaneously collects images from two different angles. The first image looks forward of nadir along the line of flight at an angle of 20° while the second image looks backwards at the same angle. The spatial resolution of HRS is 10 m and its field of view is 8°. The HRV instrument on SPOT-1 to SPOT-4 collected stereo imagery in two stages. At the first stage the HRV sensor was tilted to one side or other of the sub-satellite track to capture an off-nadir view of the area of interest. At the second, later stage a second image of the same area was collected from a later, different orbit again using the pointing capability of the instrument. The disadvantage of this two-stage approach is that surface conditions may change between the dates of acquisition of the two images, thus causing problems with the co-registering process (section 4.3.1.4) and reducing the quality of the digital elevation model (DEM) that is produced from the stereo pair. HRS uses the *along-track stereo* principle to collect two images from different angles at the same time. The ASTER sensor (section 2.3.8) uses a slightly different system, collecting simultaneous backward and nadir views in order to generate a stereo image pair. The relative (within the image) accuracy of the elevation values derived from the HRS along-track stereo image pairs is given as 5–10 m, with an absolute (compared with ground observations) accuracy of 10–15 m, making possible the generation of DEM at a map scale of 1:50 000.

This high level of geometric accuracy is due to some extent to the employment of a system for accurate orbit determination called Doppler Orbitography and Radiopositioning Integrated by Satellite (DORIS), which was designed and developed by the Centre National d'Études Spatiales, the Groupe de Recherches de Géodésie Spatiale (Space Geodesy Research Group) and the French mapping agency (Institut Géographique National, or IGN). DORIS consists of a network of transmitting stations, a receiver on board the satellite, and a control centre. A receiver on board the satellite measures the Doppler shift of the signal emitted by the transmitting stations at two frequencies (400 MHz and 2 GHz). The use of two frequencies is needed to estimate the delay in propagation of radio waves caused by the ionosphere. The data are stored in the instrument's memory, downloaded to the ground each time the satellite is within range of a receiving station, and then processed to determine the precise orbital trajectory. The orbital position of the satellite can be calculated to an accuracy of a few centimetres using this approach. A development of DORIS, called DIODE (Détermination Immédiate d'Orbite par DORIS Embarqué, or real-time orbit determination using onboard DORIS), is carried by SPOT-4 and -5; this device can, as its name suggests, provide immediate orbital information that can be downloaded to a ground station together with the image data. It is used to calculate the position of each pixel on the ground surface, and to express these results in terms of a map projection such as Universal Transverse Mercator (UTM). This is the procedure of *geometric correction* of remotely-sensed images, and it is considered in more detail in section 4.3.

The HRG instrument produces panchromatic images at a resolution of either 2.5 or 5 m. The physical resolution of the instrument is 5 m, but 2.5 m data are generated by a processing technique that the SPOT Image Company calls 'Supermode'. Supermode provides an interesting example of the application of advanced image processing methods, some of which are described in later chapters of this book. The procedure is described in Example 2.5 and Figure 2.11.

EXAMPLE 2.5: SPOT-5 SUPERMODE

The SPOT-5 satellite carries the HRG pointable imaging sensor, which produces panchromatic images at a resolution of 5 m using a CCD array with 12 000 elements. In fact, there are two such arrays. Each is programmed to collect a separate image of the area being viewed, with the second image being offset by half a pixel in the horizontal and vertical directions, as depicted in Figure 2.11(a). A new image, with a spatial resolution of 2.5 m, is derived from these overlapped cells. Example 2.5 Figure 1 shows one view of the target area taken by the first HRG sensor. Example 2.5 Figure 2 is the second 5 m resolution panchromatic image from the HRG sensor that is used in the Supermode process. This image is displaced horizontally and vertically by half a pixel with respect to the first image. This displacement produces four pixels measuring 2.5 × 2.5 m nesting within the original 5 × 5 m pixel.

Example 2.5 Figure 1 Left-hand SPOT HRG image with spatial resolution of 5 m.

(b)

Example 2.5 Figure 2 Right-hand SPOT HRG image, also with a spatial resolution of 5 m.

(c)

Example 2.5 Figure 3 Composite of Figures 1 and 2. The images in Figures 1 and 2 have a spatial resolution of 5 m, but are offset by 2.5 m. This Supermode image has been resampled and filtered to produce an image with a spatial resolution of 2.5 m.

Interpolation (resampling) is used to compute values to be placed in the 2.5 m pixels. Resampling is described in section 4.3.3. The third stage of the Supermode process uses a filtering procedure (Chapter 7) called deconvolution. SPOT Image uses a method based on the discrete wavelet transform, which is described in section 6.7. Example 2.5 Figure 3 shows the end product.

2.3.7.2 *Vegetation (VGT)*

SPOT-4 and -5 carry a new sensor, VGT, developed jointly by the European Commission, Belgium, Sweden, Italy, and France. VGT operates in the same wavebands as HRV-IR except that the swath width is 2250 km, corresponding to a field of view of 101°, with a pixel size at nadir of 1 km. This is called the 'direct' or 'regional' mode. 'Recording' or 'world-wide' mode produces averaged data with a pixel size of around 4 km. In this mode, VGT generates data sets for the region of the Earth lying between latitudes 60° N and 40° S (Arnaud, 1994). The 14 daily orbits ensure that regions of the Earth at latitudes greater than 35° are be imaged daily, whereas equatorial areas are imaged every other day. The combination of the Vegetation and HRV-IR sensors on the same platform means that data sets of significantly different spatial resolutions are obtained simultaneously.

VGT data complement those produced by the NOAA AVHRR, which is widely used to generate global data sets (Townshend and Skole, 1994). VGT has the advantage – for land applications – of three spectral bands in the optical region plus one in the infrared part of the spectrum. VGT should also be compared to the MODIS sensor that is carried by both the Terra and Aqua satellites.

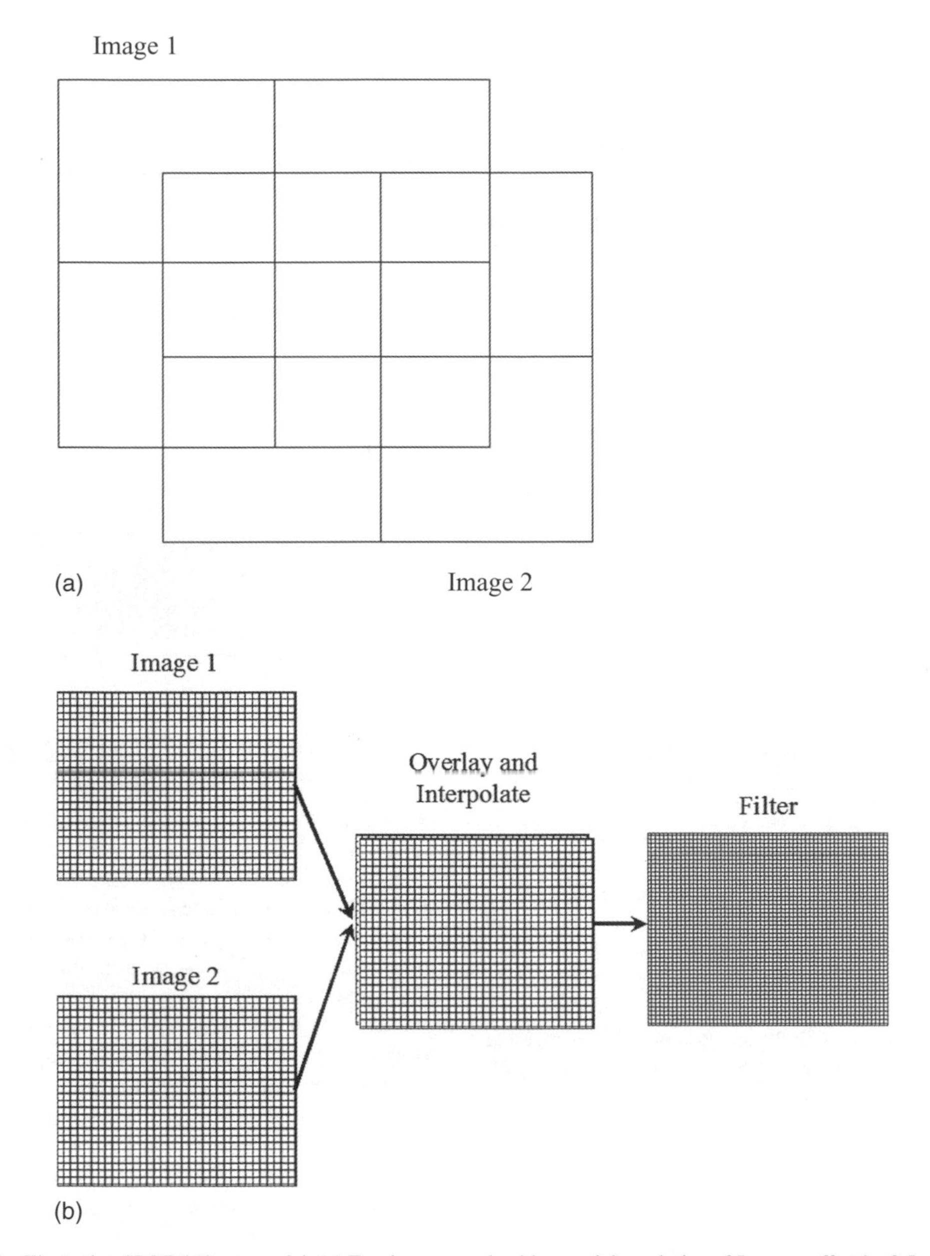

Figure 2.11 Illustrating SPOT-5 'Supermode'. (a) Two images, each with a spatial resolution of 5 m, are offset by 2.5 m in the x and y directions. (b) These two images (left side) are overlaid and interpolated to 2.5 m pixels (centre), then filtered (right) to produce a Supermode image with a resolution of 2.5 m. See Example 2.6 for more details. CNES – SPOT Image Distribution. Permission to use the data was kindly provided by SPOT Image, 5 rue des Satellites, BP 4359, F-31030 Toulouse, France (http://www.spotimage.fr).

2.3.7.3 SPOT follow-on programme

SPOT-5 is likely to be the last of the line. Its mission is planned to last until 2007. Research and development in small satellites and lightweight sensors mean that it is now more cost effective to use 'lightsats' such as TopSat, which is being developed by a consortium of UK companies with Ministry of Defence and British National Spaces Centre funding for launch in later 2003 or 2004. TopSat's scanning instrument will collect panchromatic imagery with a

Table 2.4 ASTER spectral bands. The ASTER data set is subdivided into three parts VNIR (Visible and Near Infra-Red), SWIR (Short Wave Infra-Red) and TIR (Thermal Infra-Red). The spatial resolution of each subset is: VNIR 15 m, SWIR 30 m, and TIR 90 m. The swath width is 60 km. Data in bands 1–9 are quantised using 256 levels (8 bits). The TIR bands use 12 bit quantisation. [Source: ASTER Users' Guide, Part I (General), Version 3.1, March 2001, ERSDAC, Japan, and Abrams (2000).]

Spectral region	Band number	Spectral range (μm)	Absolute accuracy	Cross-track pointing (°)
	1	0.52–0.60		
	2	0.63–0.69		
VNIR	3N	0.78–0.86	$\leq \pm 4\%$	±24
	3B	0.78–0.86		
	4	1.600–1.700		
	5	2.145–2.185		
	6	2.185–2.225		
SWIR	7	2.235–2.285	$\leq \pm 4\%$	±8.55
	8	2.295–2.365		
	9	2.360–2.430		
	11	8.475–8.825	$\leq \pm 3$ K(200–240 K)	
	12	12 8.925–9.275	$\leq \pm 2$ K(240–270 K)	
TIR	13	10.25–10.95	$\leq \pm 1$ K(270–340 K)	
	14	10.95–11.65	$\leq \pm 2$ K(340–380 K)	

spatial resolution of 2.5 m and three-band multispectral imagery at a spatial resolution of 5 m over a 15-km swath. The cost of the mission is around £10 million. The attraction of low-cost small satellites has led the French and Italian governments to agree to place a constellation of small satellites into orbit under the Pléiades programme. Readers familiar with Greek mythology will be aware that the original Pleiades were the seven daughters of Atlas who, on being pursued for nefarious purposes by Orion, were turned into doves by Zeus and transported to heaven. Orion was later sent up to heaven, but the outcome of their meeting in that place is not known. The first two Pléiades satellites will carry a high-resolution optical instrument (0.8 m resolution, 20 km swath) and an X-band SAR (c. 3 cm wavelength, 1 m spatial resolution and 10–200 km swath), respectively. Two optical and four SAR-carrying satellites are planned, with the four SAR systems being in orbit by mid-2006 and the optical systems following in 2006 and 2007. A wide field of view sensor could be developed before 2007. Joint use by civilian and military users should reduce the cost further. By 2007, it may be the case that both the Landsat and SPOT programmes have reached the end of their lives, and that moderate-resolution (20–30 m spatial resolution) imagery may no longer be as cheap and readily available as it is today (May, 2003). Further discussion of developments in high-resolution imaging can be found in section 2.3.9.

2.3.8 Advanced Spaceborne Thermal Emission and Reflection Radiometer (ASTER)

The Japanese-built Advanced Spaceborne Thermal Emission and Reflection Radiometer (ASTER) is one of the imaging sensors carried by the Terra satellite. ASTER is a high-resolution multispectral sensor, acquiring data in three spectral regions using a separate radiometer for each region. It also has a stereoscopic capability. The spectral regions in which images are acquired are the visible and near infrared (VNIR), the short-wave infrared (SWIR), and the thermal infrared (TIR) (Table 2.4). Images collected in the three spectral regions differ in terms of their spatial resolution. For the VNIR sensor, spatial resolution is 15 m. The SWIR bands have a spatial resolution of 30 m, while thermal infrared images have a 90-m spatial resolution. The band numbers, spectral ranges, and absolute accuracies of the 14 bands are shown in Table 2.4.

The ASTER instrument has a number of interesting features. First, the spatial resolution of the VNIR bands is higher than that of either the Landsat ETM+ or the SPOT HRV sensor. Like the HRV sensor, ASTER is 'pointable', that is, it can collect images of areas lying to either side of the sub-satellite track. The VNIR radiometer can scan up to ±24° off-nadir, and the other two radiometers can point up to ±8.55° off-nadir. The inclusion of several bands in both the SWIR and TIR regions makes ASTER a unique space-borne instrument. The multiple bands in these two spectral regions should prove to be valuable in geological studies.

ASTER has two spectral bands covering the 0.78–0.86 μm region. One is labelled as band 3N and the other is band 3B. Band 3N looks vertically down, while band 3B looks backwards at an angle of 27.6°. Images in these two near-infrared bands can be used to generate stereo pairs, which can be used for the production of digital elevation models. The fact that the two images are captured simultaneously in 'along track' mode reduces problems caused by changes in surface reflectance that are associated with 'across-track' mode stereo images such as those produced by SPOT -1 to SPOT-4 HRV. In across-track mode the two images making up the stereo pair are acquired on different orbits, which may be separated in time by weeks or months, depending on cloud cover and instrument availability. Welch *et al.* (1998) and Lang and Welch (1999) discuss the methods used to generate the '*Standard Data Product DEM*' from ASTER band 3B and 3N data. They cite results obtained from test sites in the USA and Mexico to show that the root mean square (RMS) error of the elevation measurements is between ±12 and ±30 m, provided that the two images can be co-registered to an accuracy of ±0.5 to ±1.0 pixels using image correlation procedures (section 4.3.2). Toutin (2002) provides details of an experiment to measure the accuracy of an ASTER DEM, and also describes the methods used by the NASA EOS Land Processes DAAC to generate ASTER DEMs (which are distributed via the USGS EROS Data Centre in Sioux Falls, SD. An example of an ASTER DEM is shown in Figure 2.12.

ASTER data products are available at different levels of processing. Level 0 is full resolution unprocessed and unreconstructed data. Level 1A is equivalent to Level 0 plus ancillary data (radiometric and geometric calibration coefficients). Radiometric and geometric corrections are applied to the Level 1B data, which are supplied in the form of quantised counts. These counts can be converted into apparent radiance using the calibration factors listed in Table 2.5. Levels 2–4 include products such as DEM derived from Band 3N/3B pairs, surface temperatures and radiances, and images derived from processing operations such as decorrelation stretch (section 6.4.3).

An account of ASTER data products is provided by Abrams (2000). The sensor is described by Yamaguchi *et al.* (1998). A number of WWW links to ASTER pages in Japan and at JPL are listed in the Optical Sensors section of the useful_links.html file that is contained the accompanying CD. Example 2.6 shows an ASTER band 3 image of the La Mancha area of central Spain.

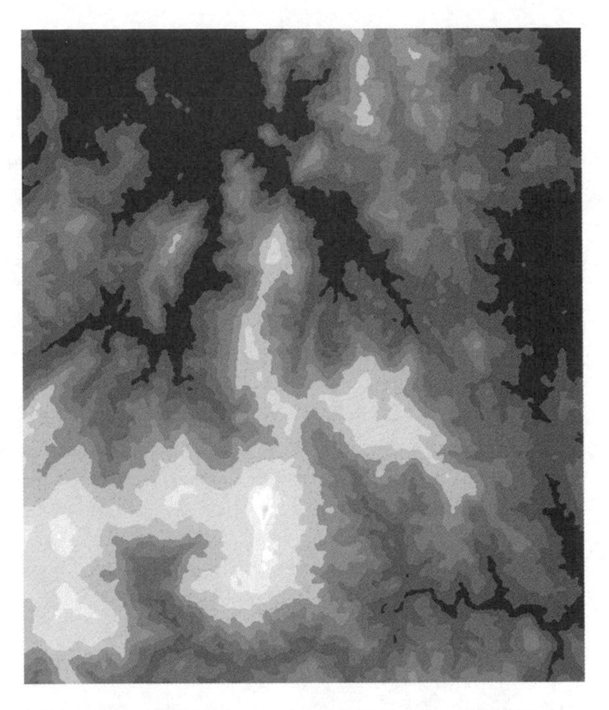

Figure 2.12 Digital elevation model derived from ASTER data. The area shown covers a part of Southern India for which conventional digital mapping is unavailable. The ASTER DEM image has been processed using a procedure called density slicing (section 5.4.1).

EXAMPLE 2.6: ASTER

This ASTER band 3N image (Example 2.6 Figure 1) shows several roads leading to the town of Tembleque (centre-right), which is located in the heart of La Mancha Alta in central Spain. The topography of this area is characterized by a generally flat, slightly undulating surface with an average altitude of 700 m, including a few scattered hills that belong to the Sierra de Algodor and Sierra de Olla mountain ranges (lower left corner of the image). The Algodor river flows to the left of these hills; it has been dammed in recent years, leading to inundation of its alluvial flood plain. The centre part of the area shown on the image is mainly agricultural land, which has been transformed from rain-fed agriculture to irrigated land. Irrigation water is mainly extracted from the underlying aquifer. The increased use of groundwater for irrigation is causing several environmental problems to this region, mainly in the form of draining nearby wetland areas, and soil salinisation.

Table 2.5 Maximum radiance for different gain settings for the ASTER VNIR and SWIR spectral bands.

	Maximum Radiance (W m^{-2} sr^{-1} μm^{-1})			
Band no	High gain	Normal gain	Low gain 1	Low gain 2
1	170.8	427	569	
2	179.0	358	477	
3N	106.8	218	290	N/A
3B	106.8	218	290	
4	27.5	55.0	73.3	73.3
5	8.8	17.6	23.4	103.5
6	7.9	15.8	21.0	98.7
7	7.55	15.1	20.1	83.8
8	5.27	10.55	14.06	62.0
9	4.02	8.04	10.72	67.0

Example 2.6 Figure 1 ASTER band 3N image of the La Mancha area of central Spain.

2.3.9 High-resolution commercial and micro-satellite systems

The release of technology from the military to the commercial sector in the United States during the 1990s has allowed a number of American private-sector companies to launch and service their own satellites, for example Space Imaging's IKONOS satellite (section 2.3.9.1). These satellites each carry a single sensor that produces either a panchromatic image, at a spatial resolution of 1 m or better, or a multispectral image, typically with four spectral bands and a spatial resolution of 4 m. Images from these sensors have brought digital photogrammetry and digital image processing much closer together. At the same time, the use of micro-satellites to carry remote sensing instruments has been pioneered by, among others, Surrey Space Technology Ltd (SSTL), a spin-off company associated with the University of Surrey, UK. The micro-satellite AlSat (Algerian Satellite), which was launched in November 2002, is the first member of a constellation of five such satellites, each carrying a multispectral scanner capable of a spatial resolution of 32 m in three spectral bands (NIR, red, green). The wide ground swath of 600 km enables a revisit of the same area almost anywhere in the world at least every four days with just a single satellite. AlSat is part of the Disaster Monitoring Constellation, an international consortium involving Algeria, Nigeria, Turkey and the UK. Figure 2.13 shows a 1024 × 1024-pixel extract from an AlSat image of the south-west United States.

Other examples of the use of micro-satellites ('lightsats') are provided in section 2.3.7.3. The UK's TopSat and the French/Italian Pléiades constellation are expected to produce panchromatic imagery at a spatial resolution of 2.5 m (TopSAT) or 0.8 m (Pléiades), plus multispectral imagery. The Pléiades system will also include small satellites carrying X-band SAR.

2.3.9.1 High-resolution commercial satellites – IKONOS

The US Space Imaging Corporation's IKONOS satellite was the first commercial high-resolution satellite to be launched successfully (the word 'icon' means 'image' or 'likeness' in Greek). High resolution is the main distinguishing characteristic of the imagery produced by

Figure 2.13 Extract from AlSAT image of the Colorado River in Arizona. This is a 1024 × 1024 pixel (33 × 33 km) extract from the full 600 × 600 km scene. AlSat is the first of a series of five microsatellites that will form Surrey's Disaster Monitoring Constellation. The original is in colour. Source: http://www.sstl.co.uk/primages/alsat-images/Coloradoriver_1024.jpg. Picture: Courtesy of CNTS, Algeria.

Figure 2.14 IKONOS panchromatic image of central London, showing bridges across the River Thames, the London Eye Ferris wheel (lower centre) and Waterloo railway station (bottom centre). The image has a spatial resolution of 1 m.

IKONOS for, in panchromatic mode, it can acquire panchromatic imagery at a spatial resolution of 1 m and multispectral imagery in four channels with a spatial resolution of 4 m. The sensor can be tilted both along and across track, and the spatial resolutions cited are valid for off-nadir pointing angles of less than 26°. The panchromatic band covers the spectral region 0.45–0.90 μm, while the four multispectral channels collect data in the blue (0.45–0.53 μm), green (0.52–0.61 μm), red (0.64–0.72 μm) and near-infrared (0.77–0.88 μm) bands. The satellite flies in a near-circular, Sun-synchronous, polar orbit at a nominal altitude of 681 km and with an inclination angle of 98.1°. The descending nodal time (when the satellite crosses the equator on a north-south transect) is 10:30. Data from the IKONOS system are quantised using an 11-bit (0–2047) scale (section 3.2.1) and are available in a variety of forms ranging from standard system-corrected to geometrically corrected (section 4.3) and stereo, for the production of DEM. The precision geocorrected imagery is claimed to have a map scale equivalent of 1:2400.

Because of their high spatial resolution, IKONOS images are used for small-area investigations, where they can replace high-altitude air photography to some extent. The fact that the sensor is pointable means that geocorrection involves more complex processing than is required for medium and low-resolution imagery from sensors such as the Landsat ETM+ and NOAA AVHRR. Expert photogrammetric knowledge is needed for these operations. An IKONOS panchromatic image of central London is shown in Figure 2.14. The detail even in this reproduction is clear, and applications in urban planning and change detection, as well as military reconnaissance, are apparent.

2.3.9.2 High-resolution commercial satellites–QuickBird

QuickBird was launched in October 2001, and is owned and operated by the American DigitalGlobe company. Like IKONOS, it carries a single instrument capable of producing panchromatic images with a spatial resolution of between 0.61 and 0.73 m, plus multispectral imagery with a spatial resolution of 2.44 and 2.88 m, depending on the angle of tilt of the sensor (which ranges up to 25° off-nadir). The sensor can be tilted along or across track, to produce stereo imagery and to ensure a revisit capability of between one and three and a half days. Imagery is available in basic (system corrected), standard (geometrically corrected to a map projection; section 4.3) and orthorectified forms. Orthorectification is a form of geometric correction that takes into account the relief of the terrain. See Toutin and Cheng (2002) for a discussion of the geometric properties of Quickbird imagery. Figure 2.15 shows a

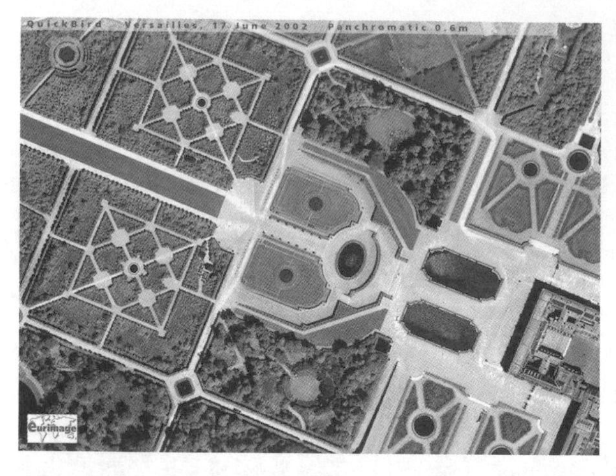

Figure 2.15 Quickbird panchromatic image of the gardens of the Palace of Versailles, near Paris, France. The proclamation of Kaiser Wilhelm I as emperor of Germany was made in the great Hall of Mirrors in the Palace of Versailles following the defeat of France in the Franco-Prussian War of 1871.

Quickbird panchromatic image of the gardens of the Palace of Versailles, near Paris.

2.4 MICROWAVE IMAGING SENSORS

As noted in section 1.2.3, the microwave region of the electromagnetic spectrum includes radiation with wavelengths longer than 1 mm. Solar irradiance in this region is negligible, although the Earth itself emits some microwave radiation. Imaging microwave instruments do not, however, rely on the detection of solar or terrestrial emissions (though passive microwave radiometers do detect terrestrial emissions). Instead, they generate their own source of energy, and are thus examples of *active* sensing devices. In this section, the properties of the SAR systems carried by the ERS-1/2, JERS-1, ENVISAT and RADARSAT satellites are presented. General details of imaging radar systems are described in sections 1.2.3 and 9.2.

SAR instruments are more complex than those operating in the optical and thermal infrared regions of the spectrum. The word *SAR* is an acronym of Synthetic Aperture Radar. In this context, an aperture is an aerial or antenna such as the ones you may have seen above the air traffic control centre at an airport, or at a military base such as Fylingdales in North Yorkshire. The first airborne radar systems carried radars that were similar in principle to these ground-based radars. The antenna was mounted on the side or on the top of the aircraft, and was pointed to one side of the aircraft at 90° to the direction of forward travel. A pulse of microwave energy was emitted by the antenna and the reflected backscatter from the target was detected by the same antenna, with the result being displayed on a cathode ray tube. This kind of radar is known as a Side-Looking Airborne Radar or SLAR. The box 'Radar History' provides a brief summary of the historical development of radar.

A BRIEF HISTORY OF RADAR

The history of the development of radar makes a fascinating story. The detection of distant targets by radio waves is an idea that dates back to the early twentieth century, when a German scientist, Huelsmayer, patented the concept. Two Americans, Taylor and Young, rediscovered Huelsmayer's idea in 1922, but nothing came of it. By the early 1930s, the British Government was becoming concerned by the prospect of aerial bombardment of cities by the Luftwaffe, and the UK Air Ministry started to investigate ways of locating incoming bombers. Some rather odd ideas emerged, such as that of the Death Ray. The UK Air Ministry offered a prize of £1 000 in 1934 for the first person to kill a sheep at 180 m using a Death Ray. The prize was never awarded; unfortunately, the number of sheep that fell in the cause of science (and patriotism) was never recorded. Thermal infrared and acoustic devices were considered, but it was Watson-Watt who, in 1935, demonstrated that he could locate an aircraft using the BBC radio transmitter near Daventry in the English Midlands. The Americans had the same idea and did, in fact, detect an aircraft using radar a month before Watson-Watt's success. Watson-Watt became a victim of his own invention when, after the Second World War, he was caught by a radar speed trap.

Early radars such as those used in the 'Battle of Britain' in August/September 1940 operated at long wavelengths (13 m) and were hampered by radio interference. A major breakthrough came in 1939–1940 with the invention of the cavity magnetron by Randall and Boot at Birmingham University. Although the USA was still neutral in 1940, the British Prime Minister, Winston Churchill, authorised the transfer of a cavity magnetron to the United States, where – in response – the Rad Lab was established at MIT, attracting scientists such as Luis Alvarez (who was later to become even more famous for his theory that dinosaurs were annihilated by a meteorite impact at about 65 MA). One American scientist described the cavity magnetron as the most valuable cargo ever to reach America's shores.

The MIT Rad Lab and the British radar research laboratories used the cavity magnetron to develop shorter-wavelength radars. The British developed the H2S radar, which operated at a 10 cm wavelength and which was fitted to bombers. This targeting radar was the first terrain-scanning radar in the world. Its effectiveness was demonstrated on 1 November 1941, when a 10-cm radar fitted to a Blenheim bomber was able to locate a town at a distance of 50 km.

Radar research and development proceeded rapidly during World War II, with naval and airborne radar being widely used by the British and American navies for detection of enemy vessels and surfaced submarines, and for gunnery control. Radars were also carried by aircraft. By 1945, radar was an essential element of the air defence system, targeting and naval operations.

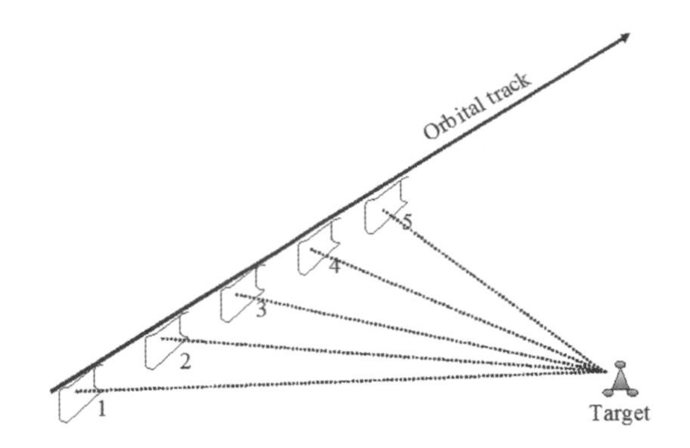

Figure 2.16 The 'synthetic aperture' is generated by the forward motion of the platform through positions 1–5, respectively. The Doppler principle is used to determine whether the antenna is looking at the target from behind or ahead.

The spatial resolution achieved by a SLAR is proportional to the length of the antenna. For satellite-borne radars, this relationship results in a practical difficulty. In order to achieve a spatial resolution of, say, 30 m from an orbital height of 700 km the antenna length would be several kilometres. If, on the other hand, it were possible to move a 10 m antenna along the orbital path, recording the backscatter from a given target at each position, then it would be possible to simulate an antenna length of several kilometres and achieve an acceptable level of spatial resolution.

One obvious difficulty in understanding how a SAR works is summarised by the question: 'How does the sensor record whether it is approaching, passing, or moving away from the target?' The answer is: it does not. Instead of recording positions relative to every possible target, which would be difficult, the SAR records the amplitude and the phase of the return signal (Figure 1.8; Figure 9.1). From these two properties of the return signals, it is possible to calculate the position of the object relative to the antenna. This calculation is based on a concept that is well within our everyday experience, namely the Doppler principle. We can tell whether a police car or an ambulance is approaching us or moving away from us by comparing the tone of the siren at one instant with that at the next instant. If the vehicle is approaching, then the siren's note appears to drop. If it is moving away from us, then the note seems to rise in pitch. The same idea is used in a SAR (the calculations are, however, quite formidable).

The Doppler principle can thus be used to increase the spatial resolution of a satellite or airborne imaging radar by allowing the calculation of the motion of the satellite platform relative to a target (Figure 2.16). In addition, the absolute distance (i.e., in kilometres) between the antenna and the target can be calculated. This distance is called the range. Microwave energy travels at the speed of light and so, if the time taken between the transmission of a pulse and the reception of the return is known, then the distance is $ct/2$ where c is the speed of light, and t is the time taken from transmission of a pulse to the reception of its return. Division by two is undertaken as the pulse travels from the antenna to the target and back again. Radar is called a ranging instrument because it can measure distance. Lidar (section 9.4) is another example of a ranging instrument. However, lidar measures only the distance from the sensor to the target. Radar can also measure some characteristics of the target as well.

We saw in Chapter 1, and in some of the examples presented earlier in this chapter, that optical sensors measure electromagnetic radiation that is reflected by the target. The chemical composition of the target (for example, chlorophyll in leaves or minerals in a soil) affects the nature of the reflected radiation in the optical region because each chemical element, or mixture of elements, absorbs some of the incoming radiation at specific wavelengths. The return signal is called the reflectance spectrum of the target. Techniques for the identification of targets by remote sensing using optical wavelengths assume that similar targets (such as specific vegetation types) have similar reflectance spectra (Chapter 8). The interaction between electromagnetic radiation and a target in the microwave region does not generate information about the types of chemical element or mixtures of elements that are present in the target, but is related to the geometry and surface roughness of that target (relative to the wavelength of the microwave energy pulse) and what are known as the 'dielectric properties' of the target, which generally are closely correlated with its moisture status.

Radar wavelength bands are described by codes such as 'L-band' or 'C-band' that came into use during World War

Table 2.6 Radar wavebands and nomenclature.

Band designation	Frequency (MHz)	Wavelength (cm)
P	300–1000	30–100
L	1000–2000	15–30
S	2000–4000	7.5–15
C	4000–8000	3.75–7.5
X	8000–12000	2.5–3.75
K_u	12000–18000	1.667–2.5
K	18000–27000	1.111–1.667
K_a	27000–40000	0.75–1.111

II for security purposes. Unfortunately, several versions of the code were used, which confused the Allies as much as the opposition. Table 2.6 shows the commonly accepted delimitation of radar wavelengths. Radar wavelength also has a bearing on the degree of penetration of the surface material that is achieved by the microwave pulses. At L-band wavelengths (approximately 23 cm), microwave radiation can penetrate the foliage of trees and, depending on the height of the tree, may reach the ground. Backscatter occurs from the leaves, branches, trunks, and the ground surface. In areas of dry alluvial or sandy soils, L-band radar can penetrate the ground for several metres. The same is true for glacier ice. Shorter-wavelength C-band radiation can penetrate the canopies of trees, and the upper layers of soil and ice. Even shorter wavelength X-band SAR mainly 'sees' the top of the vegetation canopy and the soil and ice surface.

Properties of the microwave radiation used by a SAR, other than wavelength, are important. The *polarisation* of the signal has an effect on the nature and magnitude of the backscatter. Figure 1.3 illustrates the concept of polarisation of an electromagnetic wave. In a polarised radar, the antenna can transmit and receive signals in either horizontal (H) or vertical (V) mode. If it both transmits and receives in horizontal polarisation mode it is designated as 'HH'. If both transmit and receive modes are vertical then the radar system is designated 'VV'. HV and VH modes are also used. HH and VV modes are said to be 'like polarised', whereas VH and HV modes are 'cross-polarised'. The SIR-C SAR carried by the Space Shuttle in 1994 provided polarimetric radar images in the C and L bands, plus an X-band image. The ASAR instrument on Envisat can transmit and receive in combinations of H and V polarisations, and is the only orbiting polarimetric radar at the moment, though RADARSAT-II will carry a fully polarimetric SAR. Freeman *et al.* (1994) discuss the use of multi-frequency and polarimetric radar for the identification and classification of agricultural crops (chapter 8).

Another important property of an imaging radar is the instrument's *depression angle,* which is the angle between the direction of observation and the horizontal (Figure 2.17). The angle between the direction of observation and the surface normal (a line at right angles to the slope of the Earth's surface) is the *incidence angle*, which is also shown on Figure 2.17. The local incidence angle depends on the slope of the ground surface at the point being imaged (Figure 2.17(b)). Generally, the degree of backscatter increases as the incidence angle decreases. Different depression angles are suited to different tasks; for example, ocean and ice monitoring SAR systems use lower depression angles than do SAR systems for land monitoring.

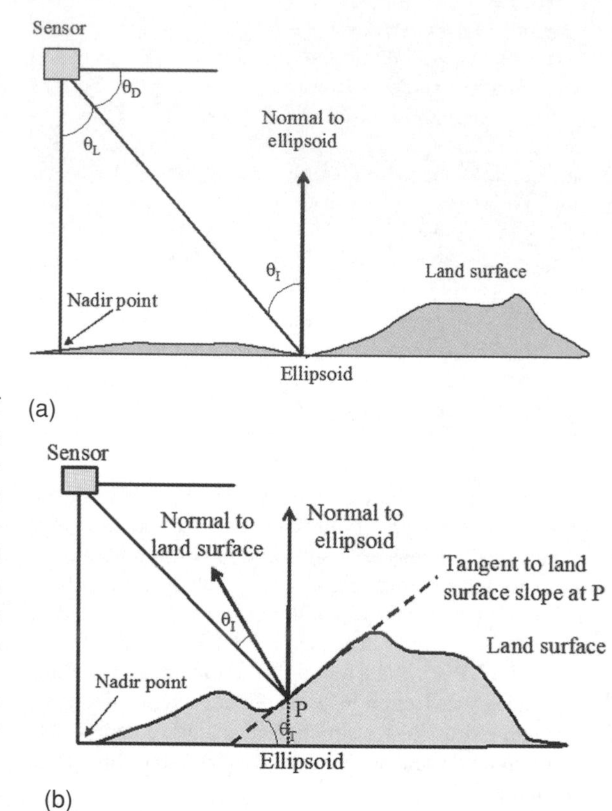

Figure 2.17 (a) Angles in imaging radar: θ_D is the depression angle (with respect to the horizontal), θ_L is the look angle, which is the complement of the depression angle, and θ_I is the incidence angle. θ_L depends on local topography and is equal to the look angle only when the ground surface is horizontal. In Figure 2.17(b) the ground slope at P is θ_T and the local incidence angle is θ_I. The normal to the land surface is the line through P that is perpendicular to the tangent to the land surface slope at P.

The view direction of a radar sensor is also important in detecting geological features with particular orientations, so that radar images of the same terrain with different 'look angles' will show different features (Blom, 1988;

Gauthier *et al.*, 1998, Koch and Mather, 1997, Lowman, 1994).

Imaging radars are side-looking rather than nadir-looking instruments, and their geometry is complicated by foreshortening (the top of a mountain appearing closer to the sensor than does the foot of the mountain) and shadow, caused by the 'far side' of a mountain or hill being invisible to the side-looking radar sensor (see Schrier, 1993, for a thorough description of the workings of a synthetic aperture radar) (Figure 2.18a, b). Furthermore, the interaction between microwave radiation and the ground surface generates a phenomenon called *speckle*, which is the result of interference resulting from the coherent integration of the contributions of all the scatterers in the pixel area (Quegan and Rhodes, 1994). Speckle magnitude is proportional to the magnitude of the back-scattered signal, and is rather more difficult to remove from the image than is additive noise. Filtering, which is used to remove unwanted features such as speckle noise, is the topic of Chapter 7. A recent development is the use of wavelets (section 6.7) to remove speckle filter noise; see, for example, Xie (2002).

A good survey of synthetic aperture radar principles and applications is contained in the EOS Instrument Panel report published by NASA (1988). See also section 1.2.3 for an introduction to imaging radar principles, and section 4.6 for a brief summary of calibration issues. Other basic sources are Leberl (1990), Ulaby *et al.* (1981–1986) and Ulaby and Elachi (1990). The textbook by Kingsley and Quegan (1991) provides a sound introduction to radar systems.

2.4.1 ERS SAR

The first space-borne imaging radar was carried by the US Seasat satellite, launched in 1978. It carried an L-band SAR, but operated for only 100 days. Thirteen years later, on 17 July 1991, the first European Remote Sensing Satellite, ERS-1, was launched into an 800 km near-polar orbit. ERS-1 carries the ATSR (section 2.3.1), which is a low-resolution optical/infrared sensor, as well as a synthetic aperture radar. A second identical ERS satellite (ERS-2) was launched in 1994, providing opportunities for 'tandem mode' operation as the two satellites orbited one behind the other to generate SAR images of the same area over a period of a few days. Such data sets are used in interferometric studies to derive digital elevation models and to measurements of small movements on the Earth's surface (section 9.2).

The ERS SAR operates at a wavelength of 5.6 cm in the C band (Table 2.6) and images an area 100 km wide to the right-hand side of the satellite track (facing the direction of motion). Microwave energy pulses are both transmitted and received in vertical polarisation mode. As noted above, radar is a side-looking sensor, transmitting pulses of electromagnetic energy in a direction to one side of the spacecraft. The *depression angle* of this beam in the case of the ERS radar is 20° for near range and 26° at the far range (Figure 2.19). Related to the radar depression angle is the *incidence angle*, which is the angle between the line connecting the SAR antenna and the target and the line perpendicular the Earth's surface to the Earth's surface at the target (the surface normal). Lillesand and Keifer (1994, p. 670; see also NASA, 1988, p. 113) note that topographic slope effects are greater than radar backscatter effects where the local incidence angle is less than 30° whereas, for incidence angles greater than 70°, topographic shadow is dominant. Between these two limits, surface roughness effects are predominant. The ERS SAR depression angle facilitates ocean and ice sheet observation, and is also good for observing some terrestrial features such as agricultural fields.

As noted earlier, the advantages of microwave radar is that it is an all-weather sensor operating independently of solar illumination. Figure 2.20 shows an ERS SAR image of the south-east coast of England from Kent to Suffolk. Areas of low radar backscatter are dark. The brighter areas are those with a high surface roughness. Bright points over the land may be features that exhibit sharp angles, such as the walls of buildings, walls around fields or steep slopes.

2.4.2 RADARSAT

The Canadian RADARSAT system is a major source of SAR imagery. The RADARSAT programme is funded through the Canadian Space Agency, with NASA and NOAA cooperation. RADARSAT is in a Sun-synchronous orbit at an altitude of 798 km, at an inclination of 98.6° to the equatorial plane. The orbit is such that the satellite should always be in sunlight, and hence power from the solar arrays is continuously available. The SAR sensor carried by RADARSAT can steer its beam so as to collect images at incidence angles in the range 20–50°, with swath widths of between 35 and 500 km, using resolutions ranging from 10 to 100 m. the radar operates in the C band and is HH polarised. RADARSAT is able to collect data in a number of modes, giving fine to coarse resolution and variable inclination angles. Figure 2.21 is an image of south central Alaska in low-resolution (ScanSAR) mode. The data are recorded on board the satellite and downlinked to ground receiving stations, of which there are three in N. America, three in S. America, one in South Africa, two in Australia and seven in Asia.

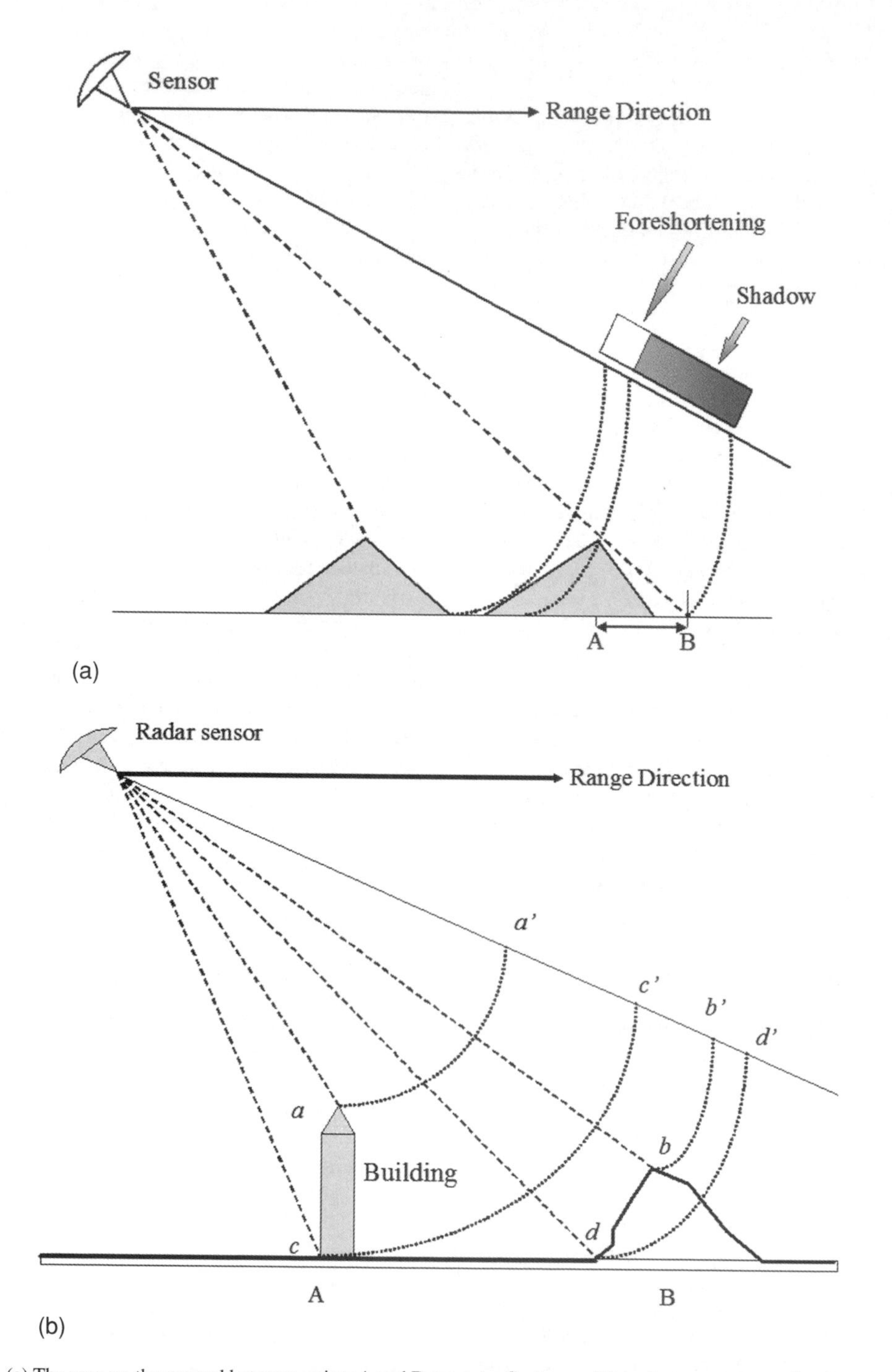

Figure 2.18 (a) The area on the ground between points A and B cannot reflect any of the microwave energy that is transmitted by the sensor, and so appears as 'radar shadow'. Also note that although the right-hand hill is fairly symmetric, the distance from hill foot to crest for the illuminated side is seen by the sensor as being considerably less than the hill foot to crest distance (AB) of the right-hand side of the hill. This effect is known as 'foreshortening'. (b) Layover is the result of the distance from the top of the building at A or the hill at B appearing to be closer to the sensor than the base of the building or the foot of the hill. This distance is called the 'slant range'.

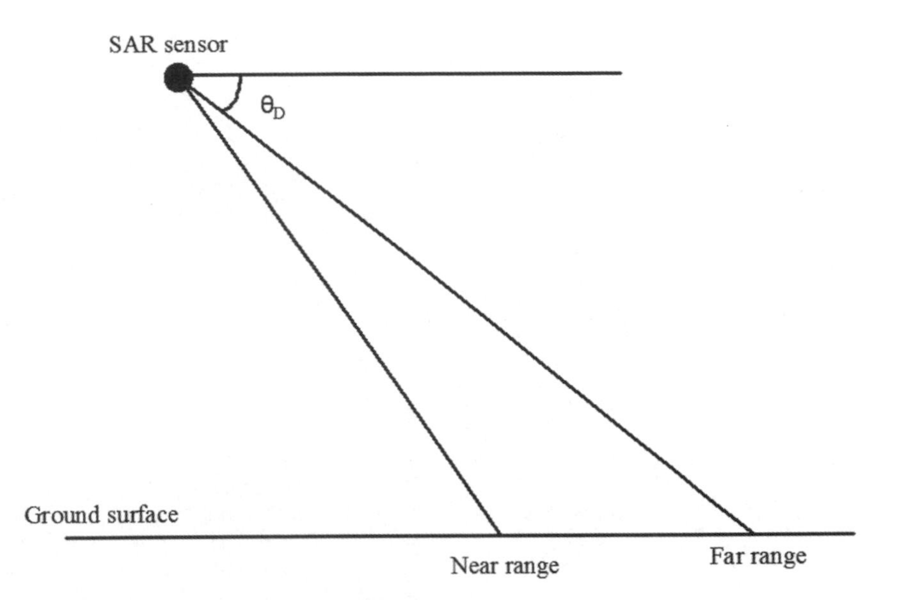

Figure 2.19 The distance from the sensor to the closest edge of the swath parallel to the azimuth direction (i.e., the flight direction) is called the near range. The distance from the sensor to the furthest part of the swath is the far range. The depression angle for the far range is shown as θ_D. Clearly, the depression angle for the near range will be greater.

Table 2.7 RADARSAT-2 modes, spatial resolutions and orbit characteristics.

RADARSAT-2 SPECIFICATIONS			
	Ultra-fine	Standard	Wide
Spatial resolution	3 m	28 m	100 m
Swath width	20 km	100 km	500 km
Revisit time		3 days	
Orbital altitude		798 km	
Nodal crossing		6:00 PM	
System life		7 years	
Orbit		Sun-synchronous	
Altitude		798 km	
Inclination		98.6°	
Period		100.7 minutes	
Repeat cycle		24 days	
Coverage access North of 70° using largest swath		Daily	
Coverage access at Equator using largest swath		Every 2–3 days	

RADARSAT-2 is scheduled for launch in 2004. It will be in an orbit similar to that used by RADARSAT-1, but with spatial resolutions of 3, 28 and 100 m in ultra-fine, normal and wide image modes (Table 2.7). The swath widths in these three modes are 20, 100 and 500 km. In addition, RADARSAT-2 will be able to collect HH-, HV- and VV-polarised images (Ulaby and Elachi, 1990).

2.5 SUMMARY

A variety of information is available from sensors carried by both satellite and aircraft platforms. Those operating in the visible and near-infrared region of the electromagnetic spectrum provide measurements that correlate with an object's colour, which is often related to the chemical or mineralogical properties of that object. Data from thermal infrared sensors are related to the temperature and thermal properties of a target, while information about surface roughness and (over the land) moisture content can be derived from data collected in the microwave (radar) wavelengths. The spectral range of remotely-sensed data available in the late 1990s from orbiting satellites covers the full range from optical to microwave. Given the number of different remote sensing programmes around the world, it is rather difficult to keep up to date and surveys such as the one provided in this chapter are only partial, and soon become outdated. Surveys such as Kramer (1994) provide a starting point, but the only way to keep informed is to join a remote sensing society and read its newsletter.

Figure 2.20 ERS SAR image of south-east England. Dark regions over the Thames estuary are areas of calm water. Bright areas over the land are strong reflections from buildings and other structures, which act as corner reflectors. Image acquired on 7 August 1991. © 1991 European Space Agency.

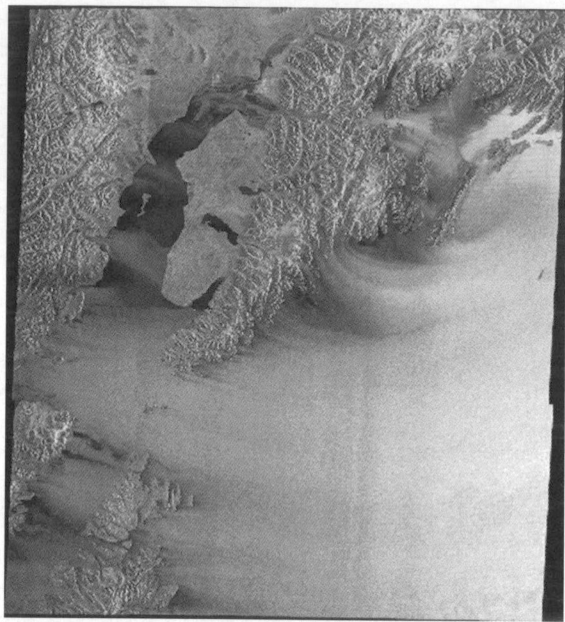

Figure 2.21 RADARSAT image of south-central Alaska using low resolution (ScanSAR) mode, covering an area of 460 km by 512 km. Kodiak Island is in the lower left of the image. The Kenai peninsula and Prince William Sound can also be seen. This is a reduced resolution, unqualified and un-calibrated image. A ScanSAR image covers approximately 24 times the area of a standard RADARSAT SAR image. © 1996 Canadian Space Agency/Agence spatiale canadienne. Received by the Alaska SAR Facility, distributed under licence by RADARSAT International.

Satellite remote sensing platforms that are currently providing image data include Landsat-7, Terra, Aqua, Envisat, SPOT, NOAA, ERS, RADARSAT, IRS, the Japanese ADEOS-2, Meteosat and other geostationary meteorological satellite systems, and high-resolution commercial systems such as IKONOS and Quickbird, with 1 m or less spatial resolution. In addition, NASA's Earth Observing-1 carries an imaging spectrometer, Hyperion, capable of resolving 220 spectral bands (from 0.4 to 2.5 μm) with a 30 m spatial resolution. The instrument images a 7.5 km by 100 km land area per image. Experience gained with airborne imaging spectrometers (Chapter 9) helps researchers to understand the problems of handling such large data volumes. Potential technical problems include the development of methods of combining multi-source (multi-sensor) data, handling large volumes of high-resolution data, and selecting the optimum combination of bands to use for a particular application. Clearly, there will be a greater need than ever in the future for research and development in the areas covered by later chapters of this book.

3

Hardware and software aspects of digital image processing

'There are 10 kinds of people in the world – those who understand the binary system and those who don't.'

(Jeremy Paxman (attrib.))

3.1 INTRODUCTION

This short chapter contains three sections, excluding this introduction. The first (section 3.2) describes and summarises the properties of digital remotely-sensed image data and the relationship between these properties and those of computer hardware, especially display and mass storage subsystems. The subjects of system processing of remotely-sensed images and the various formats of image data sets on distributable computer media such as CDs and tapes are also summarised.

The second section deals with the way in which the software employed for the processing and display of digital images uses the hardware (section 3.3). A short summary of the characteristics of the MIPS software (provided on the CD that accompanies this book) is included, so that the reader is familiar with the ways in which image data are 'invisibly manipulated' in the background by image display software in order to fit the requirements of the display hardware. Nowadays, digital images composed of numbers represented in 16- or 32-bit integer or 32-bit real form are becoming more widely available, yet the display hardware used on most PCs and workstations requires that the red, green and blue components of the image are each stored in 8-bit form. The 'invisible decisions' mentioned above relate to the scaling methods employed in order to 'make the image fit the properties of hardware'.

A short description of MIPS is provided in section 3.3. MIPS was written specifically for teaching purposes and for use by postgraduate students. It is hoped that readers will find it intuitive and easy to use. Most of the procedures described in Chapters 4–9 are implemented in MIPS, and worked examples are presented in Examples in these chapters. The software is a development of the MIPS programs that accompanied the second edition of this book in 1999. Most of the 'stand-alone' programs that were included in the original package have now been integrated into a single 'software environment'. MIPS is written mainly in Fortran 90 (using Salford Software's FTN95 compiler[1]). Some routines are written in C. These were compiled using Microsoft's Visual C++ compiler. MIPS runs under the Windows 98/Me/2000/XP operating systems. Advice on the detailed workings of MIPS is given in a number of HTML files that are included on the CD, and which can be accessed from the MIPS 'help' menu function. Consequently, this chapter is not a user's manual. Rather, it is a brief overview, which is written with the intention of showing the reader how certain key MIPS functions relate to the properties of remotely-sensed image data.

3.2 PROPERTIES OF DIGITAL REMOTE SENSING DATA

3.2.1 Digital data

Digital images are arrays of numbers, i.e. an image is represented logically as a matrix of rows and columns. These image data arrays are included in the general class of 'raster data', which means that the individual data value is not explicitly associated with a particular location on the ground. The location of each data value (or picture element, corrupted into 'pixel') is implied by its position in the array (Figure 3.1). Thus, if we know the UTM coordinates of the top left cell in the array or raster and the cell spacing in metres then we can calculate the position of any cell in the raster. The values of the numbers stored in the array elements lie in a specified range, commonly 0–255, which corresponds to the brightness range of the colour associated with that image array. The value 0 indicates lack of the associated colour (red, green or blue), and the value 255 is the brightest level at which that colour is displayed. As we will see, the two numbers at the extremities of the range (0 and 255 in this case) may be used for other purposes.

[1] http://www.salfordsoftware.co.uk/

Computer Processing of Remotely-Sensed Images: An Introduction, Third Edition. Paul M. Mather.
 ISBNs: 0-470-84918-5 (HB); 0-470-84919-3 (PB)

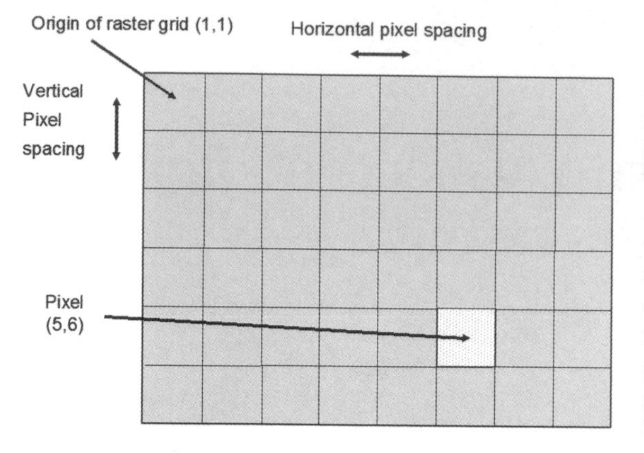

Figure 3.1 Raster data concepts. The origin of the (row, column) coordinate system is the upper left corner of the grid, at cell (row 1, column 1). The grid cell (pixel) size is usually expressed in units of ground distance such as metres. The position of any pixel can be calculated if the horizontal and vertical pixel spacing is known, and the map coordinates of a pixel can be derived if the map coordinates of pixel (1, 1) are known. Note that pixels are referenced in terms of (row, column) coordinates, so that pixel (5, 6) lies at the junction of row 5 and column 6. The horizontal and vertical pixel spacing is equal for most, but not all, remote sensing images.

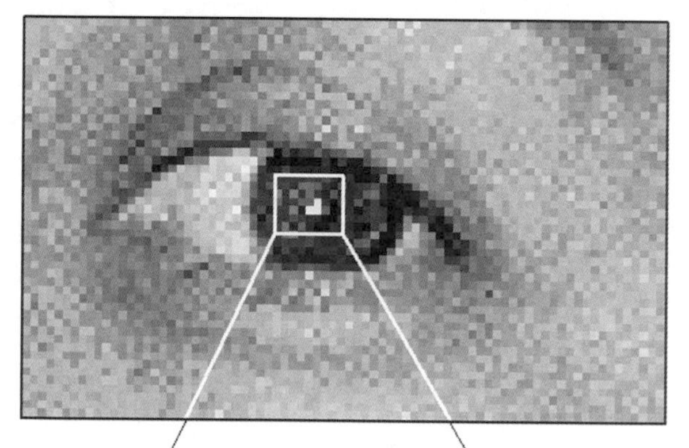

25	35	60	65	45	25
80	40	72	45	100	62
35	40	130	180	210	41
25	50	130	220	235	35
35	32	67	75	110	35
29	25	34	46	37	28

Figure 3.2 Digital image of a human eye, showing the correspondence between the grey levels of the pixels making up the image and the numerical representation of the pixel grey level in the computer's graphics memory.

Figure 3.2 shows a digital (raster) image of an eye, together with a section of the array of numbers (pixel values) corresponding to the part of the image outlined by the white rectangle. The array of pixel values is held in a special area of the computer's memory known as the *graphics memory*. Graphics memory is normally located on the graphics card and is not a part of the computer's random access memory (RAM). Figure 3.2 is a greyscale image, and so only one array of numbers is required to hold the pixel values, each of which can take on one of 256 brightness levels ranging from 0 (black) through 127 (mid-grey) to 255 (white). A greyscale image has only one component – the levels of grey – whereas a colour image has three components, these being the levels of the primary colours of light (red, green and blue) at each pixel position. The structure of the digital image in terms of individual square pixels is apparent at this scale. The correspondence between the grey levels in the image and the pixel values in the raster is also clear.

As already noted, a colour image is produced by using three raster arrays, which hold pixel values that represent the levels of the three primary colours of light (Figure 3.3). Levels 0 to 255 represent the range of each primary colour from 0 (black) to 255 (maximum intensity of red, green, or blue, hereafter RGB). Different combinations of RG and B produce the colours of the spectrum, as demonstrated by Sir Isaac Newton's famous prism experiment. The primary colours of light are 'additive' – for example, red + green = yellow. In contrast, colours used in printing are subtractive.

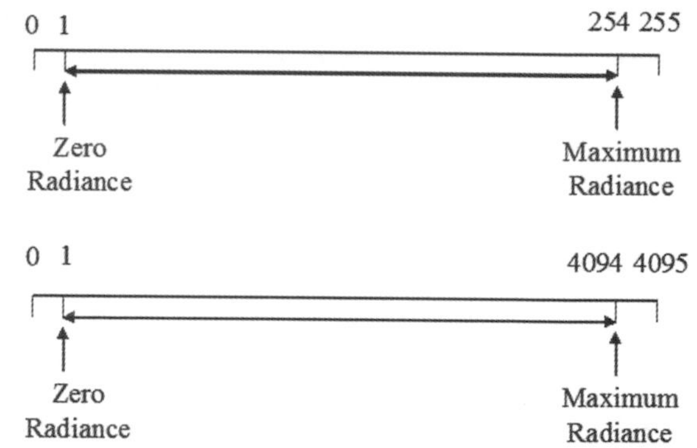

Figure 3.3 (Top) ASTER VNIR and SWIR bands: A scale of 0–255 (8 bits) is used. Quantisation level 1 is equivalent to zero radiance. Level 254 is assigned the maximum radiance level for that band and gain setting. Level 0 indicates a dummy pixel. Level 255 is used to indicate a saturated pixel. (Bottom) ASTER TIR bands: the principle is the same except that the number of quantisation levels is 4096 (12 bits). Maximum radiance equals the radiance of a 370 K blackbody at the specific waveband (10–14 μm). Based on Figure 4.4 of ASTER Users' Guide, Part II: Level 1 Data Products, Version 3.1, ERSDAC, Japan.

That is why an ink jet printer uses cyan, magenta and yellow ink. Table 3.1 lists some examples of colours generated by adding different proportions of RG and B. Note that RGB combinations in which the levels of red, green and blue are equal produce shades of grey.

Not all remotely-sensed images have pixel values that lie in the range 0–255. For example, AVHRR data (section 2.3.2) use a 0–1023 range. IKONOS pixels (section 2.3.9.1) lie in the range 0–2047, and the thermal bands of ASTER images (section 2.3.8) are measured on a 0–4095 scale. Specific use is made of the lowest and highest counts ASTER data, for example '0' and '4095'; these are used to indicate 'bad data' and 'saturated pixel', respectively (Figure 3.4).

The values stored in the cells making up a digital image (the 'pixel values' or 'pixel intensities') are represented electronically by a series of binary (base two) digits that can be thought of as 'on/off' switches, or dots and dashes in Morse code. In base two form the decimal numbers 0, 1, 2, 3, are written as 0, 1, 10, 11 . . . with each column to the left representing a successively higher power of two, rather than ten as in the everyday decimal system. If eight binary digits are used to record the value stored in each pixel, then 0 and 255 are written as 00000000 and 11111111. Thus, a total of eight *bi*nary digi*ts* (bits) is needed to represent the 256 numbers in the range 0–255. The range of pixel intensities is termed the dynamic range of the image.

Table 3.1 Combinations of the primary colours of light (red, green and blue) combine to produce intermediate colours such as purple and orange. Where the values of the three primary colours are equal, the result is a shade of grey between black and white. The intensities shown assume eight-bit representation, i.e., a 0–255 scale.

Red intensity	Green intensity	Blue intensity	Colour name
255	255	0	Yellow
0	255	255	Cyan
255	0	255	Magenta
127	0	0	Mid-red
127	127	127	Mid-grey
0	0	0	Black
255	255	255	White
241	0	171	Purple
255	155	50	Orange

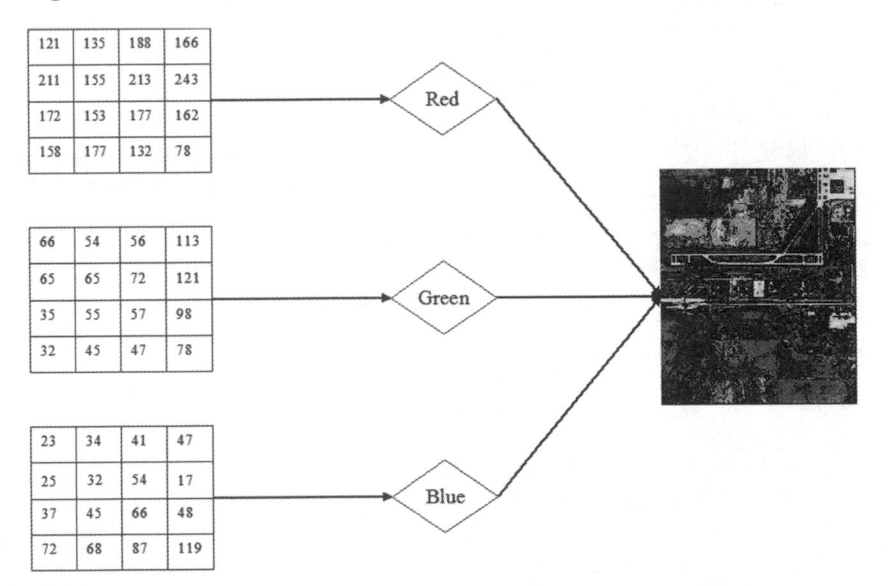

Figure 3.4 A colour image is generated on the screen using three arrays held in graphics memory. These three arrays hold numbers that show shades of red (top). The centre array shows the distribution of shades of green, and the bottom array holds the numbers corresponding to shades of blue. Each array holds integer (whole) numbers in the range of 0 (black, or lack of colour) to 255 (the brightest shade of red, green or blue). These values are converted from digital to analogue form by one of three Digital-to-Analogue Converters (DAC) (centre) before being displayed on the screen (right).

Table 3.2 Different dynamic ranges used to represent remotely-sensed image data. The number of binary digits used in the representation is proportional to the number of levels of grey or primary colour (RGB) that can be stored in the image array. Note that integer data can be signed (+ or −) or unsigned, i.e. assumed to be non-negative. Integers are represented in base two by a string of binary digits (bits), whereas real numbers are stored in computer memory as a mantissa (m) and exponent (e) in the form 'real_number = m × 2^e'. The range of a real number depends on the particular computer, whereas integer representation is common to all computers.

Number of bits	Base 10 maximum/range	Base 2 maximum
10 (integer)	1024	1111111111
16 (signed integer)	−32768 to +32767	±111111111111111
16 (unsigned integer)	0 − 65535	1111111111111111
32 (signed integer)	2147483647	±11111111 . . . 1111111
32 (real)	$\pm 3.4 \times 10^{38}$	See table caption

Because eight bits are needed to represent the range of each of the three primary colours, the resulting image is called a '24-bit image'. Other ways of representing image pixel data are shown in Table 3.2. For example, a 10-bit single-band image provides 1024 levels of grey, while a 16-bit image can represent either positive and negative data or positive data only, depending on whether signed or unsigned representation is used. Real-valued data are commonly used to store physical values rather than quantised counts. For example, the amplitude and phase of a SAR image are stored as a pair of 32-bit real numbers. Most computer monitors require an analogue (continuously varying) input signal, so the digital (discrete) values held in the graphics memory are converted to analogue form, as voltages, by a Digital to Analogue Converter (DAC), as shown in Figure 3.3. The output from the DAC is fed to the monitor input.

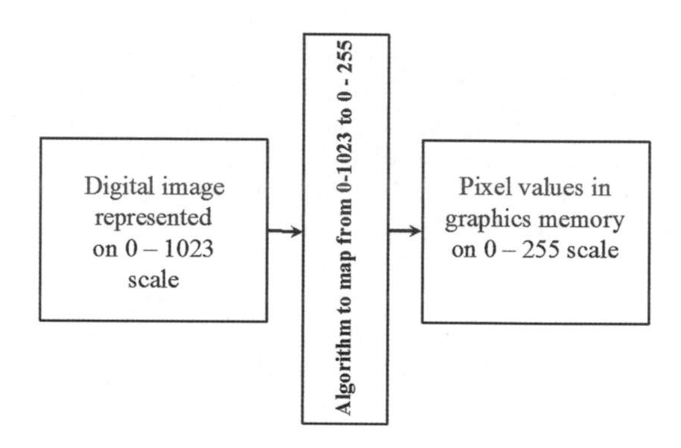

Figure 3.5 Digital images in which pixels are represented on a scale of more than eight bits (left) must be transformed onto a 0–255 scale before being transferred to the computer's graphics memory. The initial image in this example uses 10-bit representation, but it could equally well use 16 or 32 bit, or real number representation.

It is important to realise that the number of bits per pixel in the display memory is fixed at eight by the hardware. However, remotely-sensed images are provided in 10-bit, 12-bit, 16-bit or 32-bit as well as 8-bit integer form, plus 32-bit or even 64-bit real representation. Real numbers have decimal points (e.g., 2.81), whereas integers are the whole (counting) numbers. The on-screen appearance of the image is affected by the choice of method of 'mapping' (transforming) the input image to eight-bit form (Figure 3.5). The first of the two methods used by MIPS takes the numerical range of the data for each of the three colours in turn, and performs a linear mapping between input and output, using the following relationship:

$$output = \frac{(input - in_{\min})}{(in_{\max} - in_{\min})} \times 255 \qquad (3.1)$$

where *output* is a value between 0 and 255, $in_{\max}$ and $in_{\min}$ are the maximum and minimum values, respectively, in the input image, and *input* is the value of the image pixel to be converted. For example, if the input image uses real number representation and the minimum and maximum pixel values in the image are 139.76 and 2409.77, respectively, then the value placed in the 8-bit graphics memory (output in equation (3.1)) when *input* equals 1500.20 is 152 (the integer part of 152.82). Outlying (or 'rogue') values can have a substantial effect on the range of the input data and can cause this method to produce a result that lacks contrast. For example, 99% of the pixel values in an input image may lie in the range 1290–1879, with the lowest value being 5 and the highest 2009. The input value 1290 transforms to 164 using equation (3.1), while the input value 1879 is mapped to an output value of 238. Hence, 99% of the input image pixels are mapped onto the 164–238 part of the total 0–255 output brightness range, so the resulting image will look over-exposed in photographic terms, but there will be only a low contrast between the darker (164) pixels and the brighter (238) pixels. Much of the 0–255 range is left unused. However, this transform is reversible, but only approximately, as only the integer part of the *output* value is retained.

The alternative method used by MIPS to map an input image with a dynamic range greater than eight bits onto the 0–255 range is called equalisation. The input pixel values are grouped into 256 sets by amalgamating the 1024 levels into 256 classes that each contain approximately equal numbers of pixels. The 256 output classes have equal frequencies rather than equal range (Figure 3.6). This method is very similar to that of *histogram equalisation* that is described in detail in section 3.5.2. In comparison to the linear mapping approach, equalisation generally produces an output image that has greater contrast. However, since several input values can map to the same output value, the transformation is not reversible, and it is also non-linear, because the steps between the individual contiguous output classes do not correspond to equal ranges of the input values. Thus, using the figures from the example of linear mapping (above), the input values 5–1295 may correspond to output class 0, while the range 1296–1309 may correspond to the output class 1 and 1309–1311 to output class 2.

The importance of understanding the way in which the pixel brightness values that are stored in the graphics memory are generated can be illustrated by a simple example.

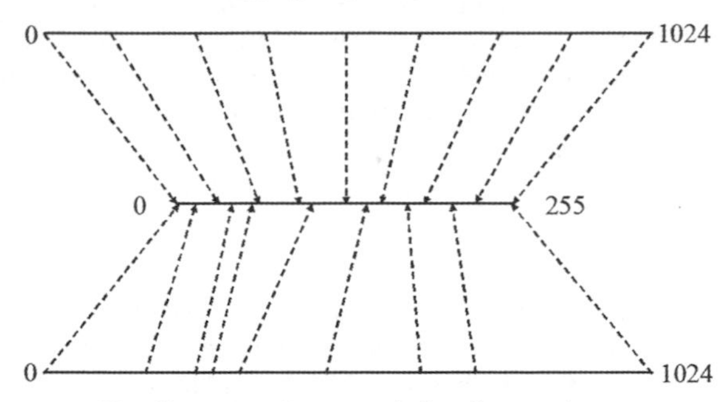

Figure 3.6 Two common methods of converting image data on to a 0–255 scale required by graphics memory. The upper part of the diagram shows that equal intervals on the 0–1024 scale map to equal intervals on the 0–255 scale. The lower part of the diagram shows how the values on the 0–1024 scale are grouped into classes of equal frequency, which map proportionally on to the 0–255 scale.

Since the values in the graphics memory are not linearly related to the pixel values in the image being displayed when the equalisation approach is used then differences and ratios calculated from the values in the graphics memory will not be proportional to the corresponding differences and ratios computed from the actual input image pixel values. In the previous paragraph it was shown that different ranges of image pixel brightness values are mapped to the 0–255 range by the equalisation procedure. If we take the difference between input image pixel brightness values of 1311 and 1296 we get 15. The difference between input image pixel brightness values of 1309 and 1308 is 1. However, input image pixel brightness values of 1311 and 1309 are both represented in the graphics memory by the value 2 (on the 0–255 range), and the input image pixel brightness values of 1296 and 1308 are placed in class 1, so in both cases the difference is calculated as 1 from the values held in graphics memory. If the linear mapping method is used then differences and ratios of the input image pixel brightness values are approximately proportional to the same computations performed on the corresponding values held in graphics memory. They are approximately proportional because only the integer part of the computation shown in equation (3.1) is stored.

Generally speaking, three different kinds of images can be stored in display memory and viewed on-screen. These types are: colour images, greyscale images and labelled or classified images. Colour images can be one of three types. The first, called natural colour, is made up of three components representing the 'real-world' colours of visible red, visible green and visible blue bands (for example, Landsat ETM+ bands 3, 2 and 1) in the red, green and blue memory banks of the display memory. Natural colour images are like ordinary colour photographs. They show colours as we would see them. If the three bands selected for display do not represent actual red, green and blue as we would see them, then the result is a false-colour image. For example, Landsat ETM+ bands 4, 3 and 2 could be stored in the red, green and blue memory banks, and displayed in RGB. The image seen on the screen would display variations in near-infrared reflectance (band 4) as shades of red, with variations in red (band 3) and blue (band 2) reflectance being seen on-screen as variations in shades of green and blue. Any three bands can be stored in the red, green and blue memory banks. The third kind of colour image is referred to as a pseudocolour image, as it is based on data that occupy a single memory bank, rather than three memory banks. This implies that the pixel values in a pseudocolour image range in value from 0 to 255. These 256 levels are associated with colours via the use of a lookup table. For example, level 0 in the display memory may be 'mapped' to the colour maximum red (with RGB components of row 0 of the lookup table set to 255, 0 and 0, respectively). The lookup table entry for level 1 may be set to brightest yellow (levels 255, 255 and 0), and so on. In this representation, the single value at a given pixel position in the image is sent to all three digital-to-analogue converters via the lookup table. In the first example given above, the red DAC converts the input '0' to the output '255' while the green and blue DACs convert the input '0' to output '0'. In the second example, the input value '1' is converted to the output value '255' by both the red and the green DACs (to produce yellow), while the blue DAC outputs '0' for an input value of '1'. Many image file formats such as BMP and TIF specify that, for a 256-level image, an array of colour values (256 × 3, i.e. 256 levels × red, green and blue) is specified and stored with the image. This array is known as either a palette or a look-up table. The distinction between natural colour, false colour and pseudocolour is shown in Figure 3.7. Refer also to Figure 3.3.

The second kind of image that can be stored in graphics memory and displayed on-screen is called a greyscale image. Like the pseudocolour image, the greyscale image has only a single input (representing a single waveband or channel). Unlike the pseudocolour image, for which the three colour DACs produce independent levels of red, green and blue, the greyscale image produces exactly the same output from the red, green and blue DACs. Recall that equal intensity values of red, green and blue combine together to form shades of grey. Thus, a greyscale pixel with a value of 127 is seen by the colour DACs as an RGB triplet with values {127, 127, 127}.

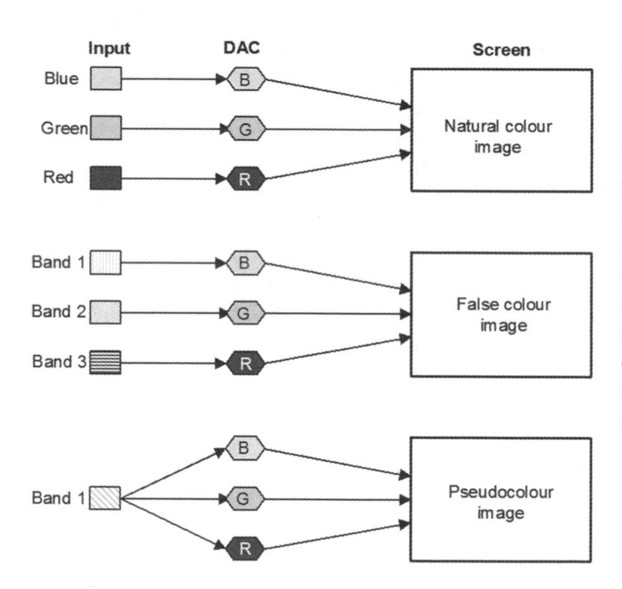

Figure 3.7 A natural colour image (top) is generated when the three input pixel values correspond to red, green and blue reflectance. If the three input pixel values represent reflectance, emittence or backscatter in three independent wavebands then a false colour image is generated (middle). If the input consists of a single band (bottom) then the DAC can be programmed to send three independent signals to the monitor, so that a pseudocolour image is generated.

A labelled or classified image is composed of pixels whose value represents a tag that indicates a property of some kind. The label itself has no numerical meaning. Chapter 8 contains a description of methods of image classification. These methods allow each pixel in an image set to be identified as belonging to a specific category, such as a specific type of land cover. These categories are described by labels such as '1', '2', '3', etc. that indicate 'water', 'broadleaved forest' or 'bare soil'. To display such an image on-screen, the labelled image is placed in graphics memory and the three DACs (R, G and B) are programmed to assign RGB values to the individual single-band pixels (just as in the case of the pseudocolour image). For example, the DAC values for label '1' ('water') could be {0, 0, 255} meaning 'no red, no green, maximum blue'. All 'water' pixels will then appear in brightest blue.

3.2.2 Data formats

One problem that the human race has consistently failed to resolve (and, sometimes, even to acknowledge) is that of standardisation. For example, car drivers in the UK, Ireland, South Africa and Japan are supposed to keep to the left, while drivers in most of the rest of the world keep to the right. Electricity is supplied at 240 volts AC in the UK, and at 110 volts AC in the United States, while US gallons and UK gallons are not the same thing at all, and the rest of the world uses litres (or even liters). It does not come as a surprise, therefore, to find that suppliers of remote sensing data provide their products in different formats. A data format describes the way in which data are written to a storage medium, such as CD. A seven-band data set, such as Landsat ETM+, may be stored in one format as:

(i) One file containing set of numerical and textual descriptions of the data (such as the number of scan lines and pixels per line in the image, the map projection used, and the latitude and longitude of the image centre),
(ii) A second file containing the pixel values in ETM+ bands 1–7, arranged band by band. For each band, the pixel values for the first scan line are written in left to right order as a single group or record, with record 2 containing the pixel values for scan line 2, and so on (Figure 3.8(a)).

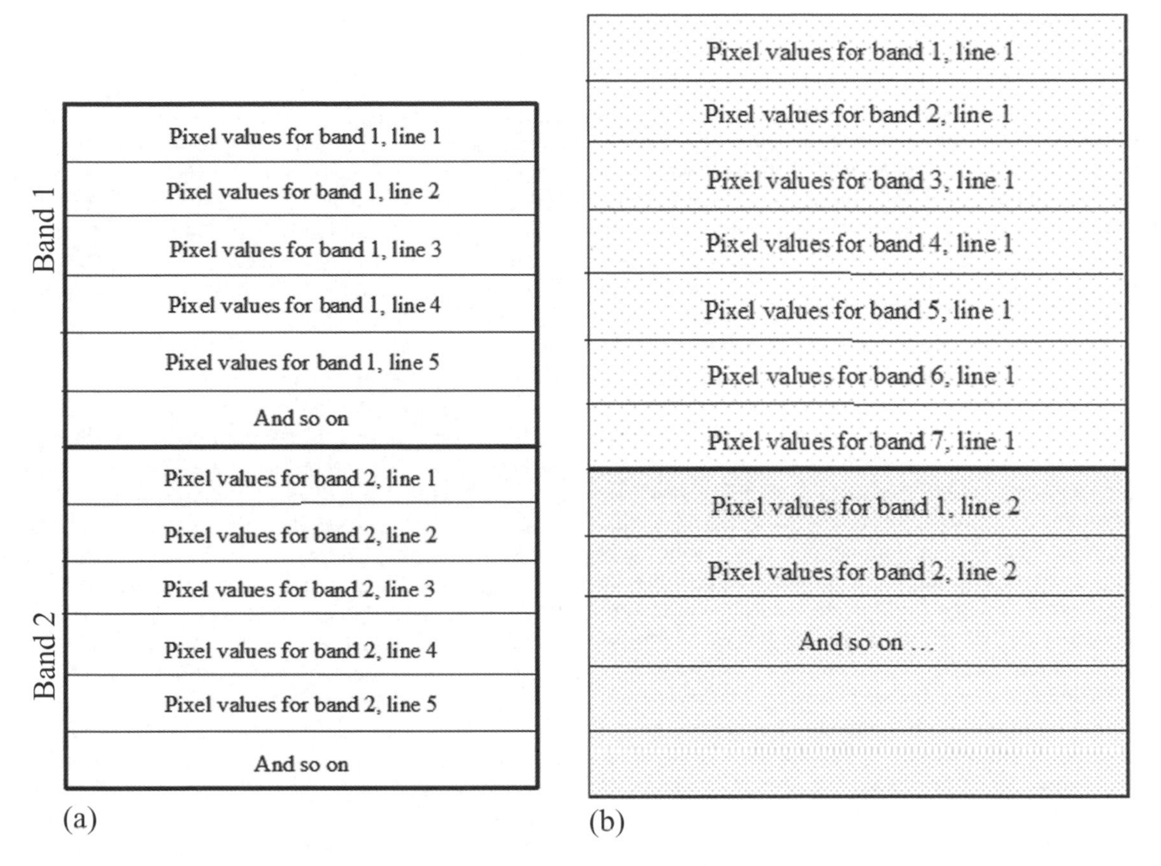

Figure 3.8 Illustrating (a) Band Sequential (BSQ) and (b) Band Interleaved by Line (BIL) formats for the storage of remotely-sensed image data. The layout in (b) assumes a seven-band image

The first of these two files contains descriptions of data held in the second file. These descriptions are known as metadata. An alternative and equally valid format may include the metadata file as before but may store the image data as sets of scan lines for all bands (thus, if the number of bands was seven, the file would hold scan line 1 for all seven bands, followed by scan line 2 for all seven bands, and so on (Figure 3.8 (b)). The first format in which the images making up the image are stored in sequential order is described as 'band sequential' or BSQ format, while the second format is known as 'Band Interleaved by Line', or BIL format.

Both of the format descriptions described above show data sets consisting of (i) metadata and (ii) image data. A number of generic data and metadata formats are in existence, as well as several proprietary formats. A generic format is one that is not restricted to a particular product, while a proprietary format is owned by a specific company or organisation. Thus, the term 'chocolate bar' is generic while 'Cadbury's Dairy Milk' or 'Hershey Bar' are proprietary, though they are chocolate bars. For example, SPOT Image provides data in a proprietary format called 'CAP'. Landsat-7 ETM+ data can be obtained in any of the 'GeoTIFF' (generic), HDF (proprietary) or Fast Landsat 7A (proprietary) formats, which are described below. Data from other instruments are provided in a range of formats, though the efforts of the CEOS (Committee on Earth Observation Satellites) to develop standardised formats are meeting with success. It should be noted that the format adopted by NASA for the Earth-Observing System (EOS) programme is the Hierarchical Data Format (HDF), which can store all of the header and image information in a single file. For example, data from the Japanese ASTER sensor (carried by the Terra satellite) is stored in a single HDF file that contains both metadata and image data.

The format in which remotely-sensed data is provided can be considered to be 'external' in the sense that everyone purchasing data from a particular company is provided with data in the same format. Image processing software producers also define their own 'internal' data formats. For example, the ENVI image processing system, marketed by Research Systems Inc., requires two files per image data set. The first is a metadata file (with the suffix '.hdr'), while the image data are held in a single file in band sequential format (with the suffix '.bsq') or in a single file in band-interleaved-by-line (BIL) format, with the suffix '.bil'. The ENVI software reads image data in an external format, such as CAP or GeoTIFF and converts from the external format to the ENVI internal format. Just to confuse matters, some agencies, such as the German Space Agency, DLR, provides DAIS imaging spectrometer data (section 9.3) in ENVI internal format.

As noted above, the ground receiving station provides two kinds of data. Image data may be thought to be the more important, but the metadata (or 'ancillary' data) describing the image data gives details of the instrument that collected the image, including calibration information, and also include information about the date and time that the image was collected, the geographical location of the image corners and centre, the size of the image in terms of rows, columns, and bands, and other information such as solar azimuth and zenith angles. In later chapters, we will see how image data can be corrected for variations in solar zenith angle, and converted to radiance units, using information provided in the metadata records. An example of the metadata that accompanies a remotely-sensed image data set is shown in Table 3.3. This table contains an extract of the metadata associated with an ASTER image set. This extract was generated by MIPS from an ASTER data set covering the La Mancha area of central Spain between –3.916 651 ° W and –2.931 887 ° E longitude, and between 39.922 905 ° N and 39.246 303 ° S latitude. The image was collected on 2 June 2002. Example 3.1 shows how Landsat ETM+ image data are read from CD-ROM, and the program output is explained.

EXAMPLE 3.1 READING A LANDSAT ETM+ IMAGE FROM CD

The MIPS *File|Import Image Set* menu provides access to a number of modules that allow you to read remotely-sensed data in one of a variety of formats. The USGS provides Landsat ETM+ data in HDF, GeoTIFF and Fast L7A formats. This example illustrates the use of Fast L7A format.

The data for Fast L7A format occupies two CDs, one of which contains the header files, which have filenames that end with 'HPN.FST', 'HRM.FST' and 'HTM.FST' and which refer to the panchromatic band (numbered 8), the visible and short-wave bands (number 1 – 5 and 7), and the thermal band (6). Two versions of the thermal band are provided, using low and high gain respectively. The data for bands 1–7 are stored on the same CD as the header files, and band 8 data is stored on the second CD. Fuller details of the use of the module can be accessed from the MIPS *Help* function. The selected image data are written to a band-sequential file, and the header information (shown below) is appended to the MIPS log file.

The header information (usually called metadata) provides details of the path and row number, the date of data acquisition, the type of resampling (CC means 'cubic convolution', which is explained in chapter 4), the image size in terms of scan lines and numbers of pixels per scan line, and the number of bits per pixel. A value of 8 bits per pixel means that the pixel values lie in the range 0–255.

The gain and bias values are used in radiometric calibration, which is covered in Chapter 4. The geometric data provide the latitude and longitude values of the image centre and corners, the map projection (UTM is Universal Transverse Mercator), and datum (WGS84). Finally, solar elevation and azimuth angles for the image centre are listed. Again, these values are used in radiometric calibration (Chapter 4).

Example Metadata Listing – Fast L7A Format.

```
READ FAST-LANDSAT 7 CD MODULE
*****************************
FILE OPENED
F:\L71201023_02320000619_HRF.FST
Header File opened. Type is Visible/Near IR
Administrative record for VNIR/SWIR Bands
========================================

REQ ID =0750008030112_0001    LOC =201/0230000      ACQUISITION DATE
=20000619
SATELLITE =LANDSAT7    SENSOR =ETM+
SENSOR MODE =NORMAL LOOK ANGLE =0.00
```

```
LOCATION =              ACQUISITION DATE =
SATELLITE =             SENSOR = SENSOR MODE = LOOK ANGLE =
LOCATION =              ACQUISITION DATE =
SATELLITE =             SENSOR = SENSOR MODE = LOOK ANGLE =
LOCATION =              ACQUISITION DATE =
SATELLITE =             SENSOR = SENSOR MODE = LOOK ANGLE =

PRODUCT TYPE =MAP ORIENTED       PRODUCT SIZE =FULL SCENE
TYPE OF PROCESSING =SYSTEMATIC  RESAMPLING =CC

VOLUME #/# IN SET = 1/ 2
PIXELS PER LINE = 8311 LINES PER BAND = 7621/ 7621
START LINE # =        BLOCKING FACTOR =
REC SIZE =63338131 PIXEL SIZE =30.00
OUTPUT BITS PER PIXEL =8 ACQUIRED BITS PER PIXEL =8
BANDS PRESENT =123457

FILENAME =L71201023_02320000619_B10.FSTFILENAME
=L71201023_02320000619_B20.FST
FILENAME =L71201023_02320000619_B30.FSTFILENAME
=L71201023_02320000619_B40.FST
FILENAME =L71201023_02320000619_B50.FSTFILENAME
=L72201023_02320000619_B70.FST
L71201023_02320000619_B10.FST
L71201023_02320000619_B20.FST
L71201023_02320000619_B30.FST
L71201023_02320000619_B40.FST
L71201023_02320000619_B50.FST
L72201023_02320000619_B70.FST

REV L7A

Number of bands referenced in the header is 6
Band identifiers are 1 2 3 4 5 7
Image filenames for VNIR/SWIR
Band 1 Filename: L71201023_02320000619_B10.FST
Band 2 Filename: L71201023_02320000619_B20.FST
Band 3 Filename: L71201023_02320000619_B30.FST
Band 4 Filename: L71201023_02320000619_B40.FST
Band 5 Filename: L71201023_02320000619_B50.FST
Band 6 Filename: L72201023_02320000619_B70.FST
Images in this file set:
Width: 8311
Depth 7621

Radiometric record
==================

GAINS AND BIASES IN ASCENDING BAND NUMBER ORDER
-6.199999809265137        0.775686297697179
-6.400000095367432        0.795686274883794
-5.000000000000000        0.619215662339154
-5.099999904632568        0.965490219639797
-1.000000000000000        0.125725488101735
-0.349999994039536        0.043725490920684
```

```
Geometric record
================

GEOMETRIC DATA MAP PROJECTION =UTM ELLIPSOID =WGS84 DATUM =WGS84
USGS PROJECTION PARAMETERS = 0.000000000000000
0.000000000000000
0.000000000000000        0.000000000000000        0.000000000000000
0.000000000000000        0.000000000000000        0.000000000000000
0.000000000000000        0.000000000000000        0.000000000000000
0.000000000000000        0.000000000000000        0.000000000000000
0.000000000000000
USGS MAP ZONE =31
UL = 0005136.6558W 540441.0389N 247500.000   5999100.000
UR = 0025703.6628E 540824.0187N 496800.000   5999100.000
LR = 0025711.8748E 520506.2184N 496800.000   5770500.000
LL = 0004051.3183W 520139.1763N 247500.000   5770500.000

CENTER = 0010525.9263E 530550.3352N 372150.000   5884800.000   4156
3811

OFFSET =-3391 ORIENTATION ANGLE =0.00
SUN ELEVATION ANGLE =57.2 SUN AZIMUTH ANGLE =147.8
```

The topic of image file formats is too large for it to be considered here in any detail. Readers should, however, be aware that some widely used file formats involve *compression* of the image data, which may result in some loss of detail. The *JPEG* (Joint Photographers Expert Group) format, for example, first separates image intensity information from colour information. These two components are then compressed separately in order to maintain the intensity (brightness) information. This is because human colour vision can deal more easily with loss of colour information than with loss of intensity information. The two components for each of a number of image sub-regions are compressed by the use of a transformation similar to the two-dimensional Fourier transform that is discussed in section 6.6. This transform expresses the image data in terms of a sequence of components of decreasing importance. The JPEG compression keeps only the first few components and discards the rest, hence losing some of the detail. Finally, repeating values such as 2, 2, 2, are replaced by counts (such as 3×2). JPEG 2000 uses a wavelet transform (section 6.7) to compress the image data. The JPEG scheme can achieve compression ratios of around 10:1, which is worthwhile when dealing with large images. For instance, a 1024×1024 image represented in red, green, and blue bands, and expressed on an 8-bit scale requires 3 Mb of disc storage. A compression ratio of 10:1 means that the image takes up only 314 573 bytes (314 Kb) on the disc. The cost is the loss of some colour information.

Other image formats are PICT (used by the Apple Macintosh), TIFF, GIF, and BMP. The BMP type is the Microsoft Windows Bitmap, which is widely used by applications programs running under the Microsoft Windows operating system. MIPS can write image files in the form of bitmaps, and these can be read by other image processing and display software (for example, to add annotation to and print the image). Users of the BMP format (referred to as a *Device-Independent Bitmap* or *DIB*) can choose whether or not to use compression. The uncompressed DIB contains a header, giving information about the size of the image and the number of bits per pixel, for example. If the image consists of a single, 8-bit, component then a colour table or palette follows. This palette maps pixel values on the 0–255 scale on to screen colours. Finally, the image pixels are listed in RGB order (24-bit DIB) or in row order (8-bit DIB), with the last scan line being stored first.

The Tagged Image File Format (TIFF) has been widely used for many years. As its name implies, the TIF format attaches information about the image data (metadata) to tags or labels. A recent development is the introduction of tags that can record specifically geographical data, such as latitude and longitude. This extension to TIFF is known as the GeoTIF format. Images stored in TIFF or GeoTIFF can be compressed or uncompressed. Normally, remotely-sensed images supplied in GeoTIFF format (such as Landsat-7 ETM+ images) are uncompressed.

Data compression is a useful way of reducing image size before transmission across a network, for example. However, care should be taken when choosing a compression method. It was mentioned above, in the discussion of the JPEG image format, that this method involved loss of

Table 3.3 Edited extract from ASTER metadata file, generated by MIPS. The definitions of the parameters are given in ASTER Level 1 Data Products Specifications (GDS Version) Version 1.3 produced by the Japanese Earth Remote Sensing Data Centre, dated 25 June 2001. Each of the metadata entities is described as a group, and each group may contain descriptive fields plus values. The portion of the metadata contained in this table refers to the Visible and Near Infrared (VNIR) band 1. The first group, VNIRBAND1DATA, contains one field, which tells us that the image consists of 4,980 scan lines each containing 4,200 pixels, represented as one byte (eight bits) per pixel. Group IMAGESTATISTICS1 provides the minimum, maximum, mean, standard deviation, mode and median of the image pixel values. Group DATAQUALITY tells us that there are no bad pixels. The PROCESSINGPARAMETERS1 group contains details of the method of relocating (resampling) the pixels during system corrections (NN indicates the Nearest Neighbour method, described in section 4.3.3), the map projection (UTM) and UTM zone (30). The projection parameters for the UTM projection are in the following order: (i) semi-major axis of the ellipsoid, (ii) semi-minor axis of the ellipsoid, (iii) scale factor at the central meridian, (iv) not used – zero, (v) longitude of the central meridian, (vi) latitude of the projection origin, (vii) false easting in the same units as the semi-major axis, and (viii) false northing in the same units as the semi-minor axis. The remaining fields are set to zero. The final group shown here is UNITCONVERSIONCOEFF1. The two values in this group are the calibration coefficients that are used to convert pixel values to radiance units. The third field of this group indicates that, after conversion, the radiance values are expressed in units of W m^{-2} sr^{-1} μm.

```
GROUP VNIRBAND1DATA

IMAGEDATAINFORMATION1                         (4980, 4200, 1)

GROUP IMAGESTATISTICS1

MINANDMAX1                                    (60, 255)
MEANANDSTD1                                   (129.753067, 25.771908)
MODEANDMEDIAN1                                (121, 157)

GROUP DATAQUALITY1

NUMBEROFBADPIXELS1                            (0, 0)

GROUP PROCESSINGPARAMETERS1

CORINTEL1                                     "N/A"
CORPARA1                                      "N/A"
RESMETHOD1                                    "NN"
MPMETHOD1                                     "UTM"
PROJECTIONPARAMETERS1                         (6378137.000000,
6356752.300000, 0.999600, 0.000000, -0.052360, 0.000000,
500000.000000, 0.000000, 0.000000, 0.000000, 0.000000,
0.000000, 0.000000)      30
UTMZONECODE1

GROUP UNITCONVERSIONCOEFF1

INCL1                                         0.676000
OFFSET1                                       -0.676000
CONUNIT1                                      "W/m2/sr/um"
```

information. It is therefore called a 'lossy' compression procedure. Compression methods that preserve all of the data are called 'lossless'. While lossy compressions may be suitable for transmission of digital TV pictures (because the human eye can tolerate some loss of colour information), in general one would choose a lossless compression scheme to encode remotely-sensed data because it is impossible to predict in advance which information can be lost without any cost being incurred. Some image transform methods, including principal components analysis (section 6.4), the discrete Fourier transform (section 6.6), and the discrete wavelet transform (section 6.7) can be used to compress images. These methods exploit the fact that some redundancy exists in a multispectral image set, and they re-express the data in such a way that large reductions in the volume of transformed data represent only small losses of information. Other methods are *run length encoding* and *Huffman coding*. Run length encoding involves the re-writing of the records of image pixels in terms of expressions of the form $(l_i, g_i)$where $l_i$is the number of pixels of value g_i that occur sequentially. Thus, a sequence of pixel values along a scan line might be 1 1 1 2 2 2 2. This sequence could be encoded as (3, 1) (4, 2) without losing any information. A special character indicates the end of a scan line. Obviously the degree of compression that results from this type of encoding depends on the existence of homogeneous sections of image – in other words, if there are no sequences of equal values then there will be no compression; in fact, if there are no sequences of equal values then there will be expansion rather than compression. Run length encoding is used in fax transmission and may be useful in compressing classified images (Chapter 8) in which individual pixel values are replaced by labels. The quadtree, which is described next, may be a better choice, as it is a two-dimensional compression scheme.

A *quadtree* is a form of two-dimensional data structure or organisation that is used in some raster GIS (Figure 3.9). Its major limitation is that the image to be encoded must be square and the side length must be a power of two. However, the image can be padded with zeros to ensure that this condition is met. The square image is firstly subdivided into four component square sub-images of size $2^{n-1} \times 2^{n-1}$. If any of these sub-images is homogeneous (meaning that all of the pixels within the sub-image have the same value) then it is not 'quartered' any further. Conversely, those sub-images that are not homogeneous are again divided into four equal parts and the process repeated (in computing terms, the procedure is recursive). When the quadtree operation is completed the individual components, which may be of differing sizes, are given identifying numbers called Morton numbers, and these Morton numbers are stored in ascending order of magnitude to form a *linear quadtree*.

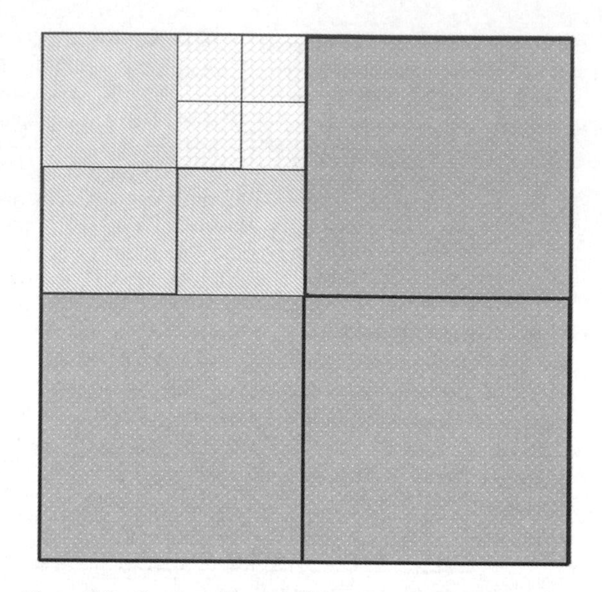

Figure 3.9 Quadtree decomposition of a raster image. The full image, each dimension of which must be a power of two, is divided into four equal parts (quads). Each quad is the subdivided into four equal parts in a recursive way. The subdivision process terminates when the pixel values in each sub-(sub-. . .)quadrant are all equal. The procedure works best for images containing large, homogeneous patches.

For images such as classified images, which generally contain significantly large homogeneous regions, the use of quadtree encoding will result in a substantial saving of storage space. If the image is inhomogeneous then the amount of storage required to store the quadtree may be greater than that required for the raw, uncompressed image. Kess *et al.* (1996) use quadtrees to compress the non-land areas of the Global Land 1-km AVHRR data set. They find that the quadtree representation produces a reduction in data volume to 6.72% of the original data size, which is better than that achieved by JPEG, GZIP, LZW and UNIX *compress* methods. More details of quadtree-based calculations can be found in Mather (1991), while the definitive reference is Samet (1990).

3.2.3 System processing

Data collected by remote sensing instruments carried by Earth-orbiting satellites are transmitted to ground receiving stations using high-bandwidth radio. These transmitted data are in raw format. They must be processed before delivery to a user. If we use the Landsat-7 ETM+ instrument as an example, we note that the image data are collected by an opto-mechanical scanner (Figure 1.2). This scanner uses a mirror to direct upwelling radiance from the ground to a set of 16 detectors, each of which records radiance values for a single scan line (recall from section 2.2.3.6.2 that the Landsat TM scanner collects data for 16 scan lines simultaneously). Each of the 16 detectors has seven components, one for each spectral band. The system operates in both forward and reverse mirror directions, so the raw format data has sets of 16 scan lines stored alternately in opposite directions. Without any pre-processing, these data would be difficult to use. Raw format data has other undesirable characteristics. The pixels forming each scan line are not properly aligned geometrically, and artefacts caused by electronic noise in the system may be present. Nor are the data calibrated to radiance units. Most remote sensing image data are delivered to the user in the form of quantised counts, often – but not always – on a scale requiring eight bits of storage (i.e., levels 0 to 255). For some applications, it is necessary to convert from quantised counts to radiance units and so this radiometric correction is a vital stage of pre-processing.

The first level of Landsat ETM+ system processing is called Level 1R. It involves removal of coherent noise, of banding effects caused by the use of 16 detectors (which can generate patterns with a period of 16 scan lines down the image) and calibration of the pixel values to radiance units. The calibration coefficients are provided in the header records of Level 1R data. However, Level 1R data are not corrected for pixel misalignments, so they must undergo a process called geometric correction (section 4.3) before they are usable. Level 1G Landsat ETM+ data are radiometrically corrected (section 4.6), like Level 1R, and the pixels are relocated or resampled (section 4.3.3) so that the image conforms to a map projection. At Level 1G, the image geometry is corrected using system information, such as orbital height of the satellite, the direction of its forward motion, and its attitude (in terms of pitch, roll and yaw). There is no geometric calibration against known ground points. However, the pixel positions are said to be accurate to within 50 m for flat areas near sea level[2]. Level 1G data are the standard Landsat-7 ETM+ product. Images derived from higher-level processing (involving the use of ground control points and digital elevation models to refine the image geometry) are offered as 'value-added' products by data providers and by third-party commercial companies.

The processing chain for SPOT HRV data is similar to that described for Landsat-7 ETM+ images. SPOT Image defines three levels of 'system corrections', beginning with Level 1A (which provides radiometric corrections to equalise detector responses), and proceeding to Level 1B, which provides a similar geometric correction to that offered by Landsat ETM+ Level 1G, in addition to radiometric equalisation of the detectors. SPOT Image also provides Level 2A pre-processing, which transforms the image data onto a cartographic projection. Like Landsat-7 ETM+ Level 1B data, neither Level 1B nor Level 2A SPOT HRV processing involves any ground control, so the accuracy of the resulting correction depends on the level of error in the determination of the satellite's orbital parameters (see section 2.3.7.1 for a description of DORIS, used to obtain accurate orbit parameters for the SPOT-5 satellite). None of these corrections includes the effects of the land surface topography, so the geometric accuracies cited by the data providers refer only to flat areas near sea level. Location errors in mountainous areas will be considerably greater.

[2] http://ltpwww.gsfc.nasa.gov/IAS/handbook/handbook_htmls/chapter13/chapter13.html#section13.1.4.

3.3 MIPS SOFTWARE

3.3.1 Installing MIPS

Details of the procedure for installing MIPS on a PC are provided in Appendix A and in the file *readme.txt* that is located in the root folder of the CD. This file can be read using Windows Notepad. In order to use the software, you need a PC running the Microsoft Windows operating system, preferably Windows 98 or later, with a CD-ROM drive. The program runs on Windows XP but has not been tested on Windows NT. Although some of the code was developed on an Intel 80486-based machine (and ran reasonably quickly), a processor speed of at least 500 MHz is recommended, together with a monitor capable of displaying 16- or 32-bit colour at a resolution of 800 *x* 600. At least 128 Mb of RAM is required, especially if you are using Windows XP, which is memory-voracious. If MIPS runs slowly when large images are being processed, ask your system administrator to check the size and location of your page file(s).

A number of 'help' files are provided with MIPS. Before you use MIPS, you should read – at least - the Help material on 'Getting Started' and the 'Frequently Asked Questions', plus Example 3.2. Help is also provided for each menu item. A set of test images is included on the CD. These images are mainly small segments (512 x 512) of Landsat MSS and TM or SPOT HRV scenes, and are intended to be used as examples that can be loaded and processed quickly as you progress through the book. The *Examples* folder on the CD-ROM contains four advanced exercises (provided by Dr Magaly Koch of Boston University). Full data sets are included with each example, giving access to multi-temporal and multi-sensor data. The *Examples* folder also includes TIFF versions of the figures used to illustrate the exercises. These TIF files are provided for the convenience

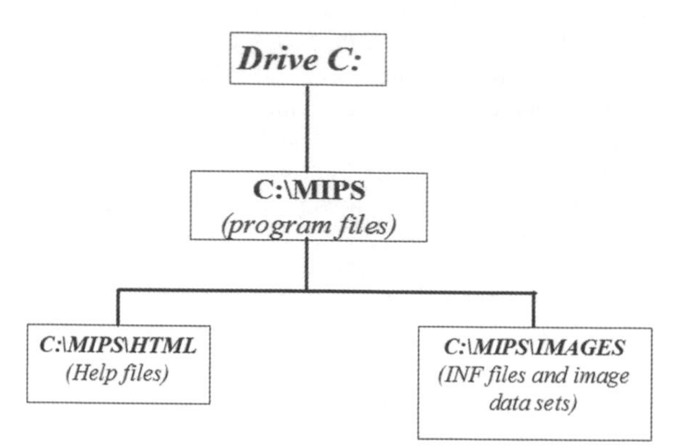

Figure 3.10 Schematic layout of MIPS folders.

of teachers and instructors, who may wish to cut and paste them into Powerpoint (or similar) presentations. The test images and the exercises should be loaded automatically when MIPS is installed (see Appendix A for more information).

The fourth item that can be installed from the CD-ROM is a set of links to WWW pages. These pages are grouped into logical classes, and should provide a valuable resource for both students and teachers. The latest version can be accessed at *http://www.geog.nottingham.ac.uk/~mather/useful_links.html*. The on-line tutorial material available via these 'useful links' complements the material contained in this book.

A small file, called *mymips.ini*, is placed in your *C:\windows* folder. The contents of this file are described in the *readme.txt* file, which is located on the CD. To uninstall MIPS, delete the folder that was created at the start of the installation procedure, and erase the file *c:\windows\mymips.ini*. The layout of the folders created by the MIPS installer program is shown in Figure 3.10.

You will find it convenient to create a shortcut to MIPS on the desktop. Right-click on an empty area of the desktop and select *New|Shortcut*. Browse to the target folder that you selected at the start of the installation process described above, and highlight *mips.exe*, followed by *Next* and *Finish*.

MIPS uses a dictionary file (called an INF file) to store details of an image set. When you create a new image set, an INF file is created so that you will not be asked to remember details of the size of the image, the number of bands, and the name(s) of the file(s) holding the images making up the set. Two types of INF files are used. The structure of the original INF file is shown in Figure 3.11. It contains details of image dimensions, number of bands, date of imaging, and other useful information. Each image is held in a separate file. The later version of the INF file is shown in

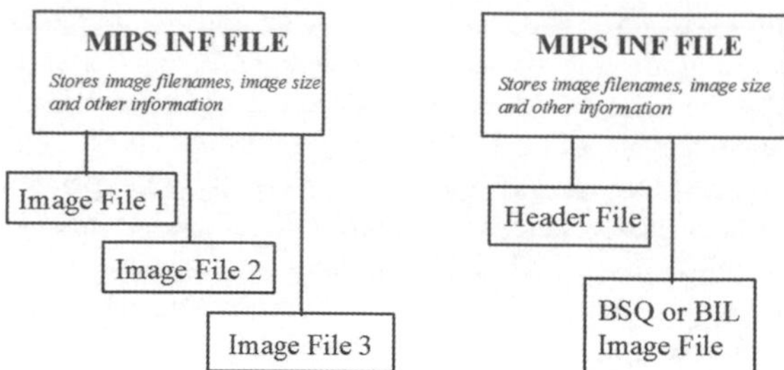

Figure 3.11 MIPS INF files. The older structure is shown on the left, with the later one on the right.

Figure 3.11. It was developed to allow the exchange of image files with the ENVI commercial software package, which stores images in band sequential (BSQ) form (see section 3.2.2 and Figure 3.8). The ENVI-style header file contains information about the image size, etc., but does not include the name of the BSQ image data file. The MIPS INF file simply holds the names of the header and image files, so that images can be accessed with a single click of the mouse button.

3.3.2 Using MIPS

You may wish to refer to Example 3.2 ('Getting Started with MIPS') before reading the remainder of this section.

EXAMPLE 3.2: GETTING STARTED WITH MIPS

The MIPS CD contains a text file called *readme.txt*. This file can be opened with Windows Notepad, and it contains all the information you need to install MIPS. In these worked examples, it is assumed that you have a directory structure as shown in Figure 3.10. MIPS also creates a file called *mymips.ini* in your Windows directory. You should never edit or delete this file. File *readme.txt* also contains instructions on how to create a short-cut icon on your desktop to access MIPS. Assuming that you have created a short-cut icon, double-click on it and MIPS will start. First, accept the licence agreement. Next, note carefully the warning that MIPS allows a maximum of eight (8) image windows to be open at any one time. Finally, use the Open File Dialogue Box to create a log file to hold details of your MIPS session. Remember that you can get help by clicking *Help|Help* on the main menu bar.

Before proceeding to display an image, we need to understand the way in which MIPS references and stores image data. The original MIPS format uses an image dictionary file to store the names (and other details) of one or more related image files (an image data set). Each of these image files is stored separately. In order to provide compatibility with the ENVI image processing software, a second storage format was added. This uses a single file to hold all of the image data (with band 1 preceding band 2, etc.). Such a file is said to be in band-sequential or BSQ format. Associated with the BSQ file is a header file with the suffix *.hdr*. Finally, a standard MIPS INF file provides access details. The two methods of storing and referencing image data sets are shown in Figure 3.11.

The MIPS toolbar is automatically switched on when you start MIPS for the first time. You can switch it on an off via the *Toolbar* menu item. MIPS remembers whether the toolbar is on or off at the end of the previous session. Some people like it; others do not, so you can customise it to suit your own tastes.

Either click the 'eye' icon on the toolbar (there is 'tooltip help' to identify the icons – just leave the cursor over the 'eye' for a second or two) or select *View|View Image*. A list of MIPS INF files is provided in the Open File Dialogue Box. Navigate to the *c:\mips\images* directory and select *litcolorado.inf*. This INF file references a traditional MIPS image set. First, provide a title for the window (such as Little Colorado River). Then select Colour Composite from the drop-down list, and finally highlight and select the three image files to be combined to form a colour composite image (this is done three times in RGB order, once for each of the three selected files). I selected bands 7, 4 and 1 in that order. There is a pause while the data are read from disc, and a window then opens with the title 'Little Colorado River', and your colour composite image is displayed. You may move the window around, or minimise it.

Now open *LaManchaETM.inf*. You will see an information message, telling you that this INF file references an ENVI 8-bit data file (8-bit means 256 levels). You can also see the size of the image file (512 lines × 512 pixels per line × 6 bands). Click *OK*. Now provide a window caption (for example 'La Mancha'), again select Colour Composite as the display type, and finally enter three file identification numbers in RGB order. These file identification numbers are the positions of the image files in the band sequential image data file. A value of 1 means the first image in the bsq file. Since there are six bands stored in the file*LaManchaETM bsq*, we cannot input a number greater than 6 or less than 1. I selected 4, 3, 2 (TM bands 4, 3 and 2) and, after a short pause while the data are read from disc, a window with the title 'La Mancha' appears, containing a false colour image.

Try to display a selection of images from the *C:\mips\images* directory. You should practise moving the windows around, closing them, minimising them, and generally familiarising yourself with starting the MIPS program and displaying both greyscale and false colour images.

Exit MIPS either by using the EXIT icon on the toolbar or clicking *File|Exit*. You do not need to close the windows containing the images before you exit, but it may sometimes be useful to close all of them at once. To do this, select *File|Close All Windows*.

When MIPS starts up, you are asked to agree to the conditions of use (which you can read by pressing the appropriate button), then to nominate a log file into which details of the operations performed during the current session are written. You can read this log file in any ASCII editor, such as Windows Notepad. The MIPS main menu bar is at the top of the screen. The menu names forming the first level in the menu hierarchy are visible (File, View, etc.). The time of day is also displayed (Figure 3.12). The time shown is updated every two seconds unless the computer is working very hard. Clicking on a main menu item produces a sub-menu in the usual way. The functions attached to each menu item are listed in the Help file.

The next few paragraphs are more technical in nature and can be skipped by readers who are not interested in the practical details of computer implementation of software for image processing.

Each image displayed in the main window occupies a 'child window' that is owned by the main window in the sense that, for example, minimising the main window will minimise all child windows. Details of the images contained within the child window or windows are stored in a database, so that clone windows can easily be generated to hold derived images that are formed when any image processing operation is executed. *Please note that a maximum of eight child windows can be displayed at any given time.* The lower toolbar should be kept visible, as it is used to convey information such as the last mouse action (e.g., the word 'Move' in Figure 3.12).

In theory, there is no limit in theory to the size of image that can be handled by MIPS, though in practice you will find that memory limitations will mean that very large images are being written to and read from disc, which inevitably slows down operations such as scrolling. As an image is read from a disc file, it is first converted (if necessary) to eight-bit representation, as described in section 3.1, which may affect the contrast and the brightness of an image that has a pixel representation of more than eight bits. The bands chosen for display in red, green and blue are written to separate arrays in computer memory. If greyscale mode is used then the selected single band is copied to all three arrays. The contents of the three arrays are converted to Windows bitmap form and transferred to the graphics memory. Whenever an image is selected, the corresponding data are transferred from the computer's physical memory (RAM)

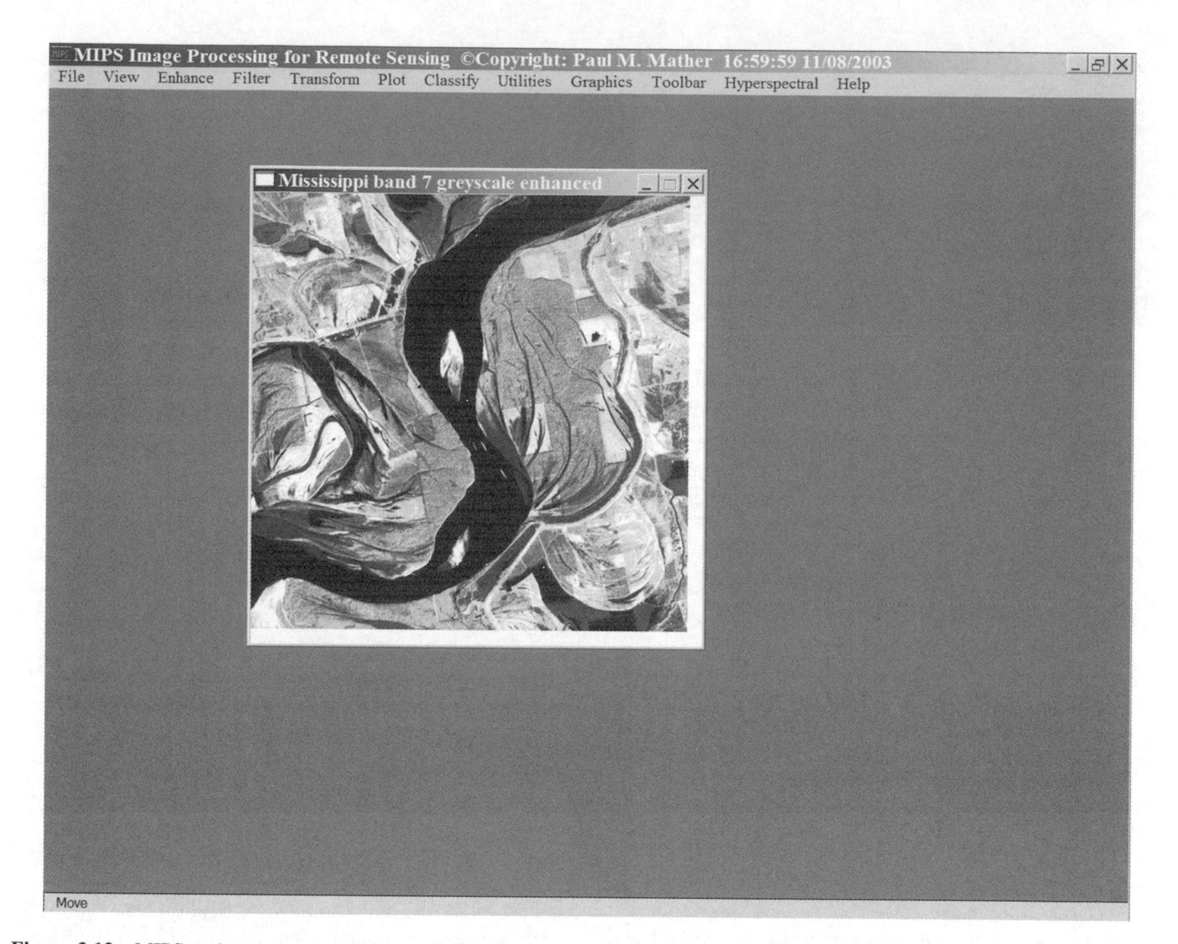

Figure 3.12 MIPS main window, showing the menu bar at the top and a child window containing a middle-infrared (Landsat TM band 7) image of the Mississippi River south-west of Memphis, Tennessee (13 January 1983). The word 'move' on the lower toolbar indicates the last mouse state.

to graphics memory. If the physical RAM is insufficient then the image arrays will be kept in virtual memory. In order to work around the 14 Mb limit on Windows bitmaps, the image is sampled before being transferred from RAM to graphics memory. Scrolling is accomplished by copying the relevant image section into the graphics memory, while zooming uses simple replication of pixels.

3.3.3 Summary of MIPS functions

The functions provided by MIPS are accessed from the main menu bar, shown in Figure 3.12. The menu items are: *File, View, Enhance, Filter, Transform, Plot. Classify, Utilities, Graphics, Toolbar, Hyperspectral* and *Help*. The following paragraphs summarise the functions associated with each menu item, and provide a navigational guide to MIPS.

The expanded *File Menu* is shown in Figure 3.13. Most of the functions relate to the import of image data in various formats, the export of image data, image printing, and INF file operations, including editing. The most-recently used INF files are listed below the horizontal separator.

The *View Menu* is shown in Figure 3.14. In alphabetical order, the functions accessed from this menu are: *Change Image Scale* (i.e. zoom, either continuously or in integral steps), *Display Cursor* (which provides the coordinates of the mouse cursor plus the pixel values at that point, read from the graphics memory), *Display Image* (which reads an INF file and allows the user to select either an image or a sub-image for display), and *Display RGB Separately*. The last of these functions decomposes a colour composite (RGB) image into its three component parts, each of which is shown as a greyscale image.

The *Enhance Menu* (Figure 3.15) references the modules that implement procedures discussed in Chapters 5 and 6. The first three procedures alter the contrast of the displayed image. The pseudocolour and density slice methods are used to convert a greyscale image into colour representation. The HSI (hue-saturation-intensity) and decorrelation

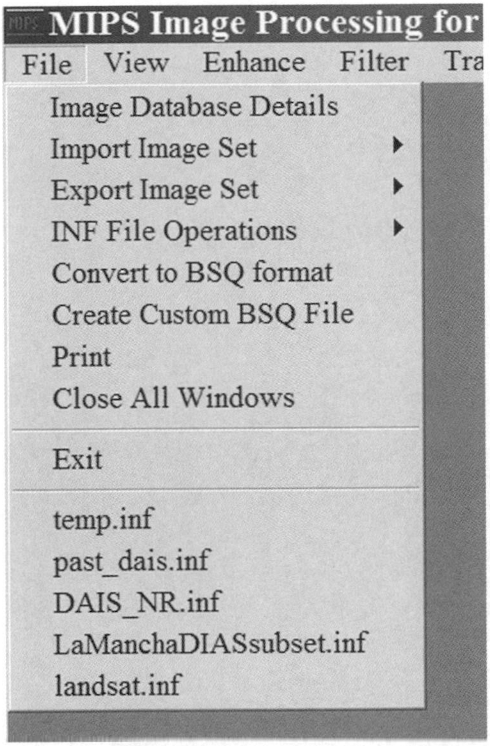

Figure 3.13 MIPS File Menu.

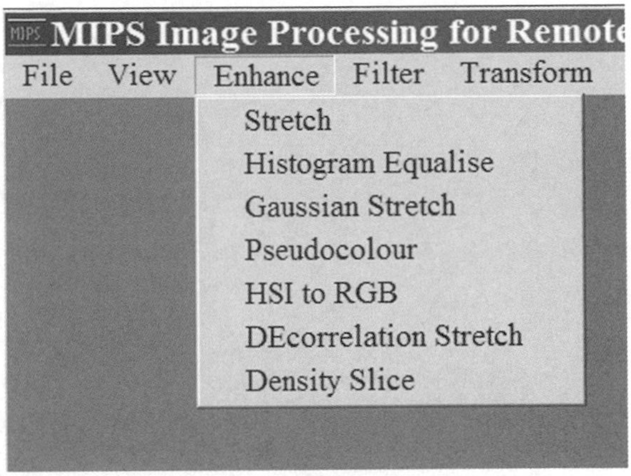

Figure 3.15 The MIPS Enhance Menu.

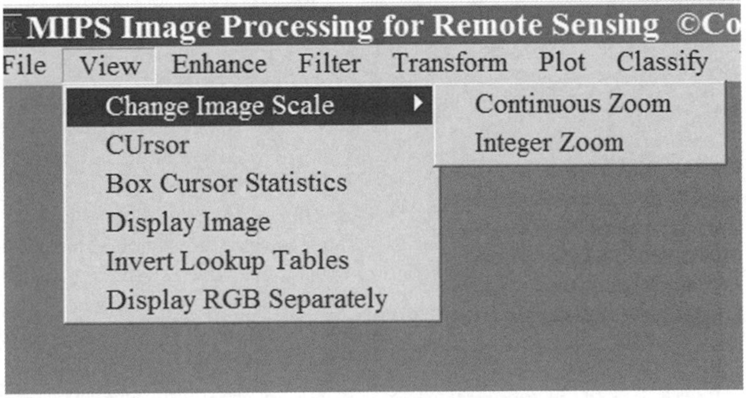

Figure 3.14 The MIPS View Menu.

stretch modules implement transformations of a colour image with the aim of broadening the dynamic range of the three component primary colours simultaneously. They are described in sections 6.5 and 6.4.3, respectively.

Filtering methods are considered in Chapter 7. The first six of the seven items on the *Filter Menu* (Figure 3.16) refer to procedures that operate directly on the image itself in order to perform high- or low-pass filter operations. The *image minus Laplacian* operation adds high-frequency information back to the image, thus sharpening it (section 7.3.2). The *Savitzky–Golay* technique is either a smoothing operation (low pass) or a high-pass filter, depending on the user's choice (section 9.3.2.2.1). The *median* filter is a low-pass operator, while both the *Sobel* and *Roberts* are high-pass filters. The *User Defined Filter* module allows the user to specify the weights for a box filter. Finally, a variety of frequency domain filters using the Fourier representation of the image, are provided. These are described in Chapters 6 and 7.

The *Transform Menu* (Figure 3.17) offers a range of procedures that are described in Chapters 4 and 6. These

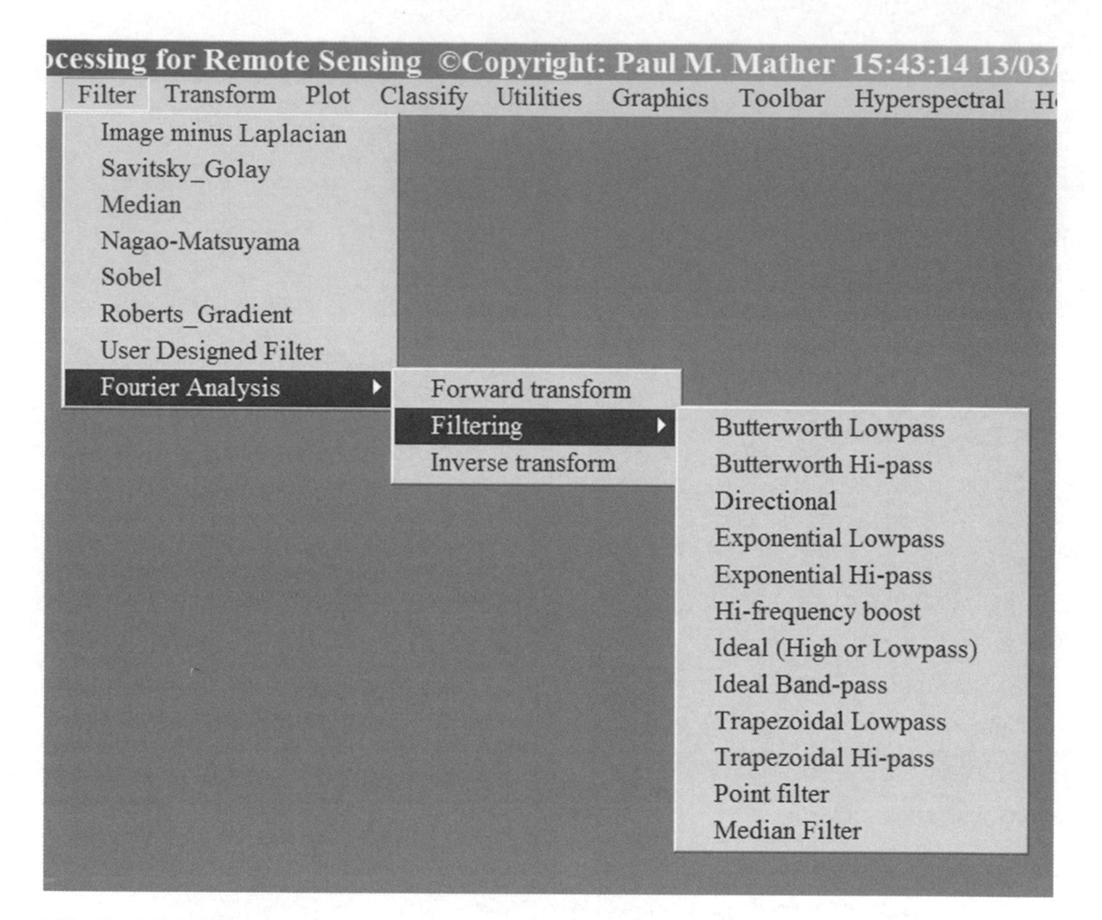

Figure 3.16 The MIPS Filter Menu.

transforms range from coordinate transformations (geometric correction) and vegetation ratios (Normalised Vegetation Index|) to statistical transforms such as Principal Components Analysis and Noise Reduction Transform. Methods of estimating sub-pixel components such as Spectral Angle Mapper and Mixture Modelling are included.

The *Plot Menu* (Figure 3.18) allows the user to view the histogram of the selected image, to plot pixel values in one band against those in a second band, to draw cross-sections on a displayed image, including lidar images, and to perform dimensional reduction analysis using Nonlinear Mapping. The Plot Derivatives module is included to demonstrate the effects of denoising a spectral reflectance curve using the wavelet transform before computing derivatives (Chapter 6 describes the wavelet transform, and the use of derivatives is discussed in Chapter 9).

The *Classify Menu* is shown in Figure 3.19. Classification is the generic name of a set of procedures for allocating labels such as 'forest', 'water', or 'grassland' to individual pixels. These procedures are described in Chapter 8.

The *Utilities Menu* (Figure 3.20) provides access to a number of procedures that do not fit conveniently into any other category. Resampling means changing the pixel size. Thus, if you have two images with pixel sizes of 10 and 20 m, respectively, you may wish to *resample* the coarser image to 10 m pixel size in order to overlay one on the other. Routine operations such as combining three greyscale images to form a colour image, changing the window caption, extraction of a sub-image, creating and applying a logical mask, adding and subtracting images, and copying images or graphics to the Windows clipboard are included as well as more specific routines for haze removal, conversion to reflectance, and cross-correlation of images. Most of these operations are described in Chapter 4.

The *Graphics Menu* and the *Toolbar Menu* are not illustrated as they each contain only two menu items. The graphics menu provides a routine for on-screen digitising

for Remote Sensing ©Copyrigh
Transform Plot Classify Utilities
Wavelet (2D) transform
Geometric Correction
HSI Pan Sharpen
Wavelet Pan Sharpen
Normalise image
Principal Components (PCA)
Tasselled Cap
NDVI
Noise Reduction Transform
Spectral Angle Mapper
Mixture Modelling

Figure 3.17 The MIPS Transform Menu.

Figure 3.18 The MIPS Plot Menu.

plus a routine that rotates a scanned image or map (for example, to ensure that the grid lines run north-south). The toolbar menu contains a toggle that switches the MIPS toolbar on or off.

The functions offered by the *Hyperspectral Menu* (Figure 3.21) are dealt with in Chapter 9. They are intended

nsing ©Copyright: Paul M.
t Classify Utilities Graphics
IsoData
Collect Training Data
Supervised Classification
Reclassify
Wavelet texture

Figure 3.19 MIPS Classify Menu.

for use with multi-channel imaging spectrometry data, such as DAIS data provided by the German Space Agency, DLR, or the Hymap data collected by the UK Natural Environment Research Council (NERC). These data sets are provided in ENVI-compatible form, in either band-sequential or band interleaved by line format. The reader may wish, at this stage, to try the *Plot Library Spectrum* function, which selects and plots a reflectance and emittence spectrum from a spectral library. The remaining functions are specialised and are discussed in Chapter 9.

The *Help Menu* accesses the MIPS help files, which are stored in the *html* subdirectory during the installation process (Figure 3.10). MIPS uses your default Internet browser

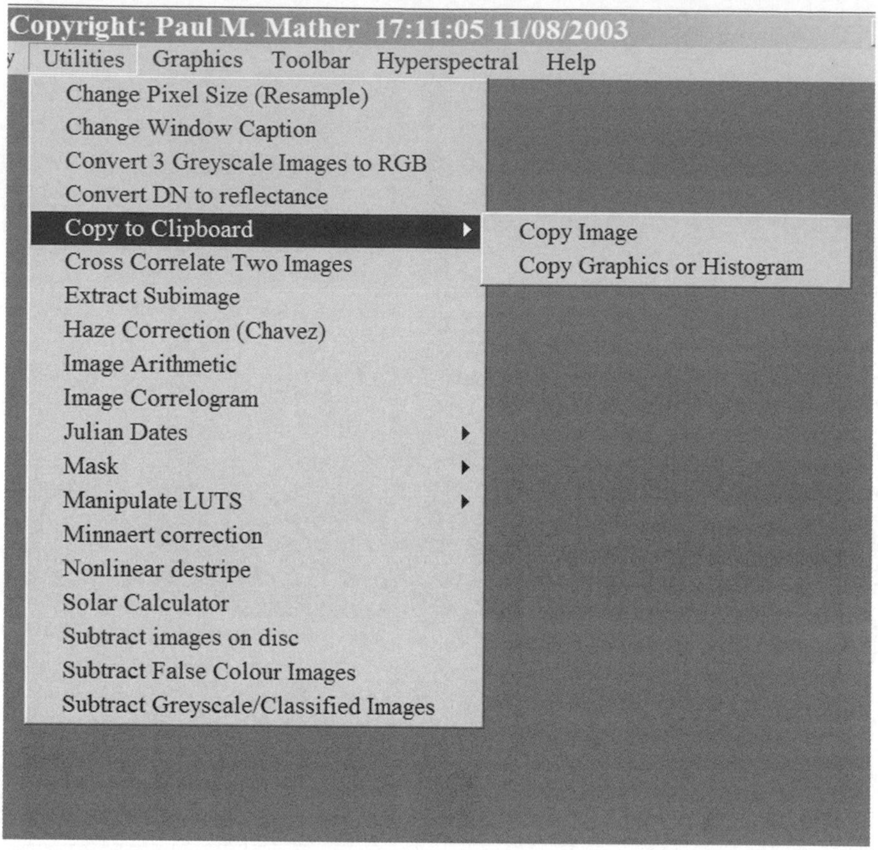

Figure 3.20 MIPS Utilities Menu.

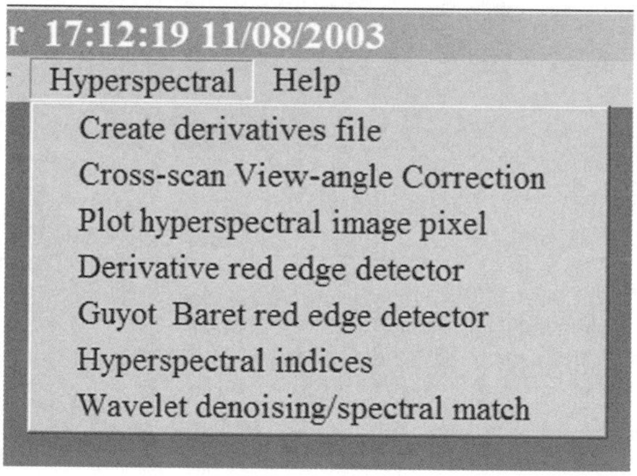

Figure 3.21 MIPS Hyperspectral Menu.

to display these help files, which can be navigated in the normal way.

Revisions and extensions to MIPS or to the help files will be placed on the WWW at http://homepage.ntlworld.com/paul.mather/ComputerProcessing3/.Readers are welcome to download these files, which will be maintained in the form of archives. You will need a utility such as *Winzip* in order to unpack the archive, which will contain a *readme_first.txt* file giving instructions on the updating procedure. To some extent, the frequency of updating will depend on the number of error reports that I receive from readers, which in turn depends on how many errors exist in the MIPS code. Note that the Help files (accessed from the Help menu item on the MIPS main menu) contains a number of items intended to guide and assist first-time users. Refer also to the list of *Frequently Asked Questions*.

3.4 SUMMARY

Appearances can be misleading, according to folk wisdom. This is nowhere more true than in the case of the manipulation and display of images. In the case of remotely-sensed images it is unlikely that deception would be employed or condoned, though anyone who is unfamiliar with the procedures used to transform image data that is represented by more than eight bits per pixel onto the 0–255 range may well be misled. The differences between false colour, true colour and pseudocolour images should also be recognised. One pitfall to be avoided by the aware reader is the use of data that has been transformed, perhaps non-linearly, onto an 8-bit scale in subsequent calculations such as band ratios and vegetation indices (section 6.2.4). A band ratio such as the Normalised Difference Vegetation Index should be computed from the image data themselves, and not from scaled values stored in a display memory.

The topic of data formats is also discussed in this chapter. Extraction of data from a storage medium such as a CD-ROM is often the first problems faced by a potential user of remotely-sensed data. Again, it is important that the user understands the characteristics of different formats and the properties of the metadata that are provided together with the image data. Differences between levels of processing must always be recognised, as should the properties of 'lossy' and 'lossless' data compression.

4

Pre-processing of Remotely-Sensed Data

4.1 INTRODUCTION

In their raw form, as received from imaging sensors mounted on satellite platforms, remotely-sensed data generally contain flaws or deficiencies. The correction of deficiencies and the removal of flaws present in the data is termed *pre-processing* because, quite logically, such operations are carried out before the data are used for a particular purpose. Despite the fact that some corrections are carried out at the ground receiving station (section 3.2.3), there is often still a need on the user's part for some further pre-processing. The subject is thus considered here before methods of image enhancement and analysis are examined in later chapters.

It is difficult to decide what should be included under the heading of 'pre-processing', since the definition of what is, or is not, a deficiency in the data depends to a considerable extent on the use to which those data are to be put. If, for instance, a detailed map of the distribution of particular vegetation types or a bathymetric chart is required then the geometrical distortion present in an uncorrected remotely-sensed image will be considered to be a significant deficiency. On the other hand, if the purpose of the study is to establish the presence or absence of a particular class of land-use (such as irrigated areas in an arid region) then a visual analysis of a suitably-processed false-colour image will suffice and, because the study is concerned with determining the presence or absence of a particular land-use type rather than its precise location, the geometrical distortions in the image will be seen as being of secondary importance. A second example will show the nature of the problem. An attempt to estimate reflectance of a specific target from remotely-sensed data will be hindered, if not completely prevented, by the effects of interactions between the incoming and outgoing electromagnetic radiation and the constituents of the atmosphere. Correction of the imagery for atmospheric effects will, in this instance, be considered to be an essential part of data pre-processing whereas, in some other case (for example, discrimination between land-cover types in an area at a particular point in time), the investigator will be interested in relative, rather than absolute, pixel values and thus atmospheric correction would be unnecessary. Measurements of change over time using multi-temporal image sets will, in the case of optical imagery, require correction for atmospheric variability, and it will also be necessary to register the images forming the multi-temporal sequence to a common geographical coordinate system. In addition, corrections for changes in sensor calibrations will be needed to ensure that like is compared with like.

Because of the difficulty of deciding what should be included under the heading of pre-processing methods, an arbitrary choice has been made. Correction for geometric, radiometric and atmospheric deficiencies, and the removal of data errors or flaws, is covered here despite the fact that not all of these operations will necessarily be applied in all cases. This point should be borne in mind by the reader. It should not be assumed that the list of topics covered in this chapter constitutes a menu to be followed in each and every application. The pre-processing techniques discussed in the following sections should, rather, be seen as being applicable in certain circumstances and in particular cases. The investigator should decide which pre-processing techniques are relevant on the basis of the nature of the information to be extracted from the remotely-sensed data.

The pre-processing techniques described in section 4.2 are concerned with the removal of data errors and of unwanted or distracting elements of the image. In reality, of course, data errors such as missing scan lines cannot be removed; the data in error are simply replaced with some other data that are felt to be better estimates of the true but unknown values. Similarly, unwanted or distracting elements of the image (such as the banding present on Landsat TM and ETM+ images, as discussed in sections 2.3.6 and 4.2.2) can only be eliminated or reduced by modifying all the data values in the image. These errors are caused by detector imbalance.

Many actual and potential uses of remotely-sensed data require that these data conform to a particular map

Computer Processing of Remotely-Sensed Images: An Introduction, Third Edition. Paul M. Mather.
 ISBNs: 0-470-84918-5 (HB); 0-470-84919-3 (PB)

projection so that information on image and map can be correlated, for example, within a Geographical Information System (GIS). Two examples will demonstrate the importance of this requirement. In Chapter 8 we see that the classification of a remotely-sensed image is best achieved by establishing the nature of ground cover categories by field work and/or by air-photo and map analysis. In order that the information so derived can be related to the remotely-sensed image, some method of transforming from the scan-line/pixel coordinate reference system of the image to the easting/northing coordinate system of the map is required. Secondly, if remotely-sensed data are to be used in association with other data within the context of a GIS then the remotely-sensed data and products derived from such data (for example, a set of classified images) will need to be expressed with reference to the geographical coordinates to which the rest of the data in the information system conform. In both these cases, there is a need for data pre-processing of a kind known as geometric correction. The same arguments can be put forward if the study involves measurements made on images produced on different dates; if information extracted from the two images is to be correlated then they must be registered, that is, expressed in terms of the same geographical coordinate system. Where an image is geometrically corrected so as to have the coordinate and scale properties of a map, it is said to be *geo-referenced*. Geometric correction and registration of images is the topic of section 4.3.

Atmospheric effects on electromagnetic radiation (due primarily to scattering and absorption) are described in section 1.2.5. These effects add to or reduce the true ground-leaving radiance, and act differentially across the spectrum. If estimates of radiance or reflectance values are to be successfully recovered from remote measurements then it is necessary to estimate the atmospheric effect and correct for it. Such corrections are particularly important (a) whenever estimates of ground-leaving radiances or reflectance rather than relative values are required, for example in studies of change over time, or (b) where the part of the signal that is of interest is smaller in magnitude than the atmospheric component. For example, the magnitude of the radiance upwelling from oceanic surfaces is generally very low, being much less than the atmospheric path radiance (the radiance scattered into the field of view of the sensor by the gaseous and particulate components of the atmosphere). If any useful information about variations in radiance upwelling from the ocean surface is to be obtained from a remotely-sensed image then the component of the signal received at the sensor that emanates from the ocean surface must be separated from the larger atmospheric component (Figure 1.23). It is fair to say that no single method of achieving this aim has yet been established, and it is also true that most of the techniques that are in use today and which produce even approximately realistic results tend to be complex in nature. In my experience, simple techniques cannot solve complex problems. The more complex techniques are well beyond the scope of this book and will thus not be considered in any detail. Section 4.4 provides an introductory review of atmospheric correction techniques.

Sections 4.5–4.7 are concerned with the radiometric correction of images. If images taken in the optical and infrared bands at different times (multi-temporal images) are to be studied then one of the sources of variation that must be taken into account is differences in the angle of the sun. A low sun-angle image gives long shadows, and for this reason might be preferred by geological users because these shadows may bring out subtle variations in topography. A high sun angle will generate a different shadow effect. If the reflecting surface is Lambertian (which is, in most cases, a considerable oversimplification) then the magnitude of the radiant flux reaching the sensor will depend on the sun and view angles. For comparative purposes, therefore, a correction of image pixel values for sun elevation angle variations is needed. This correction is considered in section 4.5. The calibration of images to account for degradation of the detectors over time is the topic of section 4.6. Such corrections are essential if multi-temporal images are to be compared, for changes in the sensor calibration factors will obscure real changes on the ground. The effects on recorded radiance levels of terrain slope and orientation are reviewed briefly in section 4.7.

The material in this chapter is introductory in scope. Research applications will require more elaborate methods of pre-processing. For example, orbital geometry models of geometric correction (section 4.3.1) may use advanced photogrammetric principles (Konecny, 2003), while the use of the more sophisticated atmospheric correction procedures (section 4.4) requires knowledge of higher-level physics. The level of presentation adopted here is intended to provide a basic level of appreciation rather than a full understanding. More elaborate treatments are provided by Slater (1980), Elachi (1987) and by various contributors to Asrar (1989).

4.2 COSMETIC OPERATIONS

Two topics are discussed in this section. The first is the correction of digital images that contain either partially or entirely missing scan lines. Such defects can be due to errors in the scanning or sampling equipment, in the transmission or recording of image data, or in the reproduction of the media containing the data, such as magnetic tape or CD-ROM. Whatever their cause, these missing scan lines are normally seen as horizontal black or white lines on the image, represented by sequences of pixel values such as zero or 255 (in an 8-bit image; Figure 3.4). Their presence intrudes upon the visual examination and interpretation of images and also affects statistical calculations based on image pixel values. Methods to replace missing values with estimates of their true (but unknown) values are reviewed in section 4.2.1. This is followed by a brief discussion of methods of 'de-striping' imagery produced by electromechanical scanners such as those carried by Landsat (TM and ETM+). As noted in Chapter 2, these scanners collect data for several scan lines simultaneously. The Landsat TM and ETM+ instruments record 16 scan lines for each spectral band on each sweep of the scanning mirror. The radiance values along each of these scan lines are recorded by separate detectors. In a perfect world, each detector would produce the same output if it received the same input. As we know, the world is far from perfect, and so over time the responses of the detectors making up the set of sixteen change at different rates. A systematic pattern is superimposed upon the image, repeating every sixteen lines. Techniques to remove this pattern are discussed in section 4.2.2. Note that they cannot be used with images recorded using solid state (push-broom) scanners such as the HRV carried by the SPOT satellites because each individual pixel across a scan line is recorded by the corresponding detector in the sensor. Hence, each column of pixels in a SPOT HRV image is recorded by the same detector. With 6000+ columns in an image, the problem of correcting for variations in the detectors is rather more severe than that presented here for electromechanical scanners.

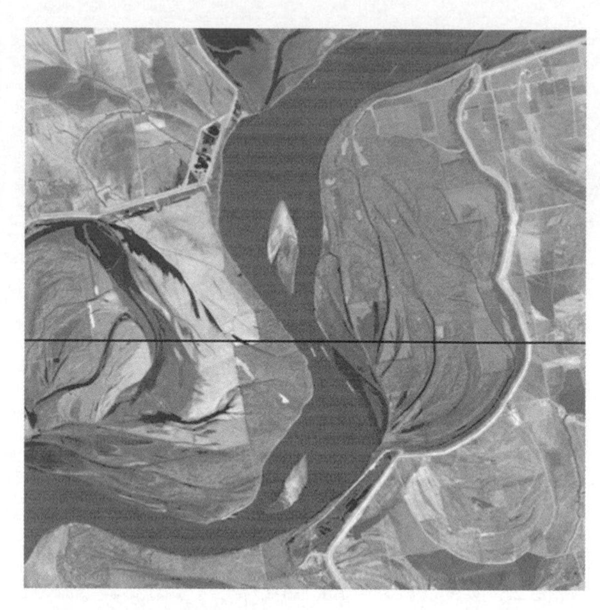

Figure 4.1 Illustrating the effect of a dropped scan line on a Landsat TM band 4 image. The effects of 16-line banding can also be seen faintly in the water (black) areas (see Figure 4.2). The 'missing scan line' is simulated in this example, in order to emphasise the phenomenon.

4.2.1 Missing scan lines

When missing scan lines occur on an image (Figure 4.1) there is, of course, no means of knowing what values would have been present had the scanner or data recorder been working properly; the missing data have gone for ever. It is, nevertheless, possible to attempt to estimate what those values might be by looking at the image data values in the scan lines above and below the missing values. This approach relies upon a property of spatial data that is called spatial autocorrelation. The word 'auto' means 'self', thus autocorrelation is the relationship between one value in a series and a neighbouring value or values in the same series. Temporal autocorrelation is usually present in a series of hourly readings of barometric pressure, for example. The value at 11 o'clock tends to be very similar to the value at 10 o'clock unless the weather conditions are quite abnormal. Spatial autocorrelation is the correlation of values distributed over a two-dimensional or geographical surface. Points that are close in geographical space tend to have similar values on a variable of interest (such as rainfall or height above sea level). The observation that many natural phenomena exhibit spatial autocorrelation is the basis of the estimation of missing values on a scan line from the adjacent values.

The simplest method (*method 1*) for estimating a missing pixel value along a dropped scan line involves its replacement by the value of the corresponding pixel on the immediately preceding scan line. If the missing pixel value is denoted by $v_{i,j}$, meaning the value of pixel i on scan line j, then the algorithm is simply:

$$v_{i,j} = v_{i,j-1}$$

Method 1 has the virtue of simplicity. It also ensures that the replacement value is one that exists in the near neighbourhood of pixel (i, j). We will consider an averaging method next, and will see that, where the assumption of positive spatial autocorrelation does not hold, the average of two adjacent pixel values will produce an estimated replacement value that is quite different from either, whereas method 1 produces an estimate that is similar to at least one of its neighbours. Method 1 will need modification whenever the missing line (j) is the first line of an image. In that instance, the value $v_{i,j+1}$ could be used.

Method 2 is slightly more complicated; it requires that the missing value be replaced by the average of the corresponding pixels on the scan lines above and below the defective line, that is:

$$v_{i,j} = (v_{i,j-1} + v_{i,j+1})/2$$

(taking the result to the nearest integer if the data are recorded as integer counts). Where the missing line is the first or last line of the image then Method 1 can be used. As indicated earlier, if nearby pixel values are not highly correlated then the averaging method can produce hybrid pixels that are unlike their neighbours on the scan lines immediately above or below. This is likely to happen only in those cases where the missing line coincides with the position of a boundary such as that between two distinct land-cover types, or between land and water.

Method 3 is the most complex. It relies on the fact that two or more bands of imagery are often available. Thus, Landsat TM produces seven bands, ETM+ produces eight (including the 15 m resolution panchromatic band) and SPOT HRV produces three bands of imagery. If the pixels making up two of these bands are correlated on a pair-by-pair basis then high correlations are generally found for bands in the same region of the spectrum. For instance, the Landsat ETM+ bands 1 and 2 in the blue-green and green wavebands of the visible spectrum are normally highly correlated. The missing pixels in band k might best be estimated by considering contributions from (i) the equivalent pixels in another, highly correlated, band and (ii) neighbouring pixels in the same band, as in the case of the two algorithms already described. If the neighbouring, highly correlated, band is denoted by the subscript r then the algorithm can be represented by:

$$v_{i,j,k} = M\{v_{i,j,r} - (v_{i,j+1,r} + v_{i,j-1,r})/2\} + (v_{i,j+1,k} + v_{i,j-1,k})/2$$

The symbol M in this expression is the ratio of the standard deviation of the pixel values in band k and the standard deviation of the pixel values in band r. This algorithm was first described by Bernstein *et al.* (1984) and is examined, together with the two algorithms outlined above, by Fusco and Trevese (1985). The conclusion of the latter authors is that the use of a second correlated band both reduces error and better preserves the geometric structures present in the image. They present some further results and elaborations of the basic algorithm, and readers wishing to go more deeply into the matter are referred to their paper.

The location of missing scan lines might not at first sight seem a topic worthy of serious consideration, for they are usually manifestly obvious when a defective image is examined visually. However, to locate such missing lines interactively using a cursor is a tedious task. The spatial autocorrelation property of images might be used as the basis for formulating a strategy that might allow missing scan lines to be located semi-automatically. If the average of the pixel values along scan line i is computed (with i running from 1 to n, where n is the number of scan lines in the image) then it might be reasonable to expect that the average of scan line i differs from the average of scan lines $i + 1$ and $i - 1$ by no more than a value e. The parameter e would be determined by looking at the frequency distribution of the scan line averages over a number of images, or for a representative (and non-defective) part of the image under consideration. Step 1 would then involve locating all those scan lines with average values that deviated by more than e from the average of the preceding scan line. The first scan line of the image could either be omitted or compared with the second scan line. At the end of step 1 we cannot be sure that the unexpectedly deviant behaviour of the scan lines picked out by this comparative method is the result of missing values. Step 2 thus involves a search along each of the scan lines picked out at step 1 for unexpected sequences of values. These unexpected sequences are most likely to be strings of extreme values, either 0 or 255 in eight-bit images. The beginnings and ends of such sequences are marked. At this stage the image can be displayed and a cursor used to mark the start of the suspect sequence. The operator is then able to check that the scan lines or portions of scan lines are indeed defective. Step 3 consists of the application of one of the three methods described earlier, which allow the defective value to be replaced by an estimate of its true but unknown value. Note that isolated aberrant values such as speckle noise on synthetic aperture radar (SAR) images are removed by the use of filters such as the Lee filter or the median filter. These methods are described in Chapter 7.

4.2.2 De-striping methods

The presence of a systematic horizontal banding pattern is sometimes seen on images produced by electromechanical scanners such as Landsat's TM (Figure 4.2). This pattern is most apparent when seen against a dark, low-radiance background such as an area of water. The reasons for the presence of this pattern, known as banding, are given in section 2.3.6.1. It is effectively caused by the imbalance between the detectors that are used by the scanner. This banding can be considered to be a cosmetic defect (like missing scan lines) in that it interferes with the visual appreciation of the patterns and features present on the image. If any statistical analysis of the pixel values is to be undertaken then the problem becomes somewhat more difficult. The pixel values recorded on a magnetic tape or CD-ROM are by no means 'raw' data, for they have been subjected to radiometric and geometric correction procedures at the ground

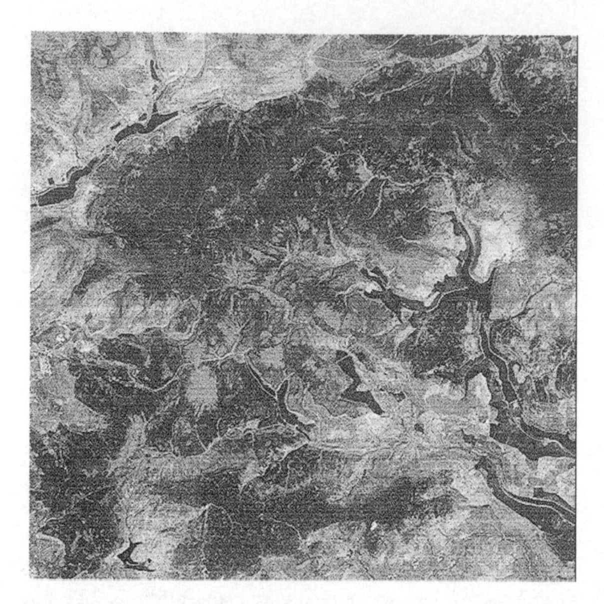

Figure 4.2 Horizontal banding effects can be seen on this Landsat-4 TM band 1 image of part of the High Peak area of Derbyshire, England. The banding is due to detector imbalance. As there are 16 detectors per band, the horizontal banding pattern repeats every 16th scan line. The image has been contrast-stretched (section 5.3) in order to emphasise the banding effect. See section 4.2.2 for more details.

receiving station, as described in section 3.2.3. Hence, there does not seem to be much force in the argument that 'raw data' should not be interfered with. If we take as our starting point the assumption that the image data should be internally consistent (that is, areas of equal ground-leaving radiance should be represented by equal pixel values in the image, assuming no other complicating factors) then some kind of correction or compensation procedure would appear to be justified. Two reasons can thus be put forward in favour of applying a 'de-striping' correction: (i) the visual appearance and interpretability of the image is thereby improved, and (ii) equal pixel values in the image are more likely to represent areas of equal ground-leaving radiance, other things being equal.

Two methods of de-striping Landsat imagery are considered in this section. For the sake of simplicity, they are illustrated with reference to MSS images, which have only six detectors per band) rather than to TM or ETM+ images, which have 16 detectors per spectral band. Both methods are based upon the shapes of the histograms of pixel values generated by each of the detectors; these histograms are calculated from lines 1, 7, 13, 19, . . . (histogram 1), lines 2, 8, 14, 20, . . . (histogram 2) and so on, in the case of a six-detector system such as the Landsat MSS.

The first method characterises the relationship between the scene radiance r_{in} that is received at the detector and the value r_{out} that is output by the sensor system in terms of a linear function. The second method is non-linear in the sense that the relationship between r_{in} and r_{out} is not characterised in terms of a single linear function; a piecewise function made up of small linear segments is used instead. Methods based on low-pass filtering (such as those described by Crippen (1989) and Pan and Chang (1992)) are mentioned in Chapter 7.

4.2.2.1 *Linear method*

The first method uses a linear expression to model the relationship between the input and output values. The underlying idea is quite simple, though it is based upon the assumption that each of the six detectors 'sees' a similar distribution of all the land-cover categories that are present in the image area. If this assumption is satisfied, and the proportion of pixels representing water, forest, grassland, bare rock, and so on, is approximately the same for each detector, then the histograms generated for a given band from the pixel values produced by the n detectors should be identical (n is the number of detectors used by the scanning instrument, for example, six for Landsat MSS and 16 for Landsat TM or ETM+). This implies that the means and standard deviations of the data measured by each detector should be the same. Detector imbalance is considered to be the only factor producing differences in means and standard deviations of the subsets of data collected by each detector. To eliminate the striping effects of detector imbalance, the means and standard deviations of the n histograms are equalised, that is, forced to equal a chosen value. Usually the means of the n individual histograms are made to equal the mean of all of the pixels in the image, and the standard deviations of the n individual histograms are similarly forced to be equal to the standard deviation of all of the pixels in the image. Example 4.1 provides a guide to the calculations that are involved.

EXAMPLE 4.1: DE-STRIPING – LINEAR METHOD

This example demonstrates the way in which the calculations involved in de-striping an image using the linear method (section 4.2.2.1) are carried out. In effect, we partition the image data into k subsets, where k is the number of detectors, and calculate the mean and standard deviation for each subset. Secondly, the values of the mean and standard deviation of the full data set are required. Given these values, a bias term and an offset term are computed for each subset. The bias term is a multiplier and the offset is a constant to be added. These biases and offsets are applied to the subsets in turn, and the net effect is that all of the subset means are equal to the overall mean and the subset standard deviations are all equal to the overall standard deviation. Given that striping results from detector imbalance (i.e. differences in mean and standard deviation from detector to detector), this equalising procedure should eliminate striping.

We first need to compute the value of the overall variance V, given by the expression:

$$V = \frac{\sum n_i \left(\bar{x}_i^2 + v_i\right)}{\sum n_i} - \bar{X}^2$$

For simplicity, we assume that the number of detectors used in the imaging instrument is 6. The columns in the table below show:

(i) detector number (1–6),
(ii) number of pixels recorded by each detector
(iii) pixel values recorded by each detector
(iv) means of the pixel values for each detector
(v) standard deviations of the pixel values for each detector
(vi) variances of the pixel values for each detector.

(i)	(ii)	(iii)	(iv)	(v)	(vi)
Detector number	Number of pixels	Pixel values	Individual means ($\bar{x}_i$)	Standard deviations (s_i)	Variances ($v_i = s_i^2$)
1	5	1 3 2 4 6	3.2	1.720	2.96
2	5	3 6 2 3 8	4.4	2.245	5.04
3	5	4 3 4 2 9	4.4	2.417	5.84
4	5	2 4 3 3 7	3.8	1.720	2.96
5	5	0 2 2 2 6	2.4	1.959	3.84
6	5	4 3 3 3 9	4.4	2.332	5.44

We need the value of the overall mean, $\bar{X}$, which appears in the numerator. This is the sum of the individual detector means (column (iv)) divided by the number of detectors (6), which gives $22.6 \div 6 = 3.76\dot{6}$. The numerator of the equation is the sum of $n_i(\bar{x}_i^2 + v_i)$ where $\bar{x}_i$ is the mean value of the five pixel values for each detector and the n_i (column (ii)) are all equal to 5. The calculation is as follows: $[(3.2^2 + 2.96) \times 5] + [(4.4^2 + 5.04) \times 5] \cdots [(4.4^2 + 5.44) \times 5] = 573$. The denominator is the sum of the n_i, which equals $5 + 5 + 5 + 5 + 5 + 5 = 30$, so the first term in the equation is equal to $530/40 = 19.1$.

The second term is the square of the overall mean ($\bar{X}^2$), that is, the square of $3.76\dot{6}$or 14.183. Finally, subtract the second term from the first (19.100–14.183) to get the value of the overall variance V, which equals 4.192. The overall standard deviation is the square of V, or 2.216. To apply the de-striping correction to the image from which the statistics were derived you must calculate the gains (b_i) and offsets (a_i), mentioned above, from:

$$a_i = \frac{S}{s_i}$$

$$b_i = \bar{X} - a_i \bar{x}_i$$

The corrected pixel values r'_{ij} are then found from the relationship $r'_{ij} = a_i r_{ij} + b_i$, where r_{ij} are the uncorrected pixel values. It is important to ensure that the gain and offset computed from the subset of data collected by detector $i$are applied to the pixels collected by that detector.

The sampling variability of the subset means and standard deviations (which measures their reliability) increases with the size of the subset, so each subset should be reasonably large. An image size of at least 1024 lines and 1024 pixels per line is suggested.

If the linear method were to be applied on a pixel-by-pixel basis to an image of any size then it would be inordinately slow. A little thought will show that, for eight-bit images collected by the Landsat TM and ETM+, there are only 256 possible values for each detector. We could build a table consisting of n columns and 256 rows. The input pixel value is the row number and the n corrected values, one per detector, form the data values for that row. The principle is simple. For a given pixel, the row number in the table is the input pixel value while the output (corrected) value is the ith value on that row, where i is the detector number ($1 \le i \le n$). The table is known as a *lookup table* or a *direct address table*. The use of lookup table methods can be very effective where 8- or 16-bit integer data are being processed, as is the case with most images, because the number of outcomes is limited and is relatively small, so that pre-calculation of results for all possible cases becomes feasible. More complicated methods using hash tables are needed if the same idea is to be applied to 32-bit integer data, which can represent integers in the range from 0 to 2 147 483 647. An example of the use of hash tables in processing remotely-sensed data is given by Mather (1985). Cormen *et al.* (2002) provide a more in-depth study of data structures and algorithms, including hash tables.

4.2.2.2 *Histogram matching*

The method of de-striping images produced by electromechanical scanners described in section 4.2.2.1 is based on the assumption that the output from a detector is a linear function of the input value according to the expression:

$$r_{\text{out}} = \text{offset} + \text{gain} \times r_{\text{in}}$$

Horn and Woodham (1979) observe that '... it appears that different gains and offsets are appropriate for different scene radiance [r_{in}] ranges. That is, the sensor transfer curves are somewhat non-linear'. In other words, the linear relationship between r_{in} and r_{out} used in section 4.2.2.1 is an over-simplification. The method described in this section uses the shape of the cumulative frequency histogram of each detector to find an estimate of the non-linear transfer function. The ideal or target transfer function is taken to be defined by the shape of the cumulative frequency histogram of the whole image, which is easily found by carrying out a class-by-class summation of the n individual detector histograms (e.g., six for Landsat MSS or 16 for Landsat TM/ETM+). The histogram for detector 1 is computed from the pixel values on scan-lines 1, 7, 13, ..., of the image, while the histogram for detector 2 is derived from the pixel values on scan lines 2, 8, 14, ..., and so on. The histograms are expressed in cumulative form (so that class 0 is the number of pixels with a value of 0, class 1 is the number of pixels with values 0 or 1, and so on). Next, each histogram class frequency is divided by the number of pixels counted in that histogram, thus ensuring that the individual histograms and the target histogram are all scaled between 0 and 1.

At this stage, we have n individual cumulative histograms and one target cumulative histogram, where n is the number of detectors. Our aim is to adjust the individual cumulative histograms so that they match the shape of the target cumulative histogram as closely as possible. This is done by adjusting the class numbers of the individual histograms. Thus, class number k in an individual histogram may be equated with class number j in the target histogram. This means that all pixels scanned by the detector corresponding to that individual histogram, and which have the value k, would be replaced in the de-striped image by the value j, which is derived from the target histogram. In order to determine the class number in the target histogram to be equated to class number k in the individual histogram, we find the first class in the target histogram whose cumulative frequency count equals or exceeds the cumulative frequency value of class k in the individual histogram. The class in the target histogram that is found is class y. An example is given in Table 4.1. The frequency value for cell 3 of an individual histogram is 0.57. This value is compared with the target histogram values until the first class with a value greater than or equal to 0.57 is found. This is class 4 of the target histogram. Class 3 of the detector histogram is thus equated to class 4 of the target histogram, and all pixel values of 3 scanned by that detector are replaced with the value 4. The procedure is applied separately to all 256 values for each of the 6 (or 16) detectors. The result is generally a reduction in the banding effect, though much depends on the nature of the image. Wegener (1990) gives a critical review of the Horn and Woodham procedure, and presents a modified form of the algorithm.

Table 4.1 Example of histogram matching for de-striping Landsat MSS and TM images. The target histogram is the cumulative histogram of the entire image or sub-image. The detector histogram is the cumulative histogram using values of pixels scanned by one of Landsat MSS's six detectors. The output pixel value to replace a given input value is found by comparison of the two histograms. For example, the detector histogram for input pixel value 3 is 0.57. The first value in the target histogram to equal or exceed 0.57 is that in row four. Hence, the pixel values in the uncorrected image that are generated by this detector are replaced by the value 4.

Input pixel value	Target histogram value	Detector histogram value	Output pixel value
0	0.09	0.08	0
1	0.18	0.11	1
2	0.33	0.18	2
3	0.56	0.57	4
4	0.60	0.66	4
5	0.76	0.78	6
6	0.95	0.95	6
7	1.00	1.00	7

The lookup table procedure described at the end of section 4.2.2.1 can be used to make the application of this method more efficient. Thus, for each detector histogram, a table can be constructed so that the output value corresponding to a given input can be easily read. The input value is the pixel value in the image being corrected while the output value is its equivalent in the de-striped image.

4.2.2.3 *Other de-striping methods*

The procedures discussed in sections 4.2.2.1 and 4.2.2.2 operate directly on the image data, which has spatial coordinates of (row, column). Hence, these procedures are said to operate in the 'spatial domain'. A number of methods of transforming an image data set from the spatial domain representation to an alternative frequency domain representation are described in Chapter 6. In particular, the Fourier transform has been widely used to determine the existence of periodicities a data series such as may be caused by a recurring 6 or 16 line horizontal pattern. More recently, the wavelet transform has been introduced as an image transform tool; it is also considered in Chapter 6.

4.3 GEOMETRIC CORRECTION AND REGISTRATION

Remotely-sensed images are not maps. Frequently, however, information extracted from remotely-sensed images is integrated with map data in a Geographical Information System (GIS) or presented to consumers in a map-like form (for example, gridded 'weather pictures' on TV or in a newspaper). If images from different sources are to be integrated (for example, multispectral data from Landsat ETM+ and SAR data from ERS, RADARSAT or ASAR) or if pairs of interferometric SAR images are to be used to develop digital elevation models (section 9.2) then the images from these different sources must be expressed on a common coordinate system. The transformation of a remotely-sensed image so that it has the scale and projection properties of a given map projection is called geometric correction or georeferencing. A related technique, called registration, is the fitting of the coordinate system of one image to that of a second image of the same area. Accurate image registration is needed if a time sequence of images is used to detect change in, for example, the land cover of an area.

A map is defined as:

> '. . . a graphic representation on a plane surface of the Earth's surface or part of it, showing its geographical features. These are positioned according to pre-established geodetic control, grids, projections and scales'
>
> (Steigler, 1978).

A map projection is a device for the representation of a curved surface (that of the Earth) on a flat sheet of paper (the map sheet). Many different map projections are in common use (see Snyder, 1982, Steers, 1962, and Frei *et al.*, 1993). Each projection represents an effort to preserve some property of the mapped area, such as uniform representation of areas or shapes, or preservation of correct bearings. Only one such property can be correctly represented, though several projections attempt to compromise by minimising distortion in two or more map characteristics. The UK Ordnance Survey uses a Transverse Mercator projection. A regular grid, graduated in metres and with its origin to the south-west of the British Isles, is superimposed on the map sheet since lines of latitude and longitude plot as complex curves on the Transverse Mercator projection.

Geometric correction of remotely-sensed images is required when the remotely-sensed image, or a product derived from the remotely-sensed image such as a vegetation index image (Chapter 6) or a classified image (Chapter 8), is to be used in one of the following circumstances (Kardoulas *et al.*, 1996):

(i) to transform an image to match a map projection,
(ii) to locate points of interest on map and image,
(iii) to bring adjacent images into registration,
(iv) to overlay temporal sequences of images of the same area, perhaps acquired by different sensors, and
(v) to overlay images and maps within a GIS.

The advent of high-resolution images obtained from instruments carried by satellites such as Quickbird, IKONOS, SPOT-5 and IRS-1C has brought the topic of geometric correction of remotely-sensed images much closer to the field of photogrammetry. For many years, photogrammetrists have used accurate camera models to perform analytical corrections on aerial photographs (Konecny, 2003; Wolf and DeWitt, 2000). The use of stellar navigation and GPS onboard satellites has meant that some of the orbital parameters required for an analytical solution are now more readily available. Finally, considerable research effort has been directed towards providing a solution to the problem of terrain or relief correction. A brief account of these developments is given in section 4.7.

The sources of geometric error in moderate spatial resolution imagery with a narrow field of view, such as the imagery produced by Landsat ETM+ and SPOT HRV are summarised in section 2.3. The main categories are (1) instrument error, (2) panoramic distortion, (3) Earth rotation and (4) platform instability (Bannari *et al.*, 1995a). Instrument errors include distortions in the optical system, non-linearity of the scanning mechanism, and non-uniform sampling rates. Panoramic distortion is a function of the angular field of view of the sensor and affects instruments with a wide angular field of view (such as the AVHRR and CZCS) more than those with a narrow field of view, such as the Landsat TM and ETM+ and the SPOT HRV. Earth rotation velocity varies with latitude. The effect of Earth rotation is to skew the image. Consider the Landsat satellite as it moves southwards above the Earth's surface. At time t, its ETM+ sensor scans image lines 1–16. At time $t + 1$, lines 17–32 are scanned. But the Earth has moved eastwards during the period between time t and time $t + 1$ therefore the start of scan lines 17–32 is slightly further west than the start of scan lines 1–16. Similarly, the start of scan lines 33–48 is slightly further west than the start of scan lines 17–32. The effect is shown in Figure 4.3. Platform instabilities include variations in altitude and attitude. All four sources of error contribute unequally to the overall geometric distortion present in an image. In this section, we deal with the geometric correction of medium-resolution digital images such as those acquired by the Landsat TM and SPOT HRV instruments. Correction of wide-angle images derived from

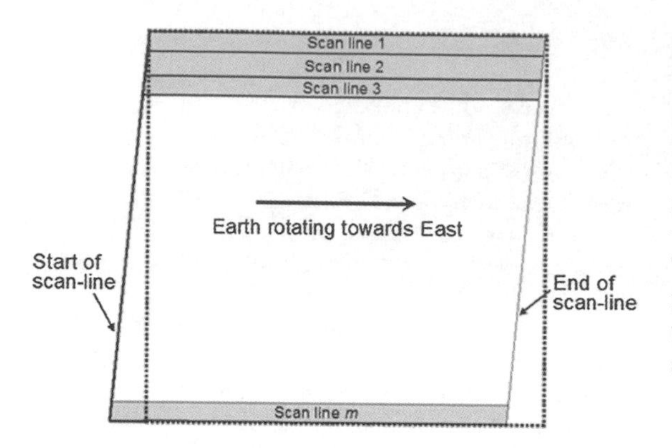

Figure 4.3 Effects of Earth rotation on the geometry of a line-scanned image. Due to the Earth's eastwards rotation, the start of each swath (of 16 scan lines, in the case of the Landsat-7 ETM+) is displaced slightly westwards. At the Equator the line joining the first pixel on each scan line (the left margin of the image) is oriented at an angle that equals the inclination angle i of the satellite's orbit. At a latitude of $(90 - i)^{\circ}$ the same line is parallel to the line of latitude $(90 - i)^{\circ}$. Thus the image orientation angle increases pole-wards. See section 4.3.1 for further details.

the NOAA AVHRR is described by Brush (1985), Crawford *et al.* (1996), Moreno and Melia (1993) and Tozawa (1983). Geocoding of SAR images is covered by Dowman (1992), Dowman *et al.* (1993), Johnsen *et al.* (1995) and Schreier (1993). The use of digital elevation data to correct images for the geometric distortion produced by relief variations is considered by Blaser and Caloz (1991), Itten and Meyer (1993), Kohl and Hill (1988), Palà and Pons (1995), Toutin (1995) and Wong *et al.* (1981). Other useful references are Fogel and Tinney (1996), Kropatsch and Strobl (1990), Kwok *et al.* (1987), Novak (1992), Shlien (1979), van Wie and Stein (1977), Westin (1990) and Wolberg (1990). Williams (1995) provides an excellent treatment of many aspects covered in this section, including geocoding of SAR and AVHRR imagery. Wolf and DeWitt (2000) is a good source of information on georeferencing of digital air photographs.

The process of geometric correction can be considered to include (i) the determination of a relationship between the coordinate system of map and image (or image and image in the case of registration), (ii) the establishment of a set of points defining pixel centres in the corrected image that, when considered as a rectangular grid, define an image with the desired cartographic properties, and (iii) the estimation of pixel values to be associated with those points. The relationship between the two coordinate systems (map and image) could be defined if the orbital geometry of the satellite platform were known to a sufficient degree of accuracy. Where orbital parameters are known, methods based upon orbital geometry give high accuracy. The DORIS system, used with the SPOT satellite, is described in section 2.3.7. Otherwise, orbital models are useful only where the desired accuracy is not high, or where suitable maps of the area covered by the image are not available. The method based on nominal orbital parameters is described in section 4.3.1, while the map-based method is covered in section 4.3.2. The extraction of the locations of the pixel centre points for the corrected image and the estimation of pixel values to be associated with these output points is considered in section 4.3.3.

4.3.1 Orbital geometry model

Orbital geometry methods are based on knowledge of the characteristics of the orbit of the satellite platform. Bannari *et al.* (1995) describe two procedures based on the photogrammetric equations of collinearity. These equations describe the properties of the satellite orbit and the viewing geometry, and relate the image coordinate system to the geographical coordinate system. They require knowledge of the geographical coordinates of a number of points on the image. Such points are known as *ground control points* or GCP.

A simple method of correcting the coordinate system of remotely-sensed images using approximate orbit parameters, described by Landgrebe *et al.* (1975), is used here to illustrate the principles involved. It is not recommended as an operational method, because it is based upon nominal rather than actual orbital parameters, which implies that the accuracy of the geometrically corrected image produced by this technique is not high. Landgrebe *et al.* (1975) suggest that the magnitude of the error is of the order of 1–2%, meaning that if the corrected image is overlaid on a map and both are aligned with reference to a well-defined point, then the error in measured coordinate positions of other points will be 1–2%. Note that the coordinate system has its origin in the top left corner, so that the x-axis runs horizontally and increasing in value to the right, with the y-axis running vertically, with values increasing downwards. Thus, the x-axis gives the pixel position across the scan line. The y-axis gives the scan line number.

(i) *Aspect ratio*

Some sensors, such as the Landsat MSS, produce images with pixels that are not square. The Landsat MSS scan lines are nominally 79 m apart, whereas the pixels along each scan line are spaced at a distance of 56 m. Since the instantaneous field of view of the MSS is 79 m there is over-sampling in the across-scan direction. As we generally require square rather than rectangular pixels, we can choose 79 m square pixels or 56 m square pixels to overcome the problem of unequal scale in the x and y directions. Because the across-scan direction is over-sampled, it is more reasonable to choose 79 m square pixels. The aspect ratio (the ratio of the $x : y$ dimensions) is 56:79 or 1:1.41. The first transformation matrix, **M1**, which corrects the image to a 1:1 aspect ratio, is, therefore:

$$\mathbf{M1} = \begin{bmatrix} 0.00 & 1.41 \\ 1.00 & 0.00 \end{bmatrix}$$

This transformation matrix is not required if the image to be geometrically transformed has square pixels.

(ii) *Skew correction*

Landsat TM and ETM+ images are skewed with respect to the North-South axis of the Earth. Landsat-1 to -3 had an orbital inclination of 99.09° whereas Landsats 4–5 and -7 have an inclination of 98.2°. The satellite heading (the direction of the forward motion of the satellite) at the Equator is therefore 9.09° and 8.2° respectively, increasing with latitude (Figure 2.1). The skew angle θ at latitude L is given (in degrees) by:

$$\theta = 90 - \cos^{-1}\frac{\sin(\theta_{\mathrm{E}})}{\cos(L)}$$

where θ_{E} is the satellite heading at the Equator and the expression $\cos^{-1}(x)$ means: the angle whose cosine is x, that is, the inverse cosine of x. Given the value of θ the coordinate system of the image can be rotated through θ degrees anticlockwise so that the scan-lines of the corrected image are oriented in an East-West direction using the transformation matrix **M2**:

$$\mathbf{M2} = \begin{bmatrix} \cos(\theta) & \sin(\theta) \\ -\sin(\theta) & \cos(\theta) \end{bmatrix}$$

The value L that is normally used in the determination of θ is the centre latitude of the image being corrected. Since the latitude of the satellite is varying continuously this value will be only an approximation.

For latitude $L = 51°$ and using the heading of Landsat-4 and -5 at the Equator, θ_{E}, the value of θ is given by:

$$\theta = 90.00° - \cos^{-1}\frac{\sin(8.20°)}{\cos(51.00°)}$$

$$= 90.00° - \cos^{-1}\frac{0.14}{0.63} = 90.00° - 76.90° = 13.10°$$

and the elements of matrix **M2** are:

$$\mathbf{M2} = \begin{bmatrix} 0.9740 & 0.2286 \\ -0.2286 & 0.9740 \end{bmatrix}$$

(iii) *Earth rotation correction*

As the satellite moves southwards over the illuminated hemisphere of the Earth, the Earth rotates beneath it in an easterly direction with a surface velocity proportional to the latitude of the nadir or sub-satellite point. To compute the displacement of the last line in the image relative to the first line we need to determine (i) the time taken for the satellite sensor to scan the image and (ii) the eastwards angular velocity of the Earth. The distance travelled by the Earth's surface can then be obtained by multiplying time by velocity. The time taken for the satellite sensor to scan the image can be found if the distance travelled by the satellite and the satellite's velocity are known. Both distance and velocity are expressed in terms of angular measure (i.e., in radians; one radian equals approximately 57°; Figure 1.6). If A is a point on the Earth's surface corresponding to the centre of the first scan line in the image, and if B is the corresponding point for the last scan line in the image, then the curve AB (the line on the Earth's surface joining points A and B) is an arc of a circle centred at the Earth's centre, O. The angle AOB can be calculated because we know that the Earth's equatorial radius (OA or OB) is approximately 6378 km, and an angle (in radian measure) is equal to arc length divided by radius (Figure 1.6(a)). The arc length AB is the distance between the centre pixels in the first and last scan lines of the image. For Landsat MSS and TM images, AB is 185 km, hence angle AOB (representing the angular distance moved by the Landsat satellite during the capture of one image) equals 185/6378 or 0.029 radians.

The orbital period (the time required for one full revolution) for Landsats 1–3 is 103.267 minutes (99 minutes for later Landsats), so the satellite's angular velocity ω_0 is $2\pi \div (103.267 \times 60)$ radians per second, or 0.001 014 r s^{-1}. The problem is now to find the time required for a satellite travelling at this angular velocity to traverse through an angle of 0.029 radians (the angular distance between the first and last scan line, see above). The answer is found by dividing the angular distance to be moved by the angular velocity, and the result of this operation is 0.029/0.001 014 = 28.6 seconds.

Now the question becomes: how far will point B (the centre of the last scan line) move eastwards during the 28.6 seconds that elapses between the scanning of the first and last scan lines of the MSS or TM image? The answer again depends on latitude. For simplicity, we will take the latitude (L) of the centre of the image. The Earth's surface velocity at latitude L is $V_E(L)$ which is defined as:

$$V_{\mathrm{E}}(L) = R\cos(L)\omega_{\mathrm{E}}$$

R is the Earth's radius, approximately 6378 km, and ω_{E} (omega-e) is the Earth's angular velocity. Since the

Earth rotates once in 23 hours, 56 minutes and 4 seconds (that is, 86 164 seconds), then its angular velocity is simply (2π/867 164) radians per second, or 0.7292×10^{-4} r s^{-1}. If L is 51° then $V_E(L)$ equals $6378 \times 10^{-3} \times 0.6293 \times 0.729721 \times 10^{-4}$ m s^{-1}, i.e., 292.7 m s^{-1}. Now that we know (i) that the time taken to scan an entire Landsat TM/ETM+ or MSS image is 28.6 seconds and (ii) the Earth's surface velocity, then the calculation of the eastward displacement of the last scan line in the image can be obtained. At 51°N the surface velocity is 292.7 m s^{-1} so the distance travelled eastwards is $292.7 \times 28.6 = 8371$ m.

These calculations assume that the line AB joining the centres of the first and last scan lines is oriented along a line of longitude whereas, in fact, the Landsat satellites (such as SPOT, Terra, Aqua and all Sun-synchronous, polar-orbiting platforms) have an orbit that is skewed relative to lines of longitude, as noted in the calculation of matrix **M2** above. The skew angle θ for 51° latitude is 14.54° (see above) so the actual eastwards displacement is $8371 \times \cos(14.54^o) = 8103$ m. These computations are summarised by the term a_{sk}:

$$a_{sk} = \frac{\omega_E \cos(L)}{\omega_O \cos(\theta)} = 0.0719 \frac{\cos(L)}{\cos(\theta)}$$

where ω_E is the Earth's angular velocity, ω_O is the satellite's angular velocity and θ and L are defined above. The transformation matrix **M3** is:

$$\mathbf{M3} = \begin{bmatrix} 1 & 0 \\ a_{sk} & 1 \end{bmatrix}$$

At 51° latitude, **M3** is:

$$\mathbf{M3} = \begin{bmatrix} 0 & 1 \\ 0.04647 & 1 \end{bmatrix}$$

Note that 'fill pixels' are added to the start of each scan-line of a Landsat MSS or TM image by some ground stations to compensate for the Earth rotation effect. If this correction is thought to be sufficient then transformation **M3** can be omitted. Alternatively, since the number of fill pixels is generally given in the header/trailer data associated with each scan-line, these fill pixels can be stripped off and the correction **M3** applied.

The three transformation matrices **M1**, **M2** and **M3** given in sub-sections (i) to (iii) above are not applied separately. Instead, a composite transformation matrix, **M**, is obtained by multiplying the three separate transformation matrices:

$$\mathbf{M} = \mathbf{M1\ M2\ M3}$$

The corrected image coordinate system is related to the raw image coordinate system by:

$$\mathbf{x}' = \mathbf{Mx}$$

where $\mathbf{x}'$ $(= x_1', x_2')$ is the vector holding the pixel and line coordinates of the corrected pixel and $\mathbf{x}$ $(= x_1, x_2)$ is the original (pixel, line) coordinate. Remember that the origin of the image coordinate system is the top left corner of the image. See Example 4.2 for more details.

EXAMPLE 4.2: ORBITAL GEOMETRY MODEL

This example illustrates the use of the orbital geometry model (section 4.3.1) in the process of geometric correction of an image. Assume that we have a square image of size 1024 lines and 1024 pixels per line. This image is shown as a square (ABCD) in Example 4.2 Figure 1 (solid line). It is assumed that the image pixels are square so that the transformation matrix $\mathbf{M}_1$ is not required. The Landsat-4 and -5 satellites have an orbital inclination angle of 8.2° so, at a latitude of 51°, angle θ is equal to 13.10°, and the

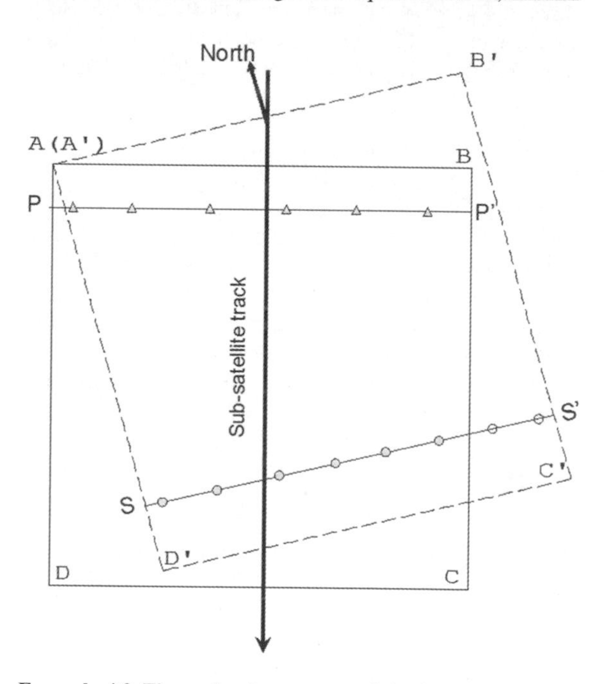

Example 4.2 Figure 1 An uncorrected (raw) square image is outlined by the solid line joining the points ABCD. After geometric transformation, the map coordinates of the image corners become A'B'C'D' (outlined by dashed line). The sub-satellite track is the ground trace of the platform carrying the scanner that collected the raw image. One scan line of the raw image is shown (line PP') with pixel centres indicated by triangles. The geometrically corrected image (A'B'C'D') has its columns oriented towards North, with rows running east-west. One scan line in the corrected image is shown (line SS') with pixel centres indicated by circles. Note that the angle θ between the sub-satellite track and North is equal to 13.0992°.

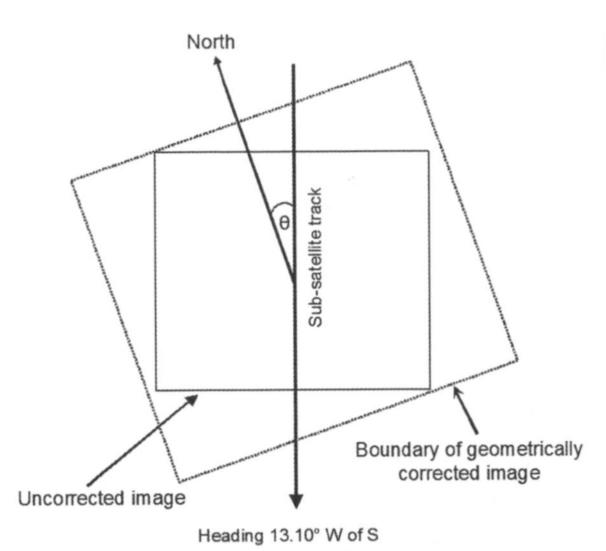

Example 4.2 Figure 2 In practical applications of geometric correction, the corrected image area is delimited by the rectangle that encloses the uncorrected image. Pixel values within the boundary of the geometrically corrected image but lying outside the area of the uncorrected image are set to zero. The angle θ measures the anti-clockwise rotation of 13.10° that is required at latitude 51° N to produce a north-oriented image.

transformation matrices $\mathbf{M}_2$ and $\mathbf{M}_3$ are given by:

$$\mathbf{M}_2 = \begin{bmatrix} 0.9740 & 0.2286 \\ -0.2286 & 0.9740 \end{bmatrix}$$

and:

$$\mathbf{M}_3 = \begin{bmatrix} 1.00000 & 0.04649 \\ 0.00000 & 1.00000 \end{bmatrix},$$

so that the final transformation matrix $\mathbf{M}$ is equal to:

$$\mathbf{M} = \begin{bmatrix} 0.9740 & 0.2719 \\ -0.2266 & 0.9634 \end{bmatrix}$$

The coordinates of the corners of the original image and their transformed equivalents are: (A (0, 0), A' (0, 0)}, {B (1024, 0), B' (997, -232)}, {C (1024, 1024), C' (1276, 754)} and {D (0, 1024), D' (278, 986)}. The second stage of the geometric correction operation is to find values for 'new' pixels located at 30 m intervals along the scan lines of the transformed image, i.e. lines that are 30 m apart and parallel to line A'B' (or D'E') in Example 4.2 Figure 1. This procedure is termed 'resampling' and is considered in section 4.3.3. Note that the values for the new pixels can only be obtained in the area where ABCD and A'B'C'D' (Example 4.2 Figure 1) overlap. In practical applications, we would generate a geometrically corrected image of a size sufficient to enclose the whole of the raw image, as shown in Example 4.2 Figure 2.

4.3.2 Transformation based on ground control points

The orbital geometry model discussed in section 4.3.1 is based on nominal orbital parameters. It takes into account only selected factors that cause geometric distortion. Variations in the altitude or attitude of the platform are not considered simply because the information needed to correct for these variations is not generally available. Some satellites such as SPOT-5 now carry instruments that provide precise orbital data (section 2.3.7), and more complex analytical models than the one described in section 4.3.1 are used to generate a geometrically corrected image.

An alternative method is to look at the problem from the opposite point of view and, rather than attempt to construct a physical model that defines the sources of error and the direction and magnitude of their effects, use an empirical method which compares differences between the positions of common points that can be identified both on the image and on a map of a suitable scale of the same area. For a Landsat ETM+ image, a map scale of at least 1:25 000 is needed, since the minimum measurable line width on a map is considered to be 1 mm, which translates to 25 m (approximately the size of one Landsat TM/ETM+ pixel) at a map scale of 1:25 000. From the differences between the distribution of the common set of points on the image and the distribution of these points on the map, the nature of the distortions present in the image can be estimated, and an empirical transformation to relate the image and map coordinate systems can be computed and applied. This empirical function should be calibrated (using ground control points), applied to the image, and then validated (using a separate test set of ground control points).

The aim of the procedures described in this section is to produce a method of converting map coordinates to image coordinates, and vice versa. Two pieces of information are required. The first is the map coordinates of the image corners. Once the image is outlined on the map, the map coordinates of the pixel centres (at a suitable scale) can be found (Figure 4.4).

The map coordinates of the image corners are found by determining an image-to-map coordinate transformation. The map coordinates of the required pixel centres are converted to image coordinates by a map-to-image coordinate transformation. Both transformations are explained in this section. The final stage, that of associating pixel values with calculated (map) pixel positions, is discussed in the next section under the heading of resampling.

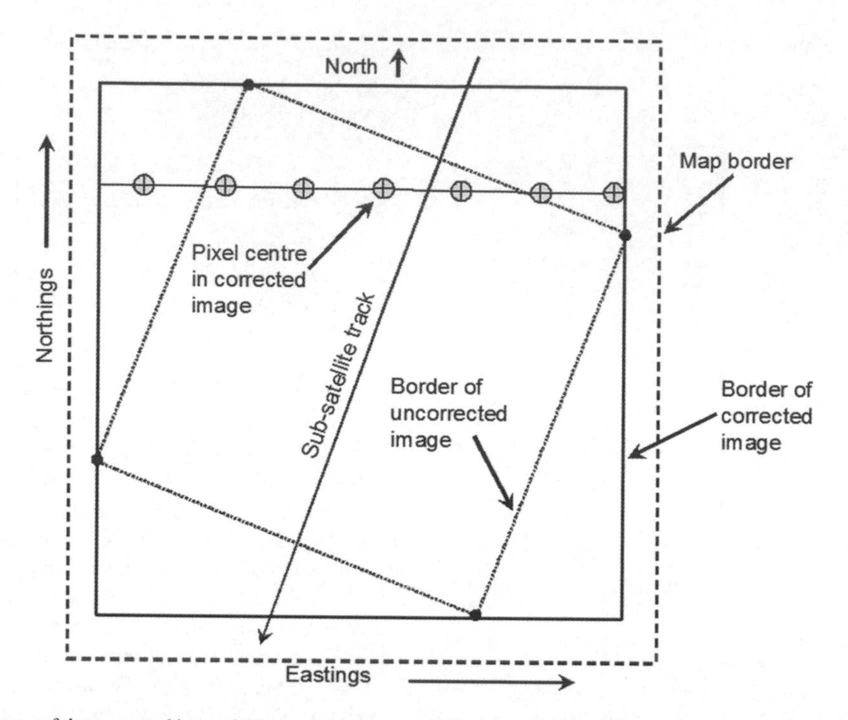

Figure 4.4 The area of the corrected image is shown by the rectangle that encloses the oblique uncorrected image. The positions of the corners of this enclosing rectangle (in map easting and northing coordinates) can be calculated from the (row, column) image coordinates of the corners of the uncorrected image using a least-squares transformation. Once the corners of the corrected image area are known in terms of map coordinates, the locations of the pixel centres in the corrected image (also in map coordinates) can be determined using simple arithmetic, noting that easting and northing map coordinates are expressed in kilometres. Finally, the pixel centre positions are individually converted to image (row, column) coordinates in order to find the image pixel value to be associated with the pixel position in the corrected image. Not all pixels in the corrected image lie within the area of the uncorrected image. Such pixels receive a zero value.

The method relies on the availability of an accurate map of the area at a suitable scale. In some parts of the world suitable maps are not available. It is thus paradoxical that accurate 'image maps' can only be produced for areas for which conventional maps are available using this method unless field surveying techniques are used. The coordinates of selected points, called ground control points, are measured on map and image. Ground control points are well defined and easily recognisable features that can be located accurately both on a map and on the corresponding image. They can be located on the ground by the use of GPS rather than by map measurement. Clavet *et al.* (1993) and Kardoulas *et al.* (1996) provide details of the use of GPS in locating ground control points for geometric correction. Cook and Pinder (1996) compare the transformation accuracy resulting from the use of control points derived from 1:24 000 US Geological Survey maps and from the use of GPS. They summarise the problems involved in the accurate measurement of control points from maps, and conclude that differential GPS provides substantially better results than map digitising. It is, however, more costly both in terms of equipment and travel time as each control point must be visited and its coordinates measured.

The symbols (x_i, y_i) refer to the map coordinates of the ith ground control point, and the symbols (c_i, r_i) refer to the image column and row image coordinates of the same point. The map coordinates can be expressed in eastings (x) and northings (y) from an arbitrary origin, or in degrees and decimal degrees of longitude and latitude. The image coordinates are expressed in terms of column (c; pixel position along the scan line) and row (r; scan line number, where scan line 1 is the first scan line of the image (Figure 4.3)). We seek a method that will allow us to convert from (x, y) to (c, r) coordinates and vice versa. To achieve this aim, the method of least squares is used.

To illustrate the method, let us assume that a sample of s_i and t_i values is available, where s and t are any two variables of interest, such as the row number of an image pixel (s) and the easting coordinate of the same point on the corresponding map (t). Furthermore, assume that values of t_i are found relatively easily whereas the values of s_i are difficult to acquire, so it would be advantageous to be able to estimate the value of s given the corresponding value of t. For this reason, we can call s the predicted variable and $t$the predictor variable. The method of least squares allows the estimation of and value of $s(s_i)$ given the corresponding value of $t(t_i)$ using a function of the form:

$$s_i = a_0 + a_1 t_i + e_i$$

Because only a single predictor variable, t, is used the expression is known as the *univariate* least-squares equation. It allows the difficult-to-measure s_i to be estimated from the value of the corresponding easier-to-measure t_i. The estimated value of s_i is written $\hat{s}$. The terms a_0 and a_1 are computed from a sample of values of s and t using the method of least squares. The criterion used in the least-squares procedure is that the sum of the squared differences between the true (and usually unavailable) values s_i and the values $\hat{s}_i$ estimated from the univariate least-squares equation (above) is the smallest for all possible value of $\hat{s}_i$. In other words, the coefficients a_0 and a_1 could, in principle, take on any values, such as a_0 = 8.999 and a_1 = –82.192. However, of all these possible values, those derived using the least-squares principle will minimise the criterion $\text{ESS} = \sum (s_i - \widehat{s}_i)^2$. The term ESS is called the 'explained sum of squares' in least-squares regression analysis. The differences between s and $\widehat{s}$ are called 'residuals', and they are represented by the term e_i. Thus, we could omit e_i from the equation and express $\hat{s}$ directly as:

$$\hat{s}_i = a_0 + a_1 t_{1i}$$

The number of predictor variables can be greater than one. For example, the *bivariate* least-squares regression equation would be used if variable s were to be predicted from variables t_1 and t_2. This equation has the form:

$$\hat{s}_i = a_0 + a_1 t_{1i} + a_2 t_{2i}$$

where s_i is the ith observation made on variable s, t_{1i} is the ith observation made on variable t_1 and t_{2i} is the ith observation made on variable t_2.

The bivariate linear least-squares function is used to find the least-squares coefficients for the following four expressions:

(i) *map* x coordinate as a function of *image* c and $r$coordinates ($x = f(c, r)$),
(ii) *map* y coordinate as a function of *image* c and r coordinates ($y = f(c, r)$),
(iii) *image* c coordinate as a function of *map* x and $y$coordinates ($c = f(x, y)$), and
(iv) *image* r coordinate as a function of *map* x and $y$coordinates ($r = f(x, y)$).

If the coefficients of each of these regression functions are known, it is possible to transform from map (x, y) to image (c, r) coordinates or from image (c, r) to map (x, y) coordinates.

The following bivariate least-squares equation relates the map x coordinate to the image r and c coordinates (item (ii) in the list above):

$$\hat{x}_i = a_0 + a_1 c_i + a_2 r_i$$

It can be interpreted as follows: the least-squares estimate of the map x coordinate of the ground control point labelled i can be found from the image column and row coordinates of that ground control point (c_i and r_i) if the least squares coefficients a_0, a_1 and a_2 are known. The values of these coefficients are determined from a sample of values of x, c and r, as described below, and are then applied to all the image pixels in order to estimate map x coordinates. The same operation is performed to find the map y coordinates of all of the pixels in the image. The steps involved are described in the following paragraphs. First, however, some of the technical details involved in the calculations need to be explained.

In mathematical terminology, the bivariate least squares equation used in the preceding paragraphs is a *first-order polynomial least-squares function*. It is first-order because neither of the predictor variables (on the right-hand side of the equation) is raised to a power greater than one. A first-order function can accomplish scaling, rotation, shearing and reflection but not warping (such as would be necessary to correct for panoramic distortion or for any similar 'bending' effect). A second- or higher-order polynomial can be used to model such distortions, though in practice it is rare for polynomials of order higher than three to be used for medium resolution satellite imagery. Where a relatively small image area is being corrected (for instance, a 1024 × 1024 segment) it should be unnecessary to correct for warping at all. Note that the polynomial method does not correct for terrain relief distortions and is therefore applicable only to relatively flat areas. See Kohl and Hill (1988) and other references listed at the end of the introduction to section 4.5 for details of a modification to the standard polynomial correction that corrects for relief-induced variations using a digital elevation model (DEM).

In general terms, a polynomial function in two variables t and u can be written concisely as:

$$\hat{s} = \sum_{j=0}^{m} \sum_{k=0}^{m-j} a_{jk} t^j u^k$$

where m is the order of the polynomial function. A third-order polynomial, written out in full, becomes:

$$\begin{aligned}\hat{s} &= a_{00}t^0u^0 + a_{01}t^0u^1 + a_{02}t^0u^2 + a_{03}t^0u^3 \\ &+ a_{10}t^1u^0 + a_{11}t^1u^1 + a_{12}t^1u^2 + a_{20}t^2u^0 \\ &+ a_{21}t^2u^1 + a_{30}t^3u^0\end{aligned}$$

(do you see the relationship between the subscripts of the coefficients a and the powers to which the corresponding t and u are raised?) The equation can be reduced to a less formidable form as:

$$\begin{aligned}\hat{s} &= a_{00} + a_{10}t + a_{01}u + a_{20}t^2 + a_{11}tu + a_{02}u^2 \\ &+ a_{30}t^3 + a_{21}t^2u + a_{12}tu^2 + a_{03}u^3\end{aligned}$$

The terms s, t and u are replaced by the appropriate terms in expressions (i)–(iv) above. If, for instance, we wished to estimate y as a function of c and r we would replace s, t and u in the polynomial expansion by y, c and r. There is thus one polynomial function for each of the four coordinate transformations. The first pair gives map coordinates (x, y) in terms of image coordinates (c, r) while the second pair give image coordinates (c, r) in terms of map coordinates (x, y).

Before we consider methods of evaluating polynomial expressions for given sets of (x, y) and (c, r) coordinates we should consider (i) the size of the sample of control points needed to give reliable estimates of the coefficients a_{ii}, (ii) the spatial distribution of the control points, and (iii) the accuracy with which they are located (Labovitz and Marvin, 1986). In mathematical terms we need to take a sample of at least n control points where n is the number of coefficients in the polynomial expression. For a first-order polynomial, n is equal to 3. For a second-order polynomial, the value of n is 6 and for a third-order equation n has the value 10. These numbers of control points are necessary purely and simply to ensure that it is *mathematically* possible to evaluate the equations defining the coefficients a_{ij}. It is important to note that the *statistical* requirement, which is concerned not so much with the theoretical feasibility of the calculations but with the reliability of the results, sets a much higher standard. Most conventional statistics texts suggest that a sample size of at least 30 is required to achieve reliable estimates, but experience suggests that 10–15 control points will give acceptable results for a first-order fit and a small image area (up to 1024^2 pixels) medium resolution imagery such as those produced by the Landsat TM/ETM+ and SPOT HRV instruments. More ground control points will be needed in areas of moderate relief, or for images produced by instruments with a wide field of view, where a second-order polynomial may be required. Results based on the use of small numbers of ground control points should be treated with caution as they may satisfy the mathematical criteria but may not satisfy the statistical ones. Mather (1995) gives details of a simulation study that emphasises the importance of adequate GCP sample size.

The second aspect of sampling that should concern users is the spatial distribution of ground control points, a topic that is treated in some detail by Mather (1995). Since the coordinate system being transformed is a two-dimensional one, it seems reasonable to suggest that the control point locations should be represented by a two-dimensional pattern. It may, for instance, be possible to take a large number of control points along a linear transect, such as a main road. The information contained in these control points refers only to one dimension. Results derived from such information could say nothing about variations in the direction perpendicular to the transect line. Another possibility is to locate control points in clusters, for example, representing road junctions in small towns that are widely separated. Again, large areas of the image would be unrepresented and results would thus be biased. In extreme cases a condition known mathematically as 'singularity' would be signalled during the least-squares computations. This is equivalent in scalar terms to trying to divide by zero. Mather (1976, p. 124–129) considers this point in detail, and also provides the technical background for two-dimensional least-squares problems. We can conclude from this brief examination not only that control points should be sufficient in number but also that they should be evenly spread, as far as possible, over the image area. This presents problems where substantial parts of the image cover sea or water areas where control points are absent. The same could be said of images covering any relatively featureless region. Unwin and Wrigley (1987) introduce the concept of the *leverage* of a control point, which measures the influence of the point on the overall fit of the polynomial function and allows the user to determine which points require most care and attention.

The accuracy with which control points are measured is also considered by Mather (1995). In his simulation experiment, increasing amounts of random 'noise' were added to the known exact locations of the ground control points, and (particularly when the distribution of control points was linear or clustered) the effects of the noise were severe. The residual error is the difference between the value of the map or image coordinate estimated from the equation and the corresponding coordinate measured on the map or image. Residual errors can be calculated for the line/pixel coordinates of the image or the easting/northing coordinates of the map. Using the appropriate polynomial function selected from the four listed above, the map easting coordinate of a control point can be estimated from the image row (line) and column values of the same control point.

The difference between the measured and estimated values of the map easting of the control point is the residual value for the map easting. This residual value is expressed in the same units as the map eastings, for example, in kilometres. A similar operation can be carried out for the map northings and the image line and pixel coordinates.

Users of some commercial software are encouraged to remove control points that are identified as 'erroneous' by the fact that their associated residual error is considered to be unacceptably high (sometimes 'high' is defined as a value more than one standard deviation distant from the mean). This may seem to be sensible step but, as Morad *et al.* (1996) show, this is not necessarily the case. Instead of eliminating control points one should seek reasons for the error, which may be the result of erroneous or inaccurate digitisation or measurement or the consequence of image distortions such as those induced by high relief. Bolstad *et al.* (1990) discuss positional uncertainty in digitising map data. Since remotely-sensed images are likely to be co-registered with maps in a Geographical Information System (GIS), it is even more important that the geometric quality of products derived from remotely-sensed image matches the requirements of the project. Overlay operations within a GIS are prone to error if mis-registration is present. Users should, therefore, take considerable care in the digitising and measurement of control point locations because locational errors are magnified by any departures from randomness of the spatial pattern of the control points, and the net result may well be unacceptable.

When using an empirical method such as polynomial least squares to carry out the geometric correction procedure, the user should always be aware of the fact that these residuals are the only measure of goodness of fit that are available. A perfect fit of the polynomial function at the control points tells us nothing about the goodness of fit in the remainder of the area. If a control point (whose location has been checked carefully) is associated with a high residual value is deleted, then the goodness of fit of the least-squares polynomial surface in the neighbourhood of that deleted point is likely to be low. Best practice is to keep some control points in reserve, and use them as check points to measure the distortion in regions of the image that are not close to those control points used to calculate the polynomial function, a procedure that is outlined below in a little more detail. You will find that the residual error values at check points will tend to fall as the order of the polynomial increases from one to two, and possibly three. Once the order of the polynomial increases beyond three it is probable that the error at check points will increase, simply because the two-dimensional surface that is represented by the polynomial function becomes increasingly flexible. Since the map and image are only being co-located at the control points, a lot of flexibility is available for distortions of the least-squares surface between the control point locations. Also, the variance of the residuals (used as an overall measure of goodness of fit) will decrease as polynomial order increases, until it becomes zero when the number of coefficients in the polynomial equals the number of control points. This does not imply a 'good fit'; rather, it indicates that the number of ground control points is too small (Example 4.3).

EXAMPLE 4.3: GROUND CONTROL POINT LOCATION USING CORRELATION METHODS

The aim of this example is to demonstrate the feasibility of automatic methods of locating ground control points (GCP) using correlation-based methods. Assume that we have a chip of pixels surrounding a known GCP in a master image, and we wish to find the coordinates of the equivalent GCP on a slave image. The master image may, for example, have been collected in 1999 and the slave image in 2003, and we do not wish to repeat the collection of GCP using manual methods. In this example, for illustrative purposes, we use the same image as master and slave.

Start MIPS and display in greyscale mode just one band of the San Joaquin image. Repeat this operation so that you have two identical images on display, the left-hand image being the master image and the right-hand one the slave, as shown in Example 4.3 Figure 1. Go to *Utilities|Correlate Two Images*. Identify the left and right images. Now you are asked to locate a GCP (normally the coordinates of the master GCP would be read from a file, but it is easier to illustrate the method using the mouse cursor). Before locating the GCP on the left (master) image, you could use *View|Change Image Scale* to zoom both left and right image by an integer factor, such as four. Once you have located the GCP on the left image you can set the chip size. The example below uses a chip size of 7 × 7 pixels. The coordinates of the point in the right image for which the correlation is highest is listed in an information window, and you are asked whether you want to carry on. Respond with NO and you will see yellow boxes on the left and right images surrounding the positions of the left (master) GCP and the candidate GCP on the right image. The log file for this example contains the following information:

```
Best correlation is 1.000
Coordinates on left-hand image are: 293 313
Coordinates on right-hand image are: 293 313
```

Not surprisingly, the correlation is exact and the right image GCP has exactly the same coordinates as the left image

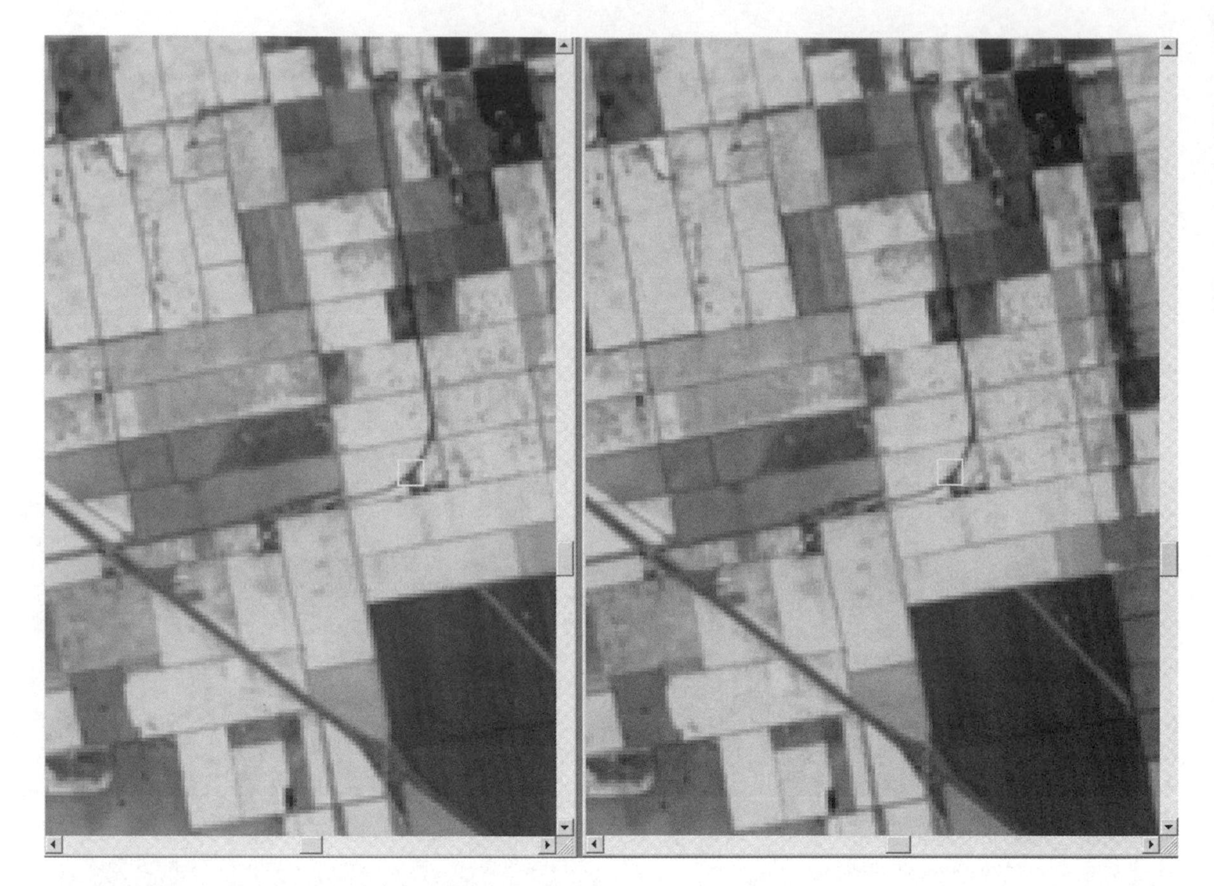

Example 4.3 Figure 1 The left- and right-hand images are zoomed by an integer scale factor of 4. Both are band 1 of the Landsat TM San Joaquin image that is included on the CD. The white boxes surrounding the master (left image) and slave GCP can be seen to the south-west of the centre of the window.

GCP that you selected using the mouse. The two images (left and right) are shown in Example 4.3 Figure 1.

Test the procedure by using another band of the San Joaquin image as the right image. Choose a number of different types of GCP, and then list what you think are good and bad GCP (for example, rectangular road junctions or field boundaries may 'work', whereas the use of random points in the centres of featureless fields may be less successful. You may also like to investigate the effects of chip size and shape. Is artificial intelligence as reliable as human experience?

We should also pay attention to the factors to be taken into consideration when control points are selected. If there is a substantial difference in time between the date of imaging and the last full map revision then it would be unwise to select as control points any feature subject to change, for example, a point on a river meander, at the edge or corner of a forest plantation or on a coastline. In tidal areas, the land/sea boundary may go back and forth – sometimes for several hundred metres – and it is thus difficult to correlate a point shown on a map as being on the coastline and the equivalent point on the image. The best control points are intersections of roads or airport runways, edges of dams, or isolated buildings and other permanent features. All are unlikely to change in position, beyond the few centimetres per century produced by continental drift, and all are capable of being located accurately both on map and image. At the 1:25 000 map scale or better, control points should be located with and error of 25 m or less, while a cursor can be used to fix the location on the image to within one resolution element. If a zoom function is used to enlarge the image, the exact position of a pair of linear features that intersect at an oblique angle can be estimated to within half a pixel (Figure 4.5).

It would be sensible to test the accuracy of the image-map and map-image coordinate transformations before using

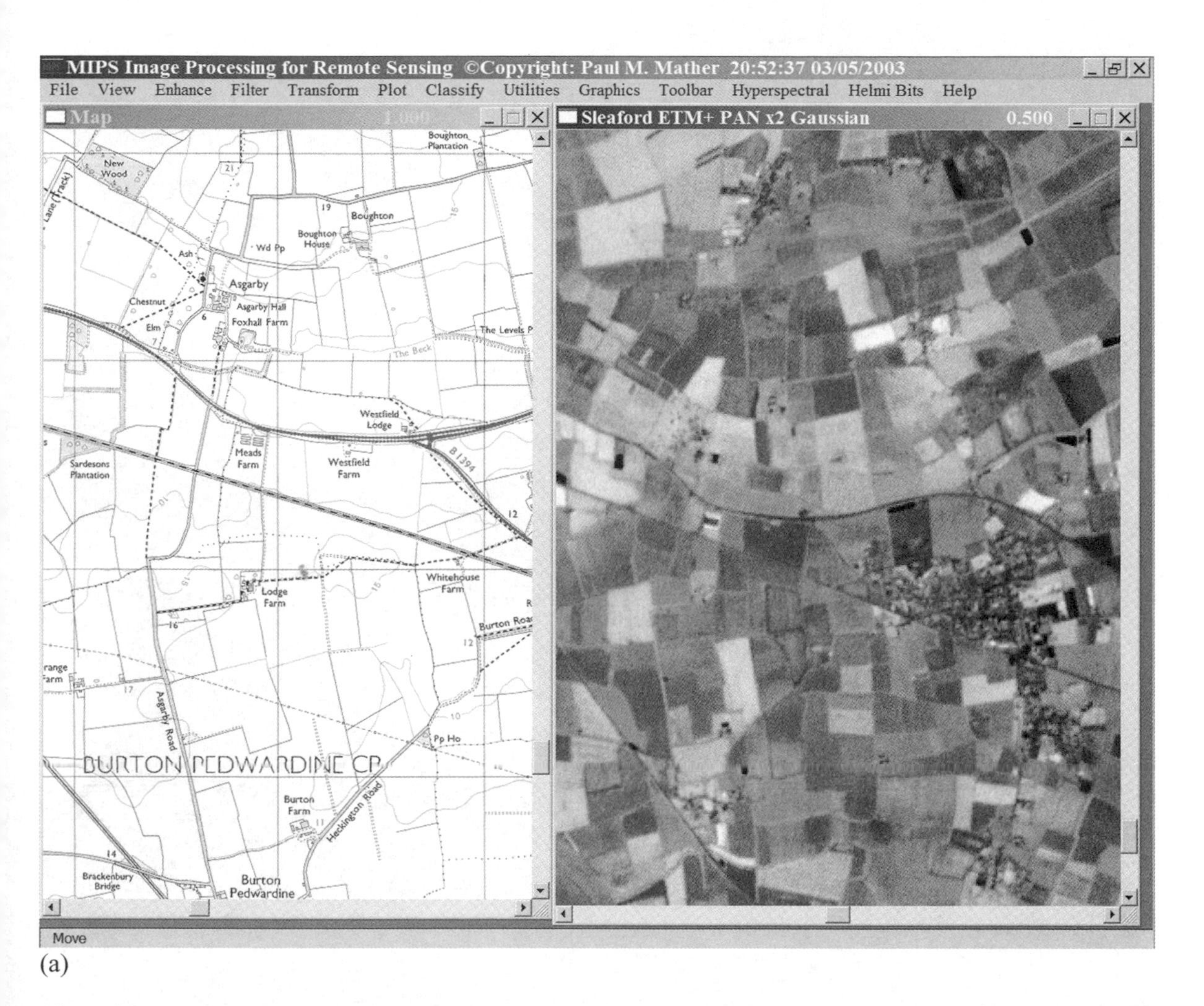

Figure 4.5 (a) Collection of ground control points from map and image using MIPS. A scanned 1:25 000-scale map of the image area is shown on the left side of the screen, with a zoomed Landsat ETM+ panchromatic image of the same area shown on the right. (b) Illustrating map-image registration. The left image is a zoomed RGB colour composite image made up of 30 m spatial resolution Landsat-7 ETM+ bands 4, 3 and 2. The right image is the corresponding 15 m resolution panchromatic band. The two images are being registered and resampled prior to 'pan-sharpening' (sections 6.5, 6.7.3). You can use this MIPS module to estimate the accuracy of your ability to locate ground control points accurately. Simply display the same image on the left and right side of the screen, as in Figure 4.2(b), and try to locate the same 6 or 7 ground control points on each image. Your log file will contain the coordinates, which should be within half a pixel (for an image zoom factor of 2) or 0.25 pixel (for an image zoom factor of 4). You may find difficulty in achieving these targets. Map reproduced by permission of Ordnance Survey on behalf of the Controller of Her Majesty's Stationery Office. © Crown copyright 100041906.

them to convert from one system to the other. It seems illogical to generate the least-squares coefficients a_i using a set of (x, y, c, r) control point coordinates and then use the same data to test the adequacy of the calculation. An independent assessment is necessary. As suggested earlier in this discussion, a subset of control points ('test points') that are not used in the least-squares estimation process should be used to assess the goodness of fit of the least-squares transformation by taking each of the test points in turn and converting its c and r image coordinates to map x and y coordinates, then calculating the residual values. The same should be done for the reverse transformation, using the least-squares coefficients listed in the MIPS log file. The standard of the map coordinate residuals and of the image coordinate residuals will give a measure of the goodness of fit. If the residual values are normally distributed, which can

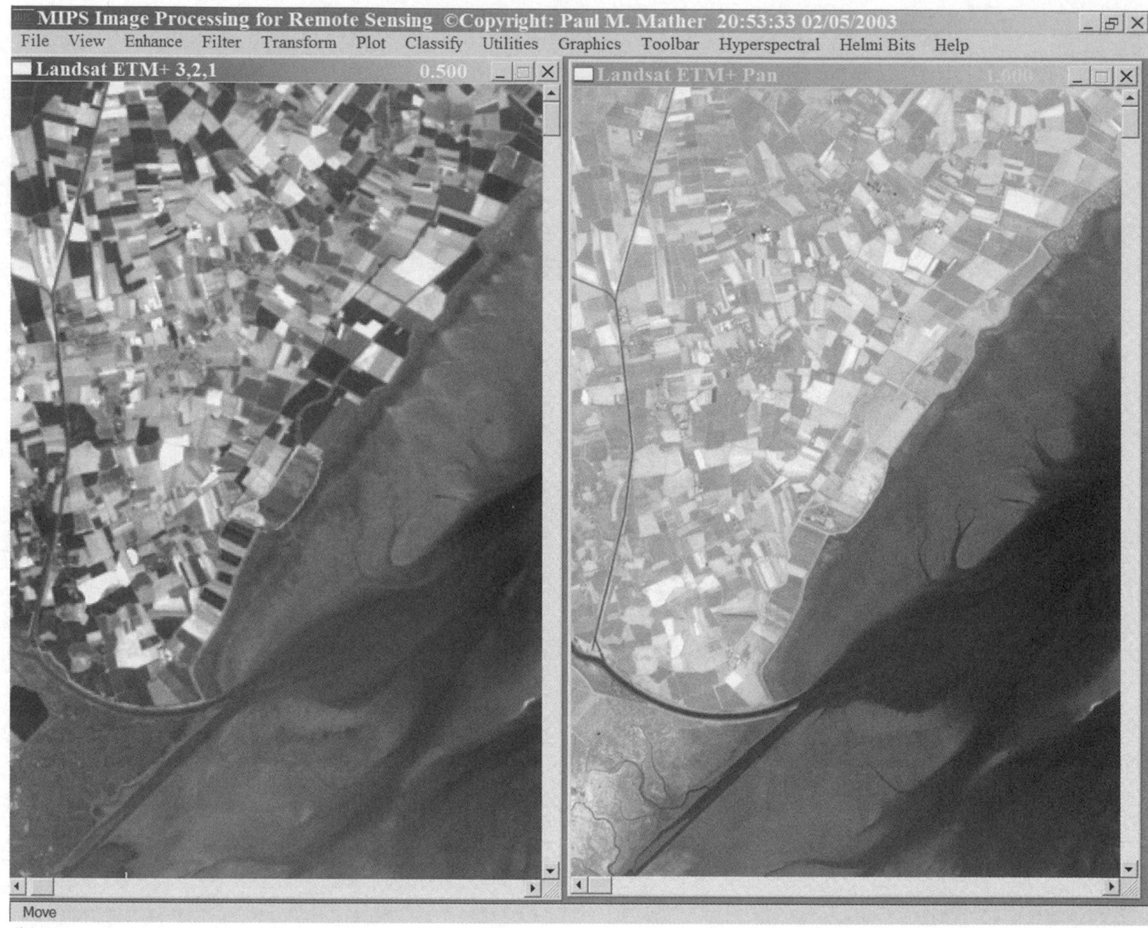

(b)

Figure 4.5 (*Cont.*)

be checked by producing a histogram of these values, then 68% of all calculated values should lie within one standard deviation of the mean. The mean residual is zero. If a root-mean-square error or standard deviation of 0.5 pixel or 10 m, for example, is cited then one should be careful not to interpret this statement to mean that *all* of the coordinate transformation results fall within the quoted range.

In some instances, control points may already be available for the required map area and for an earlier image of the region of interest. Rather than go through the tedious procedure of collecting control point information again, the map coordinates of the control points could be re-used without difficulty. However, due to fluctuations in the orbit of the satellite, it is unlikely that the control point positions on the second image would be the same as those on the first. Variation in satellite altitude and are quite small, so (a) the pixels in two separate images of the same area collected by the same satellite-borne instrument should have nearly equal sizes, and (b) the linear features whose intersection defined the control point should have approximately the same orientation from image to image of the same area. Benny (1981) made up a set of 'chips' or extracts of the first image of size 19 × 19 pixels, each chip containing a control point. The value of 19 v 19 is not a fixed, nor even a recommended one; much depends on the nature of the image, especially the contrast and the amount of detail. Larger chips are required for images lacking contrast. This procedure is illustrated in Example 4.3.

The map coordinates of the centre of the chip are known. The problem is to find the coordinates of the centre of the 19 × 19 chip on the new image. This problem is solved by a correlation procedure. Given that it is possible to estimate the position of the control point on the new image, and maximum deviations from this position in the column and row directions, then a rectangular search area can be defined. The control point is thought to be somewhere inside this search area. The 'chip' is placed over the top left 19 × 19 pixels of this search area and the correlation between the chip pixels and the pixels of the image that underlie the chip pixels is calculated and recorded, together with the row and column coordinates of the image pixel that lies below the centre pixel of the chip (Figure 4.6). The chip is then moved one column to the right and the correlation coefficient is recalculated. Once the chip has reached its furthest right position it is moved down one row and back to its leftmost position. The process continues until the chip has moved to the bottom right-hand corner of the search area.

At this stage, we have (i) the row coordinate of the chip centre, (ii) the corresponding column coordinate, and (iii) the correlation value for that position. Values (i) and (ii) could be used as conventional *xy* coordinates and isolines of correlation could be drawn on this 'map'. The most likely match would be given by the point which had the highest correlation. If the maximum correlation is low (less than about +0.4) then it might be concluded that no match has been found. Benny (1981) describes a 'spiral search' algorithm that differs from the regular search algorithm outlined above. If the regular search algorithm were coded efficiently then it would be as fast as the spiral technique.

An efficient method of conducting the regular search would be to compute the terms x_i, y_i, x_i^2, y_i^2 and $x_i y_i$ for the first (top left) chip position. The x_i are the chip pixels and the y_i are the pixels in the search area lying below the chip pixels. The label i is counted sequentially across the rows, so that for a 19 × 19 chip the label 1 refers to the first pixel on row 1, label 20 is the first pixel on row 2 and so on. The correlation coefficient r_{xy} is then computed from:

$$r_{xy} = \left(N\sum x_i y_i - \sum x_i \sum y_i\right) \Big/ \left(\left(N\sum x_i^2 - \left(\sum x_i\right)^2\right)\left(N\sum y_i^2 - \left(\sum y_i\right)^2\right)\right)^{0.5}$$

This 'short-cut' equation is not normally recommended for computer implementation. However, its use is justified on the grounds that the numbers of pixels involved at each evaluation is small (19 × 19) and so rounding errors should be negligible. The term $(N\Sigma x_i^2 - (\Sigma x_i)^2)$ needs to be computed only once for each chip as the chip pixel values are fixed – they do not change as the chip is moved over the search area. Reference to Figure 4.6 will show that as the chip is moved from position 1 to position 2 on line 1 (i.e., moves one pixel to the right) then a column of 19 image pixels enters the region covered by the chip and a column of 19 image pixels leaves that region. If we compute all the terms in the expression for r_{xy} that involve x for posi-

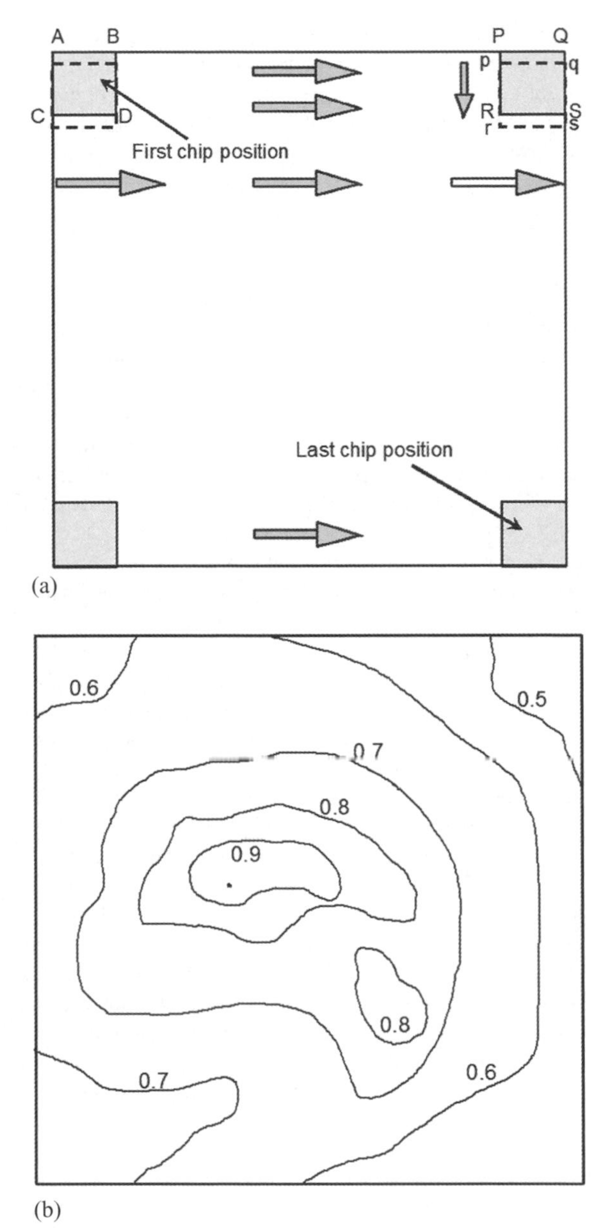

Figure 4.6 (a) Chip ABCD is covering pixels in the top left of the image. The correlation between the chip pixels and the image pixels is calculated, and the chip moves to the right by one pixel, and the procedure is repeated. The chip moves right until its right margin is coincident with the right border of the image at PQRS. The chip is moved down one line and back to the left side of the image, and the procedure is repeated. The arrows show the direction of movement of the chip. (b) Isoline map of correlations between image pixels and chip pixels, using the procedure shown in Figure 4.6(a). The star marks the point of maximum correlation.

tion 1 then we can update them by subtracting the values relating to the pixels leaving the region and by adding the values relating to the pixels entering the region. For example, we would update the calculations for the term x_i, y_i as follows:

$$x_i y_i = x_i y_i - x_{\text{out}} y_{\text{out}} + x_{\text{in}} y_{\text{in}}$$

The number of summations is reduced from $19 \times 19 = 361$ to only $(19 + 19) = 38$ for each position of the chip. If the values of the terms involving x for the start position of each horizontal sweep across the search area are 'remembered' then the corresponding values for the start of the next horizontal sweep can be computed by a similar method, this time involving the top row going out and a new bottom row coming in. Since all the calculations in the derivation of r_{xy} except the final division can be performed exactly in integer arithmetic the method is both fast and accurate. This 'moving window' technique is used in Chapter 7 when spatial filtering techniques are considered.

All the control points could be located by this correlation-based method, which requires the user to provide an estimate of the position of the control point and of the dimensions of the search area for each control point. Benny (1981) goes on to consider a method for the automatic relocation of all control points without user intervention, provided one control point can be located accurately. He estimates that, for a typical image containing 100 control points, the visual location of the control points would take about 80 hours of effort, plus 1 minute of computer time to carry out the coordinate transformation method described below. For the semi-automatic method, with the user supplying information about the approximate location of each control point, the user man-hours drop from 80 to 8, but computer time goes up to 20 minutes. Finally, the automatic procedure, in which the user supplies information relating to only one control point, the number of man-hours required falls to 0.1 (6 minutes) while the computer time requirement remains at 20 minutes. Orti *et al.* (1979) also provide a discussion of automatic control point location. Automatic identification of ground control points is examined in more detail by Ackermann (1984) and Motrena and Rebordão (1998). Automatic registration of stereo pairs of images, as a prelude to the derivation of digital elevation models, is described by Al-Rousan *et al.* (1997, 1998).

An example of the use of correlation-based methods in identifying GCP is provided as Example 4.3.

Procedures for estimating the coefficients a_{ij} in the least-squares functions (i) to (iv) above relating map and image coordinate systems are now considered. The following description assumes that we wish to estimate the map easting

EXAMPLE 4.4: POLYNOMIAL FUNCTIONS

It is not easy to visualise a polynomial function so this example provides an aid to learning in the form of an Excel worksheet (Example 4.4 Figure 1) that can be used to experiment with polynomial curve fitting in one dimension. The exercise involves fitting a polynomial function to a set of samples drawn from a sine wave. Column A of the worksheet is the sample identification number, starting from 1, while the data series (column B of the worksheet) is a simple sine wave, which completes one full oscillation every 360°. The sine wave is calculated as $b1 = a1^*$ sin *(pi()/20))*. You can plot column A as the x-axis and column B as the y-axis by selecting both columns, clicking on the Chart Wizard, selecting *'XY* Scatter', then the graph type 'scatter with data points connected by smooth line without markers'.

Now click *Chart* on the main menu bar and select *Add Trendline*. Choose the *Polynomial* option and then select the order of the polynomial (the maximum is 6). Note that the trend line type *Linear* is a first-order polynomial curve (i.e., a straight line). Example 4.2 Figure 2 shows the sine wave data with a third-order polynomial trend line superimposed. Example 4.2 Figure 3 plots the same sine wave data together with several trend lines for comparative purposes. The first-order polynomial is a straight line (no bends). The second-order polynomial has one bend, and the fifth-order function has three bends (in fact the fifth-order function could have a maximum of four bends, if the data were sufficiently scattered). The conclusion to be drawn from this experiment is that the polynomial curve becomes more flexible as the order of the polynomial function increases. In fact, a polynomial of sufficiently high order can approximate any single-valued function such as a sine wave. The bad news is that polynomial curves cannot distinguish between 'clean' data, such as the sine wave in Example 4.4 Figures 2 and 3 and data that are contaminated by error.

Consider Example 4.4 Figure 4. The dots are data points. They were generated by adding random noise to a straight line sloping up to the right, as shown in the highlighted column (C) in Example 4.4 Figure 1. The simplest polynomial – of order 1 – has no bends, and is a realistic approximation of the 'true' pattern of data points. In contrast, the sixth-order polynomial develops a number of wiggles that are responses to the random noise within the data. We could say that the sixth-order curve is 'fitting the noise in the data'.

The choice of polynomial order is thus crucial to the success of the modelling exercise. If the polynomial order is too low (such as the first-order curve in Example 4.4 Figure 3) then the model is an over-simplified representation of the true data series. If the order of the polynomial is too high, then any errors present in the data begin to influence the model.

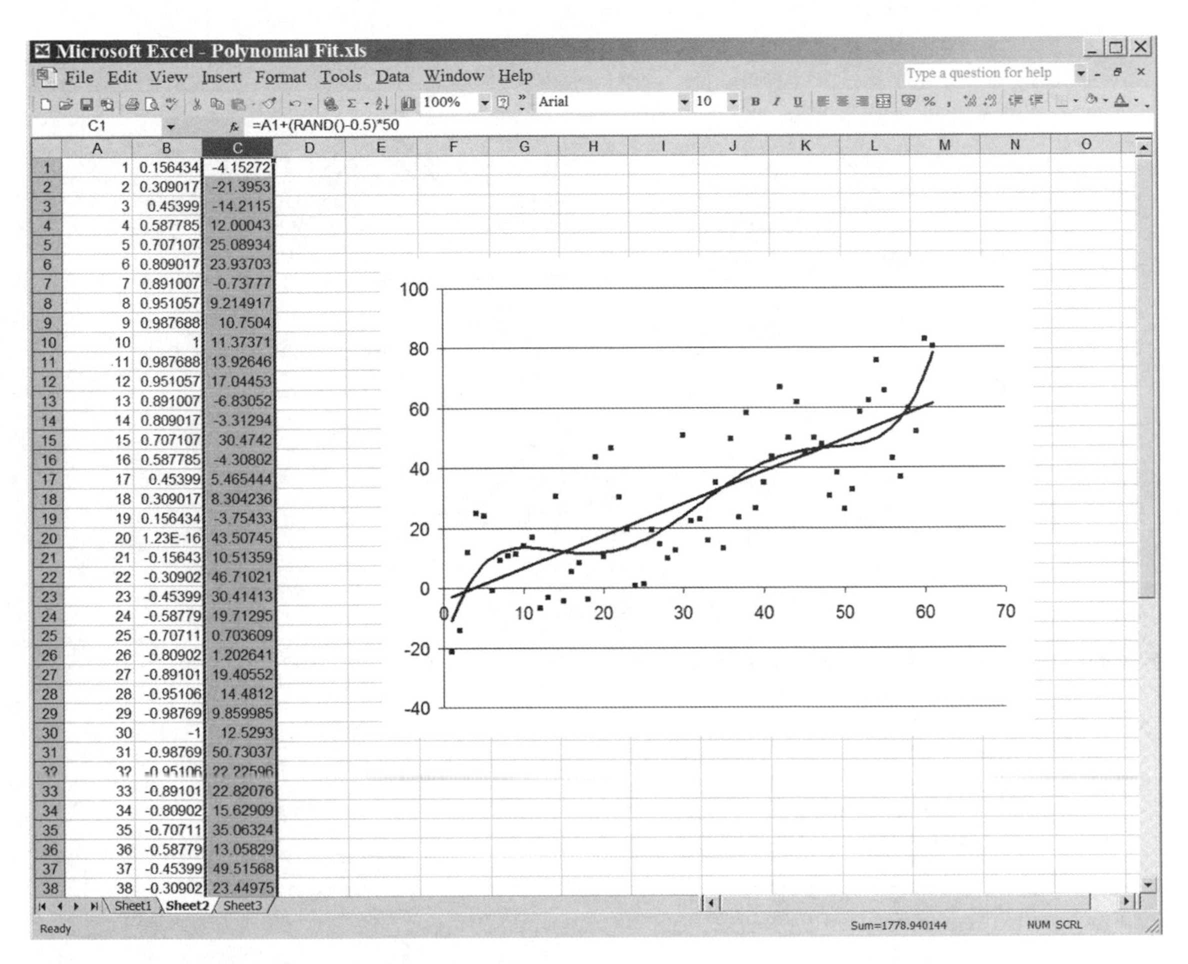

	A	B	C
1	1	0.156434	-4.15272
2	2	0.309017	-21.3953
3	3	0.45399	-14.2115
4	4	0.587785	12.00043
5	5	0.707107	25.08934
6	6	0.809017	23.93703
7	7	0.891007	-0.73777
8	8	0.951057	9.214917
9	9	0.987688	10.7504
10	10	1	11.37371
11	11	0.987688	13.92646
12	12	0.951057	17.04453
13	13	0.891007	-6.83052
14	14	0.809017	-3.31294
15	15	0.707107	30.4742
16	16	0.587785	-4.30802
17	17	0.45399	5.465444
18	18	0.309017	8.304236
19	19	0.156434	-3.75433
20	20	1.23E-16	43.50745
21	21	-0.15643	10.51359
22	22	-0.30902	46.71021
23	23	-0.45399	30.41413
24	24	-0.58779	19.71295
25	25	-0.70711	0.703609
26	26	-0.80902	1.202641
27	27	-0.89101	19.40552
28	28	-0.95106	14.4812
29	29	-0.98769	9.859985
30	30	-1	12.5293
31	31	-0.98769	50.73037
32	32	-0.95106	22.22596
33	33	-0.89101	22.82076
34	34	-0.80902	15.62909
35	35	-0.70711	35.06324
36	36	-0.58779	13.05829
37	37	-0.45399	49.51568
38	38	-0.30902	23.44975

Example 4.4 Figure 1 Spreadsheet used in Example 4.4.

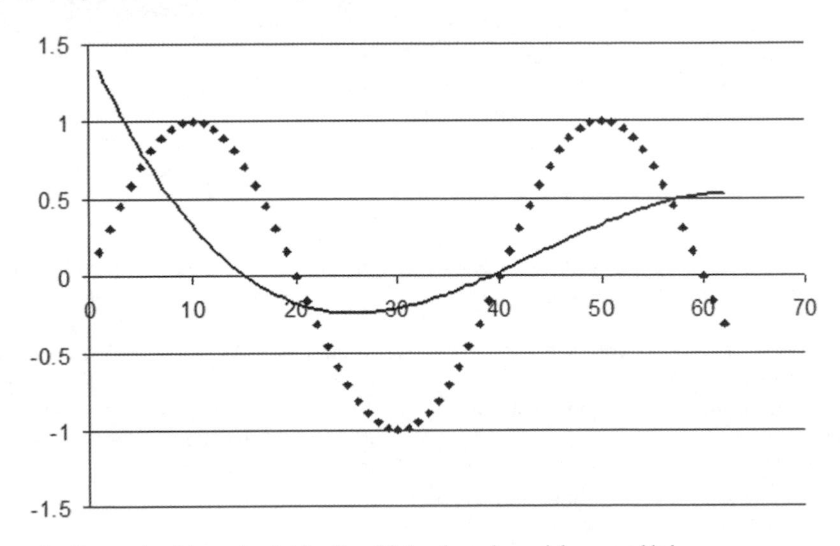

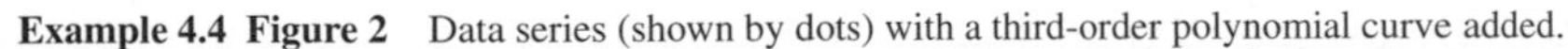

Example 4.4 Figure 2 Data series (shown by dots) with a third-order polynomial curve added.

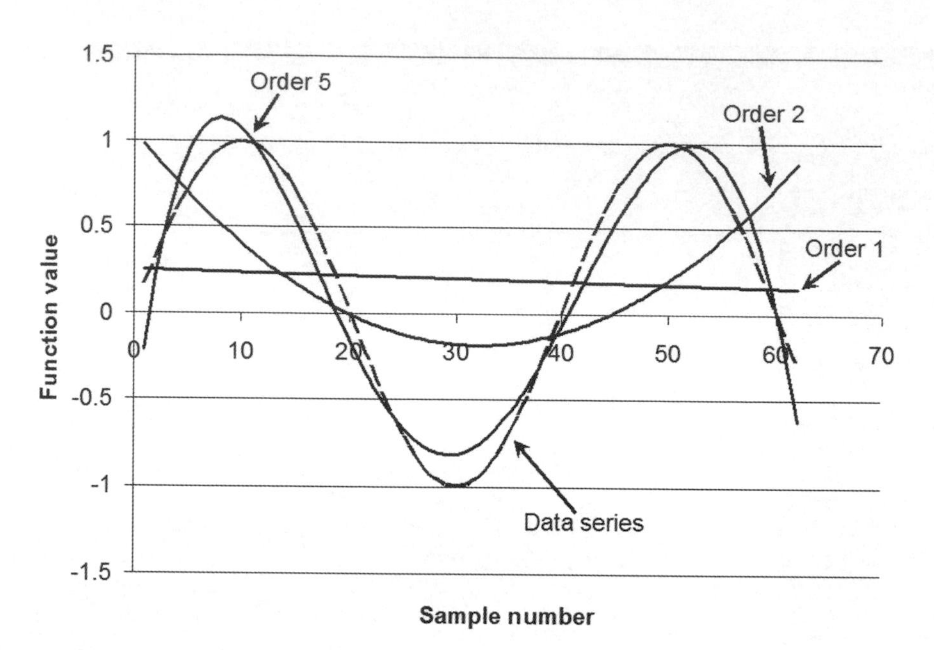

Example 4.4 Figure 3 Sine wave data with polynomial trend lines of order 1, 2 and 5.

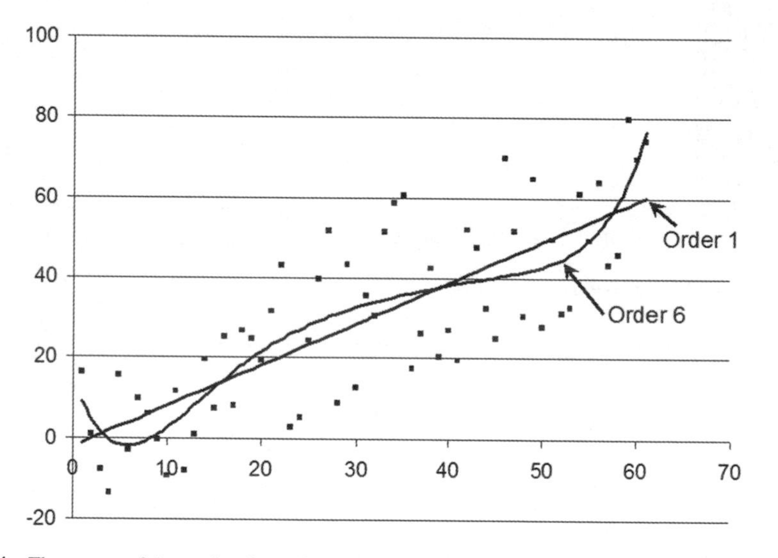

Example 4.4 Figure 4 The scatter of data points has a linear trend, sloping upwards to the right. The first-order polynomial (straight line) fits this general pattern, but the sixth-order polynomial curve is affected by random errors in the data.

e from the image column and row coordinates r and c for a set of n control points. In practice, all four functions relating map (e, n) and image (r, c) coordinates would be computed (e from r and c, n from r and c, r from e and n, and c from e and n). The set of control point map easting coordinates is denoted by the vector **e**, while the powers and cross products of the c and r values are considered to form the matrix **P**. The coefficients a_{ij} are the elements of the vector **a**. For a second-order fit we would have the system shown in Table 4.2. The method of least squares is used to find the vector of estimates **e** according to the following model:

$$\mathbf{e} = \mathbf{Pa}$$

Table 4.2 Matrix **P** and vectors **e** and **a** required in solution of second-order least-squares estimation procedure In this example the map easting vector **e** is to be estimated from the powers and cross-products of the image column (**c**) and row (**r**) vectors which form the matrix **P**. The measurements of **c**, **r** and **e** are measured at n ground control points

$$\mathbf{e} = \begin{bmatrix} e_1 \\ e_2 \\ e_3 \\ e_4 \\ \cdot \\ \cdot \\ \cdot \\ \cdot \\ e_n \end{bmatrix} \quad \mathbf{P} = \begin{bmatrix} 1 & c_1 & r_1 & c_1^2 & c_1 r_1 & r_1^2 \\ 1 & c_2 & r_2 & c_2^2 & c_2 r_2 & r_2^2 \\ 1 & c_3 & r_3 & c_3^2 & c_3 r_3 & r_3^2 \\ 1 & c_4 & r_4 & c_4^2 & c_4 r_4 & r_4^2 \\ \cdot & \cdot & \cdot & \cdot & \cdot & \cdot \\ \cdot & \cdot & \cdot & \cdot & \cdot & \cdot \\ \cdot & \cdot & \cdot & \cdot & \cdot & \cdot \\ 1 & c_n & r_n & c_n^2 & c_n r_n & r_n^2 \end{bmatrix} \quad \mathbf{a} = \begin{bmatrix} a_{00} \\ a_{10} \\ a_{01} \\ a_{20} \\ a_{11} \\ a_{02} \end{bmatrix}$$

P and **e** are explained above, while **a** is a vector of unknown coefficients which are to be estimated from the GCP data. The least-squares formula for the evaluation of **a** is:

$$\mathbf{a} = (\mathbf{P'P})^{-1}\mathbf{P'e}$$

The elements of vector **a** can be found by a standard subroutine for solving linear simultaneous equations. Such routines work efficiently with well-conditioned equations but can produce significant errors if the equations are ill conditioned. The condition of a matrix such as **P′P** in the expression above can be considered to be a measure of the degree of independence of its columns. Ill conditioning is indicated by a large determinant, or by a high value of the ratio of the largest and smallest eigenvalues of **P′P**. The degree of independence of a pair of column vectors can be visualised by analogy with the crossing point of two lines. The intersection can be measured accurately if the lines are perpendicular, whereas if the lines cross at a very acute angle the exact point of intersection is difficult to specify (Figure 4.7). In a similar way, the solutions of a set of well-conditioned equations (with the columns of the matrix **P′P** being independent, or nearly so) can be found accurately but the solution vector could be substantially in error if the matrix is badly conditioned. Various studies have indicated that the accuracy of standard procedures (using matrix inversion techniques) for the solution of the least-squares equations depends critically on the condition of the matrix. The Gram–Schmidt procedure described by Mather (1976) appears to be a more reliable algorithm, and this conclusion is confirmed by later work (Mather, 1995).

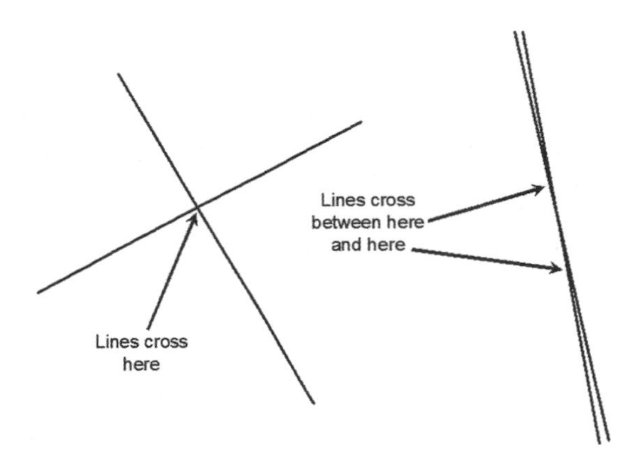

Figure 4.7 The 'condition' of a matrix in least-squares calculations can be likened to the sharpness of definition of the crossing point of two straight lines. Perpendicular lines have a sharply defined crossing point (left), while the crossing point of two near-coincident lines cannot be well defined (right).

It is likely that the matrix **P′P** will be ill-conditioned in our particular case because its columns are the powers and cross-products of two variables, c and r, e.g., c, r, c^2, cr, r^2, c^3, c^2r, cr^2 and r^3 in the case of a cubic bivariate polynomial. Hence, standard methods may well produce poor or even misleading results. These will be worsened if the spatial distribution of control points is linear or clustered or if the number of ground control points is small, as noted above. The Gram–Schmidt method is thus to be preferred, and is implemented in the MIPS software. The application of this method provides estimates of the elements of the coefficients vector **a**, which can then be used to find the map easting coordinate **e**. The coefficients of the other required functions (n estimated from c and r, c estimated from e and n, and r estimated from e and n) can be found in a similar fashion.

Note that the procedures described in this section are applicable only to images obtained from relatively stable platforms such as satellites and for areas with a low relative relief. Scanned images from aircraft contain distortions caused by rapid variations in the aircraft's attitude as measured by pitch, roll and yaw. Such imagery will contain defects such as non-parallel scan lines, which cannot easily be corrected by polynomial least-squares methods. The influence of terrain on the results obtained by the use of the methods described above are summarised earlier in this section. The correction of aircraft imagery, such as the lidar data described in Chapter 9, uses a combination of GPS and inertial navigation systems (INS) to model the position and attitude of the aircraft.

4.3.3 Resampling procedures

Once the four transformation equations relating image and map coordinate systems are known, and the results tested, the next step is to find the location on the map of the four corners of the image area to be corrected, and to work out the number of and spacing (in metres) between the pixel centres necessary to achieve the correct map scale. We can now work systematically across the map area, starting at the top left, and locate (in map coordinates e and n) the centre of each pixel in turn. Given the (e, n) location coordinates of a pixel centre on the map we can apply the transformations described in the preceding section to generate (c, r) image coordinates corresponding to the position of the pixel's centre. These (c, r) coordinates are the column and row position in the uncorrected image of the new (geometrically corrected) pixel centres (Figures 4.4 and 4.8(a)).

It is unlikely that c and r are integers; if they were, it would be possible to take the pixel value at (c, r) (pixel centre position in the uncorrected image, shown by a circle in Figure 4.4(a)) and transfer it to the corresponding pixel centre in the corrected image, shown by a plus sign in Figure 4.8(a). Non-integral values of c and r imply that the corrected pixel centre lies between the columns and rows of the uncorrected image, so that a method of interpolation is needed to estimate the pixel value at (c, r). Figure 4.8(b) illustrates this point. The coordinates of the pixel at the point (e', n') in the corrected image are transformed using the least-squares polynomial function computed from the ground control points to point (r', c') in the uncorrected image. If r' and c' were integers then the pixel values in the corrected and uncorrected image would be the same. In most cases, r' and c' are non-integral values and lie somewhere between the pixel centres in the uncorrected image.

Three methods of resampling are in common use. The first is simple – take the value of the pixel in the raw (untransformed) image that is closest to the computed (c, r) coordinates. This is called the *nearest neighbour* method. It has two advantages; it is fast and its use ensures that the pixel values in the output image are 'real' in that they are copied directly from the raw image. They are not 'fabricated' by an interpolation algorithm such as the two that are described next. On the other hand, the nearest neighbour method of interpolation tends to produce a rather blocky effect as some pixel values are repeated.

The second method of resampling is *bilinear interpolation* (Figure 4.9). This method assumed that a surface fitted to the pixel values in the immediate neighbourhood of (c, r) will be planar, like a roof tile. The four pixel centres nearest to (c, r) (i.e., points P1–P4 in Figure 4.9) lie at the corners of this tile; call their values v_{ij}. The interpolated value V at (c, r) is obtained from:

$$V = (1 - a)(1 - b)v_{i,j} + a(1 - b)v_{i,j+1} + b(1 - a)v_{i+1,j} + abv_{i+1,j+1}$$

where

$$a = c - j \qquad j = |c|$$
$$b = r - i \qquad i = |r| \text{ and}$$

$|x|$ is the absolute value of the argument x.

Note that the method breaks down if the point (c, r) is coincident with any of the four pixel centres in the uncorrected image and so a test for this should be included in any computer program. Bilinear interpolation results in a smoother output image because it essentially is an averaging process. Thus, sharp boundaries in the input image may be blurred in the output image. The computational time requirements are greater than those of the nearest neighbour method.

The third spatial interpolation technique that is in common use for estimating pixel values in the corrected image is called *bicubic* because it is based on the fitting of two third-degree polynomials to the region surrounding the point (c, r). The 16 nearest pixel values in the uncorrected image are used to estimate the value at (c, r) on the output image. This technique is more complicated than either the nearest neighbour or the bilinear methods discussed above, but it tends to give a more natural-looking image without the blockiness of the nearest neighbour or the over-smoothing of the bilinear method, though – as is the case with all interpolation methods – some loss of high-frequency information is involved. The interpolator is essentially a low-pass filter (Chapter 7). The penalty to be paid for these improvements is considerably increased computing requirements.

The cubic convolution method is described by Wolf and DeWitt (2000) as follows: first, take the 4×4 pixel area surrounding the point whose value is to be interpolated. Next, compute two vectors each with four elements. The first contains the absolute differences between the row values (in the full image) and the row value of the point to be interpolated. The second contains the absolute differences for the column values. For example, the 4×4 area may consist of rows 100, 101, 102, and 103 of the full image, with the column values being 409, 410, 411, and 412. The 4×4 set of pixel values forms a matrix $\mathbf{M}$. If the point $\mathbf{p}$ to be interpolated has row/column coordinates of (101.1, 410.5) then the row difference vector is $\mathbf{r} = [1.1, 0.1, 0.9, 1.9]$ and the column difference vector is $\mathbf{c} = [1.5, 0.5, 0.5,$ and $1.5]$.

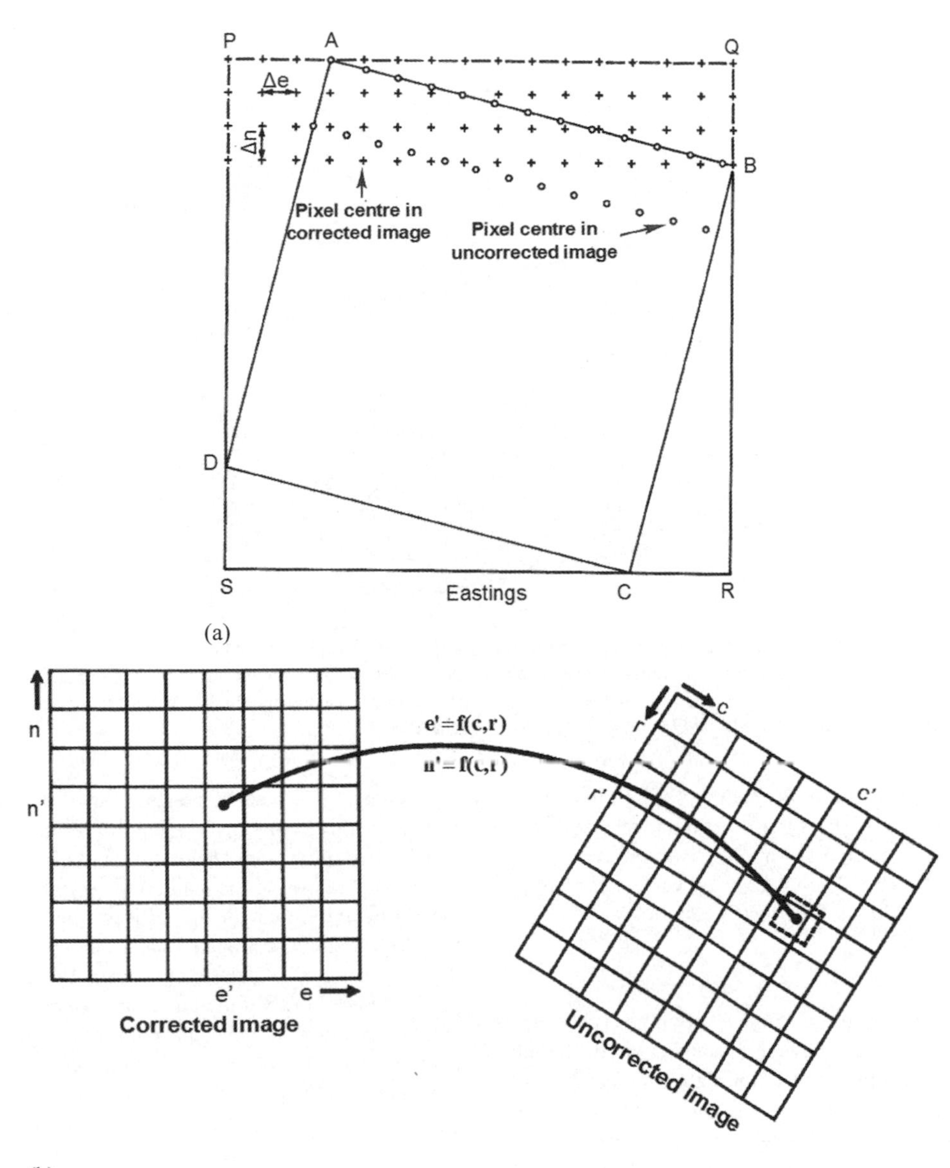

(b)

Figure 4.8 (a) Schematic representation of the resampling process. The extreme east, west, north and south map coordinates of the uncorrected image ABCD define a bounding rectangle PQRS, which is north-orientated with rows running east-west. The elements of the rows are the pixels of the geometrically corrected image. Their centres are shown by the + symbol and the spacing between successive rows and columns is indicated by Δe and Δn. The centres of the pixels of the corrected image are marked by the symbol o. See text for discussion. (b). The least-squares bivariate polynomial functions take the coordinates at point (c' r') in the uncorrected image and find the coordinates of the corresponding point (e' n') in the corrected image.

Next, a row interpolation vector and a column interpolation vector are derived from **r** and **c**. Let v represent any element of **r** or **c**. If the value of v is less than 1.0 then use function f_1 to compute the interpolation value. Otherwise, use function f_2, where f_1 and f_2 are defined as follows:

$$f_1(v) = (a+2)v^3 - (a+3)v^2 + 1$$
$$f_2(v) = av^3 - 5av^2 + 8av - 4a$$

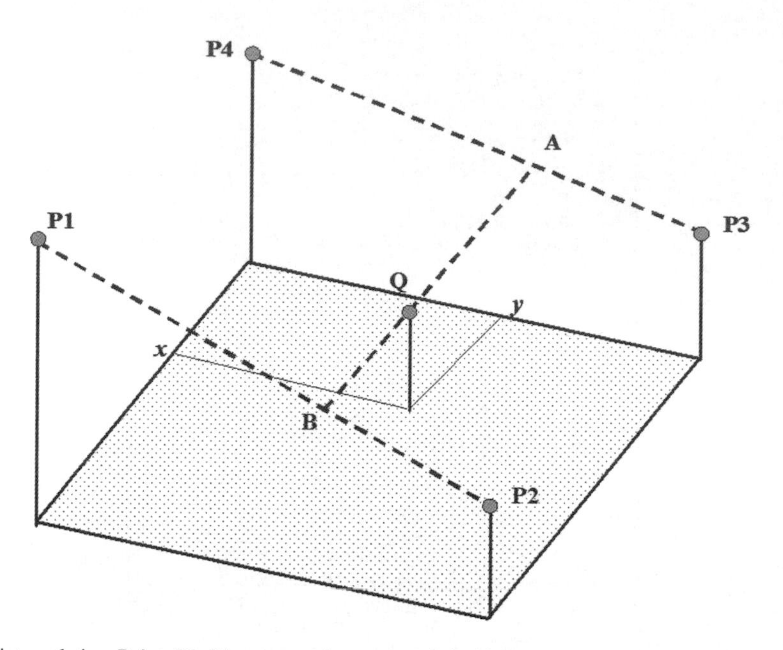

Figure 4.9 Bilinear interpolation. Points P1–P4 represent the centres of pixel in the uncorrected image. The height of the 'pin' at each of these points is proportional to the pixel value. The pixel centre in the geometrically corrected image is computed as (x, y) (point Q). The interpolation is performed in three stages. First, the value at A is interpolated along the line P4–P3, then the value at B is interpolated along the line P1–P2. Finally, the value at Q is interpolated along the line AB.

The value of the term a is set by the user, normally in the range –0.5 to –1.0. At this point the elements of **r** and **c** have been converted to row and column interpolation values, respectively. The interpolated value z at point **p** is given by the solution of the matrix equation:

$$z = \mathbf{r}'\mathbf{M}\mathbf{c}$$

The value of a in the row/column interpolation procedure is rarely mentioned, but it has a significant impact on the value returned by the cubic convolution algorithm. For example, using the same set of data as that presented by Wolf and Dewitt (2000, p. 564), different values of z can be obtained simply by changing the value of a by a small amount. Wolf and DeWitt's matrix **M** is:

$$\mathbf{M} = \begin{bmatrix} 58 & 54 & 65 & 65 \\ 63 & 62 & 68 & 58 \\ 51 & 56 & 59 & 53 \\ 52 & 45 & 50 & 49 \end{bmatrix}$$

for row numbers 618–621 and column numbers 492–495 of the full image. The row/column coordinates of point p are [619.71 493.39]. If the value of $a$is set to -0.5 then the interpolated value z at point p is 60.66, For values of a of -0.6, -0.75 and -1.0 the corresponding z values are 61.13, 61.87 and 63.15. Since the interpolated results are rounded to the nearest integer, small differences do not matter – in this example, the values 60.66 and 61.13 would be rounded to 61. Nevertheless, it would be rather odd of a professional user of the method to concern himself with the arcane algorithmic details of least-squares procedures in order to improve computational accuracy while ignoring the choice of $a$value in the cubic convolution interpolation.

The choice between the three interpolation methods of nearest neighbour, bilinear, and cubic convolution depends upon two factors - the use to which the corrected image is to be put, and the computer facilities available. If the image is to be subjected to classification procedures (Chapter 8) then the replacement of raw data values with artificial, interpolated values might well have some effect on the subsequent classification (although, if remote sensing data are to be used alone in the classification procedure, and not in conjunction with map-based data, then it would be more economical to perform the geometric correction after, rather than before, the classification). If the image is to be used solely for visual interpretation, for example in the updating of a topographic map, then the resolution requirements would dictate that the bicubic or cubic method be used. The value of the end product will, ultimately, decide whether the additional computational cost of the bicubic method is justified. Khan *et al.* (1995) consider the effects of resampling on the quality of the resulting image.

4.3.4 Image registration

Registration of images taken at different dates (multi-temporal images) can be accomplished by image correlation methods such as that described in section 4.3.2 for the location of ground control points. Although least-squares methods such as those used to translate from map to image coordinates, and vice-versa, could be used to define a relationship between the coordinate systems of two images, correlation-based methods are more frequently employed. A full account of what are termed Sequential Similarity Detection Algorithms (SSDA) is provided by Barnea and Silverman (1972) while Anuta (1970) describes the use of the Fast Fourier Transform (Chapter 6) in the rapid calculation of the inter-image correlations. Eghbali (1979) and Kaneko (1976) illustrate the use of image registration techniques applied to Landsat MSS images. Townshend *et al.* (1992) point out the problems that may arise if change detection techniques are based on multi-temporal image sets that are not properly registered. As noted earlier, any GIS overlay operation involving remotely-sensed images is prone to error because all such images must be registered to some common geographical reference frame. Misregistration provides the opportunity for error to enter the system.

4.3.5 Other geometric correction methods

The least-squares polynomial procedure described in section 4.3.2 is one of the most widely used methods for georeferencing medium-resolution images produced by sensors such as such as the Landsat ETM+, which has a nominal spatial resolution of around 30 m. The accuracy of the resulting geometrically corrected image depends, as we have already noted, on the number and spatial distribution of ground control points. Significant points to note are: (i) the map and image may not coincide in the sense of associating the same coordinates (to a specified number of significant digits) to a given point, even at the control points, because the least-squares polynomial produces a global approximation to the unknown correction function, and (ii) the method assumes that the area covered by the image to be georeferenced is flat. The effects of terrain relief can produce very considerable distortions in a geometric correction procedure based on empirical polynomial functions (Figure 4.10). The only effective way of dealing with relief effects is to use a mathematical model that takes into account the orbital parameters of the satellite, the properties of the map projection, and the nature of the relief of the ground surface. Where high-resolution images such as those produced by IKONOS and Quickbird are used, the question of accurate, relief-corrected georeferenced images becomes critical.

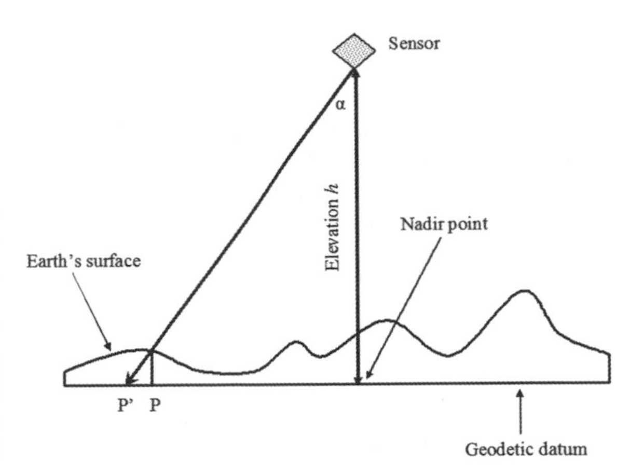

Figure 4.10 The least-squares polynomial method of geometric correction does not take terrain variations into account. The position of points on a map is given in terms of their location on a selected geodetic datum. At an off-nadir view angle of a, point P appears to be displaced to P'. The degree of displacement is proportional to the satellite altitude h and the view angle a.

The Space Imaging Company, which owns and operates the IKONOS system, does not release details of the orbital parameters of the satellite to users (though Quickbird orbital data are available). Instead, they provide the coefficients of a set of *rational polynomials* with their stereo products. The term rational means 'ratios of'. Dowman and Tuo (2000) provide a useful summary of the characteristics of this georeferencing procedure.

Geometric correction or geo-coding of SAR images is generally accomplished using orbital models. The process is complicated by the fact that SAR is a side-looking instrument, producing images in the 'slant range'. Geometric correction involves the conversion from slant range to ground range as well as scaling and north-orientation. The use of a digital elevation model in the geometric correction process allows the consideration of the effects of terrain slope and elevation. A SAR image that is corrected both for viewing geometry and terrain is said to be orthocorrected. Mohr and Madsen (2001) describe the geometric calibration of ERS SAR images. The use of SAR images to generate interferograms, which is discussed in Chapter 9, requires accurate geometric correction of the images that are used.

The derivation of Digital Elevation Models (DEM) from pairs of radar (SAR) images using interferometric methods is considered in detail in Chapter 9. It should, however, be noted here that DEM are routinely derived from stereo pairs of optical imagery produced by sensors such as SPOT HRV, ASTER, and IRS-1. All of these sensors are described in Chapter 2. Practical aspects of registering the stereo pairs prior to extraction of elevation information are covered by Al-Rousan *et al.* (1997, 1998), Dowman and Neto (1994), Giles and Franklin (1996), Theodossiou and Dowman (1990), Tokonuga and Hara (1996) and Welch *et al.* (1998). Hirano *et al.* (2003) report on validation and accuracy assessment experiments using ASTER stereo image data.

4.4 ATMOSPHERIC CORRECTION

4.4.1 Background

An introductory description of the interactions between radiant energy and the constituents of the Earth's atmosphere is given in Chapter 1. From this discussion one can conclude that a value recorded at a given pixel location on a remotely-sensed image is not a record of the true ground-leaving radiance at that point, for the magnitude of the ground-leaving signal is attenuated due to atmospheric absorption and its directional properties are altered by scattering. Figure 4.11 shows, in a simplified form, the components of the signal received by a sensor above the atmosphere. All of the signal appears to originate from the point P on the ground whereas, in fact, scattering at S_2 redirects some of the incoming electromagnetic energy within the atmosphere into the field of view of the sensor (the atmospheric path radiance) and some of the energy reflected from point Q is scattered at S_1 so that it is seen by the sensor as coming from P. This scattered energy, called 'environmental radiance', produces what is known as the 'adjacency effect'. To add to these effects, the radiance from P (and Q) is attenuated as it

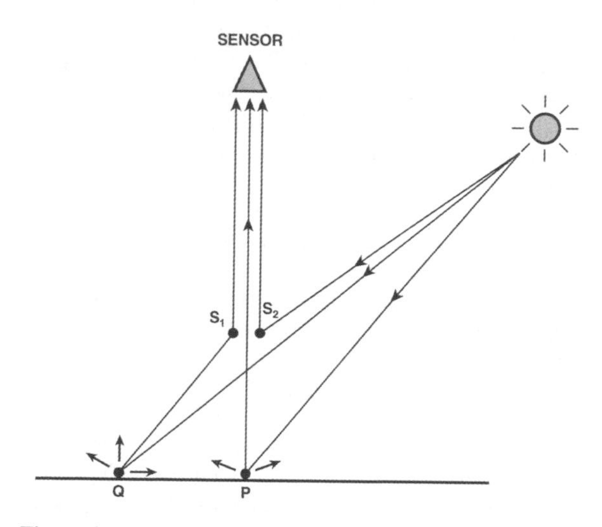

Figure 4.11 Components of the signal received by a satellite-mounted sensor. See text for explanation.

passes through the atmosphere. Other difficulties are caused by variations in the illumination geometry (the geometrical relationship between the Sun's elevation and azimuth angles, the slope of the ground and the disposition of topographic features). These problems are considered in section 4.5. If the sensor has a wide field of view, such as the NOAA AVHRR, or is capable of off-nadir viewing (for example, the SPOT HRV) then further problems result from the fact that the reflectance of a surface will vary with the view angle as well as with the solar illumination angle (this is the bi-directional reflectance property, noted in Chapter 1). Given this catalogue of problems one might be tempted to conclude that quantitative remote sensing is the art of the impossible. However, there is a variety of techniques that can be used to estimate the atmospheric and viewing geometry effects, although a great deal of research work remains to be done in this highly-complex area. A good review of terminology and principles is contained in Deschamps *et al.* (1983), while Woodham (1989) gives a lucid review of problems. The problem of the 'adjacency effect', which is mentioned above, is considered by Milovich *et al.* (1995).

Atmospheric correction might be a necessary pre-processing technique in three cases. In the first, we may want to compute a ratio of the values in two bands of a multispectral image (section 6.2.4). As noted in Chapter 1, the effects of scattering increase inversely with wavelength, so the shorter-wavelength measurements experience more scattering than the longer-wavelength data. The computed ratio will thus be a biased estimate of the true ratio. In the second situation, a research worker may wish to relate upwelling radiance from a surface to some property of that surface in terms of a physically based model. To do this, the atmospheric component present in the signal recorded by the sensor must be estimated and removed. This problem is experienced in, for example, oceanography because the magnitude of the water-leaving radiance (which carries information about the biological and sedimentary materials contained by the upper layers of the ocean) is small compared to the contribution of atmospheric effects. The third case is that in which results or ground measurements made at one time (time 1) are to be compared with results achieved at a later date (time 2). Since the state of the atmosphere will undoubtedly vary from time 1 to time 2 it is necessary to correct the radiance values recorded by the sensor for the effects of the atmosphere. In addition to these three cases, it may well be necessary to correct multispectral data for atmospheric effects even if it is intended for visual analysis rather than any physical interpretation.

The atmosphere is a complex and dynamic system. A full account of the physics of the interactions between the atmosphere and electromagnetic radiation is well beyond the scope of this book, and no attempt is made to provide a comprehensive survey of the progress that has been made in the field of atmospheric physics. Instead, techniques developed by remote sensing researchers for dealing with the problem of estimating atmospheric effects on multispectral images in the 0.4–2.4 μm reflective solar region of the spectrum are reviewed briefly. We begin by summarising the relationship between radiance received at a sensor above the atmosphere and the radiance leaving the ground surface:

$$L_s = H_{\text{tot}}\rho T + L_{\text{p}}$$

H_{tot} is the total downwelling radiance in a specified spectral band, ρ is the reflectance of the target (the ratio of the downwelling to the upwelling radiance) and T is the atmospheric transmittance. L_{p} is the atmospheric path radiance. The downwelling radiance is attenuated by the atmosphere as it passes from the top of the atmosphere to the target. Further attenuation occurs as the signal returns through the atmosphere from the target to the sensor. Some of the radiance incident upon the target is absorbed by the ground-surface material, with a proportion ρ being reflected by the target. Next, energy reflected by the ground surface from outside the target area is scattered by the atmosphere into the field of view of the sensor. Finally, the radiance reaching the sensor includes a contribution made up of energy scattered within the atmosphere; this is the path radiance term (L_{p}) in the equation. In reality, the situation is more complicated, as Figure 4.11 shows. The path radiance term L_{p} varies in magnitude inversely with wavelength for scattering increases as wavelength decreases. Hence, L_{p} will contribute differing amounts to measurements in individual wavebands. In terms of a Landsat TM or ETM+ image the blue-green band (band 1) will generally have a higher L_{p} component than the green band (band 2).

4.4.2 Image-based methods

The first method of atmospheric correction that is considered is the estimation of the path radiance term, L_{p} and its subtraction from the signal received by the sensor. Two relatively simple techniques are described in the literature. The first is the histogram minimum method, and the second is the regression method. In the first approach, histograms of pixel values in all bands are computed for the full image, which generally contains some areas of low reflectance (clear water, deep shadows or exposures of dark coloured rocks). These pixels will have values very close to zero in near-infrared bands, for example Landsat TM band 4 or SPOT HRV band 3, and should have near-zero values in the other bands in this spectral region. Yet, if the histograms of these other bands are plotted they will generally be seen to offset progressively towards higher levels. The lowest pixel values (or some combination of the lowest values) in the histograms of visible and near-infrared bands are a first approximation to the atmospheric path radiance in those bands, and these minimum values are subtracted from the respective images. Path radiance is much reduced in mid-infrared bands such as Landsat TM bands 5 and 7, and these bands are not normally corrected.

The regression method is applicable to areas of the image that have dark pixels as described above. In terms of the Landsat TM sensor, pixel values in the near-infrared band (numbered 4) are plotted against the values in the other bands in turn, and a best-fit (least-squares) straight line is computed for each using standard regression methods. The offset a on the x-axis for each regression represents an estimate of the atmospheric path radiance term for the associated spectral band (Figure 4.12).

Chavez (1988) describes an elaboration of this 'haze correction' procedure based on the 'histogram minimum' methods described above. Essentially Chavez's method is based on the fact that Rayleigh scattering is inversely proportional to the nth power of the wavelength, the value of n varying with atmospheric turbidity. He defines a number of 'models' ranging from 'very clear' to 'hazy' and each of this is associated with a value of n. A 'starting haze' value is provided for one of the short wavelength bands, and the haze factors in all other bands are calculated analytically using the Rayleigh scattering relationship. The method requires the conversion from pixel values to radiances. The gains and offsets that are used to perform this conversion

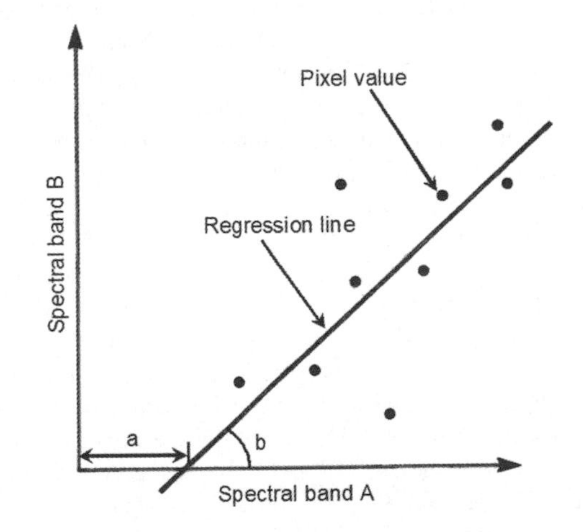

Figure 4.12 Regression of selected pixel values in spectral band A against the corresponding pixel values in band B. Band B is normally a near-infrared band (such as Landsat ETM+ band 4 or SPOT HRV band 3) whereas band A is a visible/near-infrared band.

are the pre-flight calibrations for the Landsat TM. Chavez (1996) gives details of modifications to the method which enhance its performance. A simple version of Chavez's model can be accessed in MIPS from the *Utilities* menu.

The image-based methods described above (dark object subtraction and regression) simply estimate the contribution to the radiance at a pixel of the atmospheric path radiance. The methods described next are generally used to derive estimates of the radiance reaching the sensor from the target pixel. In other words, they are normally used in conjunction with procedures to correct for illumination effects (section 4.5) and sensor calibration (section 4.6). Franklin and Giles (1995) give a systematic account of these procedures, while Aspinall *et al.* (2002) consider atmospheric correction of imaging spectrometer data sets (Chapter 9). A full radiometric correction would include conversion of pixel digital value to apparent (at-sensor) radiance, subtraction of the atmospheric contribution, topographic normalisation, and sensor calibration. The end product is an estimate of the true ground-leaving radiance at a pixel. Such values are required in environmental modelling, and in the measurement of change between two dates of imaging. The use of radiative transfer models and the empirical line method for estimating the atmospheric effect is considered next. Topographic corrections are mentioned briefly in sections 4.5 and 4.7.Sensor calibration issues are considered in section 4.6.

4.4.3 Radiative transfer model

The methods of atmospheric correction described above rely on information from the image itself in order to estimate the path radiance for each spectral band. However, a considerable body of theoretical knowledge concerning the complexities of atmospheric radiative transfer is in existence, and has found expression in numerical radiative transfer models such as LOWTRAN (Kneizys *et al.*, 1988), MODTRAN (Berk *et al.*, 1999), ATREM (Gao *et al.*, 1993) and 5S/6S (Tanré *et al.*, 1986; Vermote *at al.*, 1997)[1]. Operational (as opposed to research) use of these models is limited by the need to supply data relating to the condition of the atmosphere at the time of imaging. The cost of such data collection activities is considerable, hence reliance is often placed upon the use of 'standard atmospheres' such as 'mid-latitude summer'. The use of these estimates results in a loss of accuracy, and the extent of the inaccuracy is not assessable. See Popp (1995) for further elaborations. It is also difficult to apply radiative transfer models to archive data because of lack of knowledge of atmospheric conditions. Richter (1996) shows how spatial variability in atmospheric properties can be taken into account by partitioning the image.

The following example of the use of the 5*S* model of Tanré *et al.* (1986) is intended to demonstrate the magnitudes of the quantities involved in atmospheric scattering. Two hypothetical test cases are specified, both using Landsat TM bands 1–5 and 7 for an early summer (June 1st) date at a latitude of 51°. A standard mid-latitude summer atmosphere was chosen, and the only parameter to be supplied by the user was an estimate of the horizontal visibility in kilometres (which may be obtainable from a local meteorological station). Values of 5 km (hazy) and 20 km (clear) were input to the model, and the results are shown graphically in Figure 4.13. The apparent reflectance is that which would be detected by a scanner operating above the

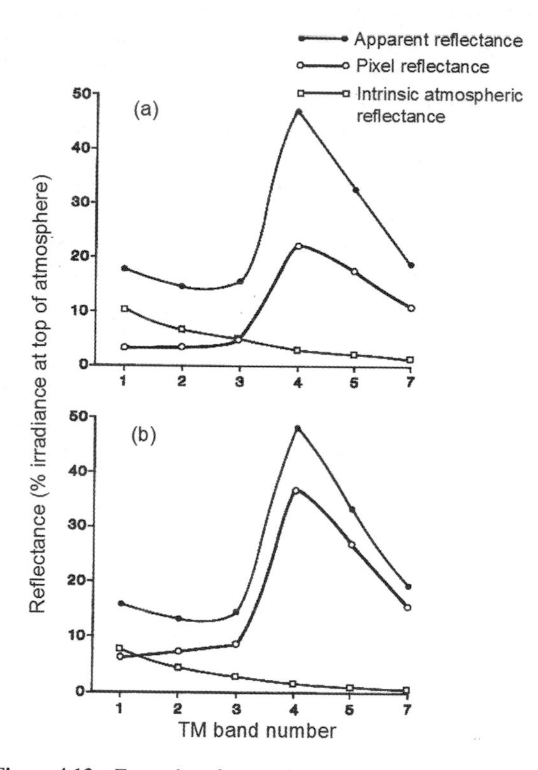

Figure 4.13 Examples of output from atmospheric model (Tanré *et al.*, 1986). Reflectance, expressed as percent irradiance at the top of the atmosphere in the spectral band, is shown for the atmosphere (intrinsic atmospheric reflectance), the target pixel (pixel reflectance) and the received signal (apparent reflectance). The difference between the sum of pixel reflectance and intrinsic atmospheric reflectance and the apparent reflectance is the background reflectance from neighbouring pixels. Examples show results for (a) 5 km visibility and (b) 20 km visibility.

[1] Fortran computer code and manuals for 6S can be freely downloaded from: *ftp://loa.univ-lille1.fr/6S/.*

atmosphere, while the pixel reflectance is an estimate of the true target reflectance which, in this example, is green vegetation. The intrinsic atmospheric reflectance is analogous to the path radiance. At 5 km visibility the intrinsic atmospheric reflectance is greater than pixel reflectance in TM bands 1 and 2. Even at the 20 km visibility level, intrinsic atmospheric reflectance is greater than pixel reflectance in TM band 1. In both cases the intrinsic atmospheric reflectance declines with wavelength, as one might expect. Note that the difference between the apparent reflectance and the sum of the pixel reflectance and the intrinsic atmospheric reflectance is equal to reflectance from pixels neighbouring the target pixels being scattered into the field of view of the sensor (see Figure 4.11). The y-axis in Figure 4.13 shows percent reflectance, that is, the proportion of solar irradiance measured at the top of the atmosphere that is reflected.

4.4.4 Empirical line method

An alternative to the use of radiative transfer models is the empirical line approach (Smith and Milton, 1999; Karpouzli and Malthus, 2003; Moran *et al.*, 2001). This is illustrated in Figure 4.14. Two targets – one light and one dark – are selected, and their reflectance is measured on the ground, using a field radiometer, to give values R on the y-axis of Figure 4.14. The radiances recorded by the sensor (shown by the x-axis, L, in Figure 4.14) are computed from the image pixel values using the methods described in section 4.6. Finally, the slope, s, and intercept, a, of the line joining the two target points are calculated. The conversion equation is $R = s(L - a)$. The term a represents the atmospheric radiance. This equation is computed for all spectral bands of interest. Smith and Milton (1999) report the results of a detailed experiment, and conclude that the method '. . . is capable of providing acceptable calibration of sensor radiance measurements to estimates of ground reflectance' (op. cit., p. 2657), though they list a number of theoretical and practical considerations that must be taken into account before reliance is placed on any results. Karpouzli and Malthus (2003) also report the results of experiments to estimate surface reflectance using the empirical line method applied to IKONOS data. The absolute differences between calculated band reflectances and corresponding reflectances measured on the ground ranged from 0 to 2.7%. The authors describe these results as '. . . highly satisfactory' (op. cit., p. 1148). Yuan and Elvidge (1996) compare these and other methods of radiometric normalisation.

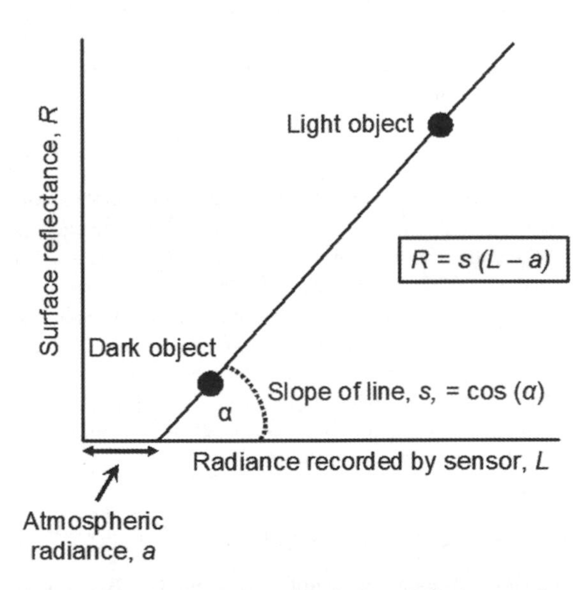

Figure 4.14 Empirical line method of atmospheric correction. Two targets (light and dark) whose reflectance (R) and at-sensor radiance (L) are known are joined by a straight line with slope s and intercept a. The reflectance for any at-sensor radiance can be computed from $R = s\ (L - a)$. Based on Figure 1 of Smith and Milton, International Journal of Remote Sensing, 1999, 20, p. 2654. © Taylor and Francis Ltd., www.tandf.co.uk/journals.

Some general advice on the need for atmospheric correction in classification and change detection studies is provided by Song *et al.* (2001). These authors suggest that atmospheric correction is not required as long as the training data and the data to be classified are measured on the same relative scale. However, if multi-temporal image data are being processed then they must be corrected for atmospheric effects to ensure that they are comparable.

4.5 ILLUMINATION AND VIEW ANGLE EFFECTS

The magnitude of the signal received at a satellite sensor is dependent on several factors, particularly:

1. reflectance of the target,
2. nature and magnitude of atmospheric interactions,
3. slope and aspect of the ground target area relative to the solar azimuth,
4. angle of view of the sensor, and
5. solar elevation angle.

Variations in the spectral reflectance of particular types of Earth-surface cover materials are discussed in Chapter 1, where a general review of atmospheric interactions is also to be found, while section 4.4 contains an introduction to the correction of image data for atmospheric effects. In this section we consider the effects of (i) the solar elevation angle, (ii) the view angle of the sensor, and (iii) the slope and aspect angles of the target.

In the absence of an atmosphere, and assuming no complicating factors, the magnitude of the radiance reflected from or emitted by a fixed target and recorded by a remote sensor will vary with the illumination angle and the angle of view of the sensor. The reflectance of the target will vary as these two angles alter, and one could therefore consider a function which described the magnitude of the upwelling radiance of the target in terms of these two angles. This function is termed the *bi-directional reflectance distribution function* (or BRDF, described in Chapter 1). When an atmosphere is present an additional complication is introduced, for the irradiance at the target will be reduced as the atmospheric path length increases. The path length (which is the distance that the incoming energy travels through the atmosphere) will increase as the solar elevation angle decreases, and so the degree of atmospheric interference will increase.

The radiance upwelling from the target also has to pass through the atmosphere. The angle of view of the sensor will control the upward path length. A nadir view will be influenced less by atmospheric interactions than would an off-nadir view; for example, the extremities of the scan-lines of a NOAA AVHRR image are viewed at angles of up to 56° from nadir, while the SPOT HRV instrument is capable of viewing angles of ±27° from nadir. Amounts of shadowing will also be dependent upon the solar elevation angle. For instance, shadow effects in row crops will be greater at low Sun elevation angles than at high angles. Ranson *et al.* (1985) report the results of a study of variations in the spectral response of soya beans with respect to illumination, view and canopy geometry. They conclude that the spectral response depended greatly on solar elevation and azimuth angles and on the angle of view of the sensor, the effect being greater in the visible red region of the spectrum than in the near-infrared. In another study, Pinter *et al.* (1983) show that the spectral reflectance of wheat canopies in MSS and TM wavebands is strongly dependent on the direction of incident radiation and its interaction with vegetation canopy properties, such as leaf inclination and size. Clearly, reflectance from vegetation surfaces is a complex phenomenon which is, as yet, not fully understood and methods for the estimation of the effects of the various factors which influence reflectance are still being evaluated at a research level. It would obviously be desirable to remove such effects by pre-processing before applying methods of pattern recognition (Chapter 8) particularly if results from analyses of images from different dates are to be compared, or if the methods are applied to images produced by off-nadir viewing or wide angle-of-view sensors.

The effects of variation in the solar elevation angle from one image to another of a given area can be accomplished simply if the reflecting surface is Lambertian (section 1.3.1). This is rarely the case with natural surfaces, but the correction may be approximate to the first order. If the solar zenith angle (measured from the vertical) is θ, the observed radiance is L and the desired view angle is x then the correction is simply:

$$L' = L\frac{\cos(x)}{\cos(\theta)}$$

This formula may be used to standardise a set of multi-temporal images to a standard solar illumination angle. If a suitable digital elevation model is available, and assuming that the image and DEM are registered, then the effects of terrain slope on the angle of incidence of the solar irradiance can be taken into account (Feng *et al.*, 2003). See Frulla *et al.* (1995) for a discussion of illumination and view angle effects on the NOAA-AVHRR imagery.

Barnsley and Kay (1990) consider the relationship between sensor geometry, vegetation canopy geometry and image variance with reference to wide field of view imagery obtained from aircraft scanners. The effect of off-nadir viewing is seen as a symmetric increase in reflectance away from the nadir point, so that a plot of the mean pixel value in each column (y-axis) against column number (x-axis) shows a parabolic shape. A first-order correction for this effect uses a least-squares procedure to fit a second-order polynomial of the form:

$$\hat{y}_x = a_0 + a_1 x$$

to the column means. The term $\hat{y}_x$ is the value on the least-squares curve corresponding to image column x. Let $\hat{y}_{\text{nad}}$ be the value on the least-squares curve at the nadir point (the centre of the scan line and the minimum point on the parabolic curve). Then the values in column x of the image can be multiplied by the value $\hat{y}_{\text{nad}}/\hat{y}_x$ in order to 'flatten' the curve so that the plot of the column means of the corrected data plot (approximately) as a straight line. This procedure can be used in MIPS – select *Hyperspectral|Cross-scan View-angle Correction.*

4.6 SENSOR CALIBRATION

Sensor calibration has one of three aims. Firstly, the user may wish to combine information from images obtained from different sensor systems such as Landsat TM and SPOT HRV. Secondly, it may be necessary in studies of change to compare pixel values obtained from images that were acquired at different times. Thirdly, remotely-sensed estimates of surface parameters such as reflectance are used in physical and biophysical models. Generally speaking, sensor calibration is normally combined with atmospheric and view angle correction (e.g., Teillet and Fedosejevs, 1995) in order to obtain estimates of target reflectance. These authors also provide details of the correction required for variations in solar irradiance over the year. A useful starting point is the review by Duggin (1985). In this section, attention is focused on the Landsat TM/ETM+ optical bands and the SPOT HRV. Calibration of the Landsat TM/ETM+ thermal band (conventionally numbered 6) is discussed by Schott and Volchok (1985) and Itten and Meyer (1993). Consideration of calibration issues relating to the NOAA-AVHRR is provided by Che and Price (1992) and by Vermote and Kaufman (1995). Gutman and Ignatov (1995) compare pre- and post-launch calibrations for the NOAA AVHRR, and show how the differences in these calibrations affect the results of vegetation index calculations (Chapter 6). A special issue of *Canadian Journal of Remote Sensing* (volume 23, number 4, December 1997) is devoted to calibration and validation issues.

Atmospheric correction methods are considered in section 4.4, while corrections for illumination and view angle effects are covered in section 4.5. In the present section, the topic of sensor calibration is reviewed. Recall from Chapter 1 that the values recorded for a particular band of a multispectral image are counts – they are the representation, usually on a 0–255 scale, of equal-length steps in a linear interpolation between the minimum and maximum levels of radiance recorded by the detector. If more than one detector is used then differences in calibration will cause 'striping' effects on the image (section 4.2.2). The term 'sensor calibration' refers to procedures that convert from counts to physical values of radiance. The numerical coefficients that are used to calibrate these image data vary over time, and consequently the relationship between the pixel value (count) recorded at a particular location and the reflectance of the material making up the surface of the pixel area will not be constant. Recall that reflectance is the ratio of the incident radiation (irradiance) and the radiance reflected or emitted by the target.

In the case of Landsat TM and ETM+, two coefficients are required. These are termed the gain and offset. The gain is the slope of the line relating counts (DN) and radiance. Offset is an additive term, as shown in Figure 4.15.

Th determination of these calibration coefficients is not an easy task. Pre-launch calibration factors (coefficients) are obtainable from ground receiving stations and from vendors of image data, and some image processing software packages incorporate these values into their sensor calibration routines. However, studies such as Thome *et al.* (1993) indicate that these calibration factors are time-varying for the Landsat-5 TM sensor, and that substantial differences

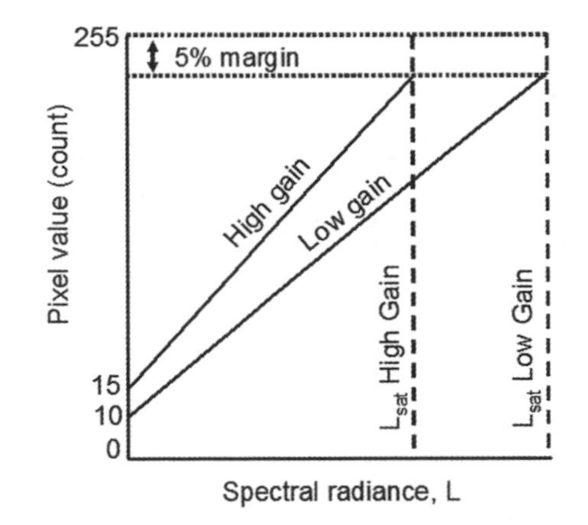

Figure 4.15 Landsat ETM+ uses one of two gain modes. The spectral radiance reaching the sensor is converted to a digital count or pixel value using high gain mode for target areas which are expected to have a maximum spectral radiance of L_{sat} (High Gain). For other target areas, the maximum radiance is specified as L_{sat} (Low Gain). Each gain setting has an associated offset, measured in counts, which is 10 for low gain and 15 for high gain. Based on Irish (2002), Figure 6.9.

in the outcome of calibration calculations depend on the values of the calibration factors that are used. The fact that a number of different processing procedures have been applied to Landsat TM data by ground stations further complicates the issue (Moran *et al.*, 1992, 1995). Thome *et al.* (1993, see also Table 2 of Teillet and Fedosjevs, 1995) provide the following expressions for Landsat-5 TM sensor gains in bands 1–5 and 7 (the offset is assumed to be constant over time):

$$G_1 = (-7.84 \times 10^{-5})D + 1.409$$
$$G_2 = (-2.75 \times 10^{-5})D + 0.7414$$
$$G_3 = (-1.96 \times 10^{-5})D + 0.9377$$
$$G_4 = (-1.10 \times 10^{-5})D + 1.080$$
$$G_5 = (7.88 \times 10^{-5})D + 7.235$$
$$G_7 = (7.15 \times 10^{-5})D + 15.63$$

D is the number of days since 1 March 1984, the date of the launch of the Landsat-5 satellite. The day number in the year is called the Julian Day. You can calculate the Julian Day relative to any year since 1753 using the *Utilities|Julian Dates* module in MIPS. The gain coefficients are used in the equation:

$$L_n^* = (PV - \text{offset})/G_n(\text{Wm}^{-2}\text{sr}^{-1}\mu\text{m}^{-1})$$

In this equation, which differs from the relationship presented in the next paragraph, uses the symbol L_n^* to denote

Table 4.3 Minimum and maximum spectral radiance at sensor (*L(min)*) and (*L(max)*) with corresponding gain (a_1) and offset (a_0) values used in the calibration of Landsat-5 TM data by the USGS EROS Data Centre. These calibration coefficients were amended on 5 May 2003 for imagery processed on and after that date. Based on Chander and Markham (undated), Table 1.

Band	Prior to 5 May 2003				5 May 2003 onwards			
Number	L(min)	L(max)	Gain	Offset	L(min)	L(max)	Gain	Offset
1	−1.52	152.10	0.602431	−1.52	−1.52	193.0	0.762824	−1.52
2	−2.84	296.81	1.175100	−2.84	−2.84	365.0	1.442510	−2.84
3	−1.17	204.30	0.805765	−1.17	−1.17	264.0	1.039880	−1.17
4	−1.51	206.20	0.814549	−1.51	−1.51	221.0	0.872588	−1.51
5	−0.37	27.19	0.108078	−0.37	−0.37	30.2	0.119882	−0.37
7	−0.15	14.38	0.056980	−0.15	−0.15	16.5	0.065294	−0.15

apparent radiance at the sensor, while *PV* is the pixel value, G_n is the sensor gain and the offsets are as follows (values in brackets): TM1 (2.523), TM2 (2.417), TM3 (1.452), TM4 (1.854), TM5 (3.423) and TM7 (2.633). Note that the values of gain and offset given in Table 4.3 refer to the procedure that is described in the following paragraphs.

Chander and Markham (undated) provide an up-to-date account of the Landsat-5 TM radiometric calibration procedures used by the USGS EROS Data Centre. This calibration procedure was revised on 5 May 2003 for all Landsat-5 TM imagery processed (not necessarily acquired) on and after this date, with the aim of improving calibration accuracy and consistency with Landsat-7 ETM+ data. The relationship between radiance at the sensor and pixel value (PV, sometimes called a digital Signal level (DSL), a digital number (DN) or a digital count (DC)) is defined for spectral band *n* of the Landsat-5 TM as follows:

$$L_n = a_0 + a_1 PV$$

where a_0 and a_1 are the offset and gain coefficients respectively, and L_n is the apparent radiance at the sensor for band *n*, measured in units of W cm^{-1}sr^{-1} μm^{-1}. The *PV* in this formula is assumed to be derived from a Landsat-5 TM image that has been subject Level 1 correction. The gain a_0 is defined as:

$$a_{0n} = \frac{L_n(max) - L_n(min)}{255}$$

$L_n(max)$ and $L_n(min)$ are the spectral radiances equivalent to pixel values of 255 and 0, respectively for band n, and a_{0n} is the gain coefficient for band *n*. Values of $L_n(max)$, $L_n(min)$, a_{0n} and a_{1n} are listed in Table 4.3.

The work of Thome *et al.* (1993) on the determination of the values of the gain and offset coefficients is noted above; their gains and offsets use the relationship between L^* and *PV* described earlier. Other references are Hall *et al.* (1991), Holm *et al.* (1989), Markhan *et al.* (1992), Moran

Table 4.4 Extract from the output from MIPS module *'File|Import Image Set|Read Fast L7A ETM+ CD'* showing the radiometric record for the visible and near-infrared bands (numbered 1–5 and 7). Column 1 contains the bias or offset values and column 2 contains the gain values used in conversion from digital counts to radiance units. The sun elevation and azimuth angles are also listed. They are expressed in degrees.

```
Radiometric record
==================
GAINS AND BIASES IN ASCENDING BAND NUMBER ORDER
-6.199999809265137     0.775686297697179
-6.400000095367432     0.795686274883794
-5.000000000000000     0.619215662339154
-5.099999904632568     0.965490219639797
-1.000000000000000     0.125725488101735
-0.349999994039536     0.043725490920684

SUN ELEVATION ANGLE =58.2 SUN AZIMUTH ANGLE =145.7
```

Table 4.5 Extract from SPOT header file showing radiometric gains and offsets. The data were generated using the MIPS module *File|Import Image Set|Read_Spot_CD*.

Scene ID:		S1H1870112102714	
Scene centre latitude:		N0434026	
Scene centre longitude:		E0043615	
Spectral mode (XS or PAN):		XS	
Preprocessing level identification:		1B	
Radiometric calibration designator:		1	
Deconvolution designator:		1	
Resampling designator:		CC	
Pixel size along line:		20	
Pixel size along column:		20	
Image size in map projection along Y-axis:		059 792	
Image size in map projection along X-axis:		075 055	
Sun calibration operation date:		19 861 115	
This is a multispectral image			
Absolute calibration gains: 00.862 62	00.798 72	00.893 10	
Absolute calibration offsets:	00.000 00	00.000 00	00.000 00

et al. (1990, 1992), Muller (1993), Olsson (1995), Price (1987, 1988), Rao and Chen (1996), Sawter *et al.* (1991), Slater *et al.* (1987), Teillet and Fedosejevs (1995), Thome *et al.* (1997) and Woodham (1989).

Calibration of Landsat-7 ETM+ image data is described in Irish (2002). The ETM+ sensor, like the previous TM instruments, operates in one of two modes, termed low and high gain. Where the expected range of upwelling radiances is small, then low-gain mode is used, otherwise high-gain mode is used. The aim of this procedure is to provide the greatest possible image contrast. Figure 4.15 shows graphs of low and high gain modes, and Table 4.3 gives minimum and maximum radiances for the two gain settings for all bands including the panchromatic band, which is labelled as band 8. Irish (2002) gives details of the gain settings used for different targets.

Conversion of Landsat ETM+ digital counts (Q) to radiance units (L) is accomplished using the following equation:

$$L_\lambda = G_\lambda Q + O_\lambda$$

The gain (G) and offset (O) values for spectral band λ should be read from the metadata provided with the image data, as shown in Table 4.4. Use of the values contained in the metadata will provide a more accurate estimate of the spectral radiance.

An alternative to the use of absolute calibration procedures for the determination of reflectance factors for Landsat data is to use one of the images in a multi-temporal image set as a reference and adjust other images according to some statistical relationship with the reference image. Thus, Elvidge *et al.* (1995) use a regression procedure to perform relative radiometric normalisation of a set of Landsat MSS images. Their procedure assumes that land cover has not changed and that the vegetation is at the same phenological stage in all images. A similar approach is used by Olsson (1993).

The corresponding expression for SPOT HRV uses calibration coefficients derived from the header data provided with the image. Table 4.5 shows a part of the output generated by the MIPS program. The header data refer to a SPOT-1 multispectral (XS) image of the Camargue area of southern France, acquired on 12 January 1987, and gain values a_i for the three multispectral channels are provided. Note that all of the offset values are zero. The apparent radiance of a given pixel is calculated from:

$$L = PV/a_1$$

where a_1 is the gain coefficient, and PV and L are as defined earlier. These coefficients are updated regularly (Begni, 1986; Begni *et al.*, 1988, Moran *et al.*, 1990). These apparent radiance values must be further corrected if imagery acquired from SPOT in off-nadir viewing mode is used (Moran *et al.*, 1990; Muller, 1993).

The procedure for conversion of ASTER visible and short wave infrared data (Table 2.4) expressed in digital counts to radiance units is similar to that described above for SPOT and ETM+ data. The sensor has a number of different gain settings, which can be selected in order to produce a high signal-to-noise ratio for different targets. Table 4.6 shows the maximum radiance values for each band for the high, normal, low 1 a1nd low 2 gain settings. In bands 1–9, pixel values 1–254 represent zero radiance and maximum radiance, respectively. The magnitude of the radiance represented by a change of unity in the quantisation

Table 4.6 Maximum radiance for different gain settings for the ASTER VNIR and SWIR spectral bands.

	Maximum radiance (W m^{-2} sr^{-1}μm^{-1})			
Band no	High gain	Normal gain	Low gain 1	Low gain 2
1	170.8	427	569	
2	179.0	358	477	
3N	106.8	218	290	N/A
3B	106.8	218	290	
4	27.5	55.0	73.3	73.3
5	8.8	17.6	23.4	103.5
6	7.9	15.8	21.0	98.7
7	7.55	15.1	20.1	83.8
8	5.27	10.55	14.06	62.0
9	4.02	8.04	10.72	67.0

level for band 1, for example, increases from 0.6751 (high gain) to 1.6877 (normal gain) to 2.2490 W m^{-2} sr^{-1}μ m^{-1}. Thus, high gain mode can be used for images of regions of low reflectance while low-gain mode is used in regions of high reflectance in order to avoid saturation (which occurs when the radiance received by the sensor exceeds the maximum radiance that can be quantised). The gain setting for any specific image is contained in the metadata. See Table 4.6 for maximum radiance values for ASTER's high, normal and low gain modes.

Given the value of radiance at the sensor (L) it is usual to convert to apparent reflectance, that is, the total reflectance (from target and atmosphere) at the sensor. This value is also known as at-satellite reflectance. Of course, if the image has been corrected for atmospheric effects then the value computed by the equation below is an estimate of actual target reflectance. Conversion to reflectance is accomplished for each individual band using the expression:

$$\rho = \frac{\pi L d^2}{E_s \cos \vartheta_s}$$

L is the radiance computed as described earlier, E_s is the exoatmospheric solar irradiance (Markham and Barker, 1987; Price, 1988; Table 4.7), d is the relative Earth–Sun distance in astronomical units (the mean distance is 1.0 AU) for the day of image acquisition, and θ_s is the solar zenith angle. The Earth-Sun distance correction factor is required because there is a variation of around 3.5% in solar irradiance over the year. The value of d is provided by the formula:

$$d = 1 - 0.01674 \cos(0.9856\,(JD - 4))$$

JD is the 'Julian day' of the year, that is, the day number counting January 1st = 1. A utility to calculate the Julian day is available via the *Utilities|Julian Dates* function in MIPS. The module actually calculates the number of days that have elapsed since a reference calendar date since the start of the first millennium, though most readers will wish to compute elapsed days since the start of a given year. The program is useful in computing the time-dependent offset and gain values from the formulae given in Thome *et al.* (1993), described above, which require the number of days that have elapsed since the launch of Landsat-5 (1 March 1984). Table 4.7 contains details of solar exo-atmospheric spectral irradiance for both the Landsat TM and ETM+ instruments. Comparisons of the radiometric characteristics of these two instruments are provided by Teillet *et al.* (2001), Masek *et al.* (2002) and Vogelmann *et al.* (2001). Thome *et al.* (1995) give details of Landsat TM radiometric calibration, while Thome (2001) discusses the absolute calibration of Landsat ETM+ data. Apparent reflectance is used by Huang *et al.* (2002) to permit comparison of images collected under different illumination conditions.

The MIPS function *Utilities|Convert DN to reflectance* allows you to convert an 8-bit Landsat TM, Landsat ETM+, or ASTER image to apparent reflectance, which is represented by 32-bit real quantities. The procedure is useful in the comparison of multitemporal images, with different solar zenith angles and possibly different irradiance inputs.

Calibration of synthetic aperture radar (SAR) imagery requires the recovery of the normalised radar cross-section (termed sigma-nought or σ_0) and measured in terms of decibels, dB). The range of σ_0 values is from +5 dB (very bright target) to −40 dB (very dark target). Meadows (1995) notes that the purpose of calibration is to determine absolute radar cross-section measurements at each pixel position, and to estimate drift or variation over time in the radiometric performance of the SAR. Calibration can be performed in three ways: by imaging external calibration targets on the ground, by the use of internal calibration data, or by examining raw data quality. Laur *et al.* (2002) give details of calibration of ERS SAR data and the derivation of σ_0.

4.7 TERRAIN EFFECTS

The corrections required to convert ETM+ and SPOT HRV data described in section 4.6 assume that the area covered by the image is a flat surface that is imaged by a narrow field of view sensor. It was noted in section 4.6 that apparent reflectance depends also on illumination and view angles, as target reflectance is generally non-Lambertian. That discussion did not refer to the commonly-observed fact that the Earth's surface is not generally flat. Variations in reflectance from similar targets will occur if these targets have a different topographic position, even if they are directly illuminated by the Sun. Therefore, the spectral

Table 4.7 Exo-atmospheric solar irradiance for (a) Landsat TM, (b) Landsat ETM+, (c) SPOT HRV (XS) bands and ASTER (Markham and Barker, 1987; Price, 1988, Teillet and Fedosejevs, 1995; Irish, 2000 and Thome, personal communication). The centre wavelength is expressed in micrometres (μm) and the exo-atmospheric solar irradiance in mW cm^{-2} sr^{-1} μm^{-1}. See also Guyot and Gu (1994), Table 2.

Landsat TM/ETM+ band number	Centre wavelength	Centre wavelength (Teillet and Fedosejevs, 1995)	Exo-atmospheric spectral irradiance	Exo-atmospheric spectral irradiance (Teillet and Fedosejevs, 1995)
1	0.486	0.4863	195.70	195.92
2	0.570	0.5706	192.90	182.74
3	0.660	0.6607	155.70	155.50
4	0.840	0.8832	104.70	104.08
5	1.676	1.677	21.93	22.075
7	2.223	2.223	7.45	7.496

(a) Landsat Thematic Mapper

Band	Bandwidth (μm)	Exo-atmospheric spectral irradiance
1	0.450 – 0.515	196.9
2	0.525 – 0.605	184.0
3	0.630 – 0.690	155.1
4	0.775 – 0.900	104.4
5	1.550 - 1750	22.57
7	2.090 – 2.350	8.207
8	0.520 – 0.900	136.8

(b) Landsat Enhance Thematic Mapper Plus (ETM+)

SPOT HRV band no	Centre wavelength	Exo-amospheric spectral irradiance
1	0.544	187.48
2	0.638	164.89
3	0.816	110.14

(c) SPOT High Resolution Visible (HRV)

ASTER Band	Bandwidth (μm)	Exo-atmospheric spectral irradiance
1	0.520 – 0.600	1846.9
2	0.630 – 0.690	1546.0
3	0.780 – 0.860	1117.6
4	0.600 – 0.700	232.5
5	2.145 – 0.2.185	80.32
6	2.185 – 2.225	74.96
7	2.235 – 2.285	69.20
8	2.295 – 2.365	59.82
9	2.360 – 2.430	57.32

(d) ASTER

reflectance curves derived from multispectral imagery for what is apparently the same type of land cover (for example, wheat or coniferous forest) will contain a component that is attributable to topographic position, and the results of classification analyses (Chapter 8) will be influenced by this variation, which is not necessarily insignificant even in areas of low relief (Combal and Isaka, 2002). Various corrections have been proposed for the removal of the 'terrain illumination effect'. See Li *et al.* (1996), Proy *et al.* (1989), Teillet *et al.* (1982), Woodham (1989) and Young and Kaufman (1986) for reviews of the problem. Danaher *et al.* (2001) and Danaher (2002) propose an empirical BRDF correction for Landsat TM and ETM+ images based on the conversion of pixel values to top-of-atmosphere reflectances, as described in the preceding section, and an empirical BRDF model.

Correction for terrain illumination effects requires a Digital Elevation Model (DEM) that is expressed in the same coordinate system as the image to be corrected. Generally, the image is registered to the DEM, as the DEM is likely to be map-based. The DEM should also be of a scale that is close to that of the image, so that accurate estimates of slope angle and slope direction can be derived for each pixel position in the image. A number of formulae are in common use for the calculation of slope and aspect 'images' from a DEM, and different results may be determined by different formulae (Bolstad and Stowe, 1994; Carara *et al.*, 1997, Hunter and Goodchild, 1997). A simple method to correct for terrain slope in areas that receive direct solar illumination is simply to use the Lambertian assumption (that the surface reflects radiation in a diffuse fashion, so that it appears equally bright from all feasible observation angles). This cosine correction is mentioned above. It involves the multiplication of the apparent reflection for a given pixel by the ratio of the cosine of the solar zenith angle (measured from the vertical) by the cosine of the incidence angle (measured from the surface normal, which is a line perpendicular to the sloping ground). Teillet *et al.* (1982, p. 88) note that this correction is not particularly useful in areas of steep terrain where incidence angles may approach 90°. Feng *et al.* (2003) describe a terrain correction for imaging spectrometer data (chapter 9) based on the assumption of Lambertian behaviour.

Non-Lambertian models include the *Minnaert correction*, which is probably the most popular method of computing a first-order correction for terrain illumination effects (though the method does not include any correction for diffuse radiation incident on a slope). Values of slope angle and slope azimuth angles are needed, so a suitable DEM is required. The Lambertian model can be written as:

$$L = L_N \cos(i)$$

where L is the measured radiance, L_N is the equivalent radiance on a flat surface with incidence angle of zero, and i is the exitance angle. The Minnaert constant, k, enters into the non-Lambertian model as follows:

$$L_N = \frac{L\cos(e)}{\cos^k(i)\cos^k(e)}$$

where i, L and L_N are defined as before and e is the angle of exitance, which is equal to the slope angle (β_t in the equation below). The value of $\cos(i)$ is found from the relationship:

$$\cos(i) = \cos(\theta_s)\cos(\beta_t) + \sin(\theta_s)\sin(\beta_t)\cos(\phi_t - \phi_t)$$

with θ_s being the solar zenith angle, β_t the slope angle, ϕ_s the solar azimuth angle and ϕ_t the slope azimuth angle. The value of the incidence angle, i, will be in the range 0–90° if the pixel under consideration receives direct illumination. If the value of i falls outside this range then the pixel lies in shadow.

The value of the Minnaert constant k is the slope of the least-squares line relating $y = \log(L\cos(e))$ and $x = \log(\cos(i)\cos(e))$. Most surfaces have k values between 0 and 1; $k = 1$ implies Lambertian reflectance while $k > 1$ implies a dominance of the specular reflectance component. Once k has been determined then the equivalent radiance from a flat surface can be calculated. However, the value of k depends on the nature of the ground cover, and so would vary over the image even in the absence of any other control. Since a sample of pixels is required in order to estimate the slope of the least-squares line relating log (L cos(e)) and log(cos(i) cos(e)) it might be necessary to segment the image into regions of similar land cover type and calculate a value of k for each type. Often, however, the purpose of performing a terrain illumination correction is to improve the identification of land cover types, hence the problem takes on circular proportions. An iterative approach might be possible, in which classification accuracy assessment (Chapter 8) is used as a criterion. However, a simpler approach would be to calculate an average k value for the whole image. Useful references include Bishop and Colby (2002), Gu *et al.* (1999) and Riaño *et al.* (2003).

Parlow (1996) describes a method for correcting terrain-controlled illumination effects using a simulation model of solar irradiance on an inclined surface. The Short Wave Irradiance Model (SWIM) computes both direct and diffuse components of irradiance for given atmospheric conditions and allows the conversion of satellite-observed radiances to equivalent radiances for a flat surface. Note that the Minnaert method, described in the preceding paragraphs, does not consider diffuse illumination. Parlow (1996) shows that correction of the image data for terrain illumination effects produces superior classification performance (Chapter 8). Further references are Conese *et al.* (1993), Egbert and Ulaby (1972), Hay and Mackay (1985), Hill *et al.* (1995), Jones *et al.* (1988), Katawa *et al.* (1988) and Smith *et al.* (1980)

4.8 SUMMARY

Methods of pre-processing remotely-sensed imagery are designed to compensate for one or more of (i) cosmetic defects, (ii) geometric distortions, (iii) atmospheric interference and (iv) variations in illumination geometry, to calibrate images for sensor degradation, and to correct image pixel values for the effects of topography. The level of pre-processing required will depend on the problem to which the processed images are to be applied. There is therefore no fixed schedule of pre-processing operations that are carried out automatically prior to the use of remotely-sensed data. The user must be aware of the geometrical properties of the image data and of the effects of external factors (such as the level of, and variations in, atmospheric haze) and be capable of selecting an appropriate technique to correct the defect or estimate the external effect, should that be necessary.

The material covered in this chapter represents the basic transformations that must be applied in order to recover estimates of ground-leaving radiance. The development of models requiring such estimates as input has expanded in recent years. Equally importantly, the use of remote sensing to measure change over time is becoming more significant in the context of global environmental change studies. Multi-temporal analysis, the comparison of measurements derived by different sensors at different points in time, and the determination of relationships between target radiance and growth and health characteristics of agricultural crops are examples of applications that require the application of corrections described in this chapter. Procedures to accomplish these corrections are not well formulated at present, and the whole area requires more research and investigation.

5

Image Enhancement Techniques

5.1 INTRODUCTION

The way in which environmental remote sensing satellite and aircraft systems collect digital images of the Earth's surface is described in Chapters 1–3. In these chapters, a remotely-sensed image is characterised as a numerical record of the radiance leaving each of a number of small rectangular areas on the ground (called pixels) in each of a number of spectral bands. The range of radiance values at each pixel position is represented (quantised) in terms of a scale which is normally eight or more bits in magnitude, depending on the type of scanner that is used and on the nature of any processing carried out at the ground station. Each pixel of a digital multispectral image is associated with a set of numbers, with one number per spectral band. For example, Landsat ETM+ provides seven bands of multispectral data, and each pixel can therefore be represented as a group (mathematically speaking, a vector) of seven elements, each expressed on the 0–255 (eight bit) range.

A digital image can thus be considered as a three-dimensional rectangular array or matrix of numbers, the x-and y-axes representing the two spatial dimensions and the z-axis the quantised spectral radiance (pixel value). A fourth dimension, time, could be added, since satellite data are collected on a routine and regular basis (every 16 days for Landsat-7, for example). The elements of this matrix are numbers in the range $0–2^n–1$, where n is the number of bits used to represent the radiance recorded for any given pixel in the image. As described in section 3.2, image data that are represented by more than $n = 8$ bits per pixel must be scaled to eight bit representation before they can be stored in the computer's display memory.

Visual analysis and interpretation are often sufficient for many purposes to extract information from remotely-sensed images in the form of standard photographic prints. If the image is digital in nature, such as the satellite and aircraft-acquired images considered in this book, a computer can be used to manipulate the image data and to produce displays that satisfy the particular needs of the interpreter. In this chapter, methods of enhancing digital images are considered. The term *enhancement* is used to mean the alteration of the appearance of an image in such a way that the information contained in that image is more readily interpreted visually in terms of a particular need. Since the choice of enhancement technique is problem-dependent, no single standard method of enhancement can be said to be 'best', for the needs of each user may differ. Also, the characteristics of each image in terms of the distribution of pixel values over the 0–255 display range will change from one area to another; thus, enhancement techniques suited to one image (for example, covering an area of forest) will differ from the techniques applicable to an image of another kind of area (for example, the Antarctic ice-cap).

There are a number of general categories of enhancement technique and these are described in the following sections. As in many other areas of knowledge, the distinction between one type of analysis and another is a matter of personal taste; some kinds of image transformations (Chapter 6) or filtering methods (Chapter 7) can, for instance, reasonably be described as enhancement techniques. In this chapter we concentrate on ways of improving the visual interpretability of an image by one of two methods:

(a) Altering image contrast, and
(b) Converting from greyscale to colour representation.

The first group of techniques consists of those methods which can be used to compensate for inadequacies of what, in photographic terminology, would be called 'exposure'; some images are intuitively felt to be 'too dark', while others are over-bright. In either case, information is not as easily comprehended as it might be if the contrast of the image were greater. In this context, contrast is simply the range and distribution of the pixel values over the 0–255 scale used by the computer's display memory. The second category includes those methods that allow the information content of greyscale image to be re-expressed in colour. This is sometimes desirable, for the eye is more sensitive to variations in hue than to changes in brightness.

Computer Processing of Remotely-Sensed Images: An Introduction, Third Edition. Paul M. Mather.
 ISBNs: 0-470-84918-5 (HB); 0-470-84919-3 (PB)

The chapter begins with a brief description of the human visual system, since the techniques covered in the following sections are fundamentally concerned with the visual comprehension of information displayed in image form.

5.2 HUMAN VISUAL SYSTEM

There are a number of theories that seek to explain the manner in which the human visual system operates. The facts on which these theories are based are both physical (to do with the external, objective world) and psychological (to do with our internal, conscious world). Concepts such as 'red' and 'blue' are an individual's internal or sensory response to external stimuli. Light reaching the eye passes through the pupil and is focused onto the retina by the lens (Figure 5.1). The retina contains large numbers of light-sensitive photoreceptors, termed rods and cones. These photoreceptors are connected via a network of nerve fibres to the optic nerve, along which travel the signals that are interpreted by the brain as images of our environment. There are around 100 million rod-shaped cells on the retina, and 5 million cone-shaped cells. Each of these cells is connected to a nerve, the junction being called a synapse. The way in which these cells respond to light is through alteration of a molecule known as a chromophore. Changes in the amount of light reaching a chromophore produce signals that pass through the nerve fibre to the optic nerve. Signals from the right eye are transmitted through the optic nerve to the left side of the brain, and vice versa.

It is generally accepted that the photoreceptor cells, comprising the rods and cones, differ in terms of their inherent characteristics. The rod-shaped cells respond to light at low illumination levels, and provide a means of seeing in such conditions. This type of vision is called *scotopic*. It does not provide any colour information, though different levels of intensity can be distinguished. Cone or photopic

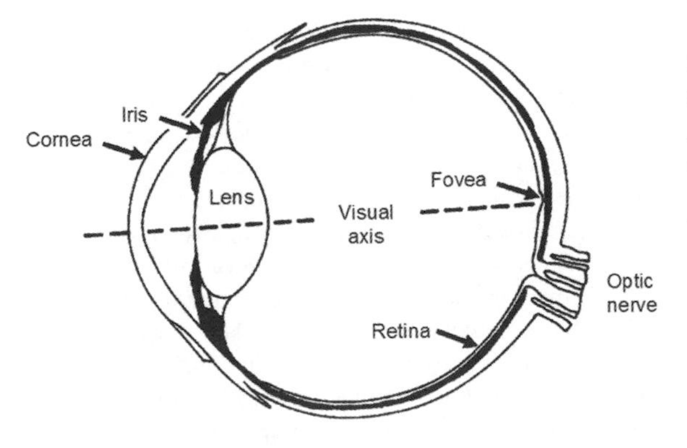

Figure 5.1 Simplified diagram of the human eye.

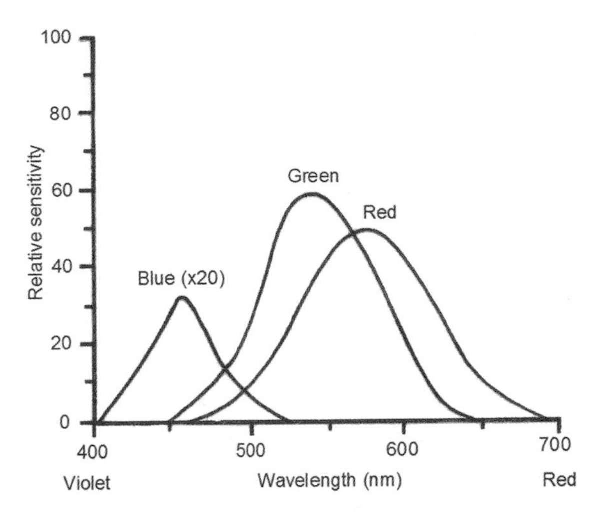

Figure 5.2 Sensitivity of the eye to red, green and blue light.

vision allows the distinction of colours or hues and the perception of the degree of saturation (purity) of each hue as well as the intensity level. However, photopic vision requires a higher illumination level than does scotopic vision. Colour is thought to be associated with cone vision because there are three kinds of cones, each kind being responsive to one of the three primary colours of light (red, green and blue). This is called the tri-stimulus theory of colour vision. Experiments have shown that the number of blue-sensitive cones is much less than the number of red- or green-sensitive cones, and that the areas of the visible spectrum in which the three kinds of cones respond do, in fact, overlap (Figure 5.2). There are other theories of colour (Land, 1977; Overheim and Wagner, 1982; Tyo *et al.*, 2003; Wasserman, 1978) but the tri-stimulus theory is an attractive one not merely because it is simple but because it provides the idea that colours can be formed by adding red, green and blue light in various combinations.

A model of 'colour space' can be derived from the idea that colours are formed by adding together differing amounts of red, green and blue light. Figure 5.3. shows a geometrical representation of the RGB (red, green, blue) colour cube. The origin is at the vertex of the cube marked 'black' and the axes are black-red, black-green and black-blue. A specific colour can be specified by its coordinates along these three axes. These coordinates are termed (R, G, B) triples. Note that white light is formed by the addition of maximum red, maximum green and maximum blue light. The line joining the black and white vertices of the cube represents colours formed by the addition of equal amounts of red, green and blue light; these are shades of grey. Colour television makes use of the RGB model of

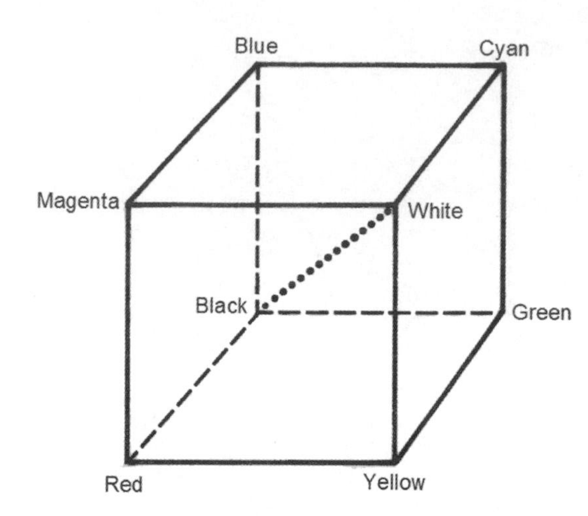

Figure 5.3 Red-green-blue colour cube.

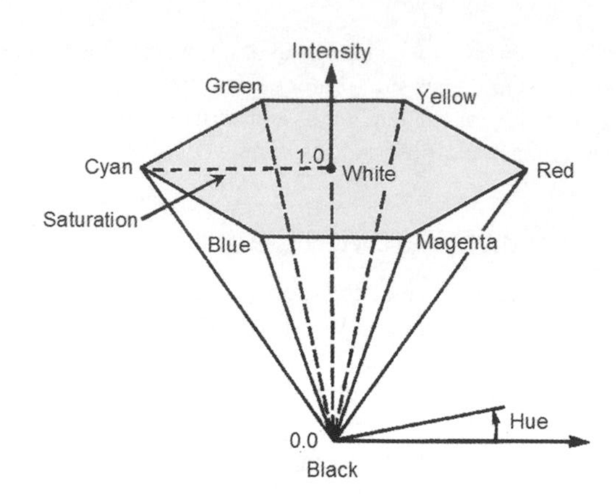

Figure 5.4 Hue-saturation-intensity (HSI) hexcone.

colour vision. A television screen is composed of an array of dots, each of which contains red-, green- and blue-sensitive phosphors. Colours on the screen are formed by exciting the red, green and blue phosphors in differing proportions. If the proportions of red, green and blue were equal at each point (but varying over the area of the screen) a greyscale image would be seen. A colour picture is obtained when the amounts of red, green and blue at each point are unequal, and so – in terms of the RGB cube – the colour at any pixel is represented by a point that is located away from the black–white diagonal line.

The RGB colour cube model links intuitively with the tri-stimulus theory of colour vision and also with the way in which a colour television monitor works. Other colour models are available which provide differing views of the nature of our perception of colour. The HSI model uses the concepts of hue (H), saturation (S) and intensity (I) to explain the idea of colour. Hue is the dominant wavelength of the colour we see; hues are given names such as red, green, orange and magenta. The degree of purity of a colour is its saturation. A pure colour is 100% saturated. Intensity is a measure of the brightness of a colour. Figure 5.4 shows a geometrical representation of the HSI model. Hue is represented by the top edge of a six-sided cone (hexcone) with red at 0°, green at 120° and blue at 240°. Pure unsaturated and maximum intensity colours lie around the top edge of the hexcone. Addition of white light produces less saturated, paler, colours and so saturation can be represented by the distance from the vertical axis of the hexcone. Intensity (sometimes called value) is shown as a distance above the apex of the hexcone, increasing upwards as shown by the widening of the hexcone. The point marked black has no hue, nor do any of the shades of grey lying on the vertical axis between the black and white points. All these shades of grey, including white and black, have zero saturation.

The RGB model of colour is that which is normally used in the study and interpretation of remotely-sensed images, and in the rest of this chapter we will deal exclusively with this model. The use of the HSI model is considered in Chapter 6 in common with other image transforms, for the representation of colours in terms of the HSI model can be accomplished by a straightforward transformation of the RGB colour coordinates. The HSI transform can be used to enhance multispectral images in terms of their colour contrast (section 5.3).

Further details of the material covered in this section can be found in Foley *et al.* (1990, 1994), Smith (1978), Williams and Becklund (1972) and Stimson (1974). Drury (1993) discusses the properties of the human visual system in relation to the choice of techniques for processing remotely-sensed images for geological applications.

5.3 CONTRAST ENHANCEMENT

The sensors mounted onboard aircraft and satellites have to be capable of detecting upwelling radiance levels ranging from low (for example, over oceans) to very high (for example, over snow or ice). For any particular area that is being imaged it is unlikely that the full dynamic range from 0 to $(2^n - 1)$ levels of the sensor will be used and the corresponding image is dull and lacking in contrast or overbright. In terms of the RGB colour cube model of section 5.2 the pixel values are clustered around a narrow section of the black-white axis (Figure 5.3). Not much detail can be seen on such images, which are either under-exposed or over-exposed in photographic terms. If the range of levels used by the display system could be altered so as to fit the full range of the black–white axis of Figure 5.3 then the contrast between the dark and light areas of the image would be improved while maintaining the relative distribution of the grey levels.

5.3.1 Linear contrast stretch

In its basic form the linear contrast-stretching technique involves the translation of the image pixel values from the observed range V_{min} to V_{max} to the full range of the display device (generally 0–255, which assumes 8-bit display memory; see chapter 3). V is a pixel value observed in the image under study, with V_{min} being the lowest pixel value in the image and V_{max} the highest. The pixel values are scaled so that V_{min} maps to a value of 0 and V_{max} maps to a value of 255. Intermediate values retain their relative positions, so that the observed pixel value in the middle of the range from V_{min} to V_{max} maps to 127. Note that we cannot map the middle of the range of the observed pixel values to 127.5 (which is exactly half way between 0 and 255) because the display system can store only the discrete levels 0, 1, 2, . . . , 255.

Some dedicated image processing systems include a hardware lookup table that can be set so that the colour that you see at a certain pixel position on the screen is a mapping or modification of the colour in the corresponding position in the display memory. The colour code in the display memory remains the same, but the mapping function may transform its value, for example by using the linear interpolation procedure described in the preceding paragraph. The fact that the colour values in the display memory are not altered can be a major advantage if the user has adopted a trial and error approach to contrast enhancement. The mapping is accomplished by the use of a lookup table that has 256 entries, labelled 0–255. In its default state, these 256 elements contain the values 0–255. A pixel value of, say, 56 in the display memory is not sent directly to the screen, but is passed through the lookup table. This is done by reading the value held in position 56 in the lookup table. In its default (do nothing) state, entry 56 in the lookup table contains the value 56, so the screen display shows exactly what is contained in the display memory. To perform a contrast stretch, we first realise that the number of separate values contained in the display memory for a given image is calculated as $(V_{max} - V_{min} + 1)$, which must be 256 or less. All lookup table (LUT) output values corresponding to input values of V_{min} or less are set to zero, while LUT output values corresponding to input values of V_{max} or more are set to 255. The range from V_{min} to V_{max} is then linearly mapped onto the range 0–255, as shown in Figure 5.5. Using the LUT shown in this figure, any pixel in the image having the value 16 (the minimum pixel value in the image) is transformed to an output value of 0 before being sent to the digital-to-analogue converter and thence to the display screen. All input values of 191 and more are transformed to output values of 255. The range of input values between 16 and 191 is linearly interpolated onto the full dynamic range of the display device, assumed in this example to be 0–255. If a colour (RGB) image is being stretched then the process is repeated separately for each of the components (R then G then B). Figure 5.6 illustrates the effects of a contrast stretch in which the lowest and highest pixel values in the image are 'stretched' to 0 and 255, respectively.

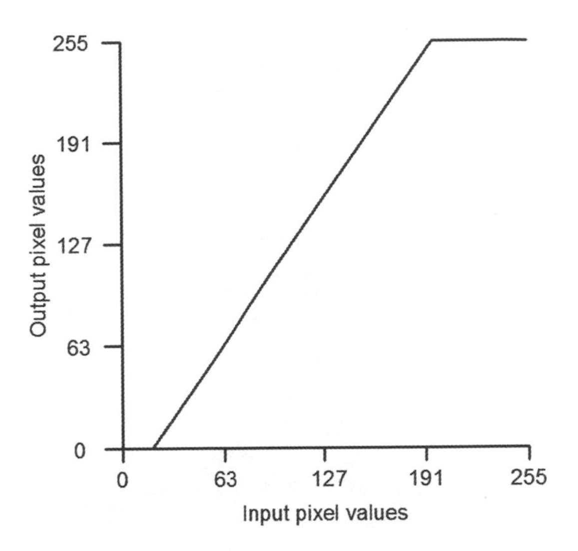

Figure 5.5 Graphical representation of lookup table to map input pixel values 16–191 on to the full intensity range 0–255. Input values less than 16 are set to 0 on output. Input values of 191 or greater are set to 255 on output. Input values between 16 and 191 inclusive are interpolated to output values 0–255.

The Microsoft Windows operating system treats images as 'bitmaps'. It does not provide for the use of lookup tables for 24-bit RGB images (i.e., eight bits for each of the red, green, and blue bands), though one is provided for 8-bit greyscale images. Rather than provide separate subprograms for greyscale and false colour images, the MIPS software treats a greyscale image as a 24-bit RGB image with the red, green and blue display memories all holding the same image. The contrast enhancement mapping described in the previous paragraph is carried out directly on the values stored in the display memory, without the use of a lookup table. Modern processors and data buses (cables connecting the different components of the computer) are now so fast that the speed penalty is not great.

A slight modification can be used to provide a little more user interaction. The basic technique, as described above, does not take into consideration any characteristic of the

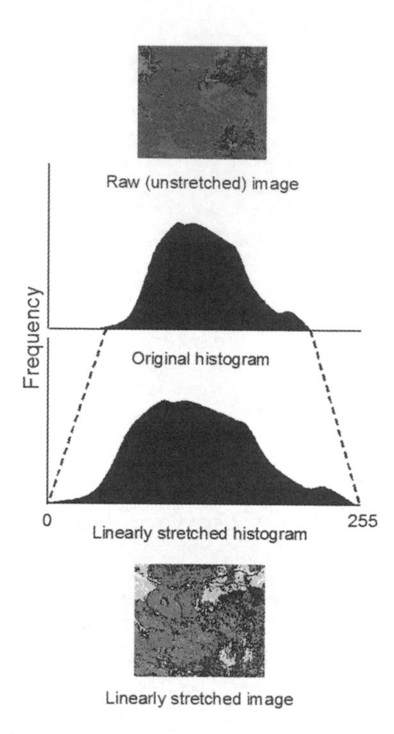

Figure 5.6 Linear contrast stretch. The extreme (highest and lowest) values of the histogram of the raw image are 'dragged' to the points 255 and 0 respectively on the x-axis of the linearly stretched histogram. The contrast stretched image (bottom) is much brighter than the original (top).

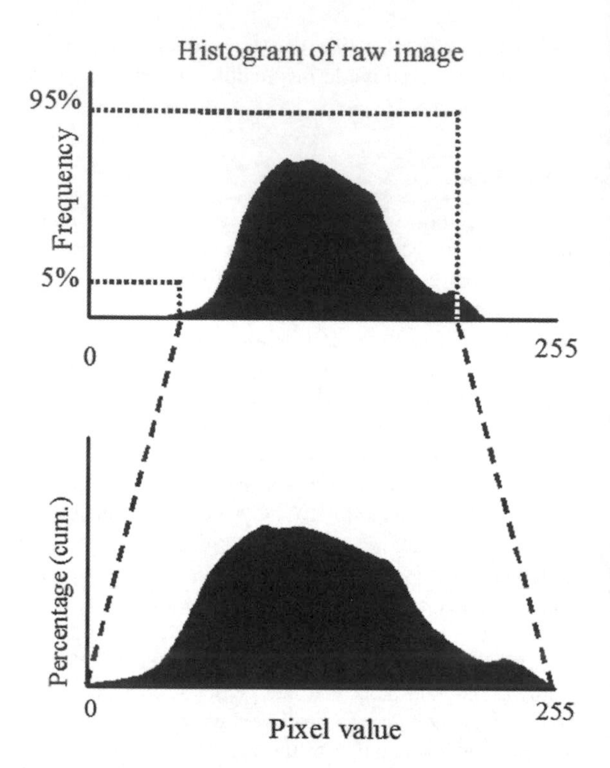

Figure 5.7 The image histogram is a plot of pixel value (0–255, x-axis) against frequency (y-axis). The raw (unstretched) image histogram is shown in the upper part of the figure. The data are stored in 8-bit form, so the theoretical range of the image pixel values is 0 to 255. The points marked 5% and 95% on the y-axis correspond to the pixel values (x-axis) for which (i) 5% of image pixel values are smaller and (ii) 5% of the image pixel values are larger (i.e., 95% are smaller), respectively. These two points are 'stretched' to points 0 and 255 on the x-axis of the lower part of the figure, thus increasing image contrast.

image data other than the maximum and minimum pixel values. These values may be outliers, located well away from the rest of the image data values. If this were the case, then it could be observed if the image histogram were computed and displayed. The image histogram for a single-band greyscale image or for one band (red, green or blue) of a false-colour image is a plot of the number of pixel values in the image having the values 0, 1, 2, ..., 255, respectively, against those class values. The histogram is an important tool in digital image processing, and we return to its uses in later sections. For the present, it is sufficient to note that it is a relatively straightforward matter to find the 5th and 95th (or any other) percentile values of the distribution of pixel values from inspection of the image histogram. The fifth percentile point is that value exceeded by 95 percent of the image pixel values, while the 95th percentile is the pixel value exceed by 5 percent of all pixel values in the image. If, instead of V_{max} and V_{min} we use $V_{95\%}$ and $V_{5\%}$ then we can carry out the contrast enhancement procedure so that all pixel values equal to or less than $V_{5\%}$ are output as zero while all pixel values greater than $V_{95\%}$ are output as 255. Those values lying between $V_{5\%}$ and $V_{95\%}$ are linearly mapped, as before, to the full brightness scale of 0 to 255. Again, this technique is applied separately to each component (red, green or blue) of a false-colour image. Of course, values other than the 5th and 95th percentiles could be used; for example, one might elect to choose the 10th and 90th percentiles, or any other pair. The chosen percentage points are usually symmetric around the 50% point, but not necessarily so. Figures 5.7 and 5.8 illustrate the application of the percentage linear contrast stretch.

Some image data providers use the value zero as a 'bad pixel' indicator. Others pad a geometrically corrected image (section 4.3) with zero pixels. These are called zero-fill pixels. We normally do not want to count these pixels in the image histogram, as their inclusion would bias the histogram. The MIPS contrast stretch module includes a check box that allows zero pixels to be ignored. Another possibility is that interest centres around a particular part of

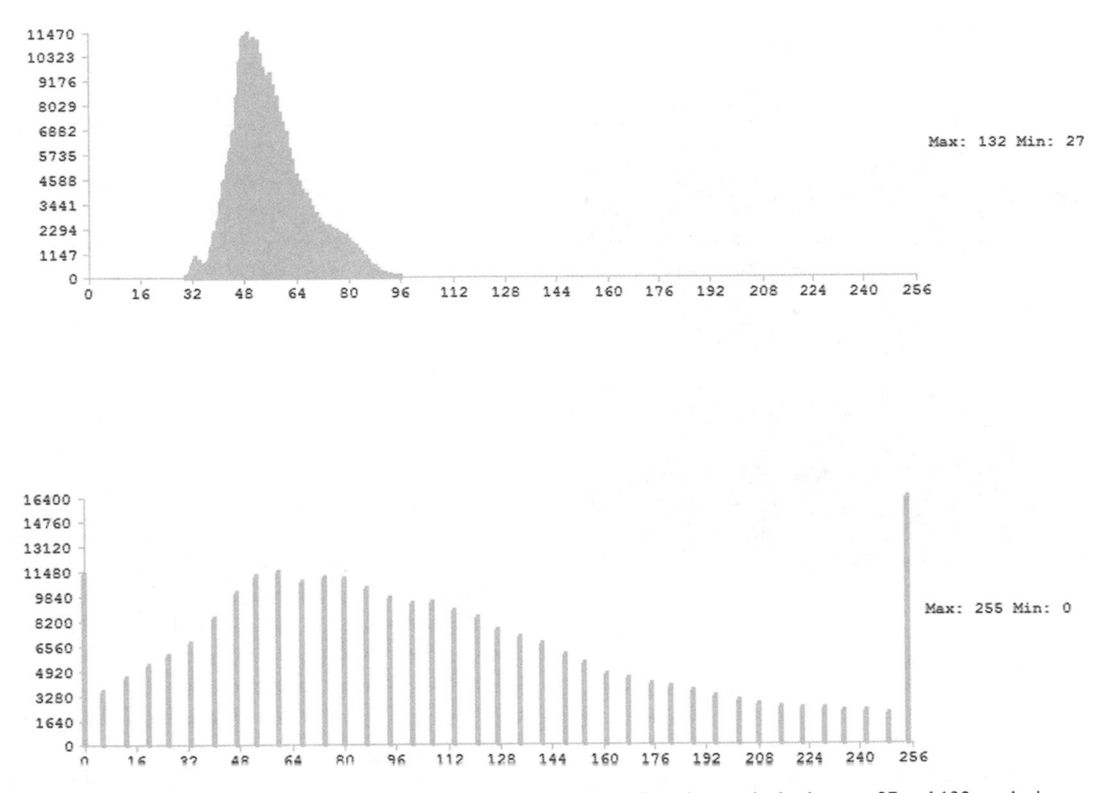

Figure 5.8 (a) Histogram of Landsat TM image. Note that the minimum and maximum pixel values are 27 and 132, so the image will appear under-exposed in photographic terms. (b) Histogram after a '5% to 95%' contrast stretch has been applied (see Figure 5.7).

the 0–255 brightness range, such as 180–250. It is possible to stretch this range so that a pixel value of 180 maps to zero, and a pixel value of 250 maps to 255 with values 181–249 being interpolated linearly. Values outside the 180–250 range remain the same. This kind of contrast stretch destroys the relationship between pixel value and brightness level, but may be effective in visualisation of a particular aspect of the information content of the image.

The use of the standard and percentage point contrast stretches is described in Example 5.1, using a Landsat TM sub-image of the Mississippi south-west of Memphis, Tennessee.

EXAMPLE 5.1 LINEAR CONTRAST STRETCH

This exercise uses the image set that is referenced by the dictionary file *missis.inf*, so you should ensure that this file (and the referenced image files) are available before you begin. The image file set consists of seven Landsat TM bands of an area of the Mississippi River south of Memphis, Tennessee. The images have 512 lines of 512 pixels.

Use *View|Display Image* to select *missis.inf* as the current image dictionary file. Display band 4 (*misstm4.img*) in greyscale. The default state of the lookup table (LUT) is to map a pixel intensity value of x (in the range of 0–255) to a display value of x, also in the range of 0–255. If the pixel values in the image are low, then the image will appear dark. Example 5.1 Figure 1 shows an image that is so dark that it may as well have been taken at night. The distribution of image pixel values can be seen by displaying the image histogram (*Plot|Histogram*), and it is immediately clear why the image is so dark – its dynamic range is very low (Example 5.1 Figure 2). The extreme values are 4 and 79, though the majority of the pixel values are in the range 10–55.

Now choose *Enhance|Stretch* and, if there is more than one image on the screen, select the appropriate window. Click the button with the caption *Automatic Stretch* and the image shown in Example 5.1 Figure 3 is generated, and displayed in a new window. The Automatic Stretch sets the lookup table so that the actual range of the data – x_{min} to x_{max} – is mapped linearly onto the output brightness

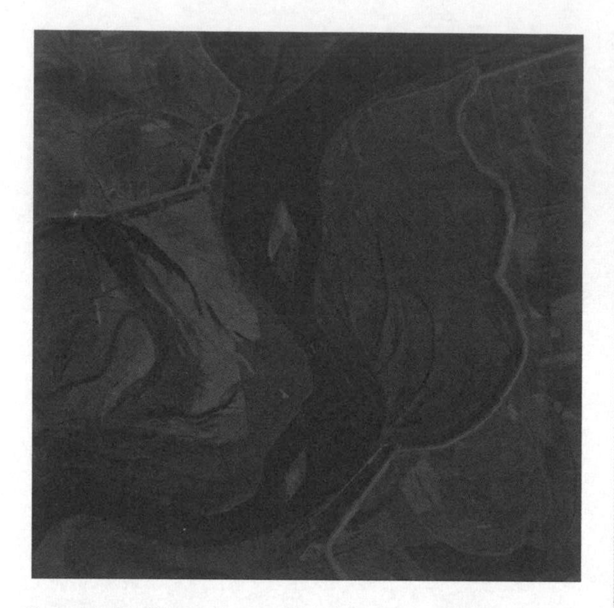

Example 5.1 Figure 1 Band 4 of Landsat-4 TM image of the Mississippi River near Memphis (details in the file *missisp.inf*). The dynamic range of the image is very low, and no detail can be seen. The histogram of this image is shown in Example 5.1 Figure 3.

range of 0–255. Use *Plot|Histogram* to view the histogram of the automatically stretched image (Example Figure 5.4).

Although the image shown in Example 5.1 Figure 3 is much easier to see than the image shown in Example 5.1 Figure 1, it is still rather dark, possibly because x_{min} and x_{max}, the extreme pixel intensity values, are outliers. Try a third experiment. Choose *Enhance|Stretch* again, and this time click *User percentage limits*. An input box will appear, with space for you to enter a lower value and an upper value. If you enter 0% and 100% then the minimum and maximum values in the data, x_{min} and x_{max}, will be used as the limits of the linear stretch, just as if you had chosen *Automatic Stretch*. What is needed is to chop off the extreme low and the extreme high pixel values. Look at the histogram of the raw image (Example 5.1 Figure 2) again. The extreme values (from visual inspection) appear to be 8 and 58, yet the lowest pixel value is 4 and the highest is 79. These relatively few low and high values are distorting the automatic stretch. If we ignore the lowest 5% and the highest 5% of the pixel values then we may see an improvement in image brightness. You actually enter 5 and 95 in the two boxes because they represent the percentages of pixels that are *lower* than the specified percentage points – in other words, we specify the 5th and 95th percentiles. The result is shown in Example 5.1 Figure 5, and the corresponding histogram is shown in Example 5.1 Figure 6.

You should relate the visual differences between the images shown in Example 5.1 Figures 1, 3 and 5 to the shape of the corresponding histograms (Example 5.1 Figures 2, 4 and 6) in order to appreciate the relationship between brightness and contrast on the one hand, and the shape and spread of the corresponding histogram.

Now try the following experiments:

- Generate stretched images from the raw image *misstm4.img* using the automatic and manual (*Set DN limits manually*) options. How would you estimate the upper and lower limits of the stretch for the manual option? What happens if you use a range narrower than the 13–42 range used by the 5%–95% option? Explain the differences between the output images.
- Try the *Histogram Equalize* and *Gaussian Stretch* options, using a range of standard deviation limits for the latter. Look at the histograms for each output image. What happens if you opt to use a specified part of the histogram rather than its full range? What happens if you

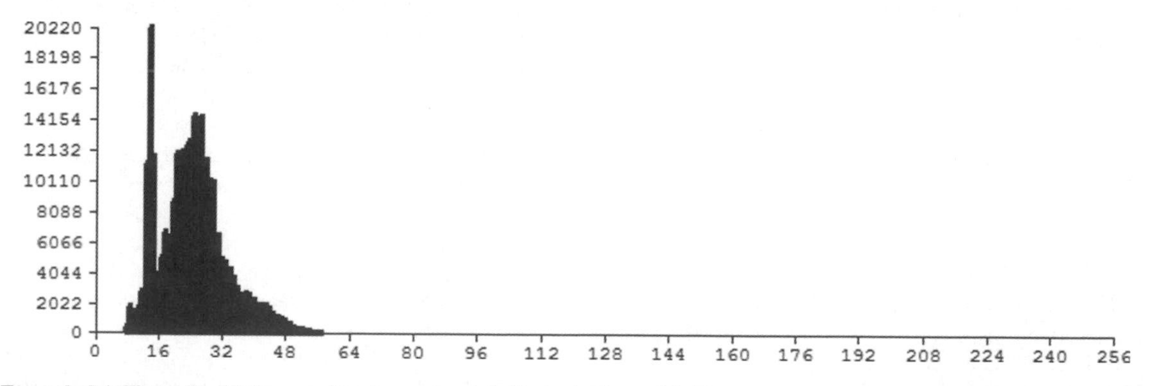

Example 5.1 Figure 2 Histogram of the image shown in Example Figure 5.1. The narrow peak at a pixel value of 14–15 represents water. The main, wider peak represents land. There are few pixel values greater than 58–60, so the image is dark. The range of pixel values is not great (approximately 8–60) and so contrast is low.

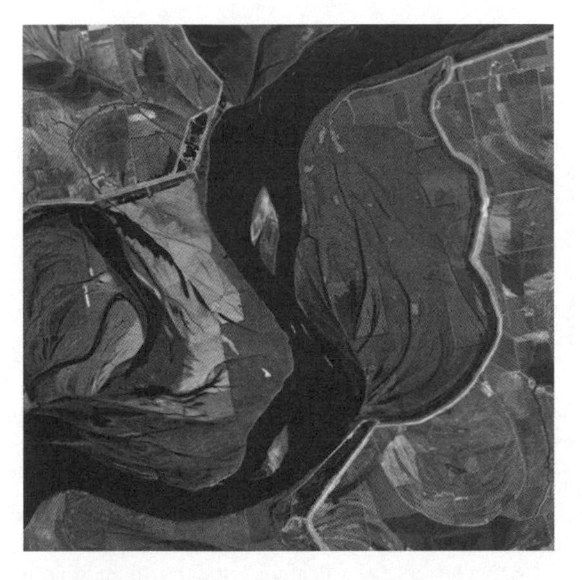

Example 5.1 Figure 3 The image shown in Figure 1 after an automatic linear contrast stretch. The automatic stretch maps the dynamic range of the image (8–60 in this case) to the dynamic range of the display (0–255). Compare the histogram of this image (Example 5.1 Figure 4) with the histogram of the raw image (Example 5.1 Figure 2).

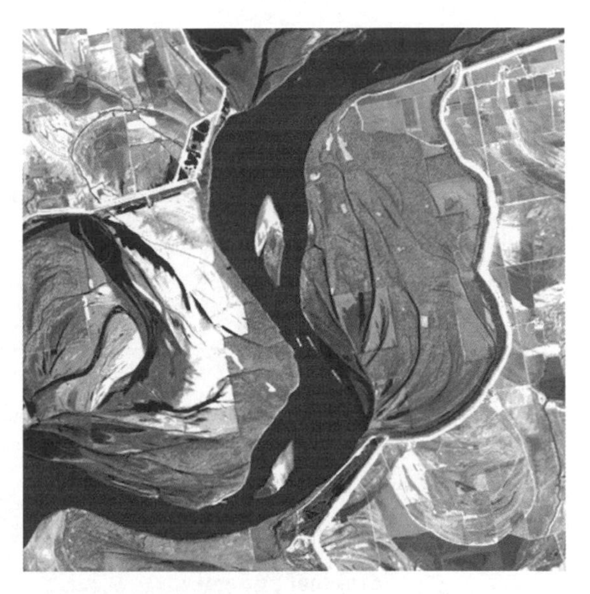

Example 5.1 Figure 5 The same image as shown in Example 5.1 Figures 1 and 3. This time, a percentage linear contrast stretch has been applied. Rather than map the lowest image pixel value to an output brightness value of zero, and the highest to 255, two pixel values are found such that 5% of all image pixel values are less than the first value and 5% are greater than the second. These two values are then mapped to 0 and 255, respectively.

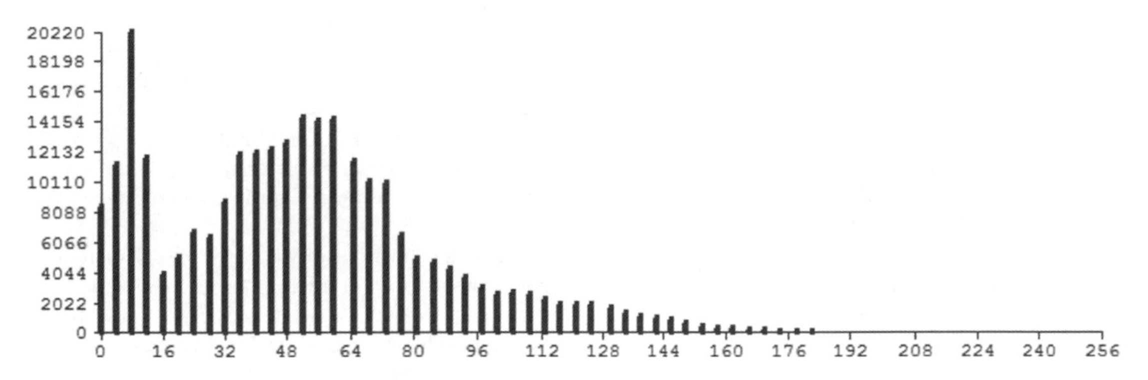

Example 5.1 Figure 4 Histogram of the contrast-stretched image shown in Example 5.1 Figure 3. Although the lower bound of the dynamic range of the image has been changed from its original value of 8 to zero, the number of pixels with values greater than 182 is relatively low. This is due to the presence of a small number of brighter pixels that are not numerous enough to be significant, but which are mapped to the white end of the dynamic range (255).

increase or decrease the number of standard deviations to either side of the mean in the Gaussian option?

- Which technique produces the 'best' output image? How would you define 'best'?
- Repeat the exercise with a different image, such as *Colo4.img* from the image set referenced by *litcolorado.inf*. Can you say that these techniques always work, or that one method is always better than the rest?

5.3.2 Histogram equalisation

The whole image histogram, rather than its extreme points, is used in the more sophisticated methods of contrast enhancement. Hence, the shape as well as the extent of the histogram is taken into consideration. The first of the two methods described here is called histogram equalisation. Its underlying principle is straightforward. It is assumed that in a well-balanced image the histogram should be such that

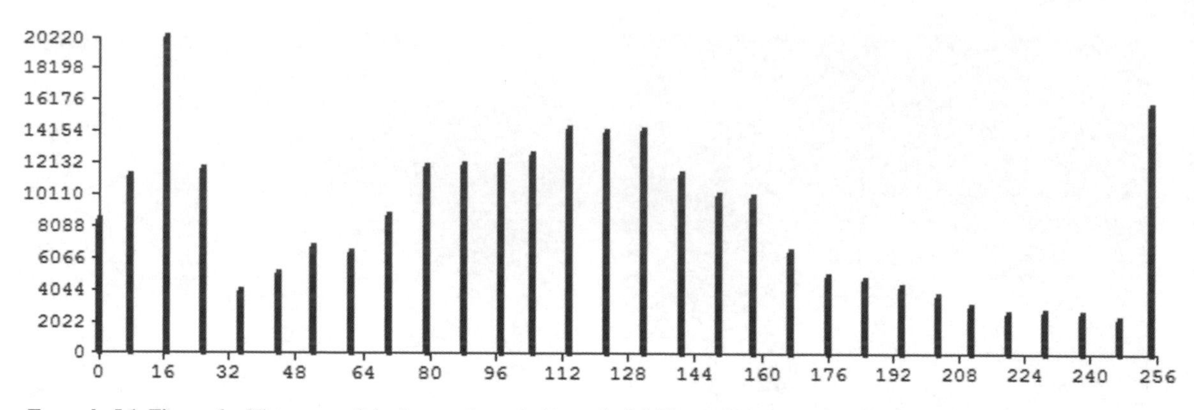

Example 5.1 Figure 6 Histogram of the image shown in Example 5.1 Figure 5. A percentage linear contrast stretch (using the 5% and 95% cutoff points) has been applied. The displayed image is now brighter and shows greater contrast than the image in Example 5.1 Figure 3.

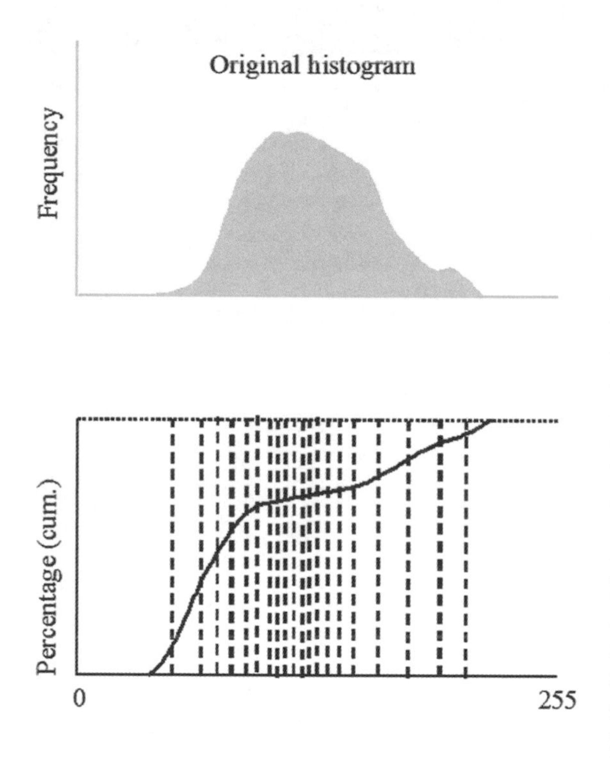

Figure 5.9 In the histogram equalisation procedure the histogram of the raw image is re-expressed in cumulative form. The cumulative histogram is divided by the vertical dashed lines into segments that each contains an approximately equal proportion of the image pixel values. The number of segments is generally less than the maximum of 256, leading to possible loss of detail, but this is compensated for by the increased contrast and average brightness.

each available brightness level contains an approximately equal number of pixel values, so that the histogram of these displayed values is almost uniform (though not all 256 classes are necessarily non-zero). If this operation, called histogram equalisation, is performed then the entropy of the image, which is a measure of the information content of the image, will be increased (section 2.2.3). Because of the nature of remotely-sensed digital images, whose pixels can take on only the discrete values 0, 1, 2, . . . , 255 it may be that there are 'too many' pixel values in one class, even after equalisation. However, it is not possible to take some of the values from that over-populated class and redistribute them to another class, for there is no way of distinguishing between one pixel value of 'x' and another of the same value. It is rare, therefore, for a histogram of the pixel values of an image to be exactly uniformly distributed after the histogram equalisation procedure has been applied.

The method itself involves, firstly, the calculation of the target number of pixel values in each class of the equalised histogram. This value (call it n_t) is easily found by dividing N, the total number of pixels in the image, by 256 (the number of histogram classes, which is the number of intensity levels in the image). Next, the histogram of the input image is converted to cumulative form with the number of pixels in classes 0 to j represented by C_j. This is achieved by summing the number of pixels falling in classes 0 to j of the histogram (the histogram classes are labelled 0–255 so as to correspond to the pixel values on an eight-bit scale):

$$C_j = n_0 + n_1 + \cdots + n_j$$

where n_j is the number of pixels taking the greyscale value j. The output level for class j is calculated very simply as C_j/n_t. See Figure 5.9.

Table 5.1 (a) Illustrating calculations involved in histogram equalisation procedure. $N = 262\,144$, $n_t = 16{,}384$. See text for explanation.

Old LUT value	Number in class	Cumulative number	New LUT value
0	1311	1311	0
1	2622	3933	0
2	5243	9176	0
3	9176	18 352	1
4	13 108	31 460	1
5	24 904	56 364	3
6	30 146	86 510	5
7	45 875	132 385	8
8	58 982	191 367	11
9	48 496	239 863	14
10	11 796	251 659	15
11	3932	255 591	15
12	3932	259 523	15
13	2621	262 144	15
14	0	262 144	15
15	0	262 144	15

(b) Number of pixels allocated to each class after the application of the equalisation procedure shown in Figure 5.1(a). Note that the smaller classes in the input have been amalgamated, reducing the contrast in those areas, while larger classes are more widely spaced, giving greater contrast. The number of pixels allocated to each non-empty class varies considerably, because discrete input classes cannot logically be split into sub-classes.

Intensity	0	1	2	3	4	5	6	7
Number	9176	22 284	0	24 904	0	30 146	0	0
Intensity	8	9	10	11	12	13	14	15
Number	45 875	0	0	58 982	0	0	48 496	22281

The method is not as complicated as it seems, as the example in Table 5.1 demonstrates. The pixel values in the raw image (16 levels in the example) are shown in column 1 of Table 5.1(a), and the number of pixels at each level is given in column 2. The cumulative number of pixels is listed in column 3. The values in column 4 are obtained by determining the target number of pixels (equal to the total number of pixels divided by the number of classes, that is, $262\,144 \div 16 = 16\,384$) and then finding the integer part of C_j divided by n_t, the target number. Thus, input levels 0 to 2 are all allocated to output level 0, input levels 3 and 4 are allocated to output level 1, and so on. Note that the classes with relatively low frequency have been amalgamated while the classes with higher frequency have been spaced out more widely than they were originally. The effect is to increase the contrast in the centre of the range while reducing contrast at the margins. Table 5.1(b) gives the numbers of pixels assigned to each output level. In this example, which uses only 16 levels for ease of understanding, the output histogram is not uniform. This is not surprising, for the number of pixels at five of the input levels considerably exceeds the target number of 16 384.

The example shows that the effect of the histogram equalisation procedure is to spread the range of pixel values present in the input image over the full range of the display device; in the case of a colour monitor. This range is normally 256 levels for each of the primary colours (red, green, blue). The relative brightness of the pixels in the original image is not maintained. Also, in order to achieve the uniform histogram the number of levels used is almost always reduced (see example, Table 5.1). This is because those histogram classes with relatively few members are amalgamated to make up the target number, n_t. In the areas of the histogram that have the greatest class frequencies the individual classes are stretched out over a wider range. The effect is to increase the contrast in the densely populated parts of the histogram and to reduce it in other, more sparsely populated areas. If there are relatively few discrete pixel values after the equalisation process then the result may be unsatisfactory compared to the simple linear contrast stretch.

Sometimes it is desirable to equalise only a specified part of the histogram. For example, if a mask is used to eliminate part of the image (for example, water areas may be set to zero) then a considerable number of pixel values of zero will be present in the histogram. If there are N zero pixels then the output value corresponding to an input pixel value of zero after the application of the procedure described above will be N/n_t which may be large; for instance, if N is equal to 86 134 and n_t is equal to 3192 then all the zero (masked) values in the original image will be set to a value of 27 if the result is rounded to the nearest integer. A black mask will thus be transformed into a dark grey one, which may be undesirable. The calculations described above can be modified so that the input histogram cells between, say, 0 and a lower limit L are not used in the calculations. It is equally simple to eliminate input histogram cells between an upper limit H and 255; indeed, any of the input histogram cells can be excluded from the calculations.

Histogram equalisation is implemented in MIPS. Two options are provided: use the entire histogram, or specify input histogram limits, to allow elimination of either the lower or the upper tail of the histogram, or both. For example, the histogram equalisation may be performed for the range of pixel values 100–199. In this case, image pixel values 0–99 and 200–255 are unaltered but values 100–199 are mapped onto the whole output brightness range. This process can lead to some unusual effects, and should therefore be used with care. The image shown in Figure 5.10 has been histogram equalised using the full range of the input image.

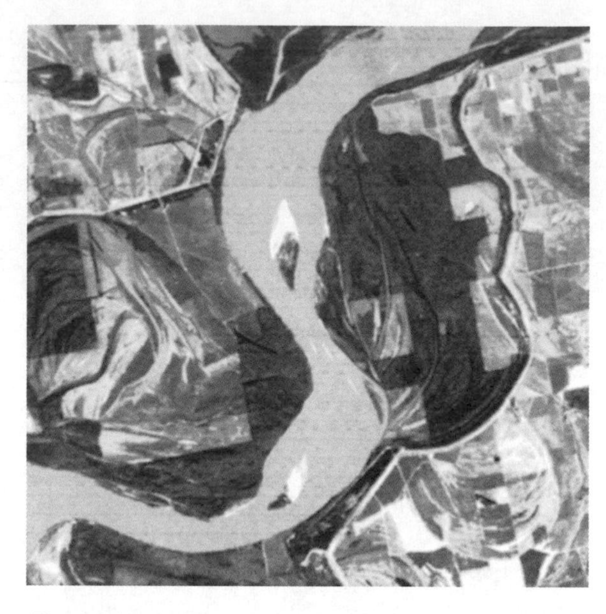

Figure 5.10 TM band 3 of the Mississippi image used in Example 5.1 following a histogram equalisation contrast stretch. The horizontal banding over the water area (section 4.2.2) is emphasised by this stretch. Can you explain why the river water in the TM band 4 image of the same area shown in Figure 1 of Example 5.1 appears black, whereas in this image (TM band 3), the river water is light grey?

5.3.3 Gaussian stretch

A second method of contrast-enhancement based upon the histogram of the image pixel values is called a *Gaussian stretch* because it involves the fitting of the observed histogram to a Normal or Gaussian histogram. A Normal distribution gives the probability of observing a value x given the mean $\bar{x}$ is defined by:

$$p(x) = \frac{1}{\sigma\sqrt{2\pi}} e^{\frac{-(x-\bar{x})^2}{2\sigma^2}}$$

The standard deviation, σ, is defined as the range of the variable for which the function $p(x)$ drops by a factor of $e^{-0.5}$ or 0.607 of its maximum value. Thus, 60.7% of the values of a normally distributed variable lie within one standard deviation of the mean. For many purposes a Standard Normal distribution is useful. This is a Normal distribution with a mean of zero and a unit standard deviation. Values of the Standard Normal distribution are tabulated in standard statistics texts, and formulae for the derivation of these values are given by Abramowitz and Stegun (1972). The shape of the Normal distribution, which will be familiar to many readers, is sketched in Figure 5.11. It is evident from this figure that the probability (or y value) decreases away from the mean in a systematic and symmetrical fashion.

An example of the calculations involved in applying the Gaussian stretch is shown in Table 5.2. The input histogram is the same as that used in the histogram equalisation example (Table 5.1). Again, 16 levels are used for the sake of simplicity. Since the Normal distribution ranges in value from minus infinity to plus infinity, some delimiting points are needed to define the end points of the area of the distribution that are to be used for fitting purposes. The range ± 3 standard deviations from the mean is used in the example. Level 1 is, in fact, the probability of observing a value of a Normally distributed variable that is 3 standard deviations or more below the mean; level 2 is the probability of observing a value of a Normally distributed variable that is between 2.6 and 3 standard deviations below the mean, and

Table 5.2 (a) Fitting observed histogram of pixel values to a Gaussian histogram. See text for discussion.

(i)	(ii)	(iii)	(iv)	(v)	(vi)	(vii)	(viii)
0	<−3.0	0.0020	530	530	1311	1311	1
1	−2.6	0.0033	868	1398	2622	3933	3
2	−2.2	0.0092	2423	3821	5243	9176	3
3	−1.8	0.0220	5774	9595	9176	18 352	4
4	−1.4	0.0448	1175	21 346	13108	31 460	5
5	−1.0	0.0779	20421	41 767	24904	56 364	6
6	−0.6	0.1156	30303	72 070	30146	86 510	7
7	−0.2	0.1465	38401	110 471	45875	132 385	8
8	0.2	0.1585	41555	152 026	58982	191 367	10
9	0.6	0.1465	38401	190 427	48496	239 863	11
10	1.0	0.1156	30303	220 730	11 796	251 659	12
11	1.4	0.0779	20421	241 151	3932	255 591	13
12	1.8	0.0448	11751	252 902	3932	259 523	14
13	2.2	0.0220	5774	258 676	2621	262 144	15
14	2.6	0.0092	2423	261 099	0	262 144	15
15	>3.0	0.0040	1045	262 144	0	262 144	15

(b) Number of pixels at each level following transformation to Gaussian model.

Intensity	(0)	(1)	(2)	(3)	(4)	(5)	(6)	(7)
Number	0	1311	0	7865	9176	13108	24904	30146
Intensity	(8)	(9)	(10)	(11)	(12)	(13)	(14)	(15)
Number	45 875	0	58 782	48 496	11 796	3932	3932	2621

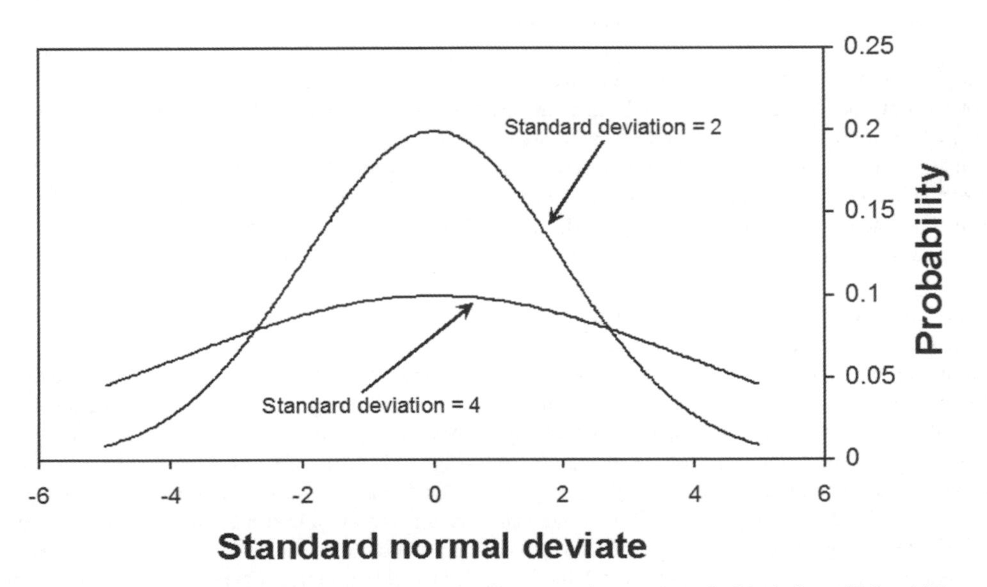

Figure 5.11 Showing two Normal (Gaussian) distributions each with a mean of zero and standard deviations of 2.0 and 4.0, respectively.

so on. These values are derived from an algorithm based on the approximation specified by Abramowitz and Stegun (1972). Column (i) of the table shows the pixel values in the original, un-enhanced image. Column (ii) gives the points on the Standard Normal distribution to which these pixel values will be mapped, while column (iii) contains the probabilities, as defined above, which are associated with the class intervals. Assume that the number of pixels in the image is $512 \times 512 = 262\,144$ and the number of quantisation levels is 16. The target number of pixels (that is, the number of pixels that would be observed if their distribution were Normal) is found by multiplying the probability for each level by the value 262 144. These results are contained in column (iv) and, in cumulative form, in column (v). The observed counts for the input image are shown by class and in cumulative form in columns (vi) and (vii). The final column gives the level to be used in the Gaussian-stretched image. These levels are determined by comparing columns (v) and (vii) in the following manner. The value in column (vii) at level 0 is 1311. The first value in column (v) to exceed 1311 is that associated with level 1, namely, 1398; hence, the input level 0 becomes the output level 1. Taking the input (cumulative) value associated with input level 1, that is, 3933, we find that the first element of column (v) to exceed 3933 is that value associated with level 3 (9595) so input level 1 becomes output level 3. This process is repeated for each input level. Once the elements of column (viii) have been determined, they can be written to the lookup table and the input levels of column (i) will automatically map to the output levels of column (viii).

The range ± 3 standard deviations as used in the example is not the only one which could have been used. A larger or smaller proportion of the total range of the Standard Normal distribution can be specified, depending on the requirements of the user. Usually, the limits chosen are symmetric about the mean, and the user can provide these limits from a terminal. An example of the Gaussian stretch is given in Figure 5.12. This image should be compared with the un-enhanced and linear stretched versions of the same image given in Example 5.1 and the image after histogram equalisation (Figure 5.10). Readers might like to select one of these images as the 'best' for visual interpretation, but it should be noted that the adequacy of any contrast stretch depends on the shape of the histogram of the un-enhanced image, and on the user's needs.

If Tables 5.1 and 5.2 are compared it will be seen that the Gaussian stretch emphasises contrast in the tails of the distribution while the histogram equalisation method reduces contrast in this region. However, at the centre of the distribution the reverse may be the case, for the target number for a central class may well be larger for the Gaussian stretch than the histogram equalisation. In the worked example the target number for each class in the histogram equalisation was 16384; note that the target numbers for classes 5 to 10 inclusive in the Gaussian stretch exceed 16384. Input classes may well have to be amalgamated in order to achieve these target numbers. Tables 5.1(b) and 5.2(b) give the number of pixels allocated to each output class after the application of the histogram equalisation and Gaussian contrast stretches, respectively. In both cases, the range of levels allocated to the output image exceeds the range of pixel values in the input image; this will result in an overall brightening of the displayed image.

The application of contrast-enhancement techniques is discussed above in terms of a single greyscale image. The techniques can be used to enhance a false colour image by applying the appropriate process to the red, green and blue channels separately. Methods of simultaneously 'stretching' the colour components of a false-colour image are dealt with elsewhere (the HSI transform in section 6.5 and the decorrelation stretch in section 6.4.3).

If an image covers two or more spectrally distinctive regions, such as land and sea, then the application of the methods so far described may well be disappointing. In such cases, any of the contrast-stretching methods described above can be applied to individual parts of the range of pixel values in the image; for instance, the histogram equalisation procedure could be used to transform the input range 0–60 to the output range 0–255, and the same could be done for the input range 61–255. The same procedure could be used whenever distinct regions occur if these regions can be identified by splitting the histogram at one or more threshold points. While the aesthetic appeal of images enhanced in this fashion may be increased, it should be noted that pixels with considerably different radiance values will be assigned to the same displayed or output colour value. The colour balance will also be quite different from that resulting from a standard colour-composite procedure.

5.4 PSEUDOCOLOUR ENHANCEMENT

In terms of the RGB colour model presented in section 5.2, a greyscale image occupies only the diagonal of the RGB colour cube running from the 'black' to the 'white' vertex (Figure 5.3). In terms of the HSI model, grey values are ranged along the vertical or intensity axis (Figure 5.4). No hue or saturation information is present, yet the human visual system is particularly efficient in detecting variations in hue and saturation, but not so efficient in detecting intensity variations. Two methods are available for converting a greyscale image to colour. The colour rendition as shown in the output image is not true or natural, for the original

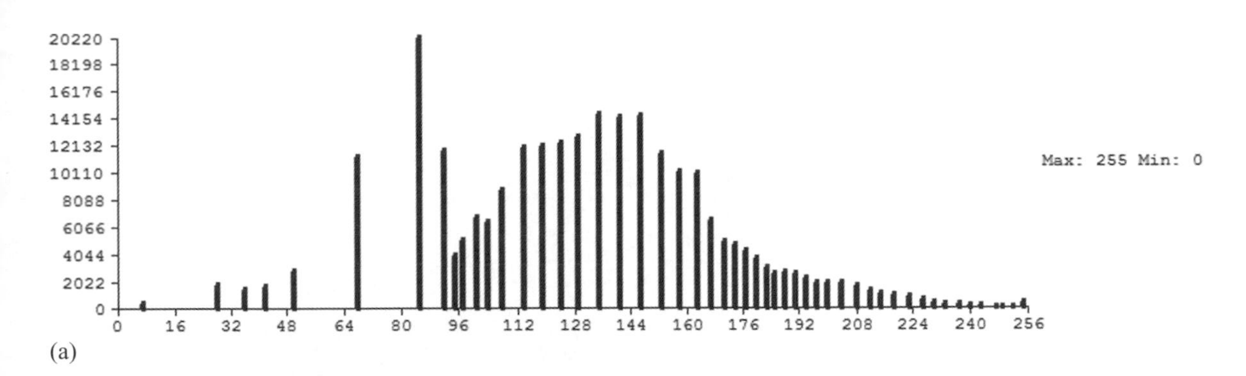

(a)

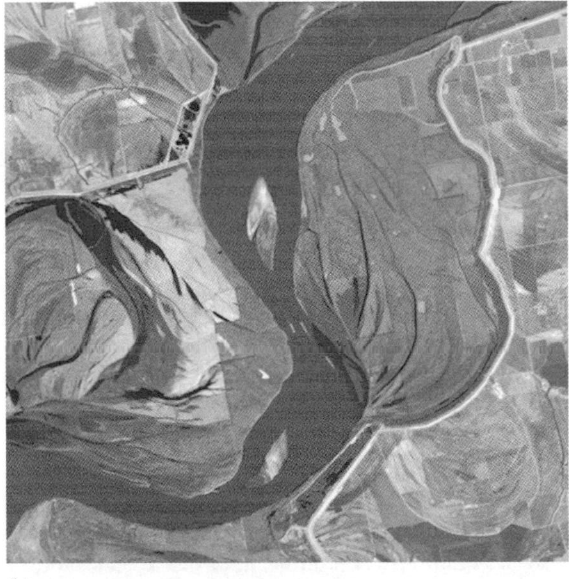

(b)

Figure 5.12 (a). Histogram of TM band 4 for the Mississippi image used in Example 5.1.A Gaussian stretch has been applied, using a cut-off of ±3 standard deviations. Note that, while adjacent histogram bins can be combined when their respective frequencies are lower than the target frequency, it is not possible to split a single, high-frequency, class such as that which occurs for the pixel value of 82, into smaller classes. The Gaussian stretch retains details in the upper and lower histogram classes. The stretched image is shown in Figure 5.12(b). (b). Landsat TM band 4 image of the Mississippi River after a Gaussian contrast stretch, described in Figure 5.12(a). Compare Figure 5.10 and the results of Example 5.1.

(input) image does not contain any colour information, and enhancement techniques cannot generate information that is not present in the input image. Nor is the colour rendition correctly described as false colour, for a false colour image is one composed of three bands of information which are represented in visible red, green and blue. The name given to a colour rendition of a single band of imagery is a pseudocolour image. Two techniques are available for converting from greyscale to pseudocolour form. These are the techniques of density slicing and pseudocolour transform. Each provides a method for mapping from a one-dimensional greyscale to a three-dimensional (RGB) colour.

5.4.1 Density slicing

Density slicing is the representation of a range of contiguous grey levels of a greyscale image by a single colour. The range of contiguous grey levels (such as 0 to 10 inclusive) is called a 'slice'. The greyscale range 0–255 is normally converted to several colour slices. It is acknowledged that conversion of a greyscale image to pseudocolour is an effective way of highlighting different but internally homogeneous areas within an image, but at the expense of loss of detail. The loss of detail is due to the conversion from a 256-level greyscale image to an image represented in terms of many fewer colour slices. The effect is (i) to reduce the number of discrete levels in the image, for several grey levels are usually mapped onto a single colour, and (ii) to improve the visual interpretability of the image if the slice boundaries and the colours are carefully selected.

In most image-processing systems the user is allowed to specify any colour for the current slice, and to alter slice boundaries in an upwards or downwards direction by means of a joystick or mouse. The slice boundaries are thus obtained by an interactive process, which allows the user to adjust the levels until a satisfactory result has been achieved. The choice of colour for each slice is important if information is to be conveyed to the viewer in any meaningful way, for visual perception is a psychological as well as a physiological process. Random colour selections may say more about the psychology of the perpetrator than about the

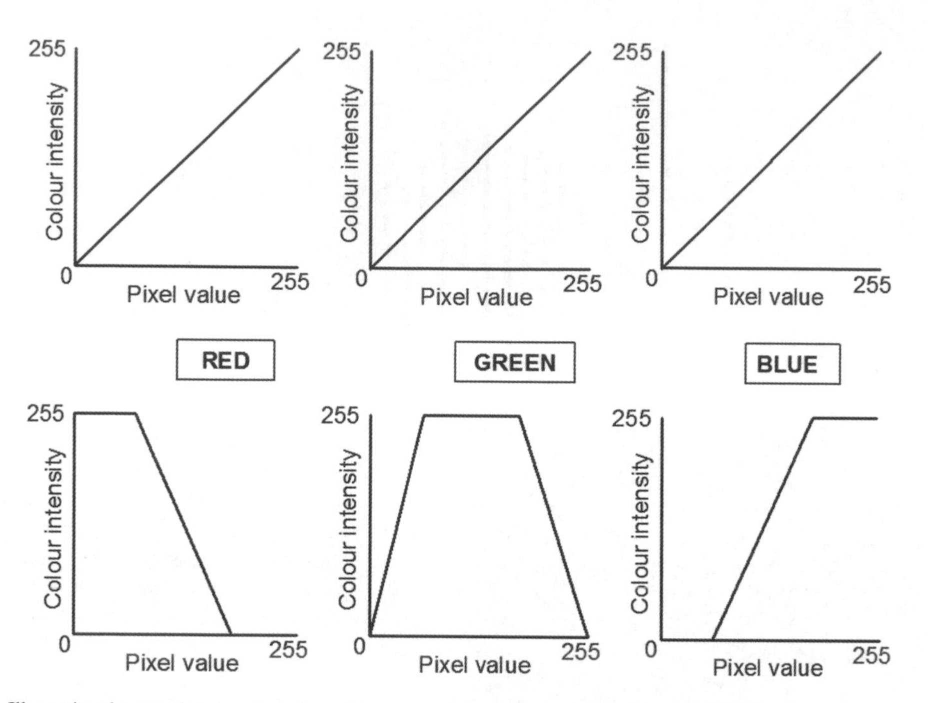

Figure 5.13 Illustrating the pseudocolour transform. The greyscale image is stored in all three (RGB) display memories, and the lookup tables for all three display memory are equivalent, sending equal RGB values to the screen at each pixel position. The pseudocolour transform treats each display memory separately, so that the same pixel value in each of the RGB display memories sends a different proportion of red, green and blue to the screen. For example, the pixel value '63' in a greyscale image would be seen on screen as a dark grey pixel. If the pseudocolour transform were to be applied, the pixel value '63' would transmit maximum red (255), maximum green (255) and no blue (0). This colour would be seen on screen as maximum yellow.

information in the image. Consider, for example, a thermal infrared image of the heat emitted by the Earth. A colour scale ranging from light blue to dark blue, through the yellows and oranges to red would be a suitable choice for most people have an intuitive feel for the 'meaning' of colours in terms of temperature. A scale taking in white, mauve, yellow, black, green and pink might confuse rather than enlighten.

Readers can use MIPS to experiment with the procedure for density slicing of a greyscale image. A single band of a suitable image should be selected, loaded and displayed, then the density slicing option used to select greyscale slices and convert them to different colours. Most readers will find that, although the technique is easily described in a textbook it is less easy to select the 'right' colours and the 'right' number of slices.

5.4.2 Pseudocolour transform

A greyscale image is can be considered to have equal red, green and blue values at each pixel position. A *pseudocolour transform* is carried out by changing the colours in the RGB display to the format shown in the lower half of Figure 5.13. The settings shown in the lower part of Figure 5.13 send different colour (RGB) information to the digital to analogue converter (and hence the screen) for the same greyscale pixel value. The result is ain image that pretends to be in colour. It is called a pseudocolour image. Unlike the density slicing method, the pseudocolour transform method associates each grey level with a discrete colour, although the difference between 90% red and 89% red may not be physically discernible in reality. If the histogram of the greyscale values in the greyscale image is not approximately uniform then the resulting pseudocolour image will be dominated by one colour, and its usefulness thereby reduced. Analysis of the image histogram along the lines of the histogram equalisation procedure (section 5.3) prior to the design of the pseudocolour LUTs would alleviate this problem.

The Fourier transform (section 6.6) can be used to split a greyscale image into its low, intermediate and high frequency components. These three components can subsequently be displayed as the red, green and blue inputs to a false-colour display. Such a display uses colour not only to enhance the visual interpretation of a greyscale image, but also to transmit information about the spatial scale components of that image.

5.5 SUMMARY

Image enhancement techniques include, but are not limited to, those of contrast improvement and greyscale to colour transformations. Other image-processing methods can justifiably be called enhancements. These include: (i) methods for detecting and emphasising detail in an image (section 7.4), (ii) noise reduction techniques, ranging from removal of banding (section 4.2.2) to filtering (Chapter 7), (iii) colour transforms based on principal components analysis, called the decorrelation stretch (section 6.4.3) and (iv) the hue-saturation-intensity (HSI) transform, section 6.5. All these methods alter the visual appearance of the image in such a way as to bring out or clarify some aspect or property of the image that is of interest to a user. The range of uses to which remotely-sensed images can be put is considerable and so, although there are standard methods of enhancement such as those described here, they should not be applied thoughtlessly but with due regard to the user's requirements and purpose.

6

Image Transforms

6.1 INTRODUCTION

An image transform is an operation that re-expresses the information content of an image or an image set. In principle, an image transform is like a good cook, who can take a set of ingredients and turn them into cakes of different types. There is nothing in each cake except the original ingredients, yet they look (and probably taste) different. There are many different ways of looking remotely-sensed images, just as there are many different ways of looking at social, economic or political questions.

A number of different transforms are considered in this chapter. The term 'transform' is used somewhat loosely, for the arithmetic operations of addition, subtraction, multiplication and division are included, although they are not strictly transforms. These operations, which are described in section 6.2, allow the generation of a derived image from two or more bands of a multispectral or multi-temporal image. The derived image may well have properties that make it more suited to a particular purpose than the original. For example, the numerical difference between two images collected by the same sensor on different days may provide information about changes that have occurred between the two dates, while the ratio of the near-infrared and red bands of a single-date image set is widely used as a vegetation index that correlates with difficult to measure variables such as vegetation vigour, biomass, and leaf area index (LAI).

Vegetation indices are also discussed in section 6.3. They are based on a model of the distribution of data values on two or more spectral bands considered simultaneously. Two examples of these transformations are the Perpendicular Vegetation Index (PVI), which uses a two-dimensional model of the relationship between vegetation and soil pixels, and the Tasselled Cap transformation, which is based on the optical (visible plus near-infrared) bands of a multispectral data set.

Section 6.4 provides an introduction to the widely-used technique of principal components analysis, which is a method of re-expressing the information content of a multispectral set of m images in terms of a set of m principal components, which have two particular properties: zero correlation between the m principal components, and maximum variance. The maximum variance property of principal components means that the components are extracted in order of decreasing variance. The first component is that linear combination of spectral bands that has the maximum variance of all possible linear combinations of the same spectral bands. The second principal component is that linear combination of spectral bands that has the maximum variance with respect to the remaining part of the data once the effects of the first principal component have been removed, and so on. The zero correlation property means that principal components are statistically unrelated, or orthogonal. It is usually found that much of the information in the original m correlated bands is expressible in terms of the first p of the full set of m principal components, where p is less than m. This property of principal components analysis is useful in generating a false-colour composite image. If the image set consists of more than three bands then the problem arises of selecting three bands for display in red, green and blue. Since the principal components of the image set are arranged in order of variance (which correlates with information) then the first three principal components can be used as the red, green and blue components of a false-colour composite image. No linear combination (i.e., weighted sum) of the original bands can contain more information than is present in the first three principal components. Another use of principal components analysis is in reducing the amount of calculation involved in automatic classification (Chapter 8) by basing the classification on p principal components rather than on m spectral bands. In addition, the p principal component images require less storage space than the m-band multispectral image from which they were derived. Hence, principal components analysis can also be considered to be a data compression transformation.

Rather than maximise the variance of the principal components, we could choose another criterion such as

Computer Processing of Remotely-Sensed Images: An Introduction, Third Edition. Paul M. Mather.
 ISBNs: 0-470-84918-5 (HB); 0-470-84919-3 (PB)

maximising the ratio of signal to noise (the signal-to-noise ratio, SNR). The standard principal components procedure, despite popular belief, does not remove noise. If we can estimate the level of noise in the data then we could extract components that are arranged in order of decreasing SNR. This modification of principal components analysis, which we will call noise-adjusted principal components analysis, is described in section 6.4.2.

Section 6.5 deals with a transformation that is concerned with the representation of the colour information in a set of three co-registered images representing the red, green and blue components of a colour image. Theories of colour vision are summarised in section 5.2 where it is noted that the conventional red-green-blue (RBG) colour cube model is generally used to represent the colour information content of three images. The hue-saturation-intensity (HSI) hexcone model is considered in section 6.5 and its applications to image enhancement and to the problem of combining images from different sources (such as radar and optical images) are described.

The transforms and operations described above act on two or more image bands covering a given area. Section 6.6 introduces a method of re-expressing the information content of a single-band greyscale image in terms of its frequency components. The discrete Fourier transform (DFT) provides for the representation of image data in terms of a coordinate framework that is based upon spatial frequencies rather than upon Euclidean distance from an origin (i.e., the conventional Cartesian or $x - y$ coordinate system). Image data that have been expressed in frequency terms are said to have been transformed from the image or spatial domain to the frequency domain. The frequency-domain representation of an image is useful in designing filters for special purposes (described in Chapter 7) and in colour coding the scale components of the image.

A related transformation, called the discrete wavelet transform, represents an attempt to bridge the gap between the spatial and frequency domains, for it decomposes the input signal (which may be one-dimensional, like a spectrum collected by a field spectrometer, or two-dimensional, like an image) in terms of wavelength (1D) or space (2D) *and* scale simultaneously. One major use of wavelets is to remove noise from (i.e., 'denoise') one- and two-dimensional signals. Wavelet-based denoising is described in Chapter 9, while the basics of the wavelet transform are considered in section 6.7.

With the increasing availability of imaging spectrometer data sets (section 9.3) that are composed of measurements in many tens or hundreds of spectral bands, methods of analysing and extracting information from the shape of spectral reflectance curves have been developed. An example is the derivative operation. The first derivative measures the slope of the spectral reflectance curve, while the second derivative measures the change in slope steepness. Both measures are useful in locating wavelengths of interest on the spectral reflectance curve, for example, the position of the 'red edge' in a vegetation spectrum. These and other advanced methods of analysing imaging spectrometer data are described in Chapter 9.

6.2 ARITHMETIC OPERATIONS

The operations of addition, subtraction, multiplication and division are performed on two or more co-registered images of the same geographical area (section 4.3). These images may be separate spectral bands from a single multispectral data set or they may be individual bands from image data sets that have been collected at different dates. Addition of images is really a form of averaging for, if the dynamic range of the output image is to be kept equal to that of the input images, re-scaling (usually division by the number of images added together) is needed. Averaging can be carried out on multiple images of the same area in order to reduce the noise component. Subtraction of pairs of images is used to reveal differences between those images and is often used in the detection of change if the images involved were taken at different dates.

Multiplication of images is rather different from the other arithmetic operations for it normally involves the use of a single 'real' image and a binary image made up of ones and zeros. The binary image is used as a mask, for those image pixels in the real image that are multiplied by zero also become zero, while those that are multiplied by one remain the same.

Division or ratioing of images is probably the arithmetic operation that is most widely applied to images in geological, ecological and agricultural applications of remote sensing, for the division operation is used to detect the magnitude of the differences between spectral bands. These differences may be symptomatic of particular land cover types. Thus, a near-infrared: red ratio might be expected to be close to 1.0 for an object which reflects equally in both of these spectral bands (for example, a cloud top) while the value of this same ratio will be well above one if the near-infrared reflectance is higher than the reflectance in the visible red band, for example in the case of vigorous vegetation.

6.2.1 Image addition

If multiple, co-registered, images of a given region are available for the same time and date of imaging then addition (averaging) of the multiple images can be used as a means of reducing the overall noise contribution. A single image might be expressed in terms of the following model:

$$G(x, y) = F(x, y) + N(x, y)$$

where $G(x, y)$ is the recorded image, $F(x, y)$ the true image and $N(x, y)$ the random noise component. $N(x, y)$ is often hypothesised to be a random Normal distribution with a mean of zero, since it is the sum of a number of small, independent errors or factors. The true signal, $F(x, y)$, is constant from image to image. Therefore, addition of two separate images of the same area taken at the same time might be expected to lead to the cancellation of the $N(x, y)$ term for, at any particular pixel position (x, y), the value $N(x, y)$ is as likely to be positive as to be negative. Image addition, as noted already, is really an averaging process. If two images $G_1(i, j)$ and $G_2(i, j)$ are added and if each has a dynamic range of 0–255 then the resulting image $G_{sum}(i, j)$ will have a dynamic range of 0–510. This is not a practicable proposition if the image display system has a fixed, 8-bit, resolution. Hence it is common practice to divide the sum of the two images by two to reduce the dynamic range to 0–255. The process of addition is carried out on a pixel-by-pixel basis as follows:

$$G_{sum}(i, j) = (G_1(i, j) + G_2(i, j))/2$$

The result of the division is normally rounded to the nearest integer. Note that if the operation is carried out directly on images stored on disc then the dynamic range may be greater than 0–255, and so the scaling factor of two will have to be adjusted accordingly. Images stored in the computer display memory have a dynamic range of 0–255. However, take care – recall from the discussion in section 5.3 that the application of a contrast stretch will cause the pixel values in the display memory to be changed, so adding together two contrast-enhanced images does not make sense. Always perform arithmetic operations on images that have not been contrast enhanced. A further difficulty is that images with a dynamic range of more than 0–255 are scaled before being written to display memory. The scaling used is automatic, and depends on the image histogram shape (section 3.2.1). Any arithmetic operation carried out on images with an extended dynamic range (i.e., greater than 0–255) that have been scaled to 0–255 for display purposes is likely to be meaningless. The operation should be carried out on the full range images stored on disc, not on their scaled counterparts.

If the algorithm described in the previous paragraph is applied then the dynamic range of the summed image will be approximately the same as that of the two input images. This may be desirable in some cases. However, it would be possible to increase the range by performing a linear contrast stretch (section 5.3.1) by subtracting a suitable offset o and using a variable divisor d:

$$G'(i, j) = (G_1(i, j) + G_2(i, j) - o)/d$$

The values of o and d might be determined on the basis of the user's experience, or by evaluating $G_{sum}(i, j)$ at a number of points systematically chosen from images G_1 and G_2. Image G' will have a stretched dynamic range in comparison with the result of the straight 'division by two'.

Other methods of removing noise from images include the use of the discrete wavelet transform to estimate a 'noise threshold' which is then applied to the data series (such as a one-dimensional reflectance spectrum or a two-dimensional image) in order to remove additive noise of the kind described above. The wavelet transform is considered further in section 6.7 and Chapter 9.

6.2.2 Image subtraction

The subtraction operation is often carried out on a pair of co-registered images of the same area taken at different times. The purpose is to assess the degree of change that has taken place between the dates of imaging (see, for example, Dale *et al.*, 1996). Image subtraction or differencing is also used to separate image components. In Chapter 9, for example, the wavelet transform is used to separate the signal (information) in an image from the noise. The difference between the original and de-noised image can be found by subtracting the de-noised image from the original image.

Image differencing is performed on a pixel-by-pixel basis. The maximum negative difference (assuming both images have a dynamic range of 0–255) is $(0 - 255 =)$ -255 and the maximum positive difference is $(255 - 0 =)$ $+255$. The problem of scaling the result of the image subtraction operation onto a 0–255 range must be considered. If the value 255 is added to the difference then the dynamic range is shifted to 0–510. Next, divide this range by 2 to give 0–255. Variable offsets and multipliers can be used as in the case of addition (section 6.2.1) to perform a linear contrast-stretch operation. Formally, the image subtraction process can be written as:

$$G_{diff}(i, j) = (255 + G_1(i, j) - G_2(i, j)/2$$

using the same notation as previously. If interest is centred on the magnitude rather than the direction of change then the following method could be used:

$$G_{absdiff}(i, j) = |G_1(i, j) - G_2(i, j)|$$

The vertical bars $|\cdot|$ denote the absolute value (regardless of sign). No difference is represented by the value 0 and the degree of difference increases towards 255. Note

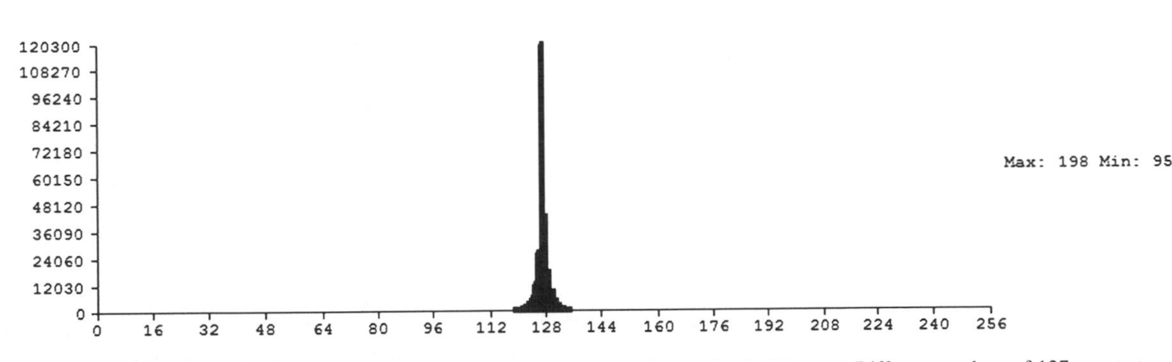

Figure 6.1 Histogram of the difference between two images, after scaling to the 0–255 range. Difference values of 127 equate to a real difference of zero. Histogram x-axis values lower than 127 indicate negative differences and values above 127 indicate positive differences.

the remarks above (section 6.2.1) on the logic of applying arithmetic operations to images with a dynamic range other than 0–255.

A difference image $G_{\text{diff}}(i, j)$ tends to have a histogram that is Normal or Gaussian in shape with the peak at a count of 127 (if the standard scaling is used), tailing off rapidly in both directions. The peak at 127 represents pixels that have not changed very much while the pixels in the histogram tails have changed substantially. The image $G_{\text{absdiff}}(i, j)$ has a histogram with a peak at or near zero and a long tail extending towards the higher values. It would be possible to use an interactive method such as the pseudocolour transform (section 5.4.2) of transforming selected parts of the dynamic range to colour, or shades of grey. If we had an image histogram such as that shown in Figure 6.1 then the areas of change could be highlighted if pixel values between 0 and 119 and between 135 and 255 were shown in red with values in the range 120–134 being represented in green (Figure 6.2). Pixels showing only a small change will be green under this scheme, while those pixels with a large difference will be red. The choice of threshold value (120 and 134) in the example is arbitrary. Normally these values would be fixed by a trial and error procedure involving inspection of the histogram of the differences shown in Figure 6.1.

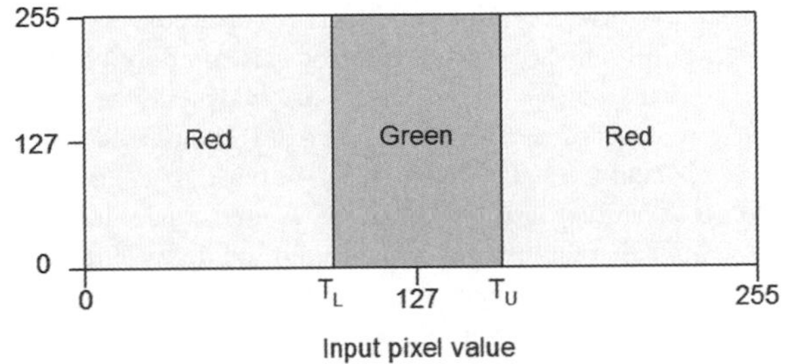

Figure 6.2 The frequency distribution of the pixel values produced by the subtraction operation is shown in Figure 6.1. We could code these values to indicate 'changed' (red) or 'unchanged' (green) using a density slicing operation. The values of the lower and upper thresholds (T_L and T_U) are selected by trial and error based on the histogram shown in Figure 6.1.

Jupp and Mayo (1982) provide an interesting example of the use of image subtraction. They use a four-band Landsat MSS image to generate a classified (labelled) image (Chapter 8) in which a single-band image is generated, with each pixel in the image being given a numerical label to indicate the land cover class to which it belongs. These labels are determined on the basis of the spectral characteristics of each pixel. For example, labels of 1, 2 and 3 could be used to indicate forest, grassland and water. The mean values of the pixels in each class in every band are computed, to produce a table of k rows and p columns, with k being the number of classes and p the number of spectral bands used in the labelling process. A new image set is then generated, with one image per spectral band. Those pixels with the label i in the classified image are given the mean value of class i. This operation results in a set of p 'class mean value' images. A third image set is then produced, consisting of p residual images which are obtained by subtracting the actual pixel value recorded in a given spectral band from the corresponding value in the 'mean' image for that pixel. Residual images can be combined for colour composite generation. The procedure is claimed to assist in the interpretation and understanding of the classified image, as it highlights pixels that differ from the mean value of all pixels allocated the same class. This could be useful in answering questions such as: 'have I omitted any significant land cover classes?' or 'why is class *x* so heterogeneous?'

Other examples of the use of change detection techniques are Collins and Woodcock (1996), Johnson and Kasischke (1998), Lambin (1996), Lambin and Strahler (1994), Siljeström and Moreno (1995) and Varjo (1996). Gong *et al.* (1992) discuss the effects of mis-registration of images (Chapter 4) on the subtraction process, and present methods to reduce these effects, which will appear on the difference image as areas of change. The book edited by Lunetta and Elvidge (1998) includes a number of contributions dealing with methods of detecting and measuring change using remotely-sensed images.

6.2.3 Image multiplication

Pixel-by-pixel multiplication of two images is rarely performed in practice. The multiplication operation is, however, a useful one if an image of interest is composed of two or more distinctive regions and if the analyst is interested only in one of these regions. Figure 6.3(a) shows a Landsat-2 MSS band 4 (green) image of part of the Tanzanian coast south of Dar-es-Salaam. Variations in reflectance over the land area distract the eye from the more subtle variations in the radiance upwelling from the upper layers of the ocean. The masking operation can eliminate variations over the distracting land region. The first step is the preparation of the mask that best separates land and water, using the near-infrared band (Figure 6.3(b)) since reflection from water bodies in the near-infrared spectral band is very low, while reflection from vegetated land areas is high, as noted in section 1.3. A suitable threshold is chosen by visual inspection of the image histogram of the near-infrared pixel values. A binary mask image is then generated from the near-infrared image by labelling with '1' those pixels that have values below the threshold (Figure 6.3(c)). Pixels whose values are above the threshold are labelled '0', so the mask image displays as a black-and-white image with the masked area appearing white. The second stage is the multiplication of the image shown in Figure 6.3(a) and the mask image (Figure 6.3(c)). Multiplication by 1 is equivalent to doing nothing, whereas multiplication by 0 sets the corresponding pixel in the masked image to 0. Using the above procedure, the pixels in the Tanzanian coast band 4 image that represent land are replaced by zero values, while 'ocean' pixels are unaltered. A linear contrast-stretch (section 5.3.1) produces the image shown in Figure 6.3(d). In practice, the pixel values in the two images (mask and band 4 in this example) are not multiplied but are processed by a simple logical function: if the mask pixel is zero then set the corresponding image pixel to zero, otherwise do nothing. Some software packages use the two extremes (0 and 255) of the range of pixel values to indicate 'less than' and 'greater than' the threshold, respectively.

6.2.4 Image division and vegetation indices

The process of dividing the pixel values in one image by the corresponding pixel values in a second image is known as ratioing. It is one of the most commonly used transformations applied to remotely-sensed images. There are two reasons why this is so. One is that certain aspects of the shape of spectral reflectance curves of different Earth surface cover types can be brought out by ratioing. The second is that undesirable effects on the recorded radiances, such as that resulting from variable illumination (and consequently changes in apparent upwelling radiance) caused by variations in topography can be reduced. Figure 6.4 shows the spectral reflectance curves for three cover types. The differences between the curves can be emphasised by looking at the gradient or slope between the red and the near-infrared bands, for example bands 3 (red) and 4 (near-infrared) in a Landsat ETM+ image, or bands 3 (near-infrared) and 2 (red) of the SPOT HRV image set. The shape of the spectral reflectance curve for water shows a decline between these two points, while that for vegetation shows a substantial increase. The spectral reflectance curve for soil increases gradually between the two bands. If a pixel value in the near-infrared band is divided by the equivalent value in the red band then the result will be a positive real number that exceeds 1.0 in magnitude. The same operation carried out on the curve for water gives a result that is less than 1.0, while the soil curve gives a value somewhat higher than 1.0. The greater the difference between the pixel values in the two bands, the greater the value of the ratio.

The two images may as well be subtracted if this were the only result to be derived from the use of ratios. Figure 6.5 shows a hypothetical situation in which the irradiance at point *B* on the ground surface is only 50% of that at *A* due to the fact that one side of the slope is directly illuminated by the Sun. Subtraction of the values in the two bands at point *A* gives a result that is double of that which would be achieved at point *B* even if both points are located on the same ground-cover type. However, the ratios of the two bands at *A* and *B* are the same because the topographic effect has been largely cancelled out in this instance. This is not always the case, as shown by the discussion below.

The two properties of the ratio, that is, reduction of the 'topographic effect' and correlation between ratio values and the shape of the spectral reflectance curve between

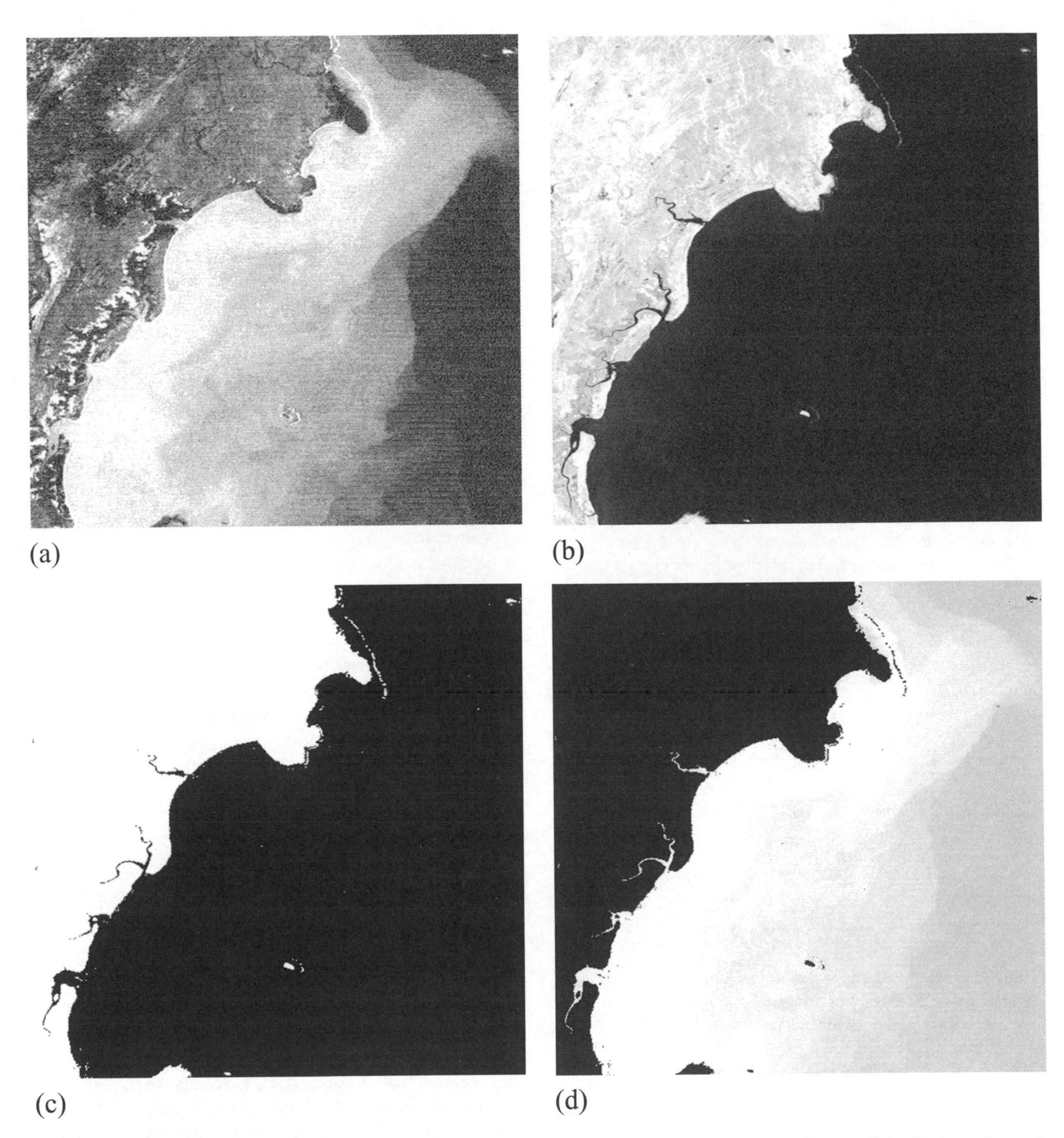

Figure 6.3 Illustrating the use of image multiplication in creating a land/sea mask that enables the full range of the display to be used to represent variations in green light penetration depth in a region of the Tanzanian coast, south of Dar es Salaam. (a) Landsat MSS Band 4 (green), (b) Landsat MSS Band 7 (near infrared), (c) Land/sea mask created from near infrared band (b). Land is shown in white. (d) Masked green band (a) with contrast stretch applied to the unmasked area.

two given wavebands, has led to the widespread use of spectral ratios in, for example, biogeography and geology. (Bannari *et al.*, 1995; Baret and Guyot, 1991; Curran, 1980, 1983; Gilabert *et al.*, 2002; Jackson, 1983; Rowan *et al.*, 1974; Pinty *et al.*, 1993; Tucker, 1979; van der Meer *et al.*, 1995). One of the most common spectral ratios used in studies of vegetation status is the ratio of the near infrared to the equivalent red band value for each pixel location.

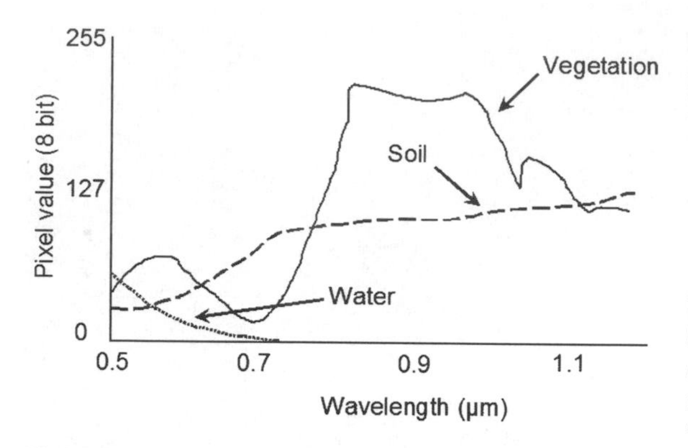

Figure 6.4 The ratio of a pixel value at near-infrared wavelengths (around 1.0 μm) to the corresponding pixel value in the red region of the spectrum (0.6–0.7 μm) will be large if the area represented by the pixel is covered by vigorous vegetation (solid curve). It will be around 1.0 for a soil pixel, but less than 1.0 for a water pixel. In effect, the IR/R ratio is measuring the slope of the spectral reflectance curve between the infrared and red wavelengths. Inspection of the curves shown in this figure shows that the curve for vegetation has a very significant slope in this region.

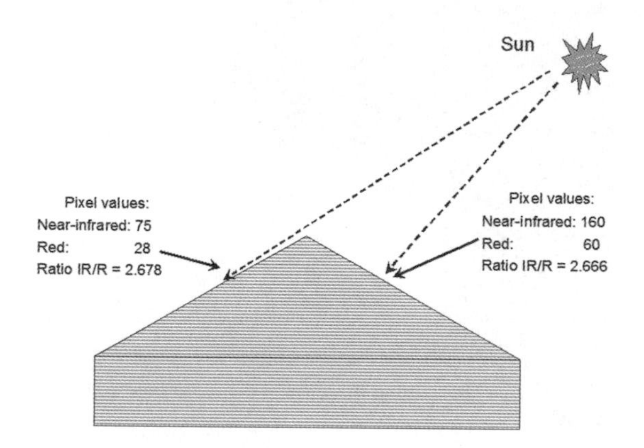

Figure 6.5 Ratio of pixel values in the near-infrared region to the corresponding pixel value in the visible red region of the spectrum. The ratios for the illuminated and shaded slopes are very similar, although the pixel values differ by a factor of more than two. Hence an image made up of IR: R ratio values at each pixel position will exhibit a much reduced shadow or topographic effect.

This ratio exploits the fact that vigorous vegetation reflects strongly in the near infrared and absorbs radiation in the red waveband (section 1.3.2.1). The result is a greyscale image which can be smoothed by a low-pass filter (section 7.2) and density sliced (section 5.4.1) to produce ain image showing variation in biomass (the amount of vegetative matter) and in green leaf area index (Bouman, 1992; Box *et al.*, 1989, Filella and Penuelas, 1994) as well as the state of health (physiological functioning) of plants. The theoretical justification of such interpretations is questioned by Myneni *et al.* (1995), who note that:

> "A central question remains unanswered: what do vegetation indices indicate?'

An example of a near-infrared:red ratio image is shown in Figure 6.6. The near-infrared band in Figure 6.6(a) is a SPOT HRV band 3 (near-infrared) image of the Nottingham area (the city centre is in the upper right, and agricultural fields can be seen in the top left corner and lower centre). The near-infrared image shows more detail in the non-urban areas than in the urban areas of the image. Figure 6.6(b) is the corresponding SPOT HRV band 2 image, showing red reflectance. Features of the urban areas are more readily seen on the HRV band 2 image. The ratio of the pixel values in the near infrared to those in the corresponding red image is shown in Figure 6.6(c) and the variation in grey level from 0 (black) to 255 (white) corresponds to the scale discussed above; water is dark and green fields are white. All three images have been contrast stretched using histogram equalisation for display purposes. The actual numerical values of the ratio could be used in any attempt to relate ratio values to biomass amounts or other indicators of vegetation health or productivity.

More complex ratios involve sums of and differences between spectral bands. For example, the Normalised Difference Vegetation Index (NDVI), defined in terms of the near infrared (NIR) and red (R) bands as:

$$\text{NDVI} = \frac{\text{NIR} - R}{\text{NIR} + R}$$

is preferred to the simple red: near-infrared ratio by many workers because the ratio value is not affected by the absolute pixel values in the near-infrared (NIR) and red (R) bands. The NDVI has been used to study global vegetation using bands 1 and 2 of the NOAA AVHRR. For example, Justice *et al.* (1985) use the NDVI in a study of vegetation patterns on a continental scale, and data produced by this method from Global Area Coverage of AVHRR are commercially available from NOAA. Figure 6.6(d) shows the NDVI image calculated for the Nottingham SPOT subimage used earlier. On the basis of visual evidence, the difference between the simple ratio and the NDVI is not great. However, the fact that sums and differences of bands are used in the NDVI rather than absolute values may make the NDVI more appropriate for use in studies where comparisons over time for a single area are involved, since the NDVI might be expected to be influenced to a lesser extent by variations in atmospheric conditions (but see below).

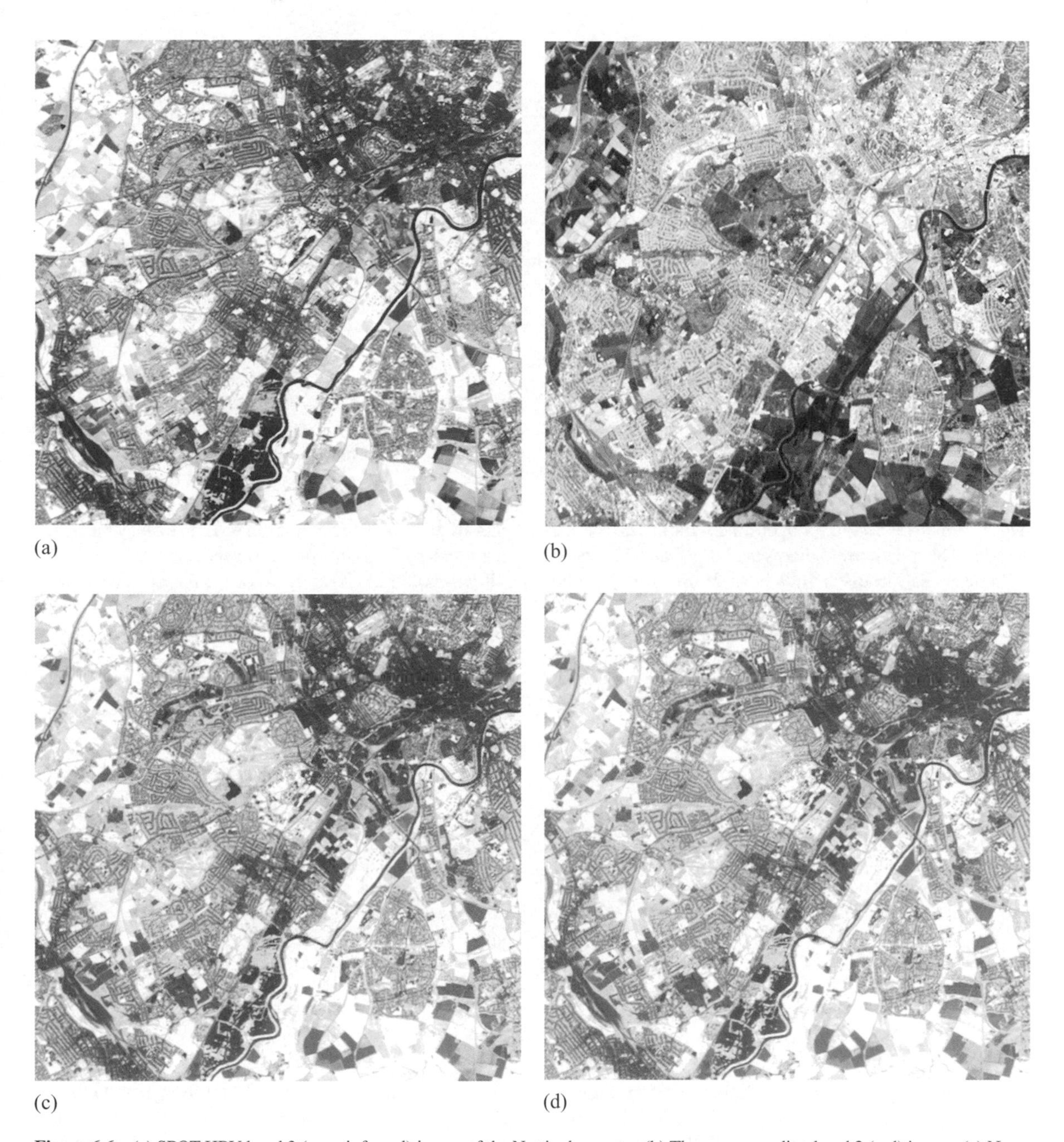

Figure 6.6 (a) SPOT HRV band 3 (near-infrared) image of the Nottingham area. (b) The corresponding band 2 (red) image. (c) Near-infrared:red ratio image. (d) Normalised Difference Vegetation Index (NDVI) image. CNES – SPOT Image Distribution. Permission to use the data was kindly provided by SPOT Image, 5 rue des Satellites, BP 4359, F-31030 Toulouse, France (*http://www.spotimage.fr*).

Sensor calibration issues, discussed in section 4.6, may have a significant influence on global NDVI calculations based on the NOAA AVHRR. Guttman and Ignatov (1995) show how the difference between pre- and post-launch calibrations lead to unnatural phenomena such as the 'greening of deserts'.

A range of different spectral band ratios is analysed by Jackson (1983), Logan and Strahler (1983) and Perry and

Lautenschlager (1984). The number of such ratios is considerable, and appears to be limited only by the imagination of the user and the degree of his or her determination to find an as-yet undiscovered combination. Fortunately, Perry and Lautenschlager (1984) find that most of these ratios are equivalent. Research into the derivation of 'better' vegetation indices is a different matter. The simple ratio and NDVI, plus other band ratios, are all affected by external factors such as the state of the atmosphere, illumination and viewing angles, and soil background reflectance. This is one reason why NDVI-composite images derived from multiple NOAA AVHRR require careful analysis. These images are produced by selecting cloud-free pixels from a number of images collected over a short period (a week or a month) to produce a single image from these selected pixels. Because of the orbital characteristics of the NOAA satellite it is probable that the NDVI values for adjacent pixels have been collected at different illumination and viewing angles. See section 1.3.1 for a discussion of the bidirectional reflectance properties of Earth surface materials. The relationship between NDVI and vegetation parameters is discussed by Chilar *et al.* (1991), Hobbs and Mooney (eds.) (1990), Pinty *et al.* (1993) and Todd *et al.* (1998). Further discussion of the problem of estimating vegetation characteristics from remotely-sensed data is contained in section 9.3, where the use of high spectral resolution (imaging spectrometer) data is discussed.

A class of indices called Soil Adjusted Vegetation Indices (SAVI) has been developed, and there are quite a number to choose from (the original soil-adjusted vegetation index, the transformed soil-adjusted vegetation index, TSAVI, the Global Environment Monitoring Index, GEMI, and a number of others). Bannari *et al.* (1995 b), Rondeaux (1995) and Rondeaux *et al.* (1996) provide an extensive review. Steven (1998) discusses the Optimised Soil-Adjusted Vegetation Index (OSAVI) and shows that the form:

$$\mathrm{OSAVI} = \frac{\mathrm{NIR} - R}{\mathrm{NIR} + R + 0.16}$$

minimises soil effects. Readers interested in pursuing this subject should refer to Baret and Guyot (1991), Huete (1989), Pinty *et al.* (1993) and Sellers (1989). Leprieur *et al.* (1996) assess the comparative value of the different vegetation indices using NOAA AVHRR data.

If ratio values are to be correlated with field observations of, for example, leaf area index (LAI) or estimates of biomass, or if ratio values for images of the same area at different times of the year are to be compared, then some thought should be given to scaling of the ratio values. Ratios are computed as 32-bit real (floating-point) values. Scaling is necessary only to convert the computed ratio values to integers on the range 0–255 for display purposes, as explained in section 3.2.1. If the user wishes to correlate field observations and ratio values from remotely-sensed images then unscaled ratio values should be used. MIPS can calculate the NDVI value at each pixel position. You can then select one of two options. First, as the NDVI is known to range between ± 1, the calculated NDVI values are arranged to that the lowest possible value (-1) is recorded as 0 (zero) and the highest ($+1$) as 255. Thus, the value zero is transformed to the 8-bit value 127. Two sets of images taken at different times can be compared directly using this scaling method, since the full range of the NDVI is transformed to the full 8-bit range of 0–255. A second MIPS option allows the user to scale the range $\mathrm{NDVI_{MIN}}$ to $\mathrm{NDVI_{MAX}}$ (the minimum and maximum NDVI values appearing in the image) onto the range 0–255. Since the values of $\mathrm{NDVI_{MIN}}$ and $\mathrm{NDVI_{MAX}}$ will normally differ from image to image, the results of this scaling operation are not directly comparable. The issue of scaling 16 and 32 bit integers and 32-bit real values onto a 0–255 range in order to display the image on a standard PC is considered in section 3.2.

It was noted above that one of the reasons put forward to justify the use of ratios was the elimination of variable illumination effects in areas of topographic slope. This, of course, is not the case except in a perfect world. Assume that the variation in illumination at point (i, j) can be summarised by a variable $c(i, j)$ so that, for the pixel at location (i, j), the ratio between channels q and p is expressed as:

$$R(i, j) = \frac{c(i, j)v(q, i, j)}{c(i, j)(v(p, i, j)} = \frac{v(q, i, j)}{v(p, i, j)}$$

where $v(q, i, j)$ is the radiance from pixel located at point (i, j) in channel q. The term $c(i, j)$ is constant for both bands q and p and therefore cancels out. If there were an additive as well as a multiplicative term then the following logic would apply:

$$R(i, j) = \frac{c(i, j)v(q, i, j) + r(q)}{c(i, j)v(p.i.j) + r(p)}$$

and it would be impossible to extract the true ratio $v(q, i, j)/v(p, i, j)$ unless the terms $r(p)$ and $r(q)$ were known or could be estimated. The terms $r(p)$ and $r(q)$ are the atmospheric path radiances for bands p and q (section 1.2.5) which are generally unknown. They are also unequal because the amount of scattering in the atmosphere increases inversely with wavelength. Switzer *et al.* (1981) consider this problem and show that the atmospheric path radiances must be estimated and subtracted from the recorded radiances before ratioing. A frequently used method, described in section 4.4.2, involves the subtraction of constants $k(p)$ and $k(q)$ which are the minimum values in the histograms of channels *p* and *q* respectively; these values might be expected to provide a first approximation to the path radiances in the two bands. Switzer *et al.* (1981) suggest that this 'histogram minimum' method over-corrects the data. Other factors, such as the magnitude of diffuse irradiance (skylight) and reflection from cross-valley slopes, also confuse the issue. It is certainly not safe to assume that the 'topographic effect' is completely removed by a ratio operation. Atmospheric effects and their removal, including the 'histogram minimum' method, are considered in Chapter 4.

The effects of atmospheric haze on the results of ratio analyses are studied in an experimental context by Jackson *et al.* (1983). These authors find that, for turbid atmospheres, the near infrared/red ratio was considerably less sensitive to variations in vegetation status and they conclude that

> 'The effect [of atmospheric conditions]on the ratio is so great that it is questionable whether interpretable results can be obtained from satellite data unless the atmospheric effect is accurately accounted for on a pixel-by-pixel basis' (p. 195).

The same conclusion was reached for the Normalised Difference Vegetation Index (NDVI). Holben and Kimes (1986) also report the results of a study involving the use of ratios of NOAA AVHRR bands 1 and 2 under differing atmospheric conditions. They find that the NDVI is more constant than individual bands.

Other problems relate to the use of ratios where there is incomplete vegetation coverage. Variations in soil reflectivity will influence ratio values, as discussed above. The angle of view of the sensor and its relationship with solar illumination geometry must also be taken into consideration if data from off-nadir pointing sensors such as the SPOT HRV or from sensors with a wide angular field of view such as the NOAA AVHRR are used (Barnsley, 1983; Wardley, 1984; Holben and Fraser, 1984). In order to make his data comparable over time and space, Frank (1985) converted the Landsat MSS digital counts to reflectances (as described in section 4.6) and used a first-order correction for solar elevation angle based on the Lambertian assumption (section 1.3.1; Figure 1.4(b); section 4.5). Other useful references are Chalmers and Harris (1981), Gardner *et al.* (1985), Kowalik *et al.* (1983), Maxwell (1976) and Misra and Wheeler (1978). As noted above, extensions to the standard ratio for the reduction of soil reflectance effects and atmospheric effects have been developed in the past few years. The general review by Rondeaux (1995) is recommended.

6.3 EMPIRICALLY BASED IMAGE TRANSFORMS

Experience gained during the 1970s with the use of Landsat MSS data for identifying agricultural crops, together with the difficulties encountered in the use of ratio transforms (section 6.2) and principal component transforms (section 6.4), led to the development of image transforms based on the observations that (i) scatter plots of Landsat MSS data for images of agricultural areas show that agricultural crops occupy a definable region of the four-dimensional space based on the Landsat MSS bands, and (ii) within this four-dimensional space the region occupied by pixels that could be labelled as 'soil' is a narrow, elongated ellipsoid. Pairwise plots of Landsat MSS bands fail to reveal these structures fully because they give an oblique rather than a 'head-on' view of the sub-space occupied by pixels representing vegetation. Kauth and Thomas (1976) propose a transformation that, by rotating and scaling the axes of the four-dimensional space, would give a more clear view of the structure of the data. They called their transform the *Tasselled Cap* since the shape of the region of the transformed feature space that was occupied by vegetation in different stages of growth appeared like a Scottish 'bobble hat'. Other workers have proposed other transforms; perhaps the best known is the *Perpendicular Vegetation Index* which was based on a similar idea to that of the Tasselled Cap, namely, that there is a definite axis in four-dimensional Landsat MSS space that is occupied by pixels representing soils, ranging from soils of low reflectance to those of high reflectance (see also Baret *et al.*, 1993). These two transformations are described briefly in the next two subsections.

6.3.1 Perpendicular Vegetation Index

A plot of radiance measured in the visible red band against radiance in the near-infrared for a partly-vegetated area will result in a plot that looks something like Figure 6.7. Bare soil pixels will lie along the line S_1–S_2, with the degree of wetness of the soil being higher at the S_1 end of the 'soil line' than at the S_2 end. Vegetation pixels will lie below and to the right of the soil line, and the perpendicular distance to the soil line was suggested by Richardson and Wiegand (1977) as a measure which was correlated with the green leaf area index and with biomass. The formula used by Richardson and Wiegand (1977) to define the Perpendicular Vegetation Index (PVI) is based on either Landsat-1–3 MSS band 7 or band 6, denoted by PVI7 and PVI6, respectively. Note that this rendition of the PVI is now of historical interest. It is examined here for illustrative purposes.

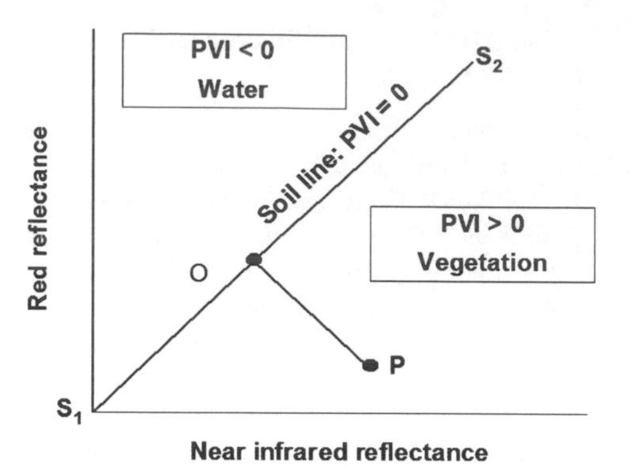

Figure 6.7 The 'soil line' S_1–S_2 joins the position of the expected red and near-infrared reflectance for wet soils (S_1) with that for dry soils (S_2). Vigorous vegetation shows high reflectance in the near-infrared (horizontal axis) and low reflectance at visible red wavelengths. Point P therefore represents a pixel that has high near-infrared and low visible red reflectance. The PVI measures the orthogonal distance from P to the soil line (shown by line PO).

Bands 6 and 7 of the Landsat MSS covered near-infrared regions and band 5 covered the visible green waveband.

$$PVI7 = \sqrt{(0.355\text{MSS7} - 0.149\text{MSS5})^2 + (0.355\text{MMSS5} - 0.852\text{MSS7})^2}$$

$$\text{PVI6} = \sqrt{(0.498\text{MSS6} - 0,487\text{MSS5} - 2.507)^2 + (2.734 + 0.498\text{MSS5} - 0.543\text{MSS6})^2)}$$

Neither of these formulae should be used on Landsat MSS images without some forethought. First, the PVI is defined as the perpendicular distance from the soil line (Figure 6.7). Richardson and Wiegand's (1977) equation for the soil line is based on 16 points representing soils, cloud and cloud shadows in Hidalgo and Willacy counties, Texas, for four dates in 1975. It is unlikely that such a small and geographically-limited sample could adequately define the soil line on a universal basis. A locally-valid expression relating 'soil' pixel values in Landsat MSS bands 5 and 7 (or 5 and 6) of the form $X5 = c + X7$ is needed. Secondly, the Richardson and Wiegand equation is based on the assumption that the maximum digital count in Landsat MSS bands 4 (green), 5 (red) and 6 (near-infrared) is 127 with a maximum of 63 in Landsat MSS band 7 (near-infrared). Landsat MSS images supplied by ESA-Earthnet are normally system corrected and the pixel values in all four Landsat MSS bands are expressed on a 0–255 scale. Landsat TM and ETM+ images are recorded on a 0–255 scale, so this problem should not arise. The PVI equations listed above do not, however, apply to ETM+ or TM images.

The PVI has been used as an index that takes into account the background variation in soil conditions, which affect soil reflectance properties. Jackson *et al.* (1983) demonstrate that the PVI is affected by rainfall when the vegetation cover is incomplete. However, they considered it to be 'moderately sensitive' to vegetation but was not a good detector of plant stress. The effects of atmospheric path radiance on the PVI were reported as reducing the value of the index by 10%–12% from a clear to a turbid atmospheric condition. This is considerably less than the 50% reduction noted for a near-infrared: red ratio.

The PVI is not now widely used. It is described here so as to introduce the concept of the 'soil line'. Nowadays, the Tasselled Cap or Kauth–Thomas transformation if generally preferred, as it can be modified to deal with data from different sensors. However, its formulation depends on the definition of the soil line using empirical data.

6.3.2 Tasselled Cap (Kauth-Thomas) transformation

The PVI (section 6.3.1) uses spectral variations in two of the four Landsat MSS bands, and relates distance from a soil line in the two-dimensional space defined by these two bands as a measure of biomass or green leaf area index. Kauth and Thomas (1976) use a similar idea except their model uses all four Landsat MSS bands. Their procedure has subsequently been extended to higher-dimensional data such as that collected by the Landsat TM and ETM+ instruments. The simpler four-band combination is considered first. In the four-dimensional feature space defined by the Landsat MSS bands, Kauth and Thomas (1976) suggest that pixels representing soils fall along an axis that is oblique with respect to each pair of the four MSS axes. A triangular region of the four-dimensional Landsat MSS feature space is occupied by pixels representing vegetation in various stages of growth. The Tasselled Cap transform is intended to define a new (rotated) coordinate system in terms of which the soil line and the region of vegetation are more clearly represented. The axes of this new coordinate system are termed 'brightness', 'greenness', 'yellowness' and 'nonesuch'. The brightness axis is associated with variations in the soil background reflectance. The greenness axis is correlated with variations in the vigour of green vegetation while the yellowness axis is related to variations in the yellowing of senescent vegetation. The 'nonesuch' axis has been interpreted by some authors as being related to atmospheric conditions. Due to the manner in which these axes are computed they are statistically uncorrelated, so that they can be represented in the four-dimensional space defined

Table 6.1 Coefficients for the Tasselled Cap functions 'brightness', 'greenness' and 'wetness' for Landsat Thematic Mapper bands 1–5 and 7.

TM band	1	2	3	4	5	7
Brightness	0.3037	0.2793	0.4343	0.5585	0.5082	0.1863
Greenness	−0.2848	−0.2435	−0.5436	0.7243	0.0840	−0.1800
Wetness	0.1509	0.1793	0.3299	0.3406	−0.7112	−0.4572

by the four Landsat MSS bands by four orthogonal lines. However, the yellowness and nonesuch functions have not been widely used and the Tasselled Cap transformation has often been used to reduce the four-band MSS data to two functions, brightness and greenness. For further discussion of the model, see Crist and Kauth (1986).

The justification for this dimensionality-reduction operation is that the Tasselled Cap axes provide a consistent, physically based coordinate system for the interpretation of images of an agricultural area obtained at different stages of the growth cycle of the crop. Since the coordinate transformation is defined *a priori* (i.e., not calculated from the image itself) it will not be affected by variations in crop cover and stage of growth from image to image over a time-series of images covering the growing season. The principal components transform (section 6.4) performs an apparently similar operation; however, the parameters of the principal components transform are computed from the statistical relationships between the individual spectral bands of the specific image being analysed. Consequently, the parameters of the principal components transform vary from one multispectral image set to another as the correlations among the bands depend upon the range and statistical distribution of pixel values in each band, which will differ from an early growing season image to one collected at the end of the growing season.

If the measurement for the jth pixel on the ith Tasselled Cap axis is given by u_j, the coefficients of the ith transformation by $\mathbf{R}_i$ and the vector of measurements on the four Landsat MSS bands for the same pixel by $\mathbf{x}_j$ then the Tasselled Cap transform is accomplished by:

$$u_j = \mathbf{R}_i'\mathbf{x}_j + c$$

In other words, the pixel values in the four MSS bands (the elements of $\mathbf{x}_j$) are multiplied by the corresponding elements of $\mathbf{R}_i$ to give the position of the jth pixel in the ith Tasselled Cap axis, $\mathbf{u}$. The constant c is an offset which is added to ensure that the elements of the vector $\mathbf{u}$ are always positive. Kauth and Thomas (1976) use a value of 32.

The vectors of coefficients $\mathbf{R}_i$ are defined by Kauth and Thomas (1976) as follows:

$$\mathbf{R}_1 = \{0.433, 0.632, 0.586, 0.264\}$$
$$\mathbf{R}_2 = \{-0.290, -0.562, 0.600, 0.491\}$$
$$\mathbf{R}_3 = \{-0.829, 0.522, -0.039, 0.194\}$$
$$\mathbf{R}_4 = \{0.223, 0.012, -0.543, 0.810\}$$

These coefficients assume that Landsat MSS bands 4 to 6 are measured on a 0–127 scale and band 6 is measured on a 0–63 scale. They are also calibrated for Landsat-1 data and slightly different figures may apply for other Landsat MSS data. The position of the $\mathbf{R}_1$ axis was based on measurements on a small sample of soils from Fayette County, Illinois. The representativeness of these soils as far as applications in other parts of the world is concerned is open to question.

Crist (1983) and Crist and Cicone (1984a and b) extend the Tasselled Cap transformation to data from the six reflective bands of Landsat TM data sets. Data from the Landsat TM thermal infrared channel (conventionally labelled band 6; see section 2.3.6) are excluded. They found that the brightness function $\mathbf{R}_1$ for the Landsat MSS Tasselled Cap did not correlate highly with the Landsat TM Tasselled Cap equivalent, though Landsat MSS greenness function $\mathbf{R}_2$ did correlate with the Landsat TM greenness function. The TM data was found to contain significant information in a third dimension, identified as wetness. The coefficients for these three functions are given in Table 6.1.

The brightness function is simply a weighted average of the six TM bands, while greenness is a visible/near infrared contrast, with very little contribution from bands 5 and 7. Wetness appears to be defined by a contrast between the mid-infrared bands (5 and 7) and the red/near infrared bands (3 and 4). The three Tasselled Cap functions can be considered to define a three-dimensional space in which the positions of individual pixels are computed using the coefficients listed in Table 6.1. The plane defined by the greenness and brightness functions is termed by Crist and Cicone (1984a) the 'plane of vegetation' while the functions brightness and wetness define the 'plane of soils' (Figure 6.8).

Several problems must be considered. The first is the now familiar problem of dynamic range compression that, in the case of the Tasselled Cap transform, assumes an added importance. One of the main reasons for supporting the use of the Tasselled Cap method against, for example, the principal components technique (section 6.4) is that the coefficients of the transformation are defined *a priori*,

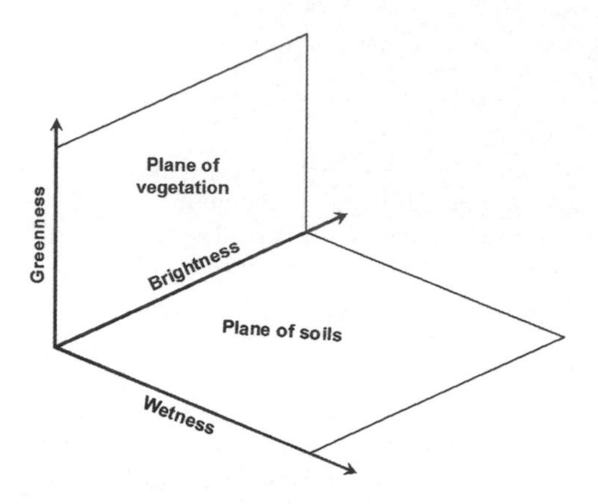

Figure 6.8 The Tasselled Cap transformation defines three fixed axes. Image pixel data are transformed to plot on these three axes (greenness, brightness and wetness) which jointly define the Plane of Vegetation and the Plane of Soils. See text for discussion. Based on Crist, E.P. and Cicone, R.C., 1986, Figure 3. Reproduced by permission of the American Society for Photogrammetry and Remote Sensing.

as noted above. However, if these coefficients are applied blindly, the resulting Tasselled Cap coordinates will not lie in the range 0–255 and will thus not be displayable on standard image processing equipment. The distribution of the Tasselled Cap transform values will vary from image to image, though. The problem is to define a method of dynamic range compression that will adjust the Tasselled Cap transform values on to a 0–255 range without destroying inter-image comparability. Because the values of the transform are scene-dependent it is unlikely that a single mapping function will prove satisfactory for all images but if an image-dependent mapping function is selected then inter-image comparison will be made more difficult. Crist (personal communication) suggests that the range of Tasselled Cap function values met with in agricultural scenes will vary between 0 and 350 (brightness), −100 and 125 (greenness) and −150 to 75 (wetness). If inter-image comparability is important, then the calculated values should be scaled using these limits. Otherwise, the functions could be evaluated for a sample of pixels in the image (e.g., 5% of the pixels could be selected) and the functions scaled using the extreme sample values according to the formula

$$y = \frac{x - x_{\min}}{x_{\max} - x_{\min}} \times 255$$

where y is the scaled value (0–255) and x is the raw value. In order to prevent over- and under-shoots, a check should be made for negative y values (which are set to zero) or values of y that are greater than 255 (these are set to 255). Problems of scaling are discussed in more detail in section 3.2.1. As the cost of disc storage has fallen substantially over the past few years, it is now more likely that the output from the Tasselled Cap procedure is represented in the form of 32-bit real values, which can be manipulated in accordance with the requirements of any particular problem.

A second problem which interferes with the comparison of multi-date Tasselled Cap images is the problem of changes in illumination geometry and variations in the composition of the atmosphere. Both of these factors influence the magnitude of the ground-leaving radiance from a particular point, so that satellite-recorded radiances from a constant target will change even though the characteristics of the target do not change. These problems are addressed in a general way in Chapter 4 (sections 4.4 and 4.5); they are particularly important in the context of techniques that endeavour to provide the means to carry out comparisons between multi-temporal images. One solution is proposed by Huang *et al.* (2002), whose use of apparent reflection is noted in section 4.6. This conversion is applied in MIPS by using the function *Utilities|Convert DN to reflectance*. Huang *et al.* (2002) use procedures similar to those described below to compute Tasselled Cap coefficients from Landsat TM apparent reflectance data.

Like the Perpendicular Vegetation Index the Tasselled Cap transform relies upon empirical data for the determination of the coefficients of the brightness axis (the soil line in the terminology of the Perpendicular Vegetation Index). It was noted above that the Kauth and Thomas (1976) formulation of the MSS Tasselled Cap was based on a small sample of soils from Fayette County, Illinois. The TM brightness function is also based on a sample of North American soils, hence applications of the transformation to agricultural scenes in other parts of the world may not be successful if the position of the brightness axis as defined by the coefficients given above does not correspond to the reflectance characteristics of the soils in the local area. Jackson (1983) describes a method of deriving Tasselled Cap-like coefficients from soil reflectance data, using Gram–Schmidt orthogonal polynomials. He uses reflectance data in the four Landsat MSS bands for dry soil, wet soil, green and senesced vegetation to derive coefficients for three functions representing brightness, greenness and yellowness. This method is implemented for the Landsat TM bands in the MIPS software, which also includes a module to perform the Tasselled Cap transformation Where possible, the coefficients of the orthogonal, Tasselled Cap-like functions should be calculated for the area of study as they may differ from those provided by Crist and Cicone (1984a and b) and listed above. A MIPS function (*Tranform|Tasselled Cap|Compute Coefficients*) is provided to allow you to do this.

6.4 PRINCIPAL COMPONENTS ANALYSIS

6.4.1 Standard principal components analysis

Adjacent bands in a multispectral remotely-sensed image are generally correlated. Multi-band visible/near-infrared images of vegetated areas will exhibit negative correlations between the near-infrared and visible red bands and positive correlations among the visible bands because the spectral characteristics of vegetation (section 1.3.2.1) are such that as the vigour or greenness of the vegetation increases the red reflectance diminishes and the near-infrared reflectance increases. The presence of correlations among the bands of a multispectral image implies that there is redundancy in the data. Some information is being repeated. It is the repetition of information between the bands that is reflected in their inter-correlations.

If two variables, x and y, are perfectly correlated then measurements on x and y will plot as a straight line sloping upwards to the right (Figure 6.9). Since the positions of the points shown along line AB occupy only one dimension, the relationships between these points could equally well be given in terms of coordinates on line AB. Even if x and y are not perfectly correlated there may be a dominant direction of scatter or variability, as in Figure 6.10. If this dominant direction of variability (AB) is chosen as the major axis then a second, minor, axis (CD) could be drawn at right angles to it. A plot using the axes AB and CD rather than the conventional x- and y-axes might, in some cases, prove more revealing of the structures that are present within the data. Furthermore, if the variation in the direction CD in Figure 6.10 contains only a small proportion of the total variability in the data then it may be ignored without too much loss of information.

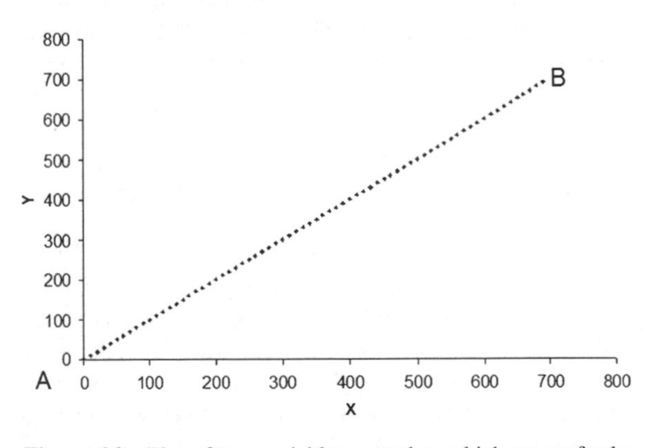

Figure 6.9 Plot of two variables, x and y, which are perfectly correlated ($r = 1.0$). The (x, y) points lie on a straight line between A and B. Although this is a two-dimensional plot, all the points lie on a one-dimensional line. One dimension is therefore redundant.

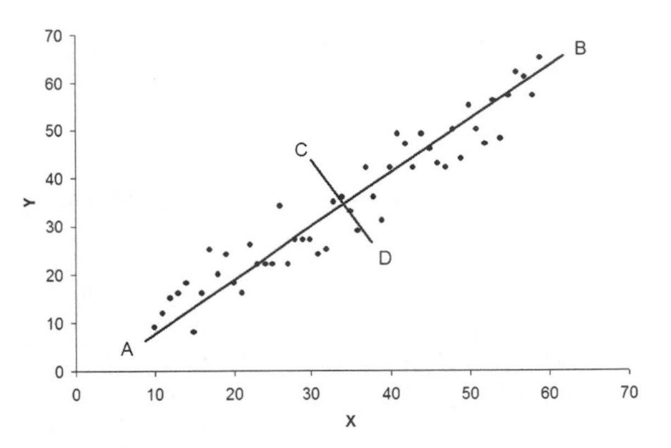

Figure 6.10 In contrast to the plot shown in Figure 6.9, this distribution of (x, y) points does not lie along a single straight line between A and B. There is some scatter in a second, orthogonal, direction CD. The distance relationships between the points would be the same if we used AB as the x-axis and CD as the y-axis, though the numerical coordinates of the points would change.

This example shows that we must draw a basic distinction between the number of variables (spectral bands) in the image data set and the intrinsic dimensionality of that data set. In Figure 6.9 the number of variables is two (x and y) but the dimensionality of the data as shown by the scatter of points is one. In Figure 6.10 the dimensionality is again effectively one, although the number of observed variables is, in fact, two. In both examples the use of the single axis AB rather than the x and y axes accomplishes two aims: (i) a reduction in the size of the data set since a single coordinate on axis AB replaces the two coordinates on the x and y axes, and (ii) the information conveyed by the set of coordinates on AB is greater than the information conveyed by the measurements on either the x or the y axes individually. In this context information means variance or scatter about the mean; it can also be related to the range of states or levels in the data, as shown in the discussion of entropy in section 2.2.3.

Multispectral image data sets generally have a dimensionality that is less than the number of spectral bands. For example, it is shown in section 6.3.2 that the four-band Landsat MSS Tasselled Cap transform produces two significant dimensions (brightness and greenness) while the six-band Landsat TM Tasselled Cap defines three functions (dimensions). The purpose of principal (not 'principle'!) components analysis is to define the number of dimensions that are present in a data set and to fix the coefficients which specify the positions of that set of axes which point in the directions of greatest variability in the data (such as axes AB and CD in Figure 6.10). These axes are always uncorrelated. A principal components transform of a multispectral image (or of a set of registered multi-temporal images, for example, Ribed and Lopez, 1995) might therefore be expected to perform the following operations:

- define the dimensionality of the data set, and
- identify the principal axes of variability within the data.

These properties of principal components analysis (sometimes also known as the Karhunen-Loève transform) might prove to be useful if the data set is to be compressed. Also, relationships between different groups of pixels representing different land cover types may become clearer if they are viewed in the principal axis reference system rather than in terms of the original spectral bands, especially as the variance of the data set is concentrated in relatively fewer principal components. The data compression property is useful if more than three spectral bands are available. A conventional RGB colour display system relates a spectral band to one of the three colour inputs (red, green and blue). The Landsat TM provides seven bands of data, hence a decision must be made regarding which three of these seven bands are to be displayed as a colour composite image. If the basic dimensionality of the TM data is only three then most of the information in the seven bands will be expressible in terms of three principal components. The principal component images could therefore be used to generate a RGB false-colour composite with principal component number 1 shown in red, number 2 in green and number 3 in blue. Such an image contains more information than any combination of three spectral bands.

The positions of the mutually perpendicular axes of maximum variability in the two-band data set shown in Figure 6.10 can be found easily by visual inspection to be the lines *AB* and *CD*. If the number of variables (spectral bands) is greater than three then a geometric solution is impracticable and an algebraic procedure must be sought. The direction of axis *AB* in Figure 6.10 is defined by the sign of the correlation between variables x and y; high positive correlation results in the scatter of points being restricted to an elliptical region of the two-dimensional space defined by the axes x and y. The line *AB* is, in fact, the major or principal axis of this ellipse and *CD* is the minor axis. In a multivariate context, the shape of the ellipsoid enclosing the scatter of data points in a p-dimensional space is defined by the variance-covariance matrix computed from p variables or spectral bands. The variance in each spectral band is proportional to the degree of scatter of the points in the direction parallel to the axis representing that variable, so that it can be deduced from Figure 6.11 that the variances of variables X and Y are approximately equal. The covariance defines the shape of the ellipse enclosing

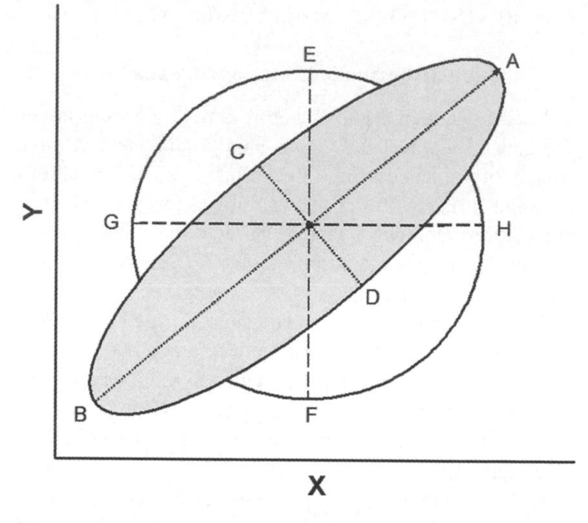

Figure 6.11 The ellipse is characterised in the two dimensional space defined by variables X and Y by long axis AB and short axis CD, which define the two orthogonal directions of maximum scatter. The circle shows equal scatter in all directions from the centre, so the positions of its axes EF and GH are purely arbitrary – there is no direction of maximum variance. In the case of the ellipse, the direction of slope of line AB indicates that there is a strong positive correlation between the two variables while its shape shows that one variable has a larger variance than the other. The lack of a preferred orientation of scatter in the case of the circle indicates a zero correlation with the two variables X and Y having equal variances.

the scatter of points. Figure 6.11 shows two distributions with the same variance. One (the ellipse) has a high positive covariance while the other (the circle) has a covariance of zero. The mean of each variable gives the location of the centre of the ellipse (or ellipsoid in a space of dimensionality higher than two). Thus, the mean vector and the variance–covariance matrix define the location and shape of the scatter of points in a p-dimensional space. The information contained in the variances and covariances of a set of variables is used in the definition of the maximum likelihood classification procedure that is described in section 8.4.2.3.

The relationship between the correlation matrix and the variance-covariance matrix sometimes leads to confusion. If the p variables making up the data set are measured on different and incompatible scales (for example, three variables may be measured in metres above sea level, barometric pressure in millibars and weight in kilograms) then the variances of these variables are not directly comparable. The importance of the variance in defining the scatter of points in a particular direction has already been stressed, so it is clear that if the variance is to be used in defining the shape of the ellipsoid enclosing the scatter of points in the p-dimensional space then the scales used to measure each variable must be comparable.

To illustrate this point, consider what would be the outcome (in terms of the shape of the enclosing ellipsoid) if the three variables mentioned earlier were measured in feet above sea level (rather than metres), inches of mercury (rather than millibars), and weight in ounces rather than kilograms. Not only would the variance of each of the variables be altered but the shape of the enclosing ellipsoid would also change. The degree of change would not be constant in each dimension, and the shape of the second ellipsoid would not be related in any simple fashion to the shape of the first ellipsoid. Consequently, the lengths and orientations of the principal axes would change. It is in these circumstances that the correlation coefficient rather than the covariance is used to measure the degree of statistical association between the spectral bands. The correlation is simply the covariance measured for standardised variables. To standardise a variable the mean value is subtracted from all the measurements and the result is divided by the standard deviation. This operation converts the raw measurements to standard scores or z-scores, which have a mean of zero and a variance of unity. The off-diagonal elements of the correlation matrix are the covariances of the standard scores and the diagonal elements are the variances of the standard scores, which are by definition always unity. Since the shape of the ellipsoid enclosing the scatter of data points is altered in a complex fashion by the standardisation procedure it follows that the orientation and lengths of the principal axes (such as *AB* and *CD* in Figure 6.10) will also change. The effects of the choice of the variance-covariance matrix **S** or the correlation matrix **R** on the results of the principal components operation are considered below, and an illustrative example is provided.

The principal components of a multispectral image set are found by algebraic methods that are beyond the scope of this book (see Mather (1976) or Richards (1993) for a more detailed discussion). In general terms, the variance-covariance matrix **S** or the correlation matrix **R** of the bands is computed. If there are p bands then each of these symmetric matrices has p rows and p columns. A set of quantities called the *eigenvalues* is found for the chosen matrix using methods of linear algebra. The magnitudes of these eigenvalues are proportional to the lengths of the principal axes of the ellipsoid whose shape is defined by the elements of **S** or **R**. The eigenvalues are measured in terms of units of variance. It follows from the previous paragraph that standardised units of variance must be used if the variables (bands) are measured on incompatible scales. If the variables are all measured in terms of the same units (such as milliwatts per square centimetre per steradian) then standardisation is unnecessary and, indeed, undesirable for it removes the effects of changes in variability between the bands.

Associated with each eigenvalue is a set of coordinates defining the direction of the associated principal axis. These coordinates are called the *eigenvectors* of the matrix **S** or **R**. The eigenvalues and eigenvectors therefore describe the lengths and directions of the principal axes. In effect the data are scaled and rotated so that the principal axes are the coordinate system in terms of which the principal component images are expressed. The eigenvectors, scaled by the square roots of the corresponding eigenvalues, can also be interpreted as correlations between the abstract principal components and the individual bands of the multispectral image. These correlations or loadings are used in the interpretation of the principal components. They relate specific spectral bands on the basis of their inter-correlations. In real-life terms, think of a supermarket bill. It shows individual items such as cereals, milk, brandy, bread, weed killer and rat poison. If you were asked what you have bought, you would not read every item from the bill. You would say, 'Some food, some drink, and a couple of things for the garden'. Thus, you would group the items on the bill in terms of their similarity or correlation, and you would use this technique to compress your answer into a single, short statement. In effect, you have done a principal components analysis on your shopping bill. The only difference is that you have decided on the names of the components ('food', 'drink', 'gardening items'). In image processing, you know the identifiers (band numbers) of the spectral bands that group together in that they have high positive or negative loadings on the same component. You also know from the principal component loadings that some bands are more closely associated than others are with whatever the principal component is picking-out from the data. You have to find a label equivalent to 'food' or 'drink' to describe that principal component. It may be a description such as 'overall brightness' or 'greenness'.

Principal component images are computed from the eigenvectors much in the same way that the Tasselled Cap features are derived from their coefficients. The eigenvector for principal component 1 provides the coefficients or weights which are used in conjunction with the observed pixel values to generate the first principal component image. The pixel values for the first principal component image are found from:

$$pc1_{ij} = a_1x_{ij1} + a_2x_{ij2} + a_3x_{ij3} + \cdots + a_mx_{ijm}$$

In this expression $pc1_{ij}$ is the value of the pixel at row i, column j in the first principal component image, the values $a_1, a_2, a_3, \ldots, a_m$ are the elements of the first scaled eigenvector and x_{ijk} is the observed pixel value at row i, column j of band $k(k = 1, m)$. The remaining principal component images are computed in the same way using the appropriate scaled eigenvectors. Scaling is performed by dividing the eigenvector by the square root of the corresponding eigenvalue.

The algebraic representation of principal component scores allows another explanation of the way in which principal components are computed. The 'maximum variance' principle is used to determine the position of the principal component axes, as described above. One could also say that there are an infinite number of coefficients $a_1, a_2, a_3, \ldots, a_m$ that could be used to define $pc1_{ij}$ in the equation above; for example, I could choose to use the vector $\{0.9, -0.3, 0.77, \ldots., -0.97\}$ or any other randomly selected set of values. If the variance of the resulting $pc1$ image were to be calculated for every possible vector of coefficients, one would have a greater variance than any of the others. That one would be the vector that consisted of the first column of eigenvectors of the matrix **S** (or **R**).

If the principal component images are to be displayed on a screen, then the values $pc1_{ij}$ (and their equivalents $pc2_{ij}$ and $pc3_{ij}$) must be scaled to fit the range 0–255. This scaling problem has already been discussed in the context of band ratios and the Tasselled Cap transform. Given that the principal component images will not generally be integers in the range 0–255, the most effective method of scaling is to store the principal component images as 32-bit real numbers, and use one of the methods of scaling described in Chapter 3 to convert the real numbers to a 0–255 range when required. When comparing principal component images produces by different computer programs, it is always wise to determine the nature of the scaling procedure used in the generation of each image.

If all available bands are input to the principal components procedure then, depending on whether the analysis is based on inter-band correlations or covariances, the information contained in a sub-set of bands may be under-represented, as a result of the spectral resolution of the sensor. For example, the Landsat ETM+ has one thermal infrared band (out of seven) whereas ASTER has four out of 13 (Table 2.4). Relative to the ASTER data set, the information content of the thermal infrared band will be under-represented in the Landsat ETM+ dataset. Siljeström *et al.* (1997) use a procedure that they call 'selective principal components analysis', which involves the division of the bands into groups on the basis of their inter-correlations. For TM data, these authors find that bands 1, 2 and 3 are strongly inter-correlated and are distinct from bands 5 and 7. Band 4 stands alone. They carry out PCA separately on bands (1, 2, 3) and bands (5, 7) and use the first principal component from each group, plus band 4, to create a false-colour image, which is then used to help in the recognition of geomorphological features.

Principal components analysis is also used in the analysis of multi-temporal images in order to identify areas of change (section 6.2.2). For example, Siljeström and Moreno (1995) use covariance-based PCA applied to a multi-temporal image set and identify specific components with areas of change. Henebry (1997) also uses PCA, applied to a multi-temporal sequence of eleven ERS SAR images over a growing season and was able successfully to identify individual principal components with specific land surface properties, such as terrain, burned/unburned prairie, and look angle variations between the eleven images.

Table 6.2(a) gives the correlation matrix for the six reflective (visible plus near- and mid-infrared) bands (numbered 1–5 and 7) of a Landsat TM image of the Littleport area shown in Figures 1.10 and 1.11. Correlations rather than covariances are used as the basis for the principal components procedure because the digital counts in each band do not relate to the same physical units (that is, a change in level from 30 to 31, for example, in band 1 does not represent the same change in radiance as a similar change in any other band. See section 4.6 for details of the differences in calibration between TM bands.). High positive correlations among all reflective bands except band 4 (near-infrared) can be observed. The lowest correlation between any pair of bands (excluding band 4) is +0.660. The correlations between band 4 and the other bands are negative. This generally high level of correlation implies that the radiances in each spectral band except band 4 are varying spatially in much the same way. The negative correlation between band 4 and the optical bands can be explained by the fact that the area shown in the image is an agricultural one, and the main land cover type is vegetation. The idealised spectral reflectance curve for vegetation (Figure 1.21) shows the contrast between the near-infrared and red channels. The vegetation ratios exploit this fact, and as the ground cover of vegetation, and the vigour (greenness) of the vegetation increase so the contrast between near-infrared and red reflectance increases as the near-infrared reflectance rises and the red reflectance falls. Hence, this negative correlation between band 4 and other bands can be partially explained. Table 6.2(a) also shows the means and standard deviations of the six bands. Note the significantly higher standard deviations for bands 4, 5 and 7 which show that variability of pixel values is greater in these bands than in bands 1, 2 and 3. The mean pixel values also differ. This is a result of the nature of the ground cover and the calibration of the TM sensor.

The eigenvalues and scaled eigenvectors, or principal component loadings, derived from the correlation matrix measure the concentration of variability in the data in six

Table 6.2 (a) Correlations among Thematic Mapper reflective bands (1–5 and 7) for the Littleport TM image. The means and standard deviations of the six bands are shown in the rightmost two columns.

TM Band	1	2	3	4	5	7	Mean	Std. Dev.
1	1.000	0.916	0.898	−0.117	0.660	0.669	65.812	8.870
2	0.916	1.000	0.917	−0.048	0.716	0.685	29.033	5.652
3	0.898	0.917	1.000	−0.296	0.757	0.819	26.251	8.505
4	−0.117	−0.048	−0.296	1.000	−0.161	−0.474	93.676	24.103
5	0.660	0.716	0.757	−0.161	1.000	0.883	64.258	18.148
7	0.669	0.685	0.819	−0.474	0.883	1.000	23.895	11.327

(b) Principal component loadings for the six principal components of the Littleport TM image. Note that the sum of squares of the loadings for a given principal components (column) is equal to the eigenvalue. The percent variance value is obtained by dividing the eigenvalue by the total variance (6 in this case because standardised components are used – see text) and multiplying by 100.

	PC 1	PC 2	PC 3	PC 4	PC 5	PC 6
TM Band 1	0.899	0.242	−0.288	0.223	0.002	−0.002
TM Band 2	0.914	0.303	−0.182	−0.143	−0.095	−0.103
TM Band 3	0.966	0.033	−0.165	−0.117	0.086	0.134
TM Band 4	−0.309	0.924	0.214	−0.006	0.076	−0.001
TM Band 5	0.871	0.019	0.470	0.038	−0.124	0.059
TM Band 7	0.904	−0.285	0.267	0.009	0.148	−0.094
Eigenvalue	4.246	1.086	0.481	0.085	0.059	0.041
% variance	70.77	18.10	8.02	1.42	0.99	0.68
Cumulative % variance	70.77	88.88	96.90	98.32	99.31	100.00

orthogonal directions (Table 6.2(b)). Over 70% of the variability in the data lies in the direction defined by the first principal component. Column 1 of Table 6.2(b) gives the relationship between the first principal component and the six TM bands; all bands except the near infrared have entries more than 0.87, while the near-infrared band has an entry of −0.309. This indicates (as was inferred from the correlation matrix) that there is considerable overlap in the information carried by the different channels, and that there is a contrast between the near infrared and the other bands. The image produced by the first principal component (Figure 6.12(a)) summarises a information that is common to all channels. It can be seen to be a weighted average of five of the six TM bands contrasted with the near-infrared band.

The second principal component of the Littleport image set (Figure 6.12(b)) is dominated by the contribution of the near-infrared band. There is a small contribution from the three visible bands. Between them, principal components 1 and 2 account for over 88% of the variability in the original six-band data set. A further 8% is contributed by principal component 3 (Figure 6.12(c)), which appears to be highlighting a contrast between the infrared and the visible bands. Visual analysis of the principal component images shown in Figure 6.12(a) - (c) appears to indicate that principal components 1 and 2 may be picking out differences between different vegetation types in the area, while principal component 3 is related to water content of the soil.

Principal components 4–6 (Figure 6.12(d)–(f)) together contain only 3.1% of the variation in the data. If the noise present in the image data set is evenly distributed among the principal components then the lower-order principal components might be expected to have a lower signal-to-noise ratio than the higher-order principal components. On these grounds it might be argued that principal components 4–6 are not worthy of consideration. This is not necessarily the case. While the contrast of these lower-order principal component images is less than that of the higher-order components, there may be patterns of spatial variability present that should not be disregarded, as Figures 6.12(d)–(f) show. The sixth principal component shown in Figure 6.12(f) is clearly spatially non-random. Townshend (1984) gives a good example of the utility of low-order principal components. His seventh principal component accounted for only 0.08% of the variability in

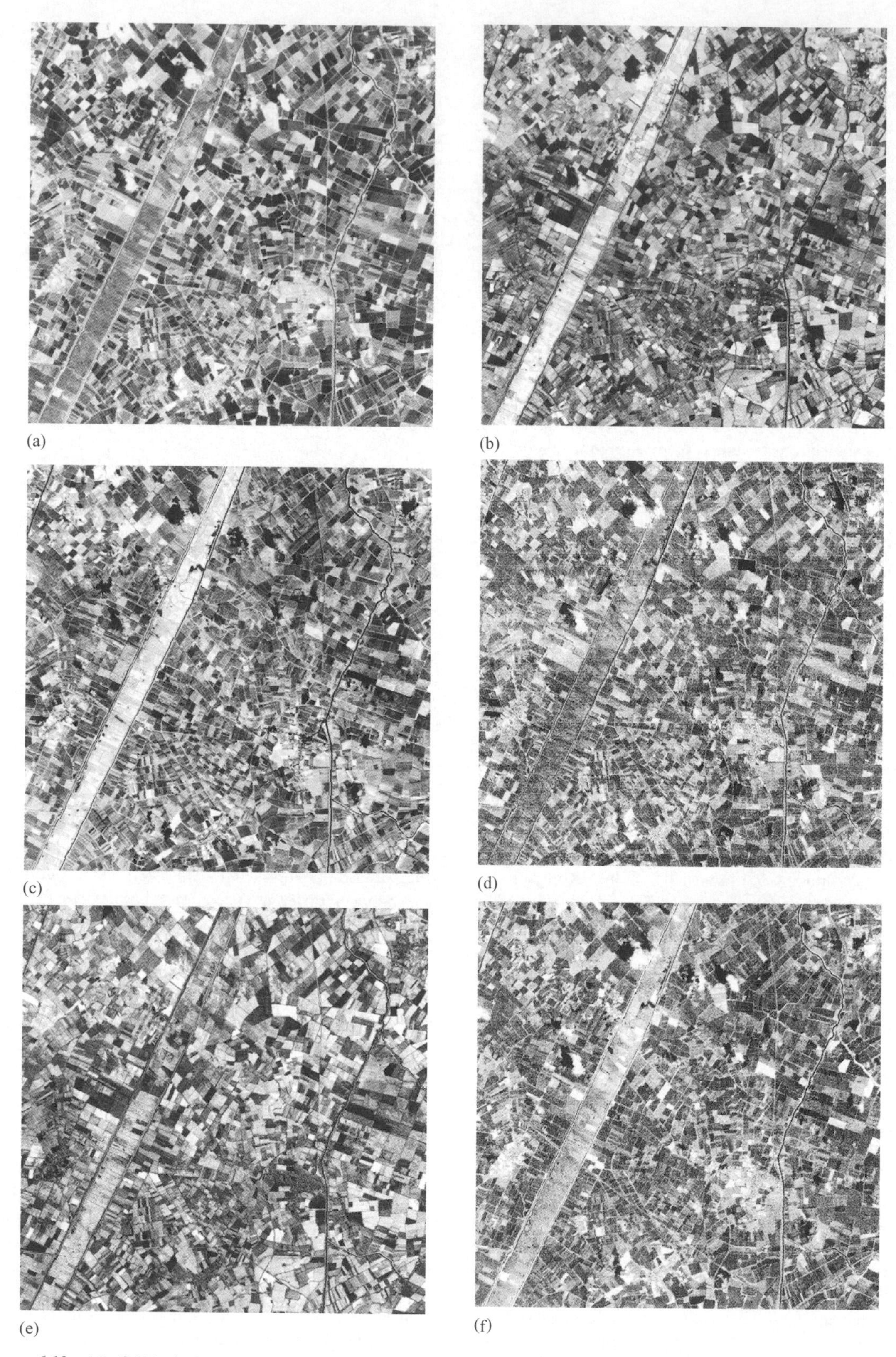

Figure 6.12 (a)–(f) Principal component images numbers 1–6 based on correlations between the Landsat TM image bands shown in Figures 1.10 and 1.11. Original data courtesy of Eurimage Ltd.

the data yet a distinction between apple and plum orchards and between peaty and silty clay soils was brought out. This distinction was not apparent on any other component or individual TM band, nor was it apparent from a study of the principal component loadings. It is important to check principal component images by eye, using one's knowledge of the study area, rather than rely solely upon the magnitudes of the eigenvalues as an indicator of information content, or on inferences drawn from the principal component loadings.

Do not be misled by the coherent appearance of the lower-order principal component images. Figure 6.12(f) shows the sixth and last principal component derived from the Littleport TM image data set. This principal component accounts for only 0.68% of the total (standardised) variance of the image data set, yet it is interpretable in terms of spatial variation in grey levels. It should be borne in mind that the information expressed by this principal component has been (i) transformed from an arbitrary 32-bit real number into a count on the 0–255 range, as described in section 3.2.1, and (ii) subjected to a histogram equalisation contrast stretch. Yet, if the first principal component were expressed on a 0–255 scale and the ranges of the other principal components adjusted according to their associated eigenvalue, then principal component 6 would have a dynamic range of 0–10.

The use of lower-order components depends on the aims of the project. If the aim is to capture as much as possible of the information content of the image set in as few principal components as possible then the lower-order principal components should be omitted. The 'eigenvalue greater than one' (or 'eigenvalue-1') criterion could be used to determine how many principal components to retain. If an individual standardised band of image data has a variance of unity then it might be possible to argue that all retained principal components should have a variance of at least one. If this argument were used then only the first two principal components of the Littleport image set would be retained, and 11% of the information in the image data set would be traded for a reduction of 66.66% in the image data volume. On the other hand, one may wish merely to orthogonalise the data set (that is, express the data in terms of uncorrelated principal components rather than in terms of the original spectral bands) in order to facilitate subsequent processing. For example, the performance of the feed-forward artificial neural net classifier, discussed in Chapter 8, may perform better using uncorrelated inputs. In such cases, all principal components should be retained. It is important to realise that the aims of a project should determine the procedures followed, rather than the reverse.

There are additional complications. The example above showed that much of the variance in a multi-band data set is concentrated in the first two or three principal components. If data compression is the primary aim of the exercise then these principal components can be used to summarise the total data set, as noted above. However, important information may be contained in the higher-order principal component. This is in part due to the fact that a multi-band data set may not comprise a single statistical population the distribution of which can be approximated by an ellipsoid (as shown in Figure 6.11). Chang (1983) has studied the problem involved in using principal components analysis when the data set is made up of two or more statistical populations. He, in fact, restricted his analyses to mixtures of two multivariate-normal distributions, though his conclusions are claimed to be more general. He states:

> 'In using principal components analysis to reduce the dimensions ... it has always been the practice to select the components with the larger eigenvalues. We disprove this practice mathematically and by the use of hypothetical and real data'.

Although Chang's (1983) study points towards this general conclusion he does not consider either the effects of non-Normal distributions or problems arising whenever the variance-covariance matrices of the populations are unequal. An earlier study by Lowitz (1978) also shows that high-order components can contain information relating to inter-group differentiation. This point should be borne in mind if principal components analysis is used as a feature-selection technique before image classification (section 8.9), and reinforces the conclusions presented in the preceding paragraph.

The example used earlier in this section is based on the eigenvalues and eigenvectors of the correlation matrix, **R**. We saw earlier that the principal components of the covariance (**S**) and correlation (**R**) matrices are not related in a simple fashion. Consideration should therefore be given to the question of which of the two matrices to use. Superficially it appears that the image data in each band are comparable, all being recorded on a 0–255 scale. However, reference to section 4.6 shows that this is not so. The counts in each band can be referred to the calibration radiances for that band, and it will be found that

(i) the same pixel value in two different bands equates to different radiance values, and
(ii) if multi-date imagery is used then the same pixel value in the same band for two separate dates may well refer to different radiance values because of differences in sensor calibration.

The choice lies between the use of the correlation matrix to standardise the measurement scale of each band or

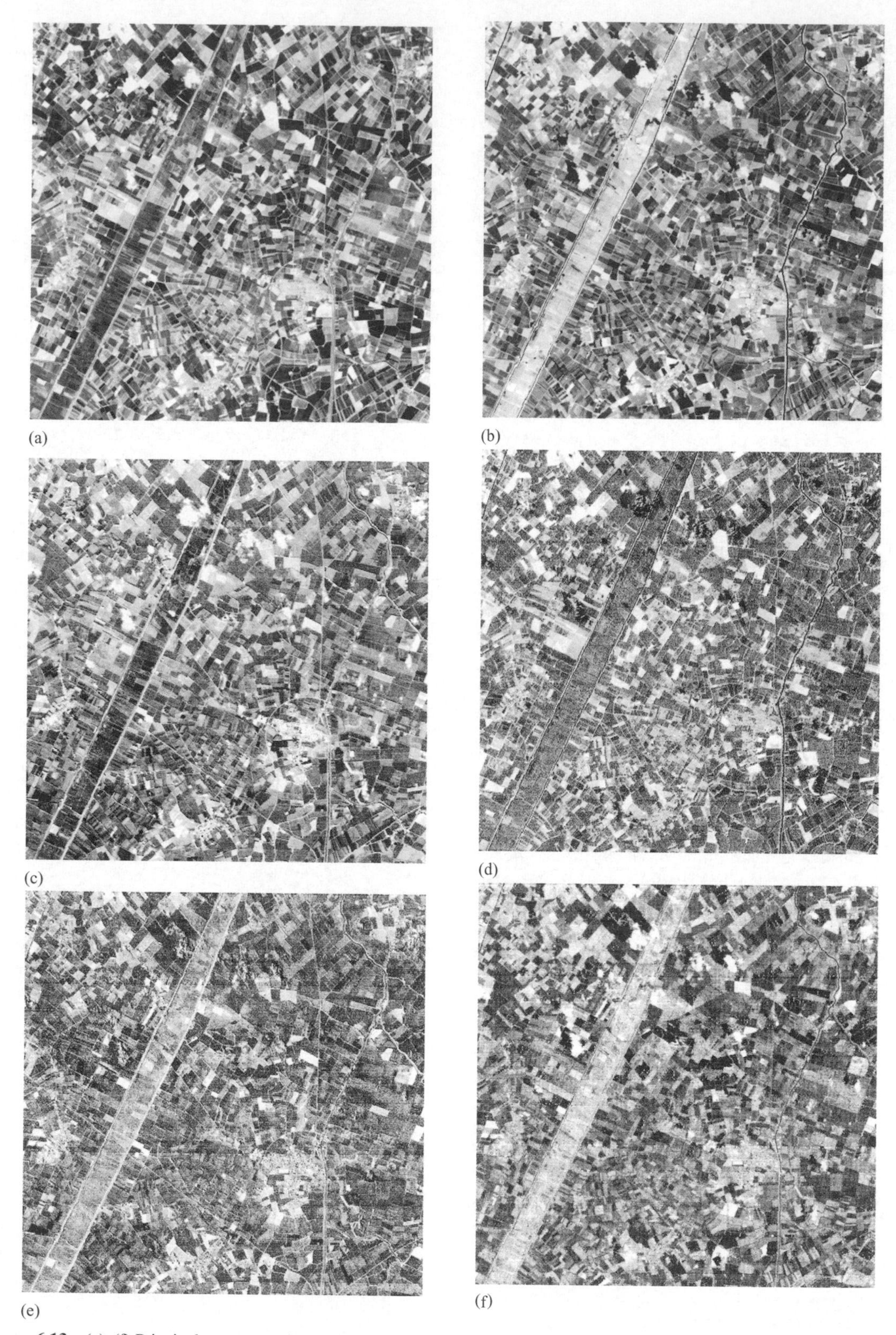

Figure 6.13 (a)–(f) Principal component images numbers 1–6 based on covariances between the Landsat TM image bands shown in Figures 1.10 and 1.11. Original data courtesy of Eurimage Ltd. Compare with Figure 6.12, which shows principal components based on inter-band correlations.

Table 6.3 (a) Variance–covariance matrix for the Littleport TM image set. The last row shows the variance of the corresponding band expressed as a percentage of the total variance of the image set. The expected variance for each band is 16.66%, but the variance ranges from 2.61% for band 2 to 47.56% for band 4.

TM Band	1	2	3	4	5	7	Mean	Std. Dev.
TM 1	78.67	45.90	67.71	−25.04	106.25	67.22	65.812	8.870
TM 2	45.90	31.94	44.09	−6.51	73.45	43.87	29.033	5.652
TM 3	67.71	44.09	72.34	−60.75	116.78	78.86	26.251	8.505
TM 4	−25.04	−6.51	−60.75	580.95	−70.32	−129.52	93.676	24.103
TM 5	106.25	73.45	116.78	−70.32	329.37	181.56	64.258	18.148
TM 7	67.22	43.87	78.86	−129.52	181.56	128.30	23.895	11.327
Percent variance	6.44	2.61	5.92	47.56	26.96	10.50	16.666	15.90

(b) Principal component loadings for the six principal components of the Littleport TM image, based on the covariance matrix shown in table 6.3 (a).

	PC 1	PC 2	PC 3	PC 4	PC 5	PC 6
TM Band 1	0.531	0.593	0.575	−0.491	−0.168	−0.018
TM Band 2	0.502	0.674	0.477	−0.034	0.139	0.215
TM Band 3	0.707	0.530	0.421	0.057	0.178	−0.085
TM Band 4	−0.834	0.552	0.000	0.019	−0.001	−0.002
TM Band 5	0.664	0.714	−0.211	−0.070	0.006	−0.003
TM Band 7	0.684	0.438	−0.064	0.233	−0.055	0.013
Eigenvalue	711.19	431.87	61.20	9.69	5.52	2.05
Percent Variance	58.22	35.35	5.01	0.79	0.45	0.17
Cumulative % variance	58.22	93.58	98.59	99.38	99.83	100.00

conversion of the image data to radiances followed by principal components analysis of the variance-covariance matrix. The differences resulting from these alternative approaches have not as yet been fully explored (Singh and Harrison, 1985). However, if multi-date imagery is used the question of comparability is of great importance.

Figure 6.13(a)–(f) and Table 6.3(a) and (b) summarise the results of a principal components transform of the Littleport TM image data set based on the variance–covariance matrix (**S**) rather than the correlation matrix (**R**). Compare the results for the analysis based on the correlation matrix, described above and summarised in Table 6.2 and Figure 6.13. The effects of the differences in the variances of the individual bands are very noticeable. Band 4 has a variance of 580.95, which is 47.56% of the total variance, and as a consequence band 4 dominates the two highest-order principal components. As in the correlation example, the first principal component is a contrast between band 4 and the other five bands, but this time the loading on band 4 is the highest in absolute terms, rather than the lowest and the percentage of variance explained by the first principal component is 58.22 rather than 70.77. Principal component two, accounting for 35.35% of the total variance rather than 18.10% in the correlation case, is now more like a weighted average. The higher-order principal components are much less important even than they are when the analysis is based on correlations, and the corresponding principal component images are noisier. Which result is 'better' depends, of course, on the user's objectives. The aim of principal components analysis should be to generate a set of images that are more useful in some way than are the un-transformed images, rather than to satisfy the pedantic requirements of statistical theory. Nevertheless, the point is clear: the principal components of **S**, the variance-covariance matrix, are quite different from the principal components of **R**, the correlation matrix. Methods of dealing with the noise that affects the lower-order principal component images are considered below.

In section 6.3.2 the relationship between the principal components and the Tasselled Cap transformations is discussed. It was noted that the principal components transform is defined by the characteristics of the inter-band correlation or covariance structure of the image, whereas the Tasselled Cap transformation is based on external data, namely, the pre-defined position of the soil brightness, greenness and wetness axes. In other words, the positions of the Tasselled Cap axes are defined mathematically by pre-determined coefficients such as those listed in Table 6.1, whereas the coefficients of the principal components transform are derived from the correlation (**R**) or variance-covariance (**S**) matrix, which varies from image to image. Because the results of a principal components analysis are image-specific it follows that principal component transformed data values cannot be directly compared between images as the principal components do not necessarily have the same pattern of principal component loadings. While principal components analysis is useful in finding the dimensionality of an image data set and in compressing data into a fewer number of channels (for display purposes, for example) it does not always produce the same components when applied to different image data sets.

Example 6.1 demonstrates the use of the principal components analysis module in MIPS.

EXAMPLE 6.1: PRINCIPAL COMPONENTS ANALYSIS (PCA)

This example uses the Littleport image set to demonstrate the use of principal components analysis in MIPS. The image set comprises 6 Landsat TM bands each with 512 lines of 512 pixels. The thermal infrared band, numbered 6, is omitted. This image set is included on the CD, and is referenced by the dictionary file *Littlept.inf*.

The principal components analysis procedure is called from the *Transform* menu. You need not have an image on display before using the PCA module, which operates directly on the data stored on disc. However, you should look at a conventional false colour composite of bands 4, 3 and 2 (displayed in red, green and blue respectively) to get some idea of the nature of the area. Littleport is a small town in Cambridgeshire in the eastern part of England. The River Ouse flows northwards towards Kings Lynn and The Wash through a flat, agricultural region that is one of the most fertile in England. A considerable amount of civil engineering work has been carried out on the drainage of the area in the past 300 years, both to prevent flooding and to drain the formerly marshy area of The Fens.

The PCA routine first asks you to locate the INF file (*Littlept.inf*) then asks you to specify the band numbers you want to include in the analysis (you are told that there are six bands available; these are the TM bands 1, 2, 3, 4, 5 and 7). Enter 1–6 to select all six bands, and verify your decision. Next, click *OK* to select the whole image area of 512 lines and 512 pixels per line. The six images are read into memory, and once this operation is complete you must choose whether to base the analysis on the inter-band correlation matrix or the inter-band covariance matrix. Read the appropriate paragraphs of section 6.4.1 so that you understand the background to this choice. Since you will run the program twice, select the correlation matrix the first time round and the covariance matrix on the second run.

Some images, for example those that have been geometrically corrected (section 4.3) will have 'fill pixels' inserted around the margins of the image (Figure 4.4). Usually, these fill pixels are given a value of zero. The next input dialogue box asks whether or not you wish to exclude zero pixels. Since there are no fill pixels on the Littleport image, answer 'no'.

The choice of the number of principal components to compute is discussed at length in the statistics literature, where the concern is to extract all principal components that contain 'significant' information. Quite often, the number of principal components is set to be equal to the number of eigenvalues of the correlation matrix that exceed 1.0 in magnitude (note that this rule is valid only when the analysis is based on correlations and not covariances). You have the choice of applying the 'eigenvalue-1' criterion or selecting the number of principal components yourself. Choose the latter option by clicking 'I will specify', as we want to view all of the principal component images, then accept the default value of 6. Finally, give the name of a file to which the principal component images are to be written, plus the names of the associated header and INF files. The six principal components are written to the output file as 32 bit real numbers, and the procedure terminates.

You can display the principal component images using *View|Display Image*. As the images are stored in the form of 32 bit real numbers, you can choose to map them onto the 0–255 range required by the display memory using either a linear or a non-linear procedure as described in Chapter 3.

Details of the calculations, including the correlation matrix and the eigenvalues and eigenvectors are written to your log file.

6.4.2 Noise-adjusted principal components analysis

The presence of noise that tends to dominate over the signal in lower-order principal component images is mentioned a number of times in the preceding section. Some researchers, for example Green *et al.* (1988) and Townshend (1984), note that some airborne thematic mapper data do not behave in this way. Roger (1996) suggests that the principal components method could be modified so as to eliminate the noise variance. This 'noise-adjusted' principal components analysis would then be capable of generating principal component images that are unaffected by noise. The method presented here and incorporated into the MIPS software is similar in principle to 'noise-adjusted principal components analysis' of Green *et al.* (1988) and Roger

(1996). It is described in more detail in Tso and Mather (2001), while Neilsen (1994) gives a definitive account.

Standard PCA extracts successive principal components in terms of the 'maximum variance' criterion. Thus, if we assume that every variable $\mathbf{x}_j$ in the principal components analysis has a mean of zero, then we could define principal component number one as that combination that maximises the expression $\sum_{i=1}^{p}\sum_{j=1}^{n} x_{ji}^2$, where x_{ji} is an element of the data matrix that consists of n observations (pixels) measured on p spectral bands. The second and subsequent principal components are defined in terms of the same criterion, but after the variance attributable to lower-order principal components has been removed.

Using this idea, principal components are seen as linear combinations of the original variables $\mathbf{x}_j$, with the coefficients of the linear combination being defined so that the maximum variance criterion is satisfied. A linear combination of variables $\mathbf{x}_j$ is simply a weighted sum of the x_{ij} and the coefficients of the linear combination are the weights. There are an infinite number of possible weights – principal components analysis determines the set that satisfies the desired criterion, that of maximum variance, as explained in section 6.4.1.

What if we use a criterion other than that of minimum variance? Could there be another set of weights that satisfies the new criterion? This new criterion could be expressed in words as 'maximise the ratio of the signal variance to the noise variance', which we will express mathematically as $\frac{\sigma_S^2}{\sigma_N^2}$, which is the ratio of signal variance (σ_S^2) to noise variance (σ_N^2). Two questions arise at this point: firstly, is there a procedure that could maximize the new criterion? Secondly, how do we calculate σ_S^2 and σ_N^2? The answer to the first question is 'yes', and it is described later in this section. The second question is, in fact, more difficult. We can easily compute the sums of squares given by the expression $\sum_{i=1}^{p}\sum_{j=1}^{n} x_{ji}^2$ because the values of $\mathbf{X}\ (= x_{ji})$ are known. However, we need a method that will separate the measurements x_{ij} into two parts, the first part representing the 'signal' part of each x (i.e., x_S) and the second representing the noise contribution (i.e., x_N). This idea could be written as $x = x_S + x_N$. If any two of these terms are known then the other can be calculated. Unfortunately, though, only x – the pixel value – is known.

One way out of this paradox is to consider the nature of 'noise'. If we can think a single-valued spatial data set (such as a digital elevation model or DEM, Figure 2.12) as consisting of the sum of spatial variability over different scales, then – for a DEM covering a hilly area at a map scale of 1:100 000 or smaller – we see patterns that relate to the *regional* disposition of hills and valleys and may be able to draw some conclusions about the structure of the area as a whole; for example, we may deduce from the fact that the valleys radiate from a central point, which is also the highest point, that the area has a dome-like structure. Alternatively, if the rivers form a trellis-like pattern, as in parts of the Appalachian Mountains in the Eastern USA, then we may conclude that the drainage pattern is controlled by the nature of the underlying rocks, which consist of alternating hard and soft bands.

In both of these instances it is the regional-scale pattern of hills and valleys on which our attention is focussed. No consideration is given to the presence of small hummocks on valley floors, or the numerous tiny gullies that are located near the heads of the main valleys. Those phenomena are too small to be of relevance at a map scale of 1:100 000, so they are ignored. They represent 'noise' whereas the regional pattern of hills and valleys represents the 'signal' or information at this spatial scale. One way of measuring the noise might be to take the difference in height between a given cell in the DEM and the cells to the north and east, respectively. If the cell of interest lies on a flat, plateau-like, area with no minor relief features, then these horizontal and vertical differences in elevation would be close to zero. In the hummocky valley bottoms and the gullied valley heads these local differences in height would be larger. Since the hummocks and gullies represent noise at our chosen scale of observation, we might think of using vertical and horizontal differences to quantify or characterise noise.

The MIPS procedure *Noise Reduction Transform* (accessed from the *Transform* menu) uses this simple idea to separate values x into the signal (x_S) and noise (x_N) components. First, the covariance matrix $\mathbf{C}$ of the full dataset (consisting of n pixels per band, with p bands) is computed. Next, the horizontal and vertical pixel differences in each of the p bands are found, and their covariance matrices calculated and combined to produce the noise covariance matrix $\mathbf{C}_N$. The covariance matrix of the signal, $\mathbf{C}_S$, is found by subtracting $\mathbf{C}_N$ from $\mathbf{C}$. Now we can define the criterion to be used in the calculation of the coefficients of our desired linear combination; it is simply 'maximise the ratio $\mathbf{C}_S/\mathbf{C}_N$', meaning: our linear combinations of the p bands will be ordered in terms of decreasing signal-to-noise ratio rather than in terms of decreasing variance (as in principal components analysis). Mathematical details of how this is done are beyond the scope of this book, but the algorithm is essentially the same as that used in the Multiple Discriminant Analysis technique, described in Mather (1976).

The outcome is a set of coefficients for p linear combinations of the p spectral bands, ranked from 1 (highest signal-to-noise ratio) to p (lowest signal to noise ratio). These coefficients are applied to the data in exactly the same way as described above for standard principal components. Now, however, the resulting images should be arranged in order of the ratio of signal-to-noise variance, rather than in

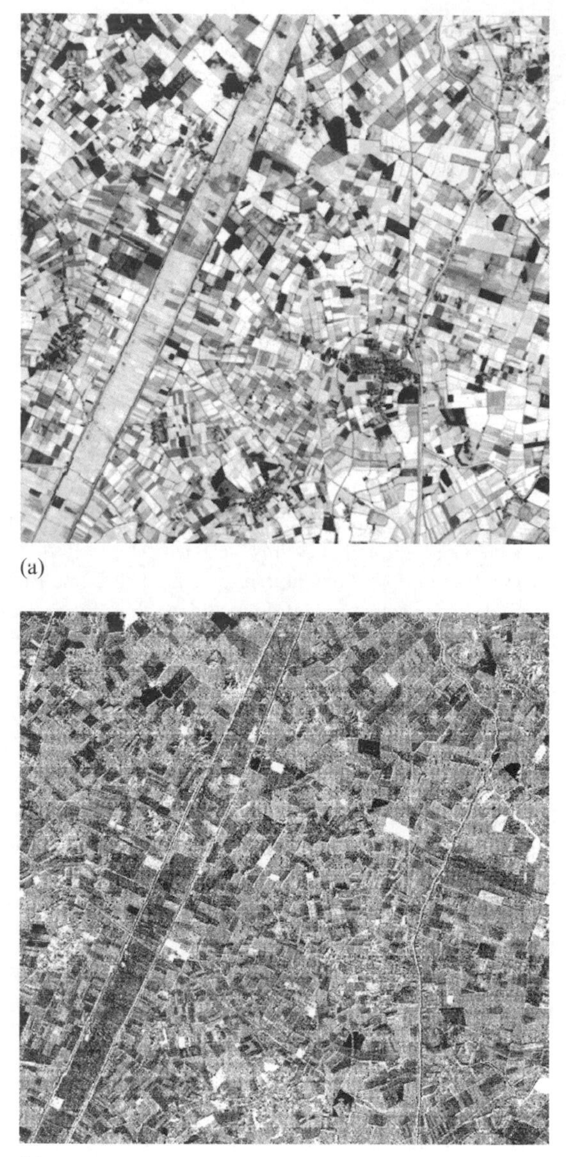

(a)

(b)

Figure 6.14 Noise-reduction transform results for the Littleport image. Figure 6.14 (a) shows the first transformed image, which has the highest signal to noise ratio. The output image with the lowest signal to noise ratio is number six, shown in Figure 6.14 (b). Compare with the results shown in Figures 6.12 and 6.13.

terms of total variance, as shown in Figure 6.14 in which the first and sixth noise-reduced principal components are displayed. To help you compare the results, the image used in this example is the same as that used at the end of section 6.4.1 (Figures 6.12 and 6.13).

6.4.3 Decorrelation stretch

Methods of colour enhancement are the subject of Chapter 5. The techniques discussed in that chapter include linear contrast enhancement, histogram equalisation and the Gaussian stretch. All of these act upon a single band of the false colour image and thus must be applied separately to the red, green and blue components of the image. As noted in section 6.4.1, Principal Components Analysis removes the correlation between the bands of an image set by rotating the axes of the data space so that they become oriented with the directions of maximum variance in the data, subject to the constraint that these axes are orthogonal. If the data are transformed by Principal Components Analysis to a 3D space defined by the principal axes, and are 'stretched' within this space, then the three contrast stretches will be at right angles to each other. In RGB space the three colour components are likely to be correlated, so the effects of stretching are not independent for each colour (Gillespie *et al.*, 1986).

Decorrelation stretching requires the three bands making up the RGB colour composite images to be subjected to a Principal Components Analysis, a stretch applied in the Principal Components space, and the result transformed back to the original RGB space. The result is generally an improvement in the range of intensities and saturations for each colour, with the hue remaining unaltered (section 6.5 provides a more extended discussion of these properties of the colour image). Poor results can be expected when the RGB images do not have approximately Gaussian histograms, or where the image covers large, homogeneous areas.

The decorrelation stretch, like Principal Components Analysis, can be based on the covariance matrix **S** or the correlation matrix **R** (section 6.4.1). Use of **R** implies that all three bands are given equal weight. If the stretch is based on **S**, each band is weighted according to its variance. The following description uses the notation **R** to indicate *either* the correlation *or* the covariance matrix. Essentially, the principal component images **y** are calculated from the eigenvector matrix **E** and the raw images **x** using the relationship $\mathbf{y} = \mathbf{Ex}$, and the inverse transform is $\mathbf{x} = \mathbf{E}^{-1}\mathbf{y}$ (Alley, 1995; Rothery and Hunt, 1990).

1. Compute the eigenvectors **E** $(= e_{ij})$ and eigenvalues $\Lambda(= \lambda_i)$ of **R**, the 3×3 matrix of inter-band correlations or covariances derived from the RGB image to be displayed.
2. Obtain the inverse, $\mathbf{E}^{-1}$, of **E** by noting that $\mathbf{E}^{-1} = \mathbf{E}'$ for an orthogonal matrix. **E'** indicates the transpose of **E**, that is, the rows of **E** are the columns of **E'**.

3. Calculate a diagonal stretching matrix **S** which has elements s_{ii} equal to $50/\sqrt{\lambda_i}$. The value 50 is arbitrary and is meant to represent the desired standard deviation in the decorrelation-stretched image. As **S** is a diagonal matrix it follows that $s_{ij} = 0$ for $i \neq j$.
4. Derive the transformation matrix $\mathbf{T} = \mathbf{E}^{-1}\mathbf{SE}$.
5. Estimate the three correction factors forming the vector **g** required to convert the transformed data back to a 0–255 scale by using the vector of means **m** (the mean value of each raw image) to determine a vector **f** that is equal to **Tm** then calculating the values required to convert each f_i to 127, the mid-point of the 0–255 scale. If $\mathbf{f} = \mathbf{Tm}$ then simply divide each f_i by 127 to obtain the correction factors g_i.
6. Apply the transformation matrix to all the {r,g,b} triples in the image. These triples, which are denoted here by the vector **x**, are scaled onto the 0–255 range via the elements of **g** calculated at step 5, i.e. $\mathbf{y} = \mathbf{Tx}$ gives the 'raw' decorrelation stretched values for a given pixel with RGB values stored in **x**. The raw y values are multiplied by the corresponding g_i to produce the scaled decorrelation stretched pixel.

The values used for the stretching parameter at step 3 require some thought (Campbell, 1996). The value of $50/\sqrt{\lambda_i}$ is suggested by Alley (1995). When combined with the use of the mean vector at step 5 to produce a vector of offsets, the result should be a distribution on each of the three bands that is centred at [127, 127, 127] and with a standard deviation of 50. This configuration does not necessarily produce optimum results. As in the case of standard PCA, and other image transforms which operate in the domain of real numbers, the scaling of the result back to the range 0–255 is logically simple but practically difficult. The method suggested above for standard PCA involves calculating the maximum and minimum of the raw transformed values, then scaling this range onto the 0–255 scale using offsets and stretching parameters. MIPS uses a rather different method. A sample of output values is taken and estimated scaling factors are derived. These estimated scaling factors are used to correct the final output, with separate transformations being used for values above and below the mean. The appearance of the final product depends crucially on the scaling method used, and there are few common-sense guidelines to help the user. The value of the decorrelation stretch is also a function of the nature of the image to which it is applied. The method seems to work best on images of semi-arid areas, and it seems to work least well where the image histograms are bimodal, for example, where the area covered by the image includes both land and sea.

MIPS provides a decorrelation stretch function on the *Enhance* menu. The visual appearance of the resulting image can often be improved if a linear, histogram equalisation or Gaussian stretch is applied following the decorrelation stretch. This may seem to defeat the object of the decorrelation stretch but essentially compensates for the inadequacy of the mapping of the decorrelation stretch output onto the range 0–255. Further consideration of the procedure is provided by Guo and Moore (1996) and Campbell (1996). The latter author presents a detailed analysis of the decorrelation stretch process. Ferrari (1992) and White (1993) illustrate the use of the method in geological image interpretation. The widespread acceptance of the decorrelation stretch procedure is shown by its inclusion as a standard product for data collected by the ASTER sensor carried by the Terra satellite.

6.5 HUE, SATURATION AND INTENSITY (HSI) TRANSFORM

Details of alternative methods of representing colours are discussed in section 5.2 where two models are described. The first is based on the red-green-blue colour cube. The different hues generated by mixing red, green and blue light are characterised by coordinates on the red, green and blue axes of the colour cube (Figure 5.3). The second representation uses the hue-saturation-intensity hexcone model (Figure 5.4) in which hue, the dominant wavelength of the perceived colour, is represented by angular position around the top of a hexcone, saturation or purity is given by distance from the central, vertical axis of the hexcone and intensity or value is represented by distance above the apex of the hexcone. Hue is what we perceive as colour (such as mauve or purple). Saturation is the degree of purity of the colour, and may be considered to be the amount of white mixed in with the colour. As the amount of white light increases so the colour becomes more pastel-like. Intensity is the brightness or dullness of the colour. It is sometimes useful to convert from RGB colour cube coordinates to HSI hexcone coordinates, and vice versa. The RGB coordinates will be considered to run from 0 to 1 (rather than 0–255) on each axis, while the coordinates for the hexcone model will consist of (a) hue expressed as an angle between 0 and 360°, and (b) saturation and intensity on a 0–1 scale. Note that the acronym IHS (intensity – hue – saturation) is sometimes used in place of HSI.

The HSI transform is useful in two ways: first, as a method of image enhancement and, second, as a means of combining co-registered images from different sources. In the former case, the RGB representation is first converted to HSI, as described by Foley *et al.* (1990), Hearn and Baker (1994) and Shih (1995). Next, separate contrast stretches are applied to the intensity (I) and saturation (S) components and the HSI representation is converted back to RGB for display purposes. It does not make sense to apply a contrast-stretch to the hue (H) component, as this would upset the colour balance of the image. Gillespie *et al.* (1986) provide further details. The HSI-based contrast stretch operation is implemented in MIPS from the *Enhance* menu item. The forward transform (RGB to HSI) is performed and, optionally, the I and S components are subjected to a linear stretch using the 95% and 5% points of their histograms, as described in the context of contrast enhancement in section 5.3.1. The stretched HSI representation can be displayed or the inverse transform applied to convert back to RGB coordinates. In practice, it is often necessary to apply a contrast enhancement procedure to the back-transformed image. See section 5.3 for more details of contrast enhancement methods.

The method can also be used to combine images from different sources (data fusion). For example, the higher-resolution panchromatic images that are collected by the Landsat ETM+ and SPOT HRV sensors can be combined with the corresponding lower-resolution multispectral images in a procedure called 'pan sharpening'. Blom and Daily (1982) combine synthetic-aperture radar (SAR) imagery with Landsat MSS imagery in a similar way (see below); see also Harris *et al.* (1990). The steps used by Blom and Daily (1982) to combine a single-band SAR and a three-band Landsat MSS images are as follows:

(i) register the SAR and Landsat MSS images (section 4.3),
(ii) transform the Landsat MSS image from RGB to HSI coordinates,
(iii) substitute the SAR image for the intensity coordinate, and
(iv) transform this combined HSI image back to RGB space.

This transformation has been found to be particularly useful in geological applications, for example Jutz and Chorowicz (1993) and Nalbant and Alptekin (1995). Further details of the HSI transform are given in Blom and Daily (1982), Foley *et al.* (1990), Green (1983), Hearn and Baker (1986), Pohl and van Genderen (1998) and Mulder (1980). Terhalle and Bodechtel (1986) illustrate the use of the transform in the mapping of arid geomorphic features, while Gillespie *et al.* (1986) discuss the role of the HSI transform in the enhancement of highly correlated images. Massonet (1993) gives details of an interesting use of the HSI transform in which the amplitude, coherence and phase components of an interferometric image (Chapter 9) are allocated to hue, saturation and intensity, respectively, and the inverse HSI transform applied to generate a false colour image that highlights details of coherent and incoherent patterns. Schetselaar (1998) discusses alternative representations of the HSI transform, and Andreadis *et al.* (1995) give an in-depth study of the transform. Pitas (1993) lists C routines for the RGB to HSI colour transform. The MIPS function *Transform*|HSI *Pan Sharpen* can be used to combine the spectral detail of an RGB colour composite image with the spatial detail of a geometrically-registered panchromatic image. Co-registration of the RGB and panchromatic images requires resampling of the coarser spatial resolution image to the spatial resolution of the panchromatic image using *Utilities|Change pixel size (resample)* followed by registration (*Transform|Geometric Correction|Register two images*). An alternative procedure for combining (fusing) a lower-resolution multispectral image and a higher spatial resolution panchromatic image using the wavelet transform is described in section 9.3.

6.6 THE DISCRETE FOURIER TRANSFORM

6.6.1 Introduction

The coefficients of the Tasselled Cap functions, and the eigenvectors associated with the principal components, define coordinate axes in the multi-dimensional data space containing the multispectral image data. These data are re-expressed in terms of a new set of coordinate axes and the resulting images have certain properties, which may be more suited to particular applications. The Fourier transform operates on a single-band (greyscale) image, not on a multispectral data set. Its purpose is to break down the image into its spatial scale components, which are defined to be sinusoidal waves with varying amplitudes, frequencies and directions. The coordinates of the two-dimensional space in which these scale components are represented are given in terms of frequency (cycles per basic interval). This is called the *frequency domain* whereas the normal row/column coordinate system in which images are normally expressed is termed the *spatial domain* (Figure 6.15). The Fourier transform is used to convert a single-band image from its spatial domain representation to the equivalent frequency domain representation, and vice versa.

The idea underlying the Fourier transform is that the greyscale values forming a single-band image can be viewed as a three-dimensional intensity surface, with the rows and columns defining two axes (x and y in Figure 6.15) and the grey level intensity value at each pixel giving the third (z) dimension. A series of waveforms of increasing frequency and with different orientations is fitted to this intensity surface and the information associated

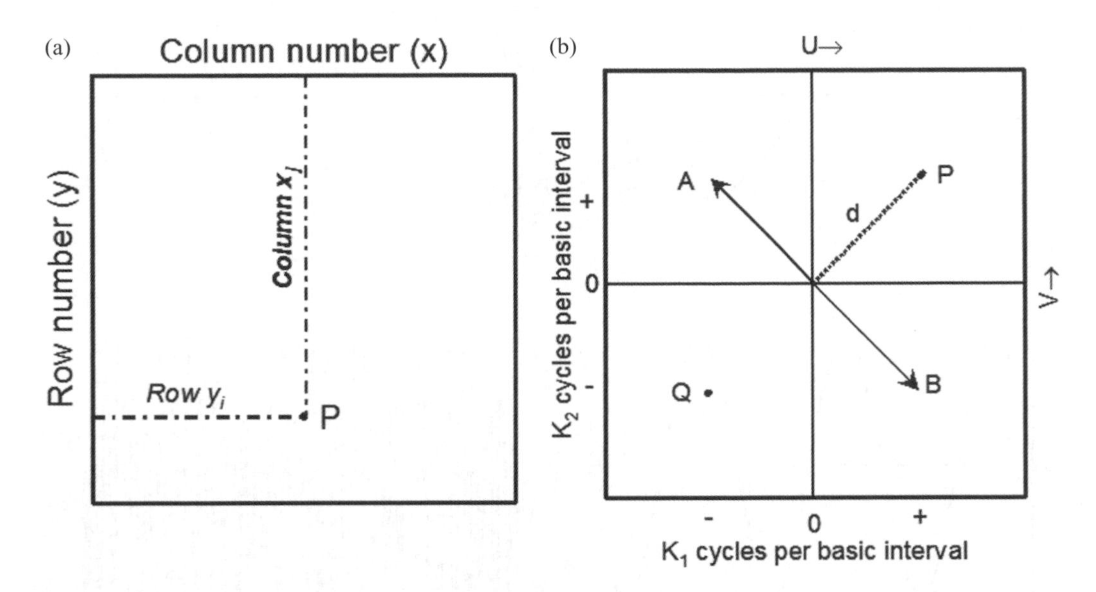

Figure 6.15 (a) Spatial domain representation of a digital image. Pixel P has coordinates (i, j) with respect to the image row and column coordinate axes, or $(i\,\Delta y, j\,\Delta x)$ metres, where Δx and Δy are the dimensions of the image pixels in the row and column direction, respectively. (b) Frequency-domain representation, showing the amplitude spectrum of the image. The value at P is the amplitude of a sinusoidal wave with frequency $k_1 = U$ and $k_2 = V$ cycles per basic interval in the u and v directions, respectively. The wavelength of this sinusoidal wave is proportional to the distance d. The orientation of the waveform is along direction AB. Point Q $(-U, -V)$ is the mirror image of point P.

with each such waveform is calculated. The Fourier transform therefore provides details of (i) the frequency of each of the scale components (waveforms) fitted to the image and (ii) the proportion of information associated with each frequency component. Frequency is defined in terms of cycles per basic interval where the basic interval in the across-row direction is given by the number of pixels on each scan line, while the basic interval in the down-column direction is the number of scan lines. Frequency could be expressed in terms of metres by dividing the magnitude of the basic interval (in metres) by cycles per basic interval. Thus, if the basic interval is 512 pixels each 20 metres wide then the wavelength of the fifth harmonic component is $(512 \times 20)/5$ or 2048 metres. The first scale component, conventionally labelled zero, is simply the mean grey level value of the pixels making up the image. The remaining scale components have increasing frequencies (decreasing wavelengths) starting with 1 cycle per basic interval, then $2, 3, \ldots, n/2$ cycles per basic interval where n is the number of pixels or scan lines in the basic interval.

This idea can be more easily comprehended by means of an example using a synthetic one-dimensional data series. This series consists of the sum of four sine waves that differ in frequency and amplitude. Sine wave 1 (Figure 6.16(a)) has an amplitude of 1.0 and completes one cycle over the basic interval of 0–6.28 radians (360°), so its frequency is 1 Hz. The second, third and fourth sine waves have amplitudes of 2.0, 2.5 and 1.75, respectively, and frequencies of 2, 3 and 32 Hz. These sine waves were calculated at 512 points along the x-axis, and Figure 6.16(e) shows the result of adding up the values of sine waves 1–4 at each of these 512 points. It is difficult to discern the components of this composite sine wave by visual analysis. However, the discrete forward Fourier transform (Figure 6.16(f) and (g)) clearly shows the frequencies (horizontal axis) and amplitudes (vertical axis) of the four component sine waves. These frequency/amplitude diagrams show the *amplitude spectrum*, which should not be confused with the electromagnetic spectrum. Other small frequency components are visible in the amplitude spectrum; these result from the fact that the series is a discrete one, measured at 512 points.

6.6.2 Two-dimensional DFT

If this simple example were extended to a function defined over a two-dimensional grid then the differences would be that (a) the scale components would be two-dimensional waveforms, and (b) each scale component would be characterised by orientation as well as amplitude. The squared amplitudes of the waves are plotted against frequency in the

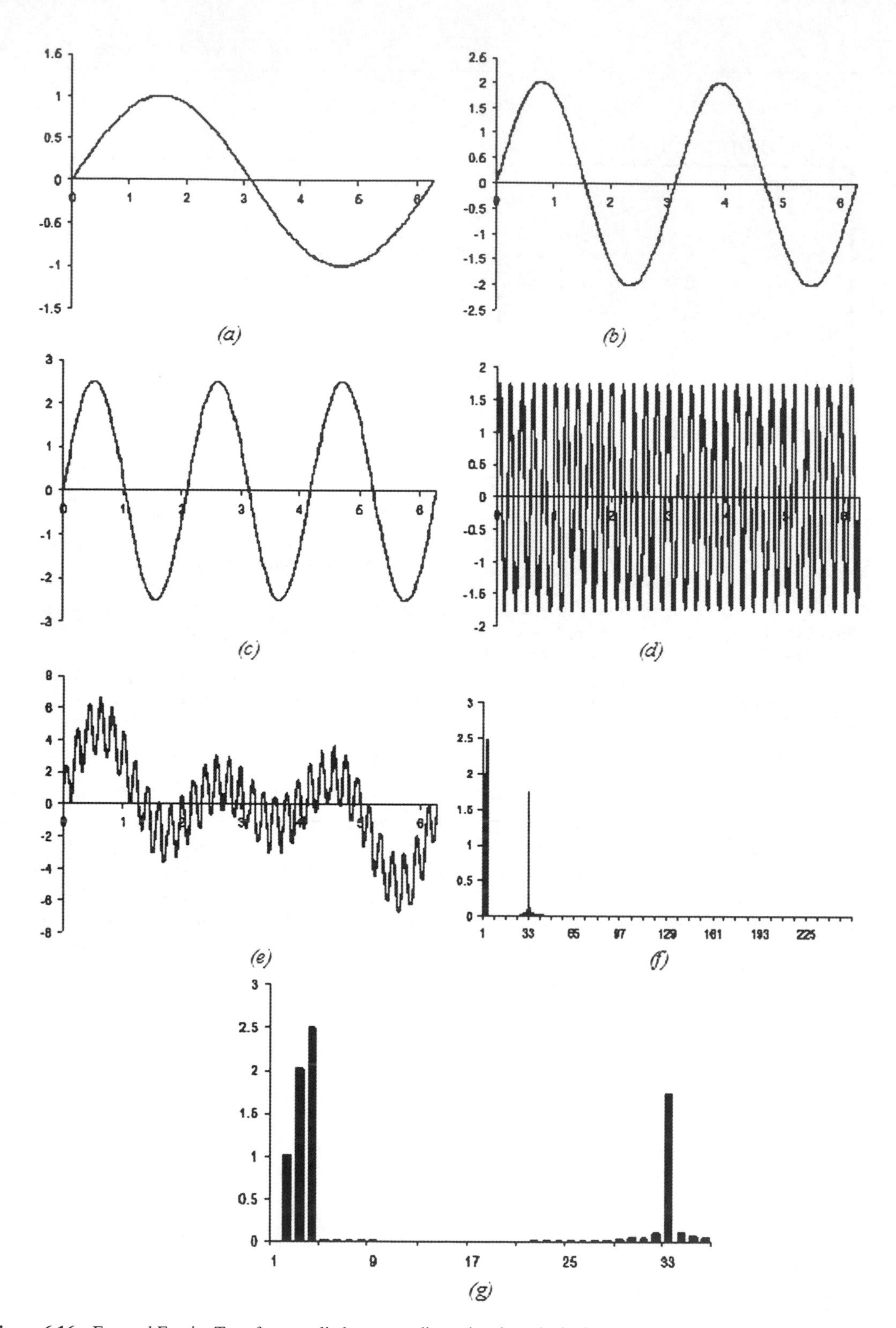

Figure 6.16 Forward Fourier Transform applied to a one-dimensional synthetic data series shown in (e). This synthetic series is the sum of the sine waves (a)–(d). The amplitude spectrum of the series is shown in (f) with the portion covering harmonics 0–34 shown in enlarged form in (g). Note that the x-axis in (f) and (g) is labelled on a scale begining at 1 rather than 0, so the point x = 33 indicates harmonic number 32. See text for explanation.

horizontal and vertical directions to give a two-dimensional amplitude spectrum, which is interpreted much in the same way as the one-dimensional amplitude spectrum in Figure 6.16(f), the major differences being

(i) the frequency associated with the point $[k_1, k_2]$ in the two-dimensional amplitude spectrum is given by:

$$k_{12} = \sqrt{(k_1^2 + k_2^2}$$

where the basic intervals given by each axis of the spatial domain image are equal, or by:

$$k_{12} = \sqrt{(k_1/n_1 \Delta t_1 + k_2/n_2 \Delta t_2)}$$

where the basic intervals in the two spatial dimensions of the image are unequal. In the latter case, $n_1 \Delta t_1$ and $n_2 \Delta t_2$ are the lengths of the two axes, n_1 and n_2 are the number of sampling points along each axis, and Δt_1 and Δt_2 are the sampling intervals. This implies that frequency is proportional to distance from the centre of the amplitude spectrum which is located at the point [0, 0] in the centre of the amplitude spectrum diagram (Rayner, 1971).

(ii) The angle of travel of the waveform whose amplitude is located at point $[k_1, k_2]$ in the amplitude spectrum is perpendicular to the line joining the point $[k_1, k_2]$ to the centre (DC) point of the spectrum ([0, 0]). This point is illustrated in Figure 6.17(b), which shows the two-dimensional amplitude spectrum of an 512 × 512 pixel image made up of a set of horizontal lines spaced 16 rows apart. These lines are represented digitally by rows of 1s against a background of 0s (Figure 6.17(b)). The amplitude spectrum shows a set of symmetric points running horizontally through the DC point, which lies at the centre of the image. The points are so close that they give the appearance of a line. They represent the amplitudes of the waveforms reconstructed from the parallel, horizontal lines which could be considered to lie on the crests of a series of sinusoidal waveforms progressing down the image from top to bottom. Since the direction of travel could be top-bottom or bottom-top the amplitude spectrum is

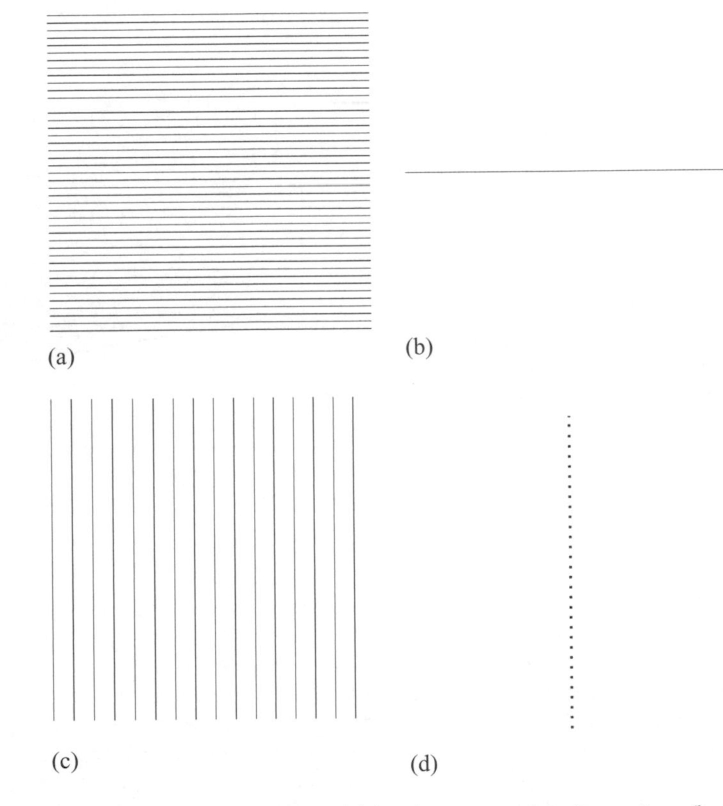

Figure 6.17 The left-hand images ((a) and (c)) show the spatial domain representation of two patterns (horizontal lines spaced 16 rows apart and vertical lines spaced 32 columns apart). The right-hand images show the frequency domain representation of the two patterns. The frequency domain representation used here is the logarithm of the two-dimensional amplitude spectrum.

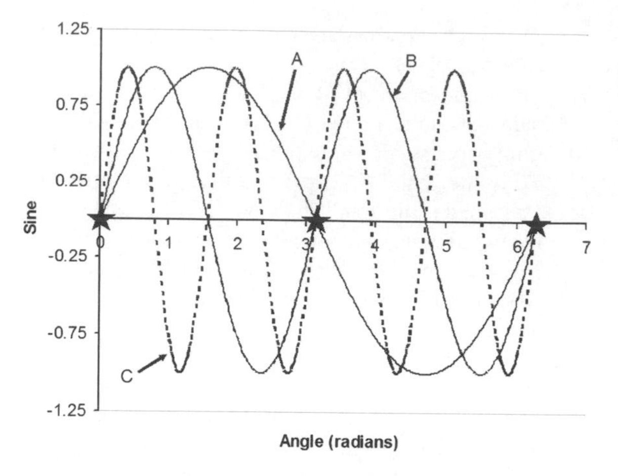

Figure 6.18 Illustrating the concept of aliasing. If we have only three values of x, at 0.0000, 3.1415 and 6.2430 then any of a number of sine waves will pass through these points. Here, three are shown. Curve A completes one cycle over the range 0–6.2430, curve B completes two cycles, and curve C completes four cycles. On the basis of the data alone, it is not possible to say whether all of these curves exist in the data or whether some are artefacts generated by aliasing (i.e., false identity).

symmetric and so the two points closest to the DC represent the amplitudes of the set of waves whose wavelength is equal to the spacing between the horizontal lines. The two points further out from the DC represent a spurious waveform, which has a wavelength equal to double the distance between the horizontal lines. Such spurious waveforms represent a phenomenon termed *aliasing* (Rosenfeld and Kak, 1982; Gonzales and Woods, 1992; Figure 6.18). Figure 6.17(c) and (d) shows a second example using a set of vertical lines as input. This time, the lines are 32 columns apart, and the amplitudes of the waveforms are seen as points. The amplitude spectrum is rotated through 90° in comparison with Figure 6.17(b) but can be interpreted in a similar way.

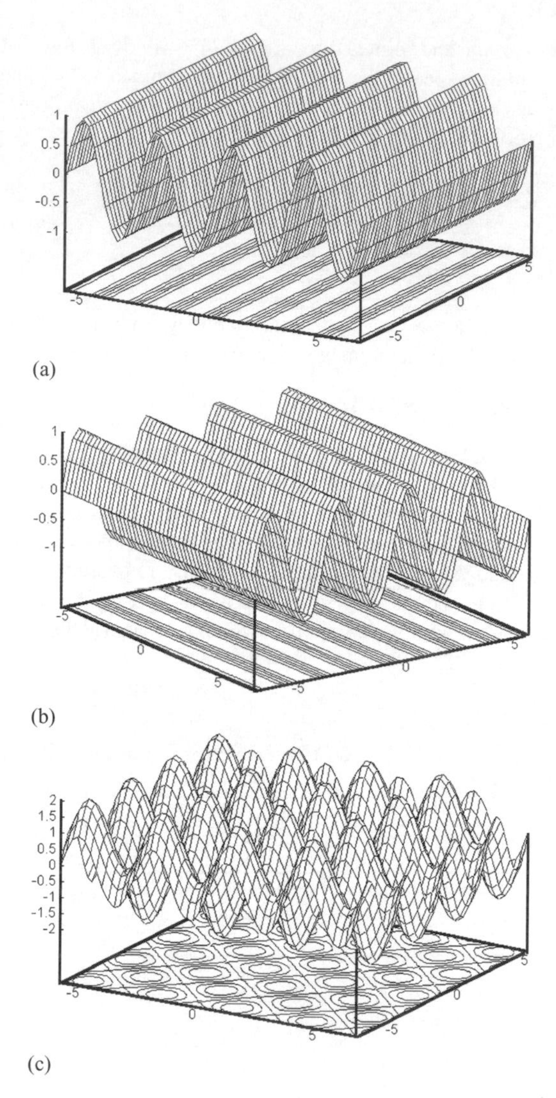

Figure 6.19 (a) Two-dimensional sine wave running parallel to the x-axis. (b) Sine wave running parallel to the y-axis. (c) Composite formed by summing the values of the sine waves shown in (a) and (b).

Figure 6.19 may help to illustrate some of these ideas. Figure 6.19(a) shows a two-dimensional sine wave orientated in a direction parallel to the x-axis of the spatial domain coordinate system. Figure 6.19(b) shows another sine wave, this time oriented parallel to the y-axis, and Figure 6.19(c) shows the result of adding these two sine waves together to generate an undulating surface.

The two-dimensional amplitude spectrum of a composite sine wave is shown in Figure 6.20(b). Figure 6.20(a) shows the pattern of high and low values (peaks and troughs – the same pattern as in Figure 6.19(c) but in the form of an image. In an ideal world, with an infinite sample (rather than the 512 × 512 grid used here) we might expect that the frequency-domain representation would show four high values located above and below and to the right and left of the DC at a distance from the DC that is proportional to the frequency of the sine waves (i.e., the number of times that the complete sine wave is repeated in the x direction (or the y direction, since the two axes are equal). The caption to Figure 6.20 explains why this is not the case, for the amplitude in the diagonal directions of Figure 6.20(a) is greater then either the vertical or horizontal amplitude.

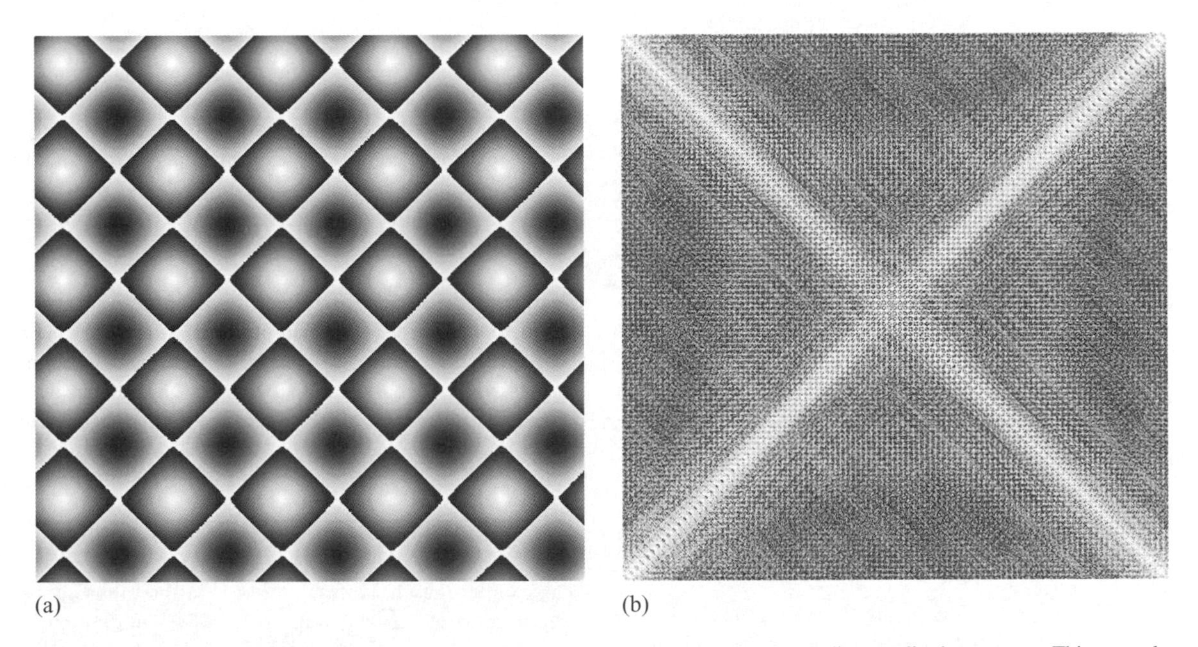

Figure 6.20 (a) Image of composite sine waves shown in Figure 6.19(c). (b) The corresponding amplitude spectrum. This example is curious in that one might think that the dominant orientations were vertical and horizontal, but not diagonal. Closer inspection of Figure 6.20(a) shows that the horizontal/vertical pattern is either dark-intermediate-dark or light-intermediate-light, whereas the diagonal pattern is dark-light-dark or light-dark-light. In other words, the amplitude or range of the diagonal pattern is greater than that of the horizontal/vertical pattern, and this is reflected in the frequency domain representation in (b). If you go back to Figure 6.19(c) you will appreciate that the diagonal pattern has a greater amplitude than the horizontal or vertical pattern, even though the image is constructed by the addition of waveforms running parallel to the x- and y-axes, respectively.

The calculation of the amplitude spectrum of a two-dimensional digital image involves techniques and concepts that are too advanced for this book. A simplified account will be given here. Fourier analysis is so called because it is based on the work of Jean Baptiste Joseph, Baron de Fourier, who was Governor of Lower Egypt and later Prefect of the Departement of Grenoble during the Napoleonic era. In his book *Theorie Analytique de la Chaleur*, published in 1822, he set out the principles of the Fourier series which has since found wide application in a range of subjects other than Fourier's own, the analysis of heat flow. The principle of the Fourier series is that a single-valued curve (i.e., one which has only a single y value for each separate x value) can be represented by a series of sinusoidal components of increasing frequency. This point is illustrated in Figure 6.16. The form of these sinusoidal components is given by:

$$f(t) = a_0 + \sum_n a_n \cos n\omega t + \sum_n b_n \sin n\omega t$$

in which $f(t)$ is the value of the function being approximated at point t, ω is equal to $2\pi/T$ and T is the length of the series. The term a_0 represents the mean level of the functions and the summation terms represent the contributions of a set of sine and cosine waves of increasing frequency. The fundamental waveform is that with a period equal to T, or a frequency of ω. The second harmonic has a frequency of 2ω, while 3ω is the frequency of the third harmonic, and so on.

The a_i and b_i terms are the cosine and sine coefficients of the Fourier series. It can be seen in the formula above that the coefficients a_i are the multipliers of the cosine terms and the b_i are the multipliers of the sine terms. Sometimes the sine and cosine terms are jointly expressed as a single complex number, that is, a number which as 'real' and 'imaginary' parts, with the form $(a_i + jb_i)$. The a part is the real or sine component, and b is the imaginary or cosine component. The term j is equal to $\sqrt{-1}$. Hence, the a_i and b_i are often called the real and imaginary coefficients. We will not use complex number notation here. The coefficients a and b can be calculated by a least-squares procedure. Due to its considerable computing time requirements this method is no longer used in practice. In its place, an algorithm called the Fast Fourier Transform (FFT) is used (Bergland, 1969, Gonzales and Woods, 1992; Lynn, 1982; Pavlidis, 1982; Pitas, 1993, Ramirez, 1985). The advantage

Table 6.4 Number of operations required to compute the Fourier transform coefficients (a) and (b) for a series of length N (column (i)) using least-squares methods (column (ii)) and the Fast Fourier Transform (FFT) (column (iii)). The ratio of column (ii) to column (iii) shows the magnitude of the improvement shown by the FFT. If each operation took 0.01 second then, for the series of length $N = 8096$, then the least-squares method would take 7 days, 18 hours and 26 minutes. The FFT would accomplish the same result in 17 minutes 45 seconds.

(i) N	(ii) N^2	(iii) $N\log_2 N$	(iv) (ii)/(iii)
2	4	2	2.00
4	16	8	2.00
16	256	64	4.00
64	4096	384	10.67
128	16 384	896	18.29
512	262144	4608	56.89
8096	67 108 864	106 496	630.15

of the FFT over the older method can be summarised by the fact that the number of operations required to evaluate the coefficients of the Fourier series using the older method is proportional to N^2 where N is the number of sample points (length of the series) whereas the number of operations involved in the FFT is proportional to $N \log_2 N$. The difference is brought out by a comparison of columns (ii) and (iii) of Table 6.4. However, in its normal implementation the FFT requires than N (the length of the series) should be a power of two. Singleton (1979 a) gives an algorithm that evaluates the FFT for a series of length N that need not be a power of two while Bergland and Dolan (1979) provide a FORTRAN program listing of an algorithm to compute the FFT in a very efficient manner.

Once the coefficients a_i and b_i are known the amplitude of the ith harmonic is computed from:

$$A_i = \sqrt{a_i^2 + b_i^2}$$

and the phase angle (the displacement of the first crest of the sinusoid from the origin, measured in radians or degrees) is defined as:

$$\theta = \tan^{-1}(b_i/a_i)$$

Generally, only the amplitude information is used.

The procedure to calculate the forward Fourier transform (it is, in fact, the forward discrete Fourier transform) for a two-dimensional series, such as a greyscale image, involves the following steps:

(i) compute the Fourier coefficients for each row of the image, storing the coefficients a_i and b_i in separate two-dimensional arrays. The coefficients form two-dimensional arrays of real numbers, or they can be considered to form a single two-dimensional complex array. The former representation is used here.

(ii) compute the Fourier transform of the columns of the two matrices composed, respectively, of the a_i and b_i coefficients to give the Fourier coefficients of the two-dimensional image. There are two sets of coefficients, corresponding to the a_i and b_i terms.

Step (ii) requires that the two coefficient matrices are transposed; this is a very time-consuming operation if the image is large. Singleton (1979b) describes an efficient algorithm for the application of the FFT to two-dimensional arrays.

Example 6.2 illustrates the use of the Fourier module in MIPS.

EXAMPLE 6.2: THE FOURIER TRANSFORM

This example has the aim of demonstrating how the Fourier amplitude spectrum is computed in MIPS, and how it is interpreted. The example is continued in section 7.5, when frequency-domain filtering is discussed.

Example 6.2 Figure 1 shows an area of the Red Sea Hills of the Sudan east of Port Sudan. It was collected by the Landsat-5 TM instrument in the middle infrared region of the spectrum (band 5). The area experiences an arid climate, so surface materials are mainly bare rock and weathering products such as sand (bright area near the lower left

Example 6.2 Figure 1 Landsat-5 TM band 5 image of part of the Red Sea Hills, Sudan.

corner). The region was tectonically active in the past and a number of fault lines are visible. The aim of this example is to show how to apply the Fourier transform to convert the representation of the image information from the spatial to the frequency domain.

First, ensure that the file *sudanhi.inf* is available. Copy it from the CD to your hard drive if necessary, and follow the steps outlined below:

1. Use *View|Display Image*, select *sudanhi.inf*, and display band 5 of this image set as a greyscale image. The dimensions of the image are 1024 × 1024 pixels.
2. Choose *Filter|Fourier Transform|Forward Transform.* The log of the amplitude spectrum of the image is displayed in a new window.

The log of the amplitude spectrum is shown above as Example 6.2 Figure 2. The origin of the frequency domain is the centre of the image, and the units of measurement are *cycles per basic interval* in the u and v (horizontal and vertical directions). The lengths of the u and v axes are both equal to 1024 pixcls, so the brightness levels of the four nearest pixels to the centre (above, below, left, and right) represent the proportion of total image information at the lowest possible spatial frequency of one cycle per basic interval (equal to a wavelength 1024 × 30 m or 30.720 km) in the horizontal and vertical directions. The second closest

Example 6.2 Figure 2 Logarithm of the Fourier amplitude spectrum of the image shown in Example Figure 2.1.

pixels in the same directions show the information present at spatial frequencies of two cycles per basic interval or a spatial wavelength of 512 × 30 m or 15.360 km.

The amplitude spectrum is interpreted in terms of spatial frequencies or wavelengths only in circumstances in which a particular frequency is to be identified. For example, the Landsat TM and ETM+ sensors gather image data in 16-line sections, and so there may be evidence of a peak in the amplitude spectrum at a spatial wavelength of $16 \times 30 = 480$ m, which corresponds to a frequency of $1024/16 = 64$ (1024 is the vertical or v axis length). Bright points positioned on the v axis at 64 pixels above and below the centre point of the amplitude spectrum would indicate the presence of a strong contribution to image variance at frequencies corresponding to a spacing of 16 lines on the image.

More generally, the shape of the amplitude spectrum can give some indication of the structures present in the image. Figure 6.17 illustrates the shape of the Fourier amplitude spectrum for two extreme cases – of vertical and horizontal lines. Hence, we would expect to see some evidence of directionality in the spectrum if the features in the image area had any preferred orientation. Another use of the amplitude spectrum is to measure the texture of the image (section 8.7.1). An image with a fine texture would have a greater proportion of high frequency information compared with an image with a coarse texture. Therefore, the ratio of high frequency components (those that are further from the origin that a given distance d_1, for example) to low frequency components (those closer to the origin than a specified distance d_2) can be used to measure the texture of an image or a sub-image. Usually, a small moving window of size 32 × 32 pixels is passed across the image, the ratio of high to low frequencies is computed, and this value is assigned to the centre pixel in the window).

The amplitude spectrum is also used as a basis for filtering the image (Chapter 7). Filtering involves the selective removal or enhancement of specific frequency bands in the amplitude spectrum, followed by an inverse Fourier transform to convert the filtered information back to the spatial domain. This procedure is illustrated in Chapter 7 using the Sudan image (above) as an example.

6.6.3 Applications of the DFT

As noted already, the main use of the Fourier transform in remote sensing is in frequency-domain filtering (section 7.5). For example, Lei *et al.* (1996) use the Fourier transform to identify and characterise noise in MOMS-02 panchromatic images in order to design filters to remove the noise. Other applications include the characterisation of particular terrain types by their Fourier transforms (Leachtenauer, 1977), and the use of measures of heterogeneity of the grey levels over small neighbourhoods based on the characteristics of the amplitude spectra of these neighbourhoods. If an image is subdivided into 32 × 32 pixel sub-images and if each sub-image is subjected to a Fourier transform then the sum of the amplitudes in the area of the amplitude spectrum closest to the origin gives the low-frequency or smoothly varying component while the sum of the amplitudes in the area of the spectrum furthest away from the origin gives the high-frequency, rapidly-changing component. These characteristics of the amplitude spectrum have also been used as measures of image texture, which is considered further in section 8.7.1. Fourier-based methods have also been used to characterise topographic surfaces (Brown and Scholz, 1985), in the calculation of image-to-image correlations during the image registration process (section 4.3.4), in analysis of the performance of resampling techniques (Shlien, 1979; section 4.3.3) and in the derivation of pseudocolour images from single-band (greyscale) images (section 5.4). De Souza Filho *et al.* (1996) use Fourier-based methods to remove defects from optical imagery acquired by the Japanese JERS-1 satellite. Temporal sequences of vegetation indices (section 6.2.4) are analysed using one-dimensional Fourier transforms by Olsson and Ekhlund (1994), and a similar approach is used by Menenti *et al.* (1993).

In the derivation of a pseudocolour image the low frequency component (Figure 7.13) is extracted from the amplitude spectrum and an inverse Fourier transform applied to give an image that is directed to the red monitor input. The intermediate or midrange frequencies are dealt with similarly and directed to the green input. The blue input is derived from the high-frequency components, giving a colour rendition of a black-and-white image in which the three primary colours (red, green and blue) represent the low, intermediate and high frequency or scale components of the original greyscale image. Blom and Daily (1982) and Daily (1983) describe and illustrate the geological applications of a variant of this method using synthetic-aperture radar images. Their method is based on the observation that grey level (tonal) variations in synthetic aperture radar images can be attributed to two distinct physical mechanisms. Large-scale features (with low spatial frequency) are produced by variations in surface backscatter resulting from changes in ground surface cover type. High spatial frequencies correlate with local slope variations, which occur on a much more rapid spatial scale of variability. The amplitude spectrum of the SAR image is split into these two components (high and low frequencies) using frequency-domain filtering methods that are described in section 7.5. The result is two filtered amplitude spectra, each of which is subjected to an inverse Fourier transform to convert from the frequency back to the spatial domain. The low-pass filtered image is treated as the hue component in HSI colour space (section 6.5) and the high-pass filtered image is treated as the intensity component. Saturation is set to a value that is constant over the image; this value is chosen interactively until a pleasing result is obtained. The authors suggest that the pseudocolour image produced by this operation is significantly easier to interpret than the original greyscale SAR image.

6.7 THE DISCRETE WAVELET TRANSFORM

6.7.1 Introduction

The idea of representing the information content of an image in the spatial domain and the frequency domain is introduced in section 6.6. These two representations give different 'views' of the information contained in the image. The discrete Fourier transform (DFT) is presented in that same section as a technique for achieving the transformation of a grey scale (single-band) image from the spatial domain of (row, column) coordinates to the frequency domain of (vertical frequency, horizontal frequency) coordinates. While the frequency domain representation contains information about the presence of different waveforms making up the grey scale image, it does not tell us *where* in the image a specific waveform with a particular frequency occurs; its is assumed that the 'frequency mix' is the same in all parts of the image. Another disadvantage of the frequency domain representation is the need to assume statistical stationarity, which requires that the mean and variance of the pixel values are constant over all regions of the image. In addition, the DFT assumes that the image repeats itself in all directions to infinity.

The DFT has a number of useful applications in the analysis of remotely-sensed images. These are elaborated in sections 6.6 and 7.5. In this section, the discrete wavelet transform (DWT) is introduced. It augments, rather than replaces, the DFT because it represents a compromise between the spatial and frequency domain representations. It is impossible to measure exactly both the frequencies present in a grey scale image *and* the spatial location of those frequencies (this is an extension of Heisenberg's Uncertainty Principle). It is, however, possible to transform an image into a representation that combines frequency bands (ranges of frequencies) and specific spatial areas. For example, the Windowed Fourier Transform generates a separate amplitude spectrum for each of a series of sub-regions of the image and thus provides some idea of the way in which the frequency content of the data series changes with time or space. The Windowed Fourier Transform does not have the flexibility of the DWT, which is generally preferred by statisticians.

An outline of the DWT is provided in section 6.7.1. Details of the use of the DWT in removing noise from the reflectance spectra of pixels (i.e., one-dimensional signals) and images (two-dimensional signals) are given in section 9.3.2. The derivation of the two-dimensional DWT is described in section 6.7.2. For a more advanced treatment, see Addison (2002), Mallat (1998) and Starck *et al.* (1998). Strang (1994) provides a gentler introduction.

6.7.2 The one-dimensional discrete wavelet transform

There are a number of ways of presenting the concept of the wavelet transform. The approach adopted in this section is more intuitive than mathematical, and it uses the idea of cascading low-pass and high-pass filters (such filters are discussed in section 7.1). Briefly stated, a low-pass filter removes or attenuates high frequencies or details, producing a blurred or generalised output. A high-pass filter removes the slowly changing background components of the input data, producing a result that contains the details without the background. Examples of low-pass and high-pass filters are shown in Figure 6.21. The wavelet transform can be considered as a sequence of (high-pass, low-pass) filter pairs, known as a filter bank, applied to a data series **x** that could, for example, be a reflectance spectrum sampled or digitised over a given wavelength range. Note that the samples are assumed to be equally spaced along the x-axis, and the number of samples (n) is assumed to be a power of 2, that is, $n = 2^j$, where j is a positive integer.

The steps involved in the one-dimensional DWT are as follows:

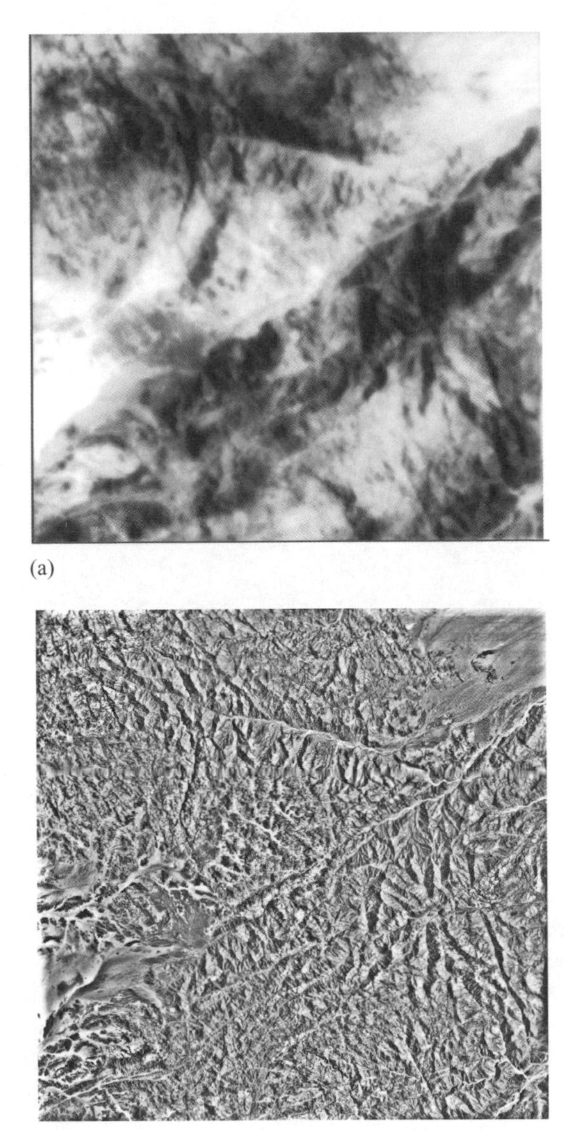

(a)

(b)

Figure 6.21 Filtered images: low-pass filtered (left) and high-pass filtered (right) images derived from the Sudan image used in Example 6.2. The filter used here is the Butterworth filter, which separates the low frequency and high frequency components of the amplitude spectrum (see Chapter 7 for details).

1. Apply a low-pass filter to the full data series **x** of length n (where n is a power of two). If $n$is not a power of two, pad the series with zeros to extend its length to the next higher power of two. Take every second element of **x** to produce a series $\mathbf{x}_1$ of length $n/2$.
2. Apply a high-pass filter to the full data series **x**, padding with zeros as in step 1 if necessary. Take every second element of the result, to produce a set of $n/2$ high-pass filter outputs comprising the first level of the detail coefficients, $\mathbf{d}_1$, of length $n/2$.
3. Take the result from step 1, which is a data series $\mathbf{x}_1$ of length $n/2$ that has been smoothed once. Apply a low-pass filter. Take every second element of the result to produce a series $\mathbf{x}_2$ of length $n/4$.
4. Apply a high-pass filter to the output from step 1. Take every second element of the result, to give a vector $\mathbf{d}_2$of length $n/4$. Vector $\mathbf{d}_2$ forms the second level of detail coefficients.
5. Repeat the low-pass/high-pass filter operation on the output of step 3. Continue filtering and decimating until the length of the resulting data series is one.

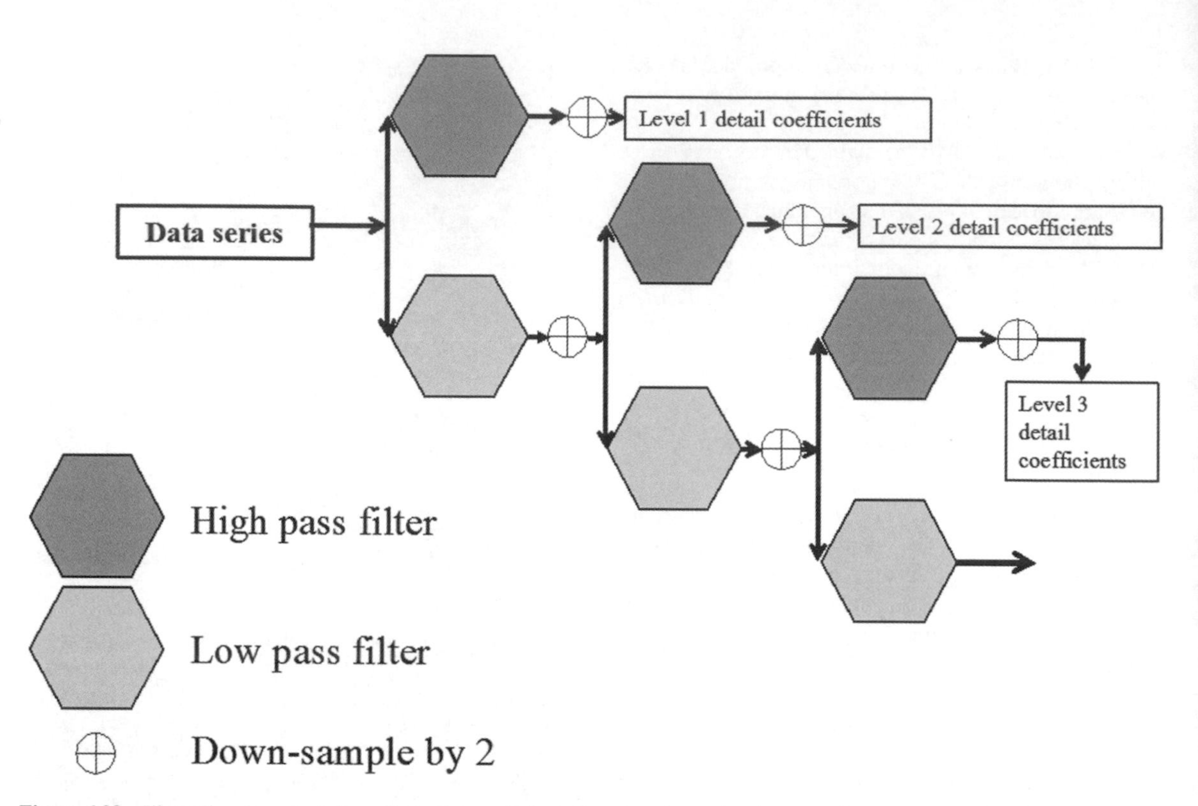

Figure 6.22 Illustrating the calculation of wavelet coefficients for a one-dimensional series of measurements. Levels 1, 2 and 3 are shown but the procedure continues until down-sampling by two results in a series containing only a two elements.

These operations are sometimes termed *sub-band coding* and the sequence of high pass and low pass filters is known as a *quadrature mirror* filter. The operations are shown schematically in Figure 6.22.

The wavelet or detail coefficients formed at steps 2, 4, and so on (i.e., vectors $\mathbf{d}_1, \mathbf{d}_2, \mathbf{d}_3, \ldots, \mathbf{d}_{n/2}$ using the notation introduced above) can be interpreted as follows. The vector $\mathbf{d}_1$ represents the detail (plus noise) in the original image. Vector $\mathbf{d}_2$ characterises the detail (plus noise) in the once-smoothed image. Vectors $\mathbf{d}_3, \mathbf{d}_4, \ldots, \mathbf{d}_n$ contain the detail (plus noise) derived by filtering the twice-smoothed, thrice-smoothed, and n-times smoothed image. There are $n/2$ elements in $\mathbf{d}_1$, $n/4$ elements in $\mathbf{d}_2$, $n/8$ elements in $\mathbf{d}_3$, and so on, as the series length is halved at each step. Thus, for an original series length n of 32, the number of detail coefficients is $16 + 8 + 4 + 2 + 2$. The last two coefficients are somewhat different from the others, but that need not concern us here. Each subset of the ($n = 32$)-point raw data sequence, with a length of 16, 8, 4 and 2, is derived from an increasingly smoothed series, the effect of down-sampling being to make the series sampled further and further apart. If the location of the samples is plotted, a time-scale diagram such as that shown in Figure 6.23 is produced. If the input data series is formed of the elements of a reflectance spectrum, then the time dimension is replaced by wavelength.

In a data series of length 512, the 256 coefficients at level one are derived from the original (raw) series. The 128 coefficients at level two come from the series after one smoothing, the 64 coefficients at level three are extracted from the twice-smoothed series, and so on. As progressive smoothing implies a continual reduction in high-frequency content, it follows that the representation of the data with the greatest high frequency content is level one, and that with the lowest high-frequency content is level nine. The higher frequencies are sampled at a higher rate than the low frequencies, which is logical as high frequencies vary more quickly in time and space than do low frequencies. An alternative way of representing the wavelet detail coefficients is to place them in the order shown in Figure 6.24.

The sampling pattern in Figure 6.24 is termed dyadic (from the Greek word for 'two') because the sample size is reduced by a factor of two at each level. Some authors

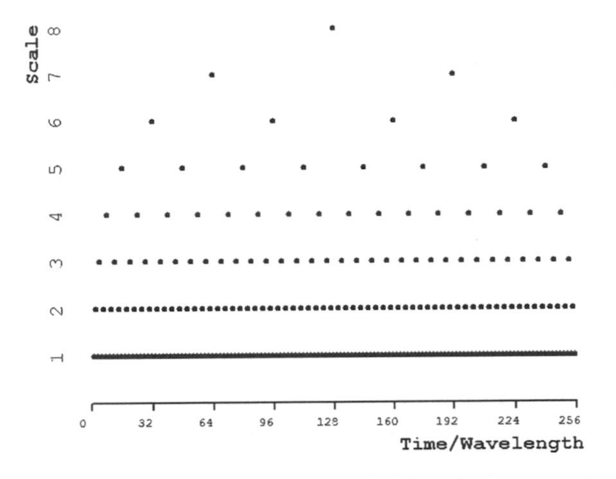

Figure 6.23 Time-scale graph for a one-dimensional series of length 256 points. The raw data series can be considered to be level 0. Dyadic sampling is applied, so that at level one, the number of detail coefficients is 128, reducing to 64 at level two, and so on. See Figure 6.24 for an alternative representation of this process of down-sampling. Note that when remotely-sensed data are used, the horizontal axis represents wavelength.

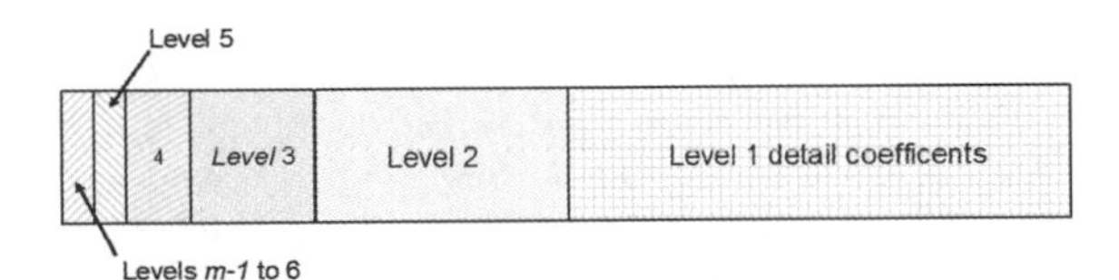

Figure 6.24 Assume that the data series length is $n = 2^m$. There are m-1 sets of detail coefficients. The $n/2$ highest-level (1) detail coefficients are shown on the right. The number of detail coefficients decreases by a factor of two at each level, so that the number of level 2 coefficients is $n/4$, the number of level 3 coefficients is $(n/8)$ and so on.

refer to this process as decimation[1]. The magnitudes of the dyadic samples can be represented in a time-scale diagram by placing the wavelet detail coefficients in 'layers' from top to bottom, as shown in Figure 6.25.

[1] Decimation implies a reduction by a ratio of 1:10 (as in the ancient Roman military punishment, revived in 71 AD by the Roman general Crassus after his men has fled before Spartacus's army of slaves, and described by Plutarch in *The Life of Crassus*: '*...five hundred that were the beginners of the flight, he divided into fifty tens, and one of each was to die by lot, thus reviving the ancient Roman punishment of decimation...*' - had he killed one in every two he would have had no army left).

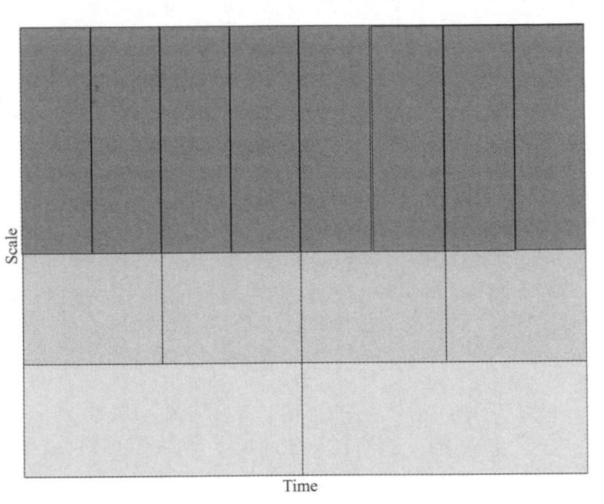

Figure 6.25 Wavelet detail coefficients are arranged in the time-scale diagram to show their extent (in time, horizontally) and in scale (vertically). Each box corresponds to one coefficient, with the number of coefficients being reduced by a factor of two at each level. The highest-level coefficients have the greatest sampling frequency in time, but cover the greatest extent in scale.

Probably the best way to discover what wavelets are, and how they work, is to look at some examples. We begin with a very simple sine wave, shown in Figure 6.26. This sine wave repeats 40 times over the basic interval of 512 samples so its frequency is 40 Hz. The coefficients resulting from a DWT applied to the data of Figure 6.26 are shown in Figure 6.27. The number of levels for $n = 512$ is nine. All levels of detail coefficients show a symmetry that mirrors the symmetry of the sine wave.

Figure 6.28 shows the Discrete Fourier Transform (DFT) of the same sine wave data shown in Figure 6.26. Since there is only a single frequency (40 Hz) present in the data, the Fourier amplitude spectrum – not unexpectedly – shows a single spike centred at a frequency of 40 Hz. The narrow spread of values around 40 Hz in the Fourier amplitude spectrum is the result of the fact that a continuous function (a sine wave) has been sampled at a specific spacing.

The second example is an extension of the first. Figure 6.29 shows a composite sine wave, formed by adding together two sine waves with frequencies of 40 and 80 Hz, respectively. The DFT of the data shown in Figure 6.29 is given in Figure 6.30. Again, the Fourier representation is adequate, as it correctly shows two spikes centred at 40 and 80 Hz. The result of the DWT is not shown, because it also represents the data adequately. A more complicated signal is generated by the Chirp function, which is a sine wave with frequency increasing with time (Figure 6.31), unlike the sine wave data, which are uniform, symmetric,

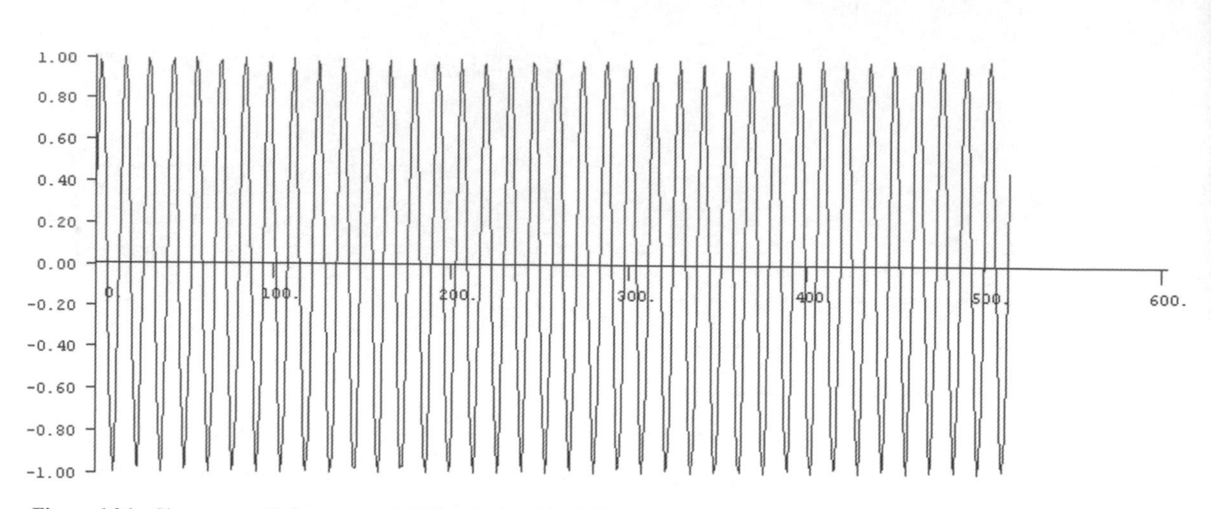

Figure 6.26 Sine wave with frequency of 40 Hz calculated for 512 points. The amplitude of the sine wave is given by the vertical axis, and sample number is shown on the horizontal axis.

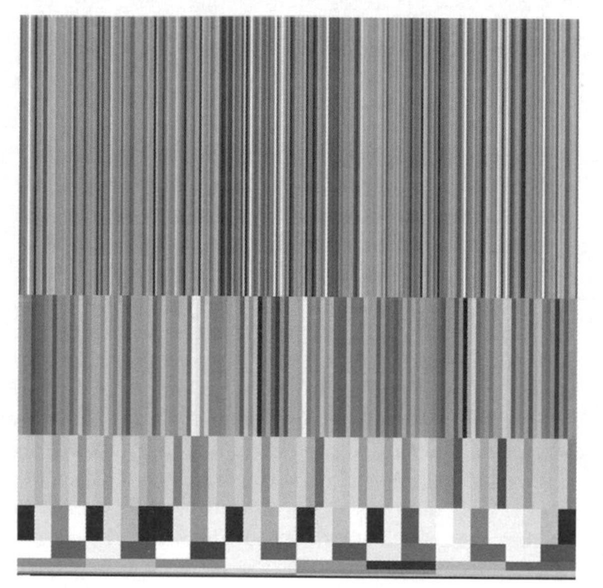

Figure 6.27 Time (horizontal axis) – scale (vertical axis) display of wavelet detail coefficients for a sine wave of frequency 40 Hz, shown in Figure 6.26. The sine wave was sampled at 512 points, so there are nine levels of detail coefficients.

and repeat to infinity. The Fourier amplitude spectrum simply indicates a number of frequency components in the approximate range 1–50 Hz. It does not pick out the change in frequency of the sine waves as time (represented by the x-axis) increases (Figure 6.32). However, the higher levels of detail coefficients generated by the DWT (Figure 6.33) show a pattern of increasing frequency from left to right, which gives a more realistic visual representation of the variation in the data set than does the DFT.

If we take the two sine waves shown in Figure 6.29 but, rather than adding them together at each of the 512 sample points, we sample the 40 Hz sine wave at points 1–256 and sample the 80 Hz sine wave at points 257–512, then the result is a separated pair of sine waves (Figure 6.34). Interestingly, the amplitude spectrum of the series shown in Figure 6.34 that is produced by the Fourier transform is identical to that shown in Figure 6.30, which was generated from the data shown in Figure 6.29. This result shows that the Fourier transform cannot define *where* in the data series a particular frequency occurs. It uses a universal basis function (i.e. based on the full data series) composed of sine and cosine waves repeating to infinity, whereas the DWT uses a localised basis function (as the filters deal with the data 2, 4, 8, 16, ..., points at a time). The DWT is thus said to have compact as opposed to universal support. The DWT result for the data of Figure 6.34 is shown in Figure 6.36, and it is apparent that the higher-level DWT coefficients show a clear distinction between the left and right halves of the data series.

6.7.3 Two-dimensional discrete wavelet transform

The two-dimensional DWT is calculated in the same way as the two-dimensional DFT, that is, by performing a one-dimensional DWT along the rows of the image to give an intermediate matrix, and then applying the one-dimensional DWT to the columns of this intermediate matrix. The size of the image is m rows and n columns, where both m and n are powers of two. The highest-level detail coefficients form the right-hand end of the one-dimensional

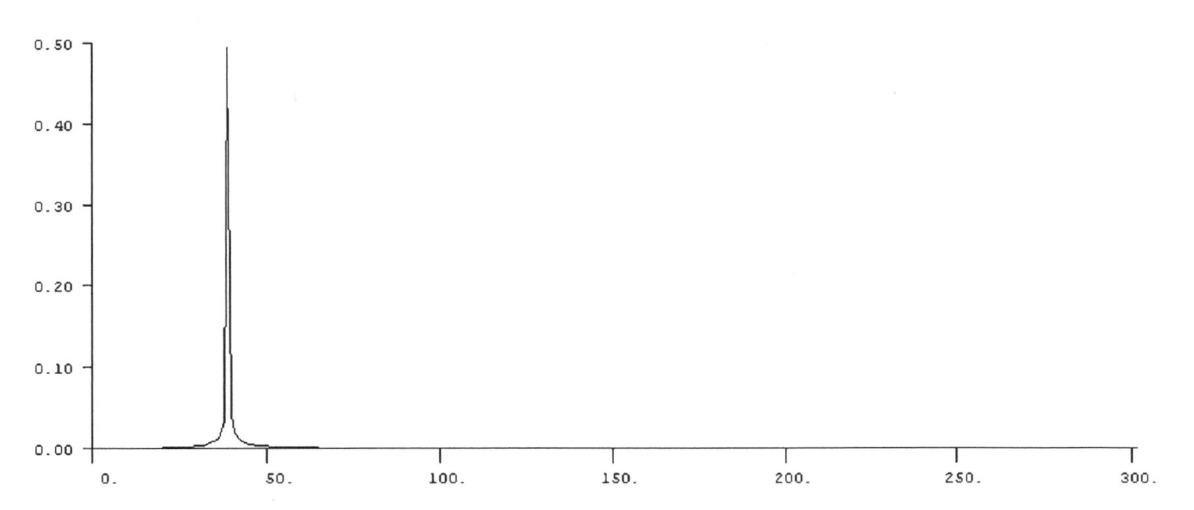

Figure 6.28 Fourier amplitude spectrum of the sine wave shown in Figure 6.26. The presence of a periodic component with a frequency of 40 Hz is apparent. The horizontal axis is time, and the vertical axis measures variance.

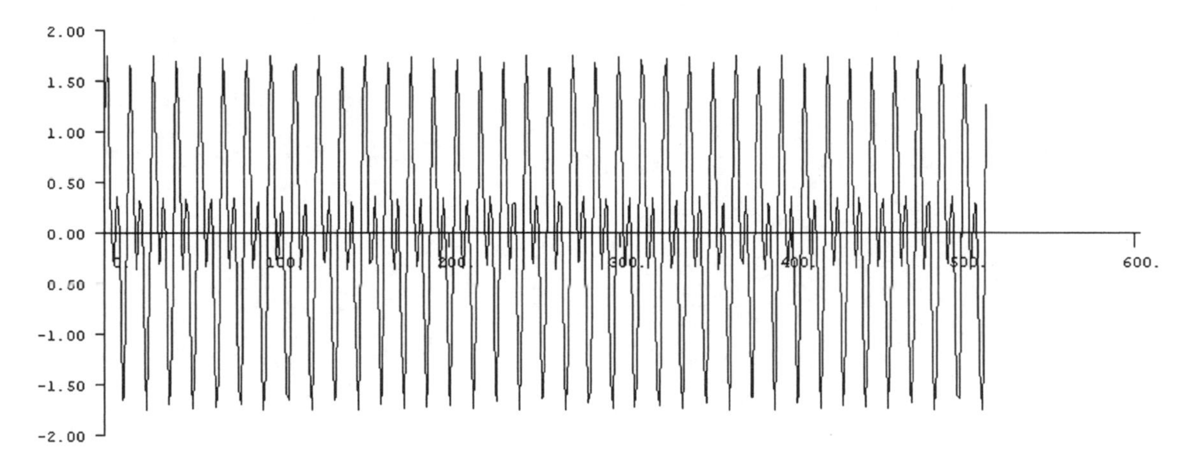

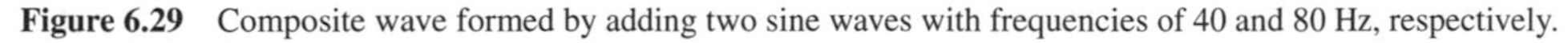

Figure 6.29 Composite wave formed by adding two sine waves with frequencies of 40 and 80 Hz, respectively.

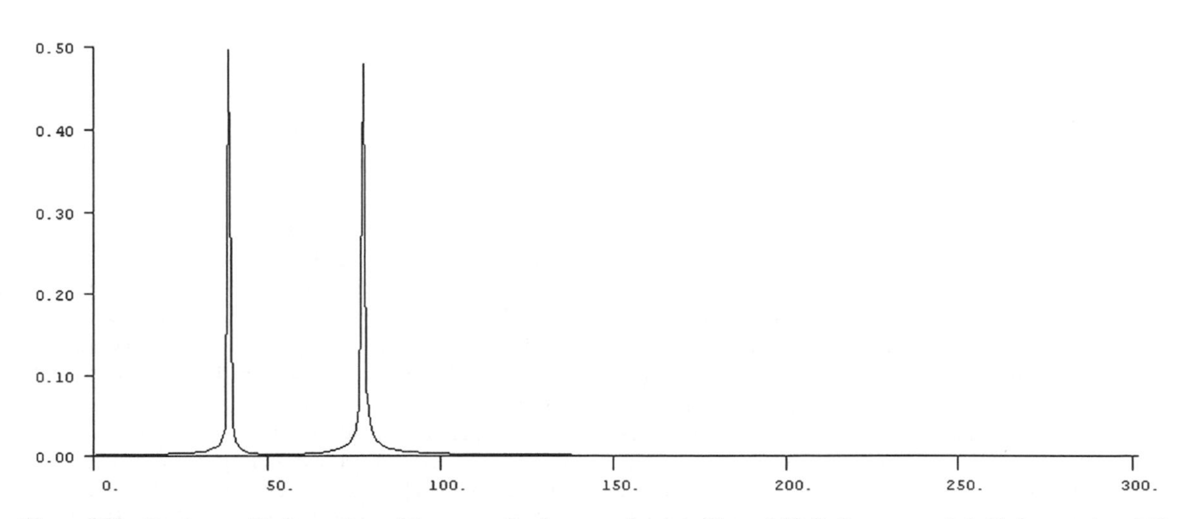

Figure 6.30 Fourier amplitude spectrum of the composite sine wave shown in Figure 6.29. Both components (with frequencies of 40 and 80 Hz, respectively) are identified.

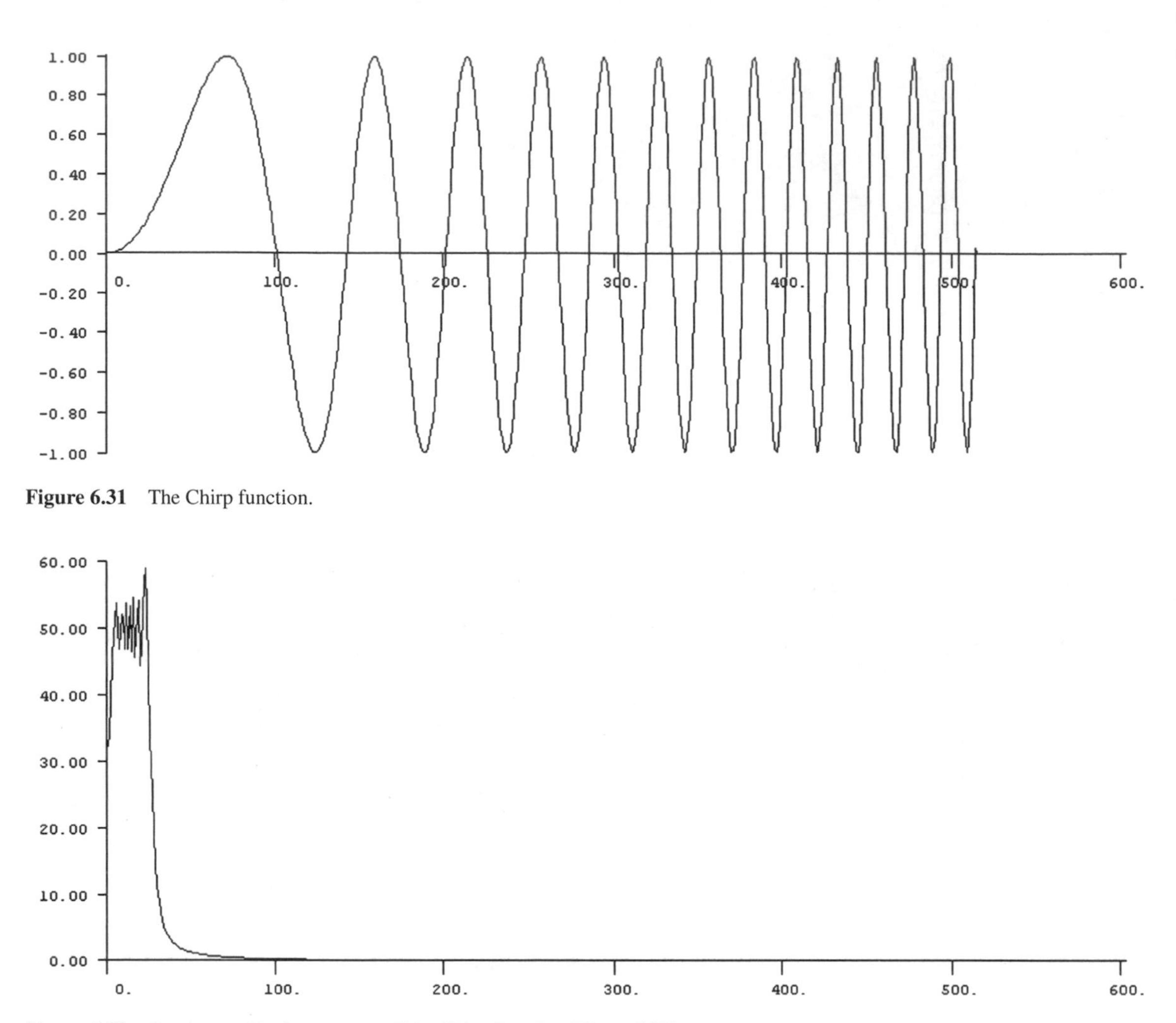

Figure 6.31 The Chirp function.

Figure 6.32 Fourier amplitude spectrum of the Chirp function (Figure 6.31).

DWT output vector (Figure 6.37) while the lowest level detail coefficients are located at the left-hand end of the one-dimensional DWT output. If you think about m such output vectors lying horizontally and being intersected by n vectors lying vertically then the highest level detail in the vertical direction forms the lower part of the output image while the highest level detail in the horizontal direction forms the lower part of the output image. The result is that the highest level of 'horizontal' detail coefficients dominates the upper right quadrant of the output image, the highest level 'vertical' detail coefficients dominates the lower left quadrant of the output image, and highest level horizontal and vertical coefficients interact in the lower right quadrant of the output image. The 'once-smoothed' image (reduced in resolution by a factor of two) is located in the upper left quadrant of the output, where the lowest-level horizontal and vertical detail coefficients interact (Figure 6.37).

The second-level decomposition acts on the coefficients contained in the upper right quadrant of the first level decomposition. This upper right quadrant is converted by row-column operations to a set of sub-quadrants – again LL, LH, HH and HL. This process can be repeated until the sub-quadrants are only one pixel in size, or it can be stopped at any point.

The importance of the two-dimensional DWT is that the direction (horizontal, vertical or diagonal) of the detail coefficients at each hierarchical level are clearly defined. The highest-level detail coefficients (indicated by HH, HV, and

Figure 6.33 Time-scale plot of wavelet transform detail coefficients for Chirp function (Figure 6.31). Time is represented by the x-axis and scale by the y-axis, with the finest scale at the top and the coarsest scale at the bottom.

VV in Figure 6.36) can be extracted and processed for particular purposes. For example, noise present in the image can be estimated and removed using procedure described in section 9.3.2.2.2. The information present in the higher-level detail coefficients can be used to characterise the texture present in an image. Texture can range from smooth to rough, and is a measure of the local variability of grey levels (section 8.7.1).

Applications of wavelets in image processing are given by Ranchin and Wald (1993) and Prasad and Iyengar (1997). Du *et al.* (2002) use wavelets to remove haze in high-resolution images, while wavelets are used to smooth interferograms (section 9.2) by Braunich *et al.* (2000). The use of wavelets in a cosmetic procedure (removal of striping, section 4.2.2) is described by Torres and Infante (2001). An interesting application uses wavelets to combine panchromatic and multispectral images (a technique known as 'pan sharpening'; see also section 6.5, where the use of the HSI transform in image fusion is described). Three bands forming an RGB colour composite image and a co-registered panchromatic image of the same area are required. The RGB composite may, for instance, be formed from three of the six reflective Landsat ETM+ bands with a spatial resolution of 30 m (Chapter 2) and these bands are fused with the corresponding panchromatic band, which has a spatial resolution of 15 m. These two images (RGB composite and panchromatic) are co-registered but have different spatial resolutions, so the 30 m bands are resampled to 15 m to give a pixel-to-pixel match. Next, the two-dimensional DWT is used to determine the level 1 decomposition of each image. The H_1, D_1 and V_1 components (Figure 6.36) of the panchromatic band are substituted for those of the RGB bands, and the three images formed from this operation are transformed back to the spatial domain using the inverse DWT. Some normalisations are required to ensure that the various components are matched. See Garguet-Duport *et* al. (1996), Yocky (1996) and Zhou *et al.* (1999) for more details. The use of two-the dimensional DWT in image 'de-noising' is discussed in Chapter 9.

6.8 SUMMARY

A range of image transform techniques is considered in this chapter. Arithmetic operations (addition, subtraction, multiplication and division) have a utilitarian use – for example, image subtraction is used routinely in the separation of the high-frequency component of an image during the filtering process (section 7.3.1) while addition is used in the method of lineament detection described in section 7.4. Image division, or ratioing, is one of the most common transformations applied to remotely-sensed images in both geological and agricultural studies for simple band ratios reflect differences in the slopes of the spectral reflectance curves of Earth surface materials. Problems experienced with the use of ratios include the difficulty of separating the effect of atmospheric path radiances, and the choice of dynamic range compression technique. Nevertheless ratio images are a widely used and valuable tool. The empirical transformations considered in section 6.3 were developed for use with images of agricultural areas. One of the special problems here is that the database on which these transformations (the Perpendicular Vegetation Index and the Tasselled Cap transformation) are based is limited and site-specific. Their unrestricted use with images from parts of the world other than those regions of the United States where they were developed is questionable.

The principal components analysis (Karhunen–Loève) transformation has a long history of use in multivariate statistics. Even so it is not as well understood by the remote sensing community as it might be. Like most parametric statistical methods it is based on a set of assumptions that must be appreciated, if not completely satisfied, if the methods are to be used successfully. The final transformation techniques covered in this chapter are the Fourier transform and the wavelet transform. The level of

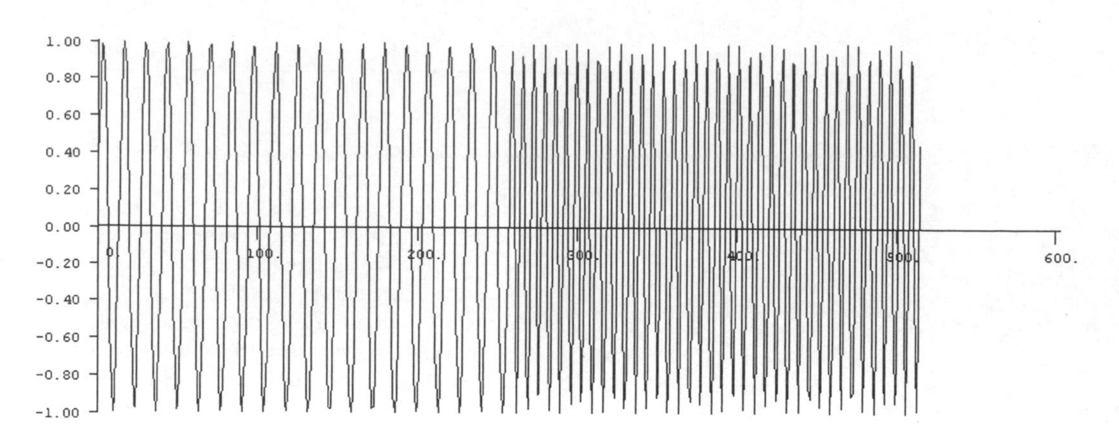

Figure 6.34 The same two sine waves as in Figure 6.29. However, in this example, the first sine wave forms the left half of the series (samples 1–256) and the second sine wave is on the right (samples 257–512).

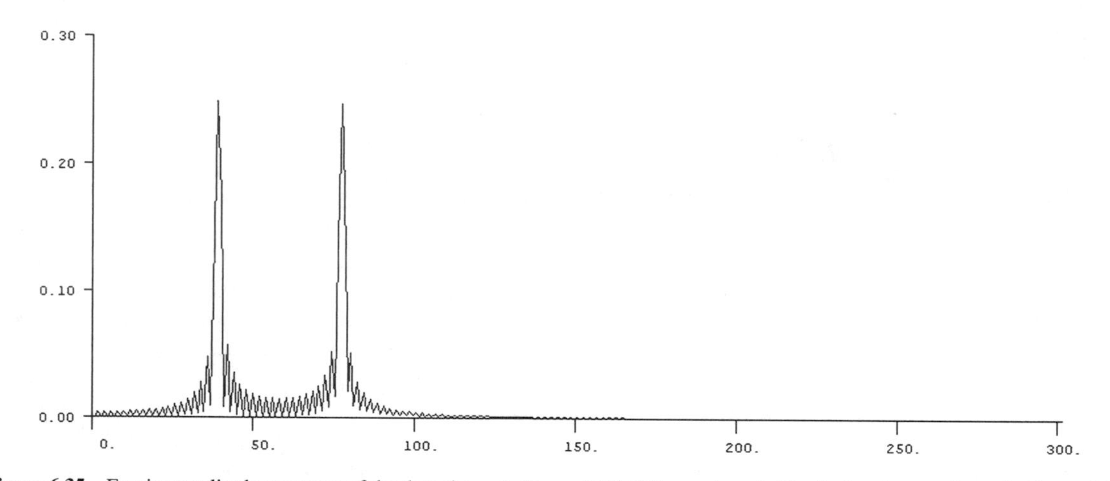

Figure 6.35 Fourier amplitude spectrum of the data shown in Figure 6.34. This spectrum is identical to the one shown in Figure 6.30, though the input data are different.

Figure 6.36 Wavelet time-scale diagram of the two sine waves shown in Figure 6.34.

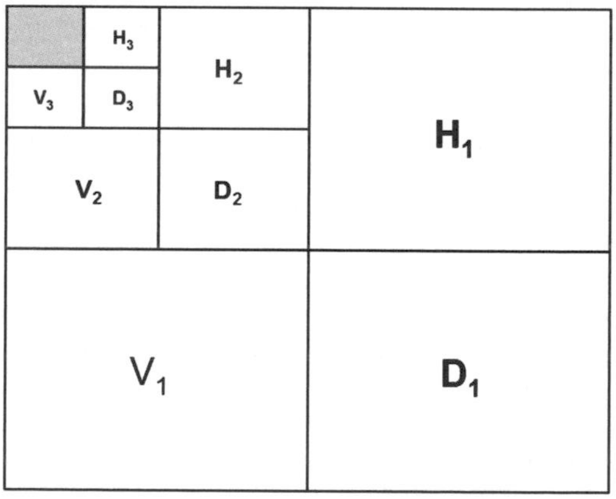

H_3 V_3 D_3 H_2 V_2 D_2 H_1 V_1 D_1

Figure 6.37 Disposition of 'horizontal' (H), 'vertical'(V) and 'diagonal' (D) wavelet detail coefficients in the two-dimensional DWT. Three levels of decomposition are shown, indicated by subscripts 1, 2 and 3. The thrice-smoothed image occupies the top left cell of the third level.

mathematics required to understand a formal presentation of these methods is generally well above that achieved by undergraduate students in geography, geology and other Earth sciences. The intuitive explanations given in sections 6.6 and 6.7 might serve to introduce such readers to the basic principles of the method and allow a fuller understanding of the frequency-domain filtering techniques described in section 7.5 and the de-noising procedures applied to reflectance spectra and images in section 9.2.

The reader should appreciate that the presentation in the latter parts of this chapter is largely informal and non-mathematical. Many pitfalls and difficulties are not covered. These will, no doubt, be discovered serendipitously by the reader in the course of project work.

7

Filtering Techniques

7.1 INTRODUCTION

The image enhancement methods discussed in chapter 5 change the way in which the information content of an image is presented to the viewer, either by altering image contrast or by coding a greyscale image in pseudocolour so as to emphasise or amplify some property of the image that is of interest to the user. This chapter deals with methods for selectively emphasising or suppressing information at different spatial scales present in an image. For example, we may wish to suppress the high-frequency noise pattern caused by detector imbalance that is sometimes seen in Landsat MSS and TM images and which results from the fact that the image is electro-mechanically scanned in groups of six lines (MSS) or 16 lines (TM/ETM+) (sections 2.3.6 and 4.2.2). On the other hand, we may wish to emphasise some spatial feature or features of interest, such as curvilinear boundaries between areas that are relatively homogeneous in terms of their tone or colour, in order to sharpen the image and reduce blurring. The techniques operate selectively on the image data, which are considered to contain information at various spatial scales. The idea that a spatial (two-dimensional) pattern, such as the variation of grey levels in a grey scale image, can be considered as a composite of patterns at different scales superimposed upon each other is introduced in section 6.6 in the context of the Fourier transform. Large-scale background or regional patterns, such as land and sea, are the basic components of the image. These large-scale patterns can be thought of as 'background' with 'detail' being added by small-scale patterns. Noise, either random or systematic, is normally also present.

In abstract terms, the information contained in an image can be represented by the following model:

$$\begin{aligned}\text{image data} &= \text{regional pattern} + \text{local pattern} + \text{noise}\\ &= \text{background} + \text{foreground (detail)} + \text{noise}\\ &= \text{low frequencies} + \text{high frequencies} + \text{noise}\end{aligned}$$

There is no reason to suppose that noise affects only the foreground or detail, though noise is often described as a high-frequency phenomenon. Noise can be either random or periodic. An example of random noise is the speckle pattern on SAR images. Periodic noise can be the result of a number of factors, such as the use of an electromechanical scanner, or the vibration from an aircraft engine.

The representation of the spatial variability of a feature in terms of a regional pattern with local information and noise superimposed has been widely used in disciplines that deal with spatially distributed phenomena. Patterns of variation are often summarised in terms of generalisations. For example, a geographer might note that, in Great Britain, 'Mean annual rainfall declines from west to east' in the knowledge that such a statement describes only the background pattern, upon which is superimposed the variations attributable to local factors. In both geography and geology, the technique of trend surface analysis has been found useful in separating the regional and local components of such spatial patterns (Davis, 1973; Mather, 1976).

By analogy with the procedure used in chemistry laboratories to separate the components of a suspension, the techniques described in this chapter are known as *filtering*. A digital filter can be used to extract a particular spatial scale component from a digital image. The slowly varying background pattern in the image can be envisaged as a two-dimensional waveform with a long wavelength or low frequency; hence a filter that separates this slowly varying component from the remainder of the information present in the image is called a *low-pass filter*. Conversely, the more rapidly varying detail is like a two-dimensional waveform with a short wavelength or high frequency. A filter to separate out this component is called a *high-pass filter*. These two types of filter are considered separately. Low-frequency information allows the identification of the background pattern, and produces an output image in which the detail has been smoothed or removed from the original (input) image (hence low-pass filtering can be thought of as a form of blurring the image). High-frequency information allows us either to isolate or to amplify the local detail. If the high-frequency detail is amplified by adding back to the image some multiple of the high-frequency component

Computer Processing of Remotely-Sensed Images: An Introduction, Third Edition. Paul M. Mather.
 ISBNs: 0-470-84918-5 (HB); 0-470-84919-3 (PB)

extracted by the filter then the result is a sharper, de-blurred image. Anyone who listens to the organ music of J. S. Bach will be able to identify the low frequency components as the slowly changing low bass notes played by the foot pedals, while the high frequency detail consists of the shorter and much more numerous notes played on the manuals. Bach's music and remotely sensed images have something in common, for both combine information at different scales (temporal scales in music, spatial scales in images).

Three approaches are used to separate the scale components of the spatial patterns exhibited in a remotely sensed image. The first is based upon the transformation of the frequency domain representation of the image into its scale or spatial frequency components using the Discrete Fourier Transform (section 6.6), while the second method is applied directly to the image data in the spatial domain. A third, more recent, development is that of the discrete wavelet transform, which uses both frequency (scale) and spatial representations of the data. The principles of the wavelet transform are discussed in section 6.7 and applications are summarised in chapter 9. Fourier-based filtering methods are considered in section 7.5. In the following two sections the most common spatial-domain filtering methods are described. There is generally a one-to-one correspondence between spatial and frequency-domain filters. However, specific filters may be easier to design in the frequency domain but may be applied more efficiently in the spatial domain. The concept of spatial and frequency-domain representations is shown in Figure 6.21. Whereas spatial-domain filters are generally classed as either high-pass (sharpening) or as low-pass (smoothing), filters in the frequency domain can be designed to suppress, attenuate, amplify or pass any group of spatial frequencies. The choice of filter type can be based either on spatial frequency or on direction, for both these properties are contained in the Fourier amplitude spectrum (section 6.6).

7.2 SPATIAL DOMAIN LOW-PASS (SMOOTHING) FILTERS

Before the topic of smoothing a two-dimensional image is considered, we will look at a simpler expression of the same problem, which is the smoothing of a one-dimensional pattern. Figure 7.1 shows a plot of grey levels along a cross-section from the top left corner (0, 0) to the bottom right corner (511, 511) of the TM band 7 image shown in Figure 1.11(b). Figure 7.1(a) shows the cross-section for the unfiltered image, while Figure 7.1(b) shows the same cross-section after the application of a low pass (smoothing) filter. Clearly, the level of detail has been reduced and the cross-section curve is more generalised, though the main peaks are still apparent. Figure 7.2 displays another plot showing grey level value (vertical axis) against position across a scan-line of a digital image. The underlying pattern is partially obscured by the presence of local patterns and random noise. If the local variability, and the random noise, were to be removed then the overall pattern would become more clearly apparent and a general description of trends in the data could then be more easily made. The solid line in Figure 7.2 is a plot of the observed pixel values against position along the scan line, while the dotted line and the broken line represent the output from median and moving-average filters, respectively. These filters are described below. Both produce smoother plots than the raw data curve, and the trend in the data is more easily seen. Local sharp fluctuations in value are removed. These fluctuations represent the high-frequency component of the data and may be the result of local anomalies or of noise. Thus, low-pass filtering is used by Crippen (1989), Eliason and McEwen (1990), and Pan and Chang (1992) to remove banding effects on remotely sensed images (section 4.2.2), while Dale *et al.* (1996) use a low-pass filter in an attempt to smooth away the effects of image-to-image misregistration. Example 7.1 illustrates the way in which image cross-sections such as those used in this section can be generated using MIPS.

EXAMPLE 7.1: IMAGE CROSS SECTIONS

Cross-sectional plots such as those shown in Figures 7.1 and 7.2 can be produced using the MIPS software, which is supplied with this book. To use the MIPS cross-section procedure, firstly display an image and then select the *Plot|Cross-sections* menu item and follow the instructions. You can specify the start and end point of the cross section in one of two ways: using the mouse (left-click at the start point, then hold the left mouse button down, move to the end point and click the right button) or by entering coordinates manually. Manual coordinate entry is useful if you wish to repeat the cross section using another image, such as a filtered image. The procedure plots the displayed pixel values, so that results will be different if an image enhancement procedure (such as a contrast stretch, described in Chapter 5) has been applied. If a false colour image is displayed, then three cross sections are drawn in red, green and blue respectively, representing the pixel values stored in the red, green and blue memory banks.

Readers are encouraged to familiarise themselves with cross-sections from different images in order to gain a 'feel' for the way in which the shape of a cross section can be estimated visually from the image. As individual filters are introduced in the following pages, cross sections of filtered images can be compared to the original cross-section, as is done in Figure 7.1(a) and (b). The plots are automatically written to the Windows clipboard, so that you can paste them into another application, such as a word processor.

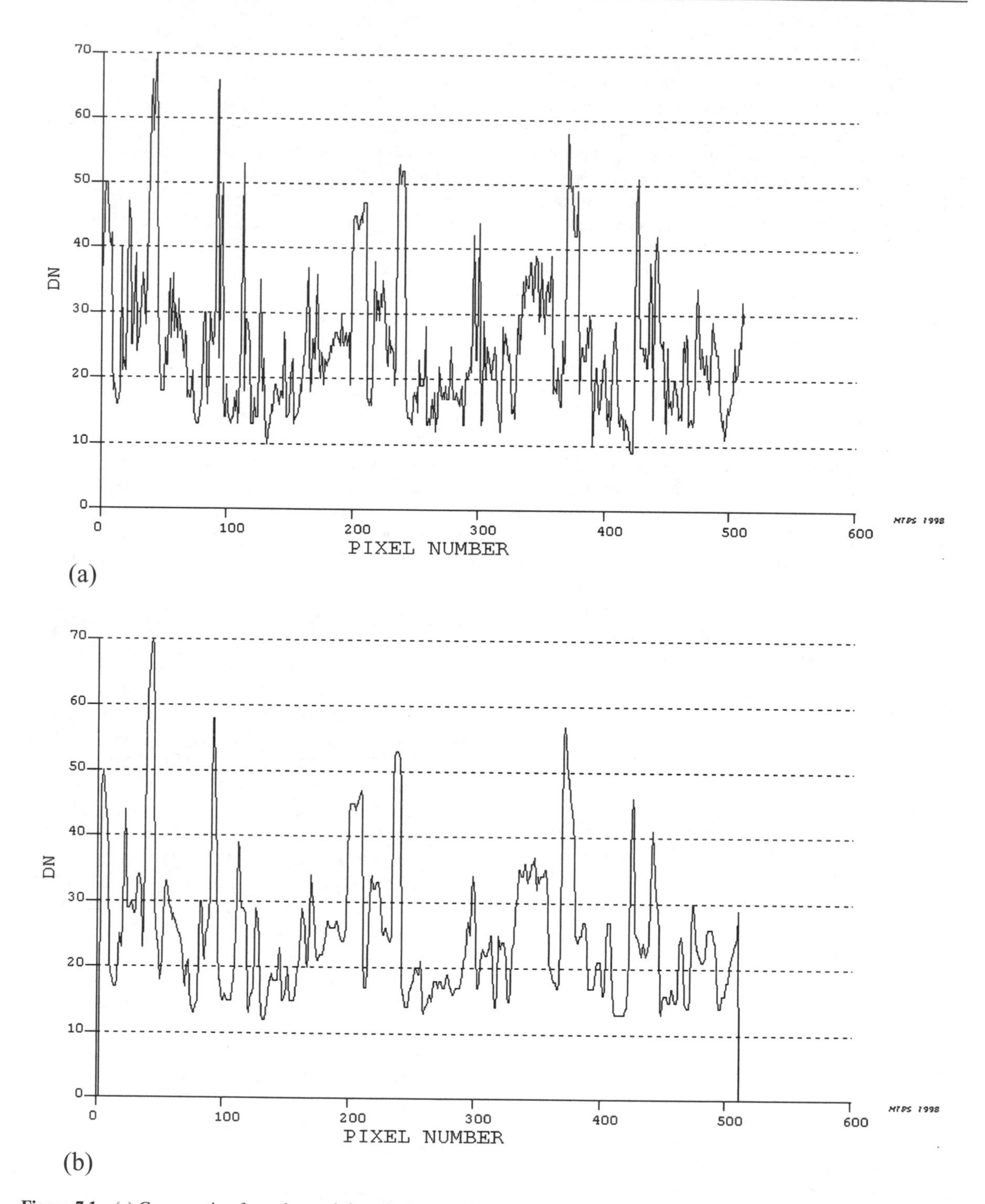

Figure 7.1 (a) Cross section from the top left to the bottom right-hand corner of the Littleport TM band 7 image shown in Figure 1.11(b). (b) Cross section between the same points as used in Figure 7.1(a) after the application of a smoothing filter (a 5 × 5 median filter was used to generate this cross section, as described in section 7.2.2). The reduction in detail is clearly apparent.

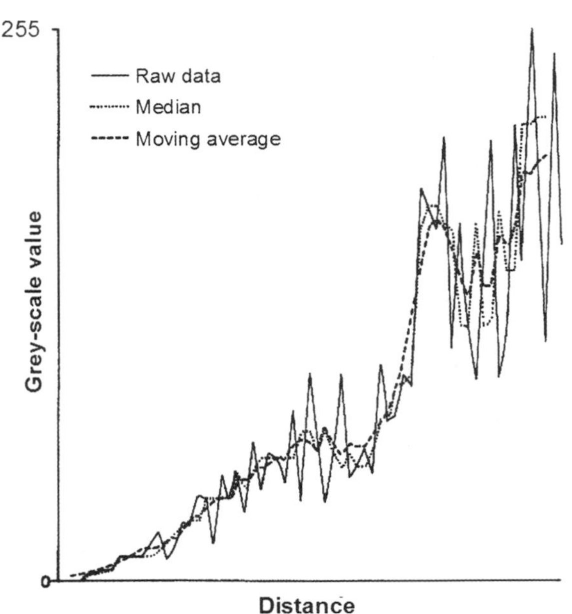

Figure 7.2 One-dimensional data series showing the effect of median (low-pass) filter and moving average (low-pass) filtering.

7.2.1 Moving average filter

The moving average filter simply replaces a data value by the average of the given data point and a specified number of its neighbours to the left and to the right. If the coordinate on the horizontal axis of Figure 7.2 is denoted by the index j then the moving-average filtered value at any point j is x'_j. The procedure for calculating x'_j depends on the number of local values around the data point to be filtered that are used in the calculation of the moving average. This number is always an odd, positive integer so that there is a definite central point (thus, the central value in the sequence 1, 2, 3 is 2 whereas there is no specific central value in the sequence 1, 2, 3, 4). The broken line in Figure 7.2 is based on a five-point moving average, defined by:

$$x'_j = (x_{j-2} + x_{j-1} + x_j + x_{j+1} + x_{j+2})/5$$

Five raw data values represented by the vector x and centred on point x_j are summed and averaged to produce one output value (x'_j). If a three-point moving average had been used then three raw data values centred on point j (i.e., points $j - 1$, j, and $j + 1$) would be summed and averaged to give one output value at x'_j.

If the number of data elements included in the averaging process is n, then $[n/2]$ values at the beginning of the input series and $[n/2]$ values at the end of the input series do not have output values associated with them, because some of the input terms x_{j-1}, x_{j-2} and so on will not exist for $j < [n/2]$, just as some of the terms $x_{j+1}, x_{j+2} \ldots$ will not exist for $j > N - [n/2]$ (the symbol [.] indicates the integer part of the given expression and N is the total number of raw data values in the (input) series that is being filtered). The filtered (output) series is thus shorter than the input series by n-1 elements, where n is the length of the filter (three point, five point, etc.). Thus, a moving average curve, such as that shown in Figure 7.2, will have no values at points x_1 and x_2 or at x_{n-1} and x_n. A 5 × 5 filter applied to an image will leave an unfiltered margin of 2 pixels around the four sides of the image. These marginal pixels are usually set to zero.

In calculating a five-point moving average for a one-dimensional series the following algorithm might be used: add up the input (x) values 1–5 and divide their sum by 5 to give x'_3, the first filtered (output) value. Note that filtered values x'_1 and x'_2 cannot be calculated; the reason is given in the preceding paragraph. Next, add raw data values x_2 to x_6 and divide their sum by 5 to give x'_4. This procedure is repeated until output value x'_{N-2} has been computed, where N is the number of input values (again, x'_{n-1} and x'_n are left undefined). This algorithm is rather inefficient, for it overlooks the fact that the sum of x_2 to x_6 is easily obtained from the sum of x_1 to x_5 simply by subtracting x_1 from the sum of x_1–x_5 and adding x_6. The terms x_2, x_3 and x_4 are present in both summations, and need not be included in the second calculation. If the series is a long one then this modification to the original algorithm will result in a more time-efficient program.

A two-dimensional moving average filter is defined in terms of its horizontal (along-scan) and vertical (across-scan) dimensions. Like the one-dimensional moving average filter, these dimensions must be odd, positive and integral. However, the dimensions of the filter need not be equal. A two-dimensional moving average is described in terms of its size, such as 3 × 3. Care is needed when the filter dimensions are unequal to ensure that the order of the dimensions is clear; the Cartesian system uses (x, y) where x is the horizontal and y the vertical coordinate, with an origin in the lower left of the positive quadrant. In matrix (image) notation, the position of an element is given by its row (vertical, y) and column (horizontal, x) coordinates, and the origin is the upper left corner of the matrix or image. The central element of the filter, corresponding to the element x'_j in the one-dimensional case described earlier, is located at the intersection of the central row and column of the $n \times m$ filter window. Thus, for a 3 × 3 window, the central element lies at the intersection of the second row and second column. To begin with, the window is placed

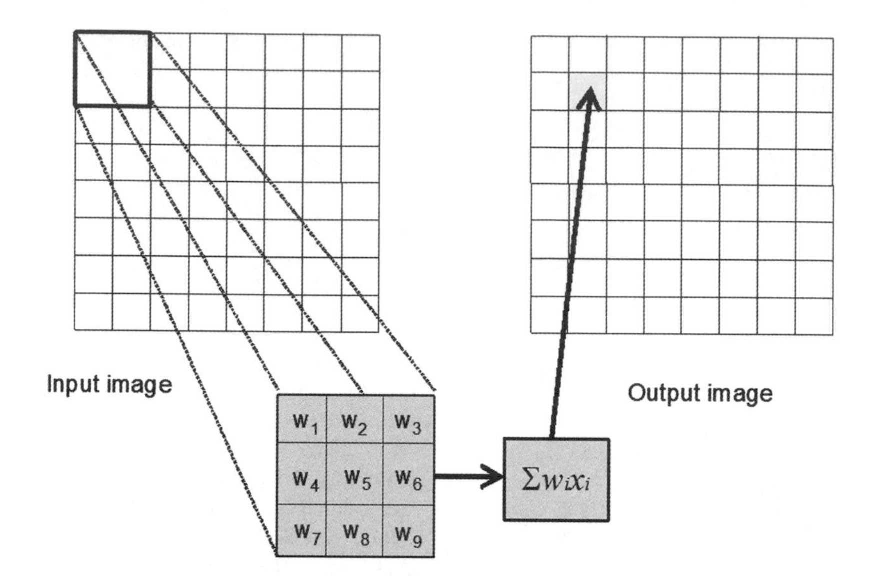

Figure 7.3 Illustrating the operation of a spatial domain filter. In this example, the filter size is 3 rows by 3 columns. The filter size is always an odd, positive integer so that there is a 'central' pixel (on row 2, column 2 in this case). The 3 × 3 matrix of filter weights ($w_1 \ldots w_9$) is placed over a 3 × 3 patch of image pixels and the weight is multiplied by the associated pixel value. These products are summed and the result is placed in the position in the output image that corresponds to the position of the central cell of the weight matrix. In the case of a 3 × 3 moving average filter, with nine elements, the weights w_i are all equal to 1/9.

in the top left corner of the image to be filtered (Figure 7.3) and the average value of the elements in the area of the input image that is covered by the filter window is computed. This value is placed in the output image at the point in the output image corresponding to the location of the central element of the filter window. In effect, the moving average filter window can be thought of as a matrix with all its elements equal to 1; the output from the convolution of the window and the image is the sum of the products of the corresponding window and image elements divided by the number of elements in the window. If $\mathbf{F}$ $(= f_{ij})$ is the filter matrix, $\mathbf{G}$ $(= g_{ij})$ is the input image and $\mathbf{O}$ $(= o_{ij})$ the output (filtered) image then:

$$o_{ij} = \left\{ \sum_{p=-b}^{b} \sum_{q=-c}^{c} g_{p+i,q+j} f_{r+p,s+q} \right\} \Big/ mn$$

where

- b integer part of $n/2$,
- c integer part of $m/2$,
- n number of rows in filter matrix,
- m number of columns in filter matrix,
- r central row in filter matrix $(= [n/2])$,
- s central column in filter matrix $(= [m/2])$,
- i, j image pixel underlying element (r, s) of filter matrix (coordinates in row/column order).

For example, given a 5 × 3 filter matrix the value of the pixel in the filtered image at row 8, column 15 is given by:

$$o_{8,15} = \left\{ \sum_{p=-2}^{2} \sum_{q=-1}^{1} g_{8+p,15+q} f_{3+p,2+q} \right\} \Big/ 15$$

with $b = 2$, $c = 1$, $r = 3$ and $s = 2$. Note that the indices i and j must be in the range $b < i < (N - b + 1)$ and $c < j < (M - c + 1)$ if the image has N rows and M columns numbered from 1 to N and 1 to M, respectively. This means that there are b empty rows at the top and bottom of the filtered image and c empty columns at either side of the filtered image. This unfiltered margin can be filled with zeros or the unaltered pixels from the corresponding cells of the input image can be placed there.

The initial position of the filter window with respect to the image is shown in Figure 7.3. Once the output value from the filter has been calculated, the window is moved one column (pixel) to the right and the operation is repeated. The window is moved rightwards and successive output values are computed until the right-hand edge of the filter window hits the right margin of the image. At this point, the filter window is moved down one row (scan-line) and back to the left-hand margin of the image. This procedure is repeated (Figure 7.4). The window is moved rightwards and successive output values are computed until

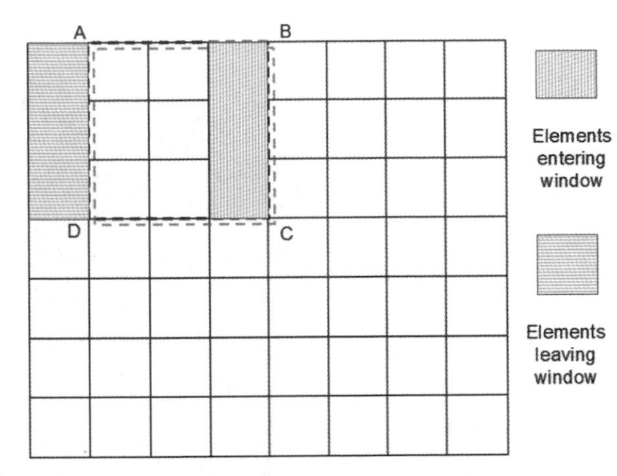

Figure 7.4 The filter window ABCD has moved one column to the right from its initial position (position 1) in the top left corner of the image and now is in position 2. The elements of the column to be subtracted from the sum calculated at position 1 are indicated by horizontal shading. Those elements to be added to the sum calculated at position 1 are indicated by vertical shading.

the filter window reaches the right-hand edge of the image. At this point, the filter window is moved down one row (scan line) and back to the left-hand margin of the image. This procedure is repeated until the filter window reaches the bottom right-hand corner of the input image. The output image values form a matrix that has fewer rows and columns than the input image because it has an unfiltered margin corresponding to the top and bottom rows and the left and right columns of the input matrix that the filter window cannot reach. Generally, these missing rows and columns are filled with zeroes in order to keep the input and output images the same size.

The effect of the moving average filter is to reduce the overall variability of the image and lower its contrast. At the same time those pixels that have larger or smaller values than their neighbourhood average (think of them as grey level peaks and troughs) are respectively reduced or increased in value so that local detail is lost. Noise components, such as the banding patterns evident in line-scanned images, are also reduced in magnitude by the averaging process, which can be considered as a smearing or blurring operation. In cases where the overall pattern of grey level values is of interest, rather than the details of local variation, neighbourhood grey level averaging is a useful technique The moving average filter described in this section is an example of a general class of filters called box filters; refer to McDonnell (1981) for a survey.

The moving average filter can be implemented in MIPS using the *filter|user_defined* menu option (see also Example 7.2). Examples of 3 × 3 and 5 × 5 moving average filter

1/9	1/9	1/9
1/9	1/9	1/9
1/9	1/9	1/9

(a)

1/25	1/25	1/25	1/25	1/25
1/25	1/25	1/25	1/25	1/25
1/25	1/25	1/25	1/25	1/25
1/25	1/25	1/25	1/25	1/25
1/25	1/25	1/25	1/25	1/25

(b)

Figure 7.5 Moving average filter weights for (a) a 3 × 3 filter and (b) a 5 × 5 filter. The same effect can be achieved by using filter weights of 1, so that the weighted summation in Figure 7.3 becomes a simple addition. The sum is then divided by the product of the two dimensions of the filter, i.e. 9 in the case of (a) and 25 for (b).

weights are given in Figure 7.5. The input to MIPS takes the form of an array of integers with a divisor that is equal to the product of the two dimensions of the box. Thus, in Figure 7.5(a) the divisor is 9, while in Figure 7.5(b) it is 25. If you apply these two filters, you will see that increasing the window size of a moving average filter results in a greater degree of smoothing, since more pixels are included in the averaging process.

EXAMPLE 7.2: MOVING AVERAGE FILTERS

The purpose of this example is to evaluate the effects of changing the window size of the two-dimensional moving average filter. Start MIPS and display a greyscale or false colour image of your choice. In this example, In this example, I am using a Landsat ETM+ band 4 image of the south-west corner of The Wash on the east coast of England. The original (contrast stretched) image is shown in Example 7.2 Figure 1. Next, select *Filter|User Defined*

Example 7.2 Figure 1 Contrast-stretched Landsat ETM+ image of the south-west corner of The Wash in eastern England.

Example 7.2 Figure 2 The image shown in Figure 1 after the application of a 3 × 3 moving average filter.

Filter. Select the default vertical and horizontal size of the filter window (i.e., 3 × 3), the default weights (all ones), and the default divisor (9). These weights and divisor define a 3 × 3 moving average filter. The filtered image (Example 7.2 Figure 2) is shown in a new window. Finally, choose *Filter|User Defined Filter* again, and change the filter size

Example 7.2 Figure 3 The image shown in Figure 1 after the application of a 7 × 7 moving average filter.

to 7 × 7. Accept the default weights and divisor, and a 7 × 7 moving average filtered image appears on the screen (Example 7.2 Figure 3). You can try other moving average window sizes (they do not have to be square; 3 v 1 or 7 × 3 are acceptable, for example). How would you describe (a) the general effect of the moving average filter, and (b) the specific effect of window size?

7.2.2 Median filter

An alternative smoothing filter uses the median of the pixel values in the filter window (sometimes called the neighbourhood) rather than the mean. The median filter is generally thought to be superior to the moving average filter, for two reasons. Firstly, the median of a set of n numbers (where n is an odd integer) is always one of the data values present in the set. Secondly, the median is less sensitive to errors or to extreme data values. This can be demonstrated by a simple, one-dimensional, example. If the nine pixel values in the neighbourhood of, and including the point (x, y) are $\{3, 1, 2, 8, 5, 3, 9, 4, 27\}$ then the median is the central value (the fifth in this case) when the data are ranked in ascending or descending order of magnitude. In this example the ranked values are $\{1, 2, 3, 3, 4, 5, 8, 9, 27\}$ giving a median value of 4. The mean is 6.88, which would be rounded up to a value of 7 for display purposes, as most display systems use an integer scale of grey levels, such as 0–255. The value 7 is not present in the original data, unlike the median value of 4. Also, the mean value is larger than six of the nine observed values, and may be thought to be unduly influenced by one extreme data value (27), which is three times larger than the next highest value in the set. Thus, the median filter removes isolated extreme pixel values or spikes, such as the value 27 in the example, which might represent isolated noise pixels. It also follows that the median filter preserves edges better than a moving-average filter, which blurs or smooths the grey levels in the neighbourhood of the central point of the filter window. Figure 7.2 shows (a) a one-dimensional sequence of values, (b) the result of applying a moving average of length five to the given data, and (c) the result of applying a median filter also of length five.

It is clear that, while both the median and the moving average filters remove high-frequency oscillations, the median filter more successfully removes isolated spikes and better preserves edges, defined as pixels at which the gradient or slope of grey level value changes markedly. Synthetic Aperture Radar (SAR) images often display a noise pattern called *speckle* (section 2.4). This is seen as a random pattern of bright points over the image. The median filter is sometimes used to eliminate this speckle without unduly blurring the sharp features of the image (MacFarlane and Thomas, 1984; Blom and Daily, 1982).

The mean is relatively easily computed; it involves a process of summation and division, as explained earlier, and considerable savings of computer time can be achieved by methods described in section 7.2.1. In contrast, the median requires the ranking of the data values lying within the $n \times m$ filter window centred on the image point that is being filtered. The operation of ranking or sorting is far slower than that of summation, for $[n/2] + 1$ passes through the data are required to determine the median ($[n/2]$ indicates 'the integer part of the result of dividing n by 2'). At each pass, the smallest value remaining in the data must be found by a process of comparison. Using the values given in the example in the preceding paragraph, the value 1 would be picked out after the first pass, leaving eight values. A search of these eight values gives 2 as the smallest remaining value, and so on for ($[n/2] + 1 = [9/2] + 1 = 4 + 1 = 5$) passes. The differences between the summation and ranking methods are amplified by the number of times the operation is carried out; for a 3×3 filter window and a 512×512 image the filter is evaluated $510 \times 510 = 260,100$ times. Thus, although the median filter might be preferred to the moving average filter for the reasons given earlier, it might be rejected if the computational cost was too high.

Fortunately, a less obvious but faster method of computing the median value for set of overlapping filter windows is available when the data are composed of integers. This fast algorithm begins as usual with the filter window located in the top left-hand corner of the image, as shown in Figure 7.3. A histogram of the $n \times m$ data points lying within the window is computed and the corresponding class frequencies are stored in a one-dimensional array. The median value is found by finding that class (grey level) value such that the cumulative frequency for that class equals or exceeds $[n/2] + 1$. Using the data given earlier, the cumulative class frequencies for the first few grey levels are (0) 0; (1) 1; (2) 2; (3) 4 and (4) 5. The value in brackets is the grey level and the number following the bracketed number is the corresponding *cumulative* frequency, so that no pixels have values of zero, whereas 4 pixels have values of 3 or less. Since n is equal to 9, the value of ($[n/2] + 1$) is 5 and no further calculation is necessary; the median is 4. This method is considerably faster than the obvious (brute force) sorting method because histogram calculation does not involve any logical comparisons, and the total histogram need not be checked in order to find the median. Further savings are achieved if the histogram is updated rather than recalculated when the filter window is moved to its next position, using a method similar to that illustrated in Figure 7.4. Firstly, the cumulative frequencies are reduced as necessary to take account of the left-hand column of the window, which is moving out of the filter, and then the cumulative frequencies are incremented according to the values of the new right-hand column of pixel values, which is moving into the window. This part of the procedure is similar to the updating of the sum of pixel values as described in section 7.2.1 in connection with the moving average filter. If the fast algorithm is used then the additional computational expense involved in computing the median is not significant.

The concept of the median filter was introduced by Tukey (1977) and its extension to two-dimensional images is discussed by Pratt (1978). The fast algorithm described above was reported by Huang *et al.* (1979), who also provide a number of additional references. See also Brownrigg (1984) and Danielsson (1981). Blom and Daily (1982) and Rees and Satchell (1997) illustrate the use of the median filter applied to SAR images.

7.2.3 Adaptive filters

Both the median and the moving average filter apply a fixed set of weights to all areas of the image, irrespective of the variability of the grey levels underlying the filter window. Several authors have considered smoothing methods in which the filter weights are calculated anew for each window position, the calculations being based on the mean and variance of the grey levels in the area of the image underlying the window. Such filters are termed adaptive filters. Their use is particularly important in the attenuation of the multiplicative noise effect known as speckle, which affects synthetic aperture radar images. As noted in section 7.2.2, the median filter has been used with some success to remove speckle noise from SAR images. However, more advanced filters will, in general, produce superior results in the sense that they are capable of removing speckle without significantly degrading the high-frequency component of the SAR image.

One of the best-known speckle suppression filters is the *sigma filter*, proposed by Lee (1983a and b). This filtering method is based on the concept of the Normal distribution. Approximately 95% of the values of observations belonging to a Normal distribution with mean μ and standard deviation σ fall within $\pm 2\sigma$ of the mean value. Lee's method assumes that the grey level values in a single-band SAR image are Normally distributed and, for each overlapping, rectangular window, computes estimates of the local mean $\bar{x}$ and local standard deviation s from the pixels falling within the window. A threshold value is computed from those pixels whose values lie within $\pm 2s$ of the window mean $\bar{x}$. Pixels outside this range are not included in the calculation. The method breaks down when only a few of the pixels in the window have values that are within $\pm 2s$ of the window mean. A parameter k is used to control the procedure. If fewer than k pixels are selected by the threshold (i.e., fewer than k pixels lie in the range $\bar{x} \pm 2s$) then the procedure is aborted, and the filtered pixel value to the left of the current position is used. Alternatively, the average of the four neighbouring pixels replaces the window centre pixel. This modification can cause problems in the first case if the pixel concerned is located on the left margin of the image. In the second case, the filtered values for the pixels to the right of and below the current pixel will need to be calculated before the average can be obtained. Other developments in the use of the sigma filter are summarised by Smith (1996), who describes two simple modifications to the standard sigma filter to improve its computational efficiency and preserve fine features. Reviews of speckle filtering of SAR images are provided by Desnos and Matteini (1993) and Lee *et al.* (1994). Wakabayeshi and Arai (1996) discuss a new approach to speckle filtering using a chi-square test. Martin and Turner (1993) consider a weighted method of SAR speckle suppression, while Alparone *et al.* (1996) present an adaptive filter using local order statistics to achieve the same objective. Order statistics are based on the local grey level histogram, for example the median. More advanced methods of speckle filtering using simulated annealing are described by White (1993). Other references are Beauchemin *et al.* (1996), who use a measure of texture (the contrast feature derived from the Grey Level Co-occurrence Matrix, described in section 8.7.1) as the basis of the filter, and Lopes *et al.* (1993). More recent developments in the suppression of speckle noise are based on the discrete wavelet transform (section 6.7 and 9.3.2.2).

0	0	0	0	0
0	1	1	1	0
0	1	1	1	0
0	1	1	1	0
0	0	0	0	0

0	1	1	1	0
0	1	1	1	0
0	0	1	0	0
0	0	0	0	0
0	0	0	0	0

0	0	0	0	0
0	0	0	1	1
0	0	1	1	1
0	0	0	1	1
0	0	0	0	0

0	0	0	0	0
0	0	0	0	0
0	0	1	0	0
0	1	1	1	0
0	1	1	0	0

0	0	0	0	0
1	1	0	0	0
1	1	1	0	0
1	1	0	0	0
0	0	0	0	0

1	1	0	0	0
1	1	1	0	0
0	1	1	0	0
0	0	0	0	0
0	0	0	0	0

0	0	0	1	1
0	0	1	1	1
0	0	1	1	0
0	0	0	0	0
0	0	0	0	0

0	0	0	0	0
0	0	0	0	0
0	0	1	1	0
0	0	1	1	1
0	0	0	1	1

0	0	0	0	0
0	0	0	0	0
0	1	1	0	0
1	1	1	0	0
1	1	0	0	0

Figure 7.6 Nagao and Matsuyama filters. The 1s and 0s can be interpreted as a logical mask, with 1 meaning 'true' and 0 meaning 'false'. For a given window position, the variance of the pixels in the 'true' positions is calculated. The pixel value transferred to the output image is the mean of the 'true' pixels in the window with the lowest variance.

The idea of edge-preserving smoothing, as used in the sigma filter, is also the basis of a filtering method proposed by Nagao and Matsuyama (1979). This method attempts to avoid averaging pixel values that belong to different 'regions' that might be present in the image. The boundary between two regions contained within a window area might be expected to be represented by an 'edge' or sharp discontinuity in the grey level values. Hence, Nagao and Matsuyama suggest that a bar be rotated around the centre of the window and the bar at the position with the smallest standard deviation of the pixels' grey scale values be selected as the 'winner', since a small standard deviation indicates the absence of any edges. The centre pixel value is replaced by the average of the pixel values in the winning bar (Figure 7.6). The Nagao-Matsuyama filter is implemented in MIPS (under the *Filter* menu item). Figure 7.7 shows the output of the filter for the Wash image used in Example 7.2.

7.3 SPATIAL DOMAIN HIGH-PASS (SHARPENING) FILTERS

The process of imaging or scanning involves blurring, as noted in the discussion of the Point Spread Function (PSF) in Chapter 2. High frequencies are more heavily suppressed than are the low-frequency components of the image. It

Figure 7.7 Output from the Nagao–Matsuyama filter for the image used in Example 7.2.

might therefore seem likely that the visual quality of an image might be improved by selectively increasing the contribution of its high-frequency components. Since the low-pass filters discussed in section 7.2 involve some form of averaging (or spatial integration) then the use of the 'mathematical opposite' of averaging or integrating, namely the derivative function, might seem to be suited to the process of sharpening or de-blurring an image. However, a simpler way of performing an operation that is equivalent to high-pass filtering is considered before derivative-based methods are discussed.

7.3.1 Image subtraction method

According to the model described in sections 6.6 and 7.1 an image can be considered to be the sum of its low and high frequency components, plus noise. The low-frequency part can be isolated by the use of a low-pass filter as explained in section 7.2. This low-frequency image can be subtracted from the original, unfiltered, image leaving behind the high frequency component. The resulting image can be added back to the original, thus effectively doubling the high-frequency part or, as in the case of Thomas *et al.* (1981), a proportion of the low-pass filtered image is subtracted from the original image to give a high-frequency image. Thomas *et al.* (1981) use the model

$$R^* = R - fR' + C$$

in which R^* is the filtered pixel value at the centre of a 3×3 window, R is the original value, R' is the average of the 3×3 window, f is a proportion between 0 and 1 (Thomas *et al.* (1981) use 0.8) and C is a constant whose function is to ensure that R^* is always positive. This subtractive box filter is used by Thomas *et al.* to enhance a circular geological feature.

The addition and subtraction operations must be done with care (section 6.2). The sum of any two pixel values drawn from images each having a dynamic range of 0–255 can range from 0 to 510, so division by two is needed to keep the sum within the 0–255 range. The difference between two pixel values can range from –255 to +255; if the result is to be expressed on the range 0–255 then (i) 255 is added to the result and (ii) this value is divided by two. In a difference image, therefore, 'zero' has a greyscale value of 127.

7.3.2 Derivative-based methods

Other methods of high-pass filtering are based on the mathematical concept of the derivative, as noted earlier. The derivative of a continuous function at a specified point is the rate of change of that function value at that point. For example, the first derivative of position with respect to time (the rate of change of position over time) is velocity, assuming direction is constant. The greater the velocity of an object the more rapidly it changes its position with respect to time. The velocity can be measured at any time after motion commences. The velocity at time t is the first derivative of position with respect to time at time t. If the position of an object were to be graphed against time then the velocity (and hence the first derivative) at time t would be equal to the slope of the curve at the point time $= t$. Hence, the derivative gives a measure of the rate at which the function is increasing or decreasing at a particular point in time or, in terms of the graph, it measures the gradient of the curve.

In the same way that the rate of change of position with time can be represented by velocity so the rate of change of velocity with time can be found by calculating the first derivative of the function relating velocity and time. The result of such a calculation would be acceleration. Since acceleration is the first derivative of velocity with respect to time and, in turn, velocity is the first derivative of position with respect to time then it follows that acceleration is the second derivative of position with respect to time. It measures the rate at which velocity is changing. When the object is at rest its acceleration is zero. Acceleration is also zero when the object reaches a constant velocity. A graph of acceleration against time would be useful in determining those times when velocity was constant or, conversely, the times when velocity was changing.

In terms of a continuous grey scale image, the analogue of velocity is the rate of change of greyscale value over space. This derivative is measured in two directions – one with respect to x, the other with respect to y. The overall first derivative (with respect to x and y) is the square root of the sum of squares of the two individual first derivatives. The values of these three derivatives (in the x direction, y direction and overall) tell us (i) how rapidly the greyscale value is changing in the x direction, (ii) how rapidly it is changing in the y direction and (iii) the maximum rate of change in any direction, plus the direction of this maximum change. All these values are calculable at any point in the interior of a continuous image. In those areas of the image that are homogeneous, the values taken by all three derivatives (x, y and overall) will be small. Where there is a rapid change in the grey scale values, for example at a coastline in a near-infrared image, the gradient (first derivative) of the image at that point will be high. These lines or edges of sharp change in grey level can be thought of as being represented by the high-frequency component of the image for, as mentioned earlier, the local variation from the overall background pattern is due to high-frequency components (the background pattern is the low-frequency component). The first derivative or gradient of the image therefore identifies the high-frequency portions of the image.

What does the second derivative tell us? Like the first derivative it can be calculated in both the x and y directions and also with respect to x and y together. It identifies areas where the gradient (first derivative) is constant, for the second derivative is zero when the gradient is constant. It could be used, for example, to find the top and the bottom of a 'slope' in grey level values.

Images are not continuous functions. They are defined at discrete points in space, and these points are usually taken to be the centres of the pixels. It is therefore not possible to calculate first and second derivatives using the methods of calculus. Instead, derivatives are estimated in terms of differences between the values of adjacent pixels in the x and y directions, though diagonal or corner differences are also used. Figure 7.8 shows the relationship between a discrete, one-dimensional function (such as the values along a scan line of a digital image, as shown in Figure 7.1) and its first and second derivatives estimated by the method of differences. The first differences (equivalent to the first derivatives) are:

$$\Delta x\, p(i, j) = p(i, j) - p(i - 1, j)$$
$$\Delta y\, p(i, j) = p(i, j) - p(i, j - 1)$$

in the x (along-scan) and y (across-scan) directions respectively, while the second difference in the x direction

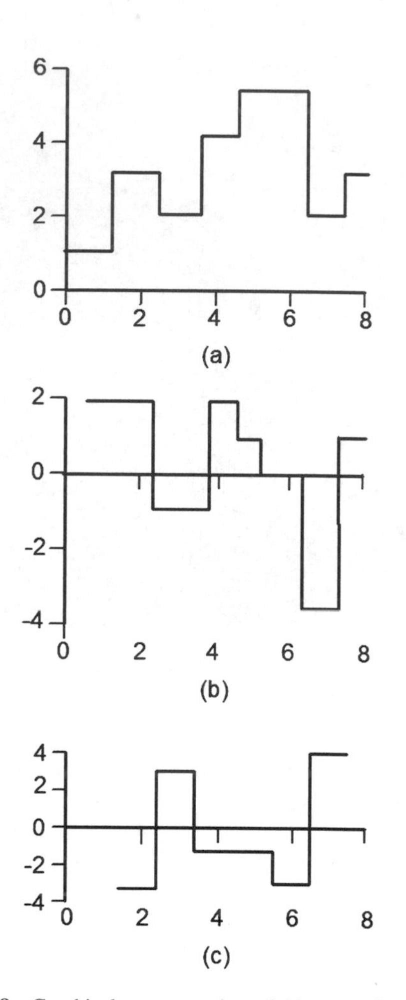

Figure 7.8 Graphical representation of (a) a one-dimensional data series with corresponding (b) its first and (c) second-order differences.

is:

$$\begin{aligned}\Delta x_2\, p(i, j) &= \Delta x\, p(i + 1, j) - \Delta x p(i, j)\\ &= [p(i + 1, j) - p(i, j)]\\ &\quad - [p(i, j) - p(i - 1, j)]\\ &= p(i + 1, j) + p(i - 1, j) - 2p(i, j)\end{aligned}$$

Similarly, the second difference in the y direction is:

$$\Delta y_2\, p(i, j) = p(i, j + 1) + p(i, j - 1) + 2p(i, j)$$

The calculation and meaning of the first and second differences in one dimension are illustrated in Table 7.1. A discrete sequence of values, which can be taken as pixel values along a scan line, is shown in the top row of Table 7.1 and the first and second differences are shown in rows two and three. The first difference is zero where the rate

Table 7.1 Relationship between discrete values (f) along a scan line and the first and second differences ($\Delta(f)$, $\Delta_2(f)$)). The first difference (row 2) indicates the rate of change of the values of f shown in row 1. The second difference (row 3) gives the points at which the rate of change itself alters. The first difference is compute from $\Delta(f) = f_i - f_{i--1}$, and the second derivative is found from $\Delta_2(f) = \Delta(\Delta(f)) = f_{i+1} + f_{i-1} - 2f_i$.

f	0	0	0	1	2	3	4	5	5	5	5	4	3	2	1	0
$\Delta(f)$		0	0	1	1	1	1	1	0	0	0	−1	−1	−1	−1	−1
$\Delta_2(f)$		0	1	0	0	0	0	−1	0	0	−1	0	0	0	0	

of change of greyscale value is zero, positive when 'going up' a slope and negative when going down. The magnitude of the first difference is proportional to the steepness of the 'slope' of the greyscale values, so steep 'slopes' (where greyscale values are increasing or decreasing rapidly) are characterised by first differences that are large in absolute value. The second difference is zero where the first difference values are constant, negative at the foot of a 'slope' and positive at the top of the 'slope'. The extremities of a 'slope' are thus picked out by the second difference.

The computation of the magnitude of the maximum first difference or gradient of a digital image can be carried out by finding Δx and Δy as above and then determining the composite gradient, given by:

$$\Delta xy\, p(i, j) = \sqrt{\{[\Delta x\, p(i, j)]^2 + [\Delta y\, p(i, j)]^2\}}$$

and the direction of this composite gradient is:

$$\Delta = \tan^{-1}[\Delta y\, p(i, j)/\Delta x\, p(i, j)]$$

Other gradient measures exist. One of the most common is the Roberts Gradient, ΔR. It is computed in the two diagonal directions rather than in the horizontal and vertical directions from:

$$\Delta R = \sqrt{\{[p(i, j) - p(i+1, j+1)]^2 + [p(i, j+1) - p(i+1, j)]^2\}}$$

or

$$\Delta R = |\, p(i, j+1) - p(i+1, j+1)\,| + |\, p(i, j+1) - p(i+1, j)\,|$$

The second form is sometimes preferred for reasons of efficiency as the absolute value (|.|) is more quickly computable then the square root, and raising the inter-pixel difference values to the power of two is avoided. The Roberts Gradient function is implemented in the MIPS program, via the *Filter* main menu item. Figure 7.9(a) shows the histogram of the Landsat ETM+ sub-image of the south-west corner of The Wash used in Example 7.2 after the application of the Roberts gradient operator. Most of the pixels in the Roberts Gradient image are seen to have values less than 40, so a linear contrast stretch was applied, setting the lower and upper limits of the stretch to 0 and 40, respectively. The result is shown in Figure 7.9(b), in which the grey level values are proportional to ΔR.

In order to emphasise the high-frequency components of an image a multiple of the gradient values at each pixel location (except those on the first and last rows and columns) can be added back to the original image. Normally the absolute values of the gradient are used in this operation. The effect is to emphasise those areas where the grey scale values are changing rapidly. Another possibility is to define a threshold value by inspection of the histogram of gradient values. Where the gradient value at a pixel location exceeds this threshold value the pixel value is set to 255, otherwise the gradient is added back as before. This will over-emphasise the areas of greatest change in grey level.

The second difference function of a digital image is given by:

$$\begin{aligned}\Delta xy_2 p(i, j) &= \Delta x_2 p(i, j) + \Delta y_2 p(i, j)\\ &= [p(i+1, j) + p(i-1, j) + p(i, j+1)\\ &\quad + p(i, j-1)] - 4p(i, j)\end{aligned}$$

In image processing, this function is called the Laplacian operator. Like its one-dimensional analogue shown in Table 7.1 this operator takes on a negative value at the foot of a greyscale 'slope' and a positive value at the crest of a 'slope' (Figure 7.10). The magnitude of the value is proportional to the gradient of the 'slope'. If absolute values are taken, then Laplacian operator will pick out the top and the bottom of 'slopes' in greyscale values. Alternatively, the signed values (negative at the foot, positive at the crest) can be displayed by adding 127 to all values, thus making 127 the 'zero' point on the grey scale. Negative values of the Laplacian will be shown by darker shades of grey, positive values by lighter grey tones. Like the gradient image, the Laplacian image can be added back to the original image though, as noted below, it is more sensible to subtract the Laplacian. The effect is sometimes quite dramatic, though much depends on the 'noisiness' of the image. Any adding-back of high-frequency information to an already noisy image will inevitably result in disappointment. Figure 7.11(b) shows the result of subtracting the Laplacian image from the original, un-enhanced image of the Painted Desert area of Arizona (Figure 7.11(a)). The result is much less 'hazy' than the original, and the effects of the linear stretch are considerable.

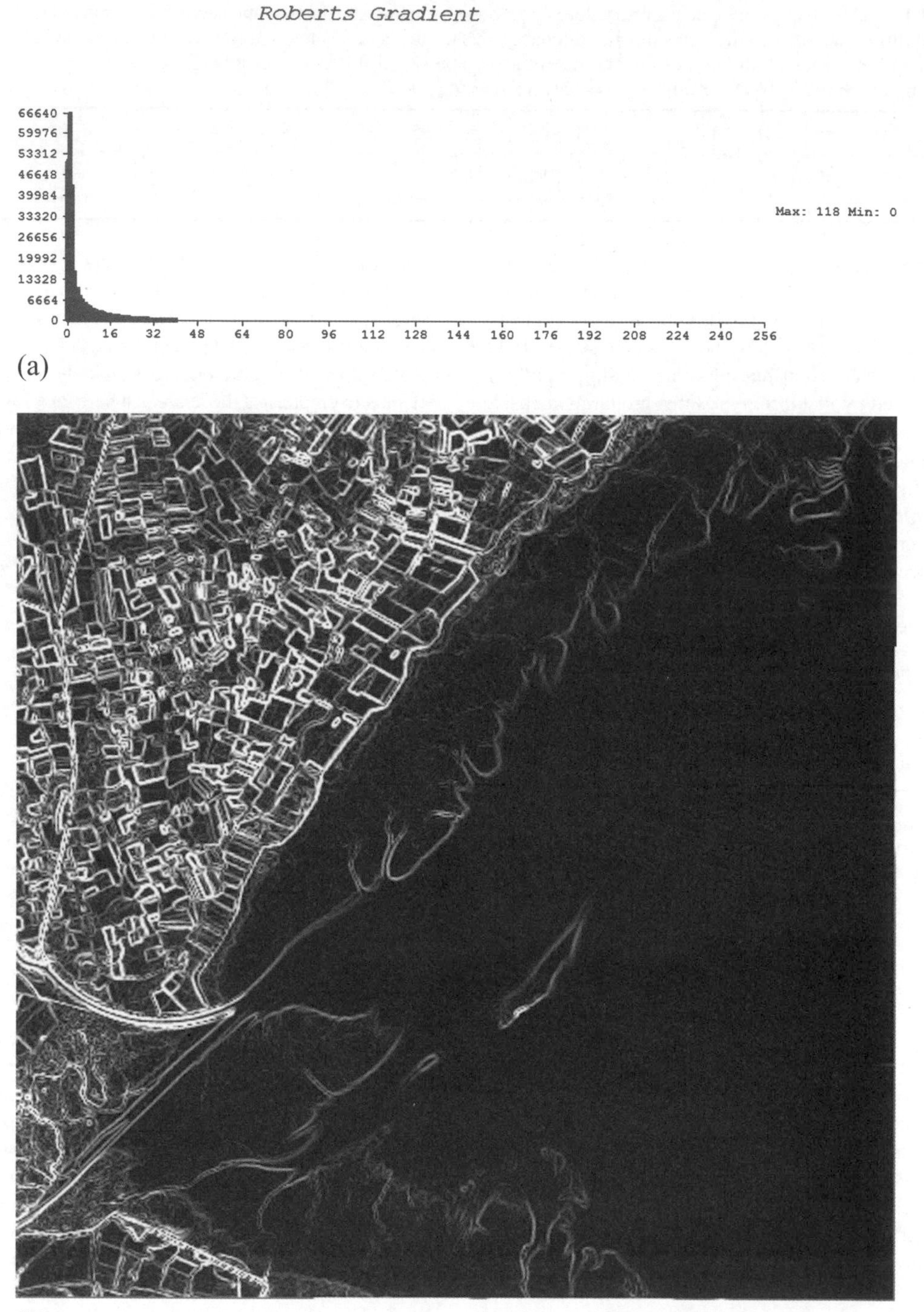

Figure 7.9 **(a)** Histogram of the image shown in Figure 7.9(b). **(b)** Landsat *ETM+* band 4 image of the south-west corner of The Wash in eastern England, after the application of the Roberts Gradient filter and a linear contrast stretch using specified lower and upper limits of 0 and 40 (see the image histogram, Figure 7.9(a)). The field boundaries are brought out very clearly, as is the mouth of the river Witham near Boston, Lincolnshire (lower left).

Table 7.2 (a) Weight matrix for the Laplacian operator. (b) These weights subtract the output from the Laplacian operator from the value of the central pixel in the window.

(a)

0	1	0
1	−4	1
0	1	0

(b)

0	−1	0
−1	5	−1
0	−1	0

Figure 7.10 Section across a greyscale image. The Laplacian operator outputs a positive value at points where the grey level curve reaches a minimum, and negative values at those points where the grey level curve shows a maximum. Subtraction of the output from the Laplacian operator from the grey level curve would over-emphasise both the dark minima at the base of a slope and the bright maximum at the top of the slope.

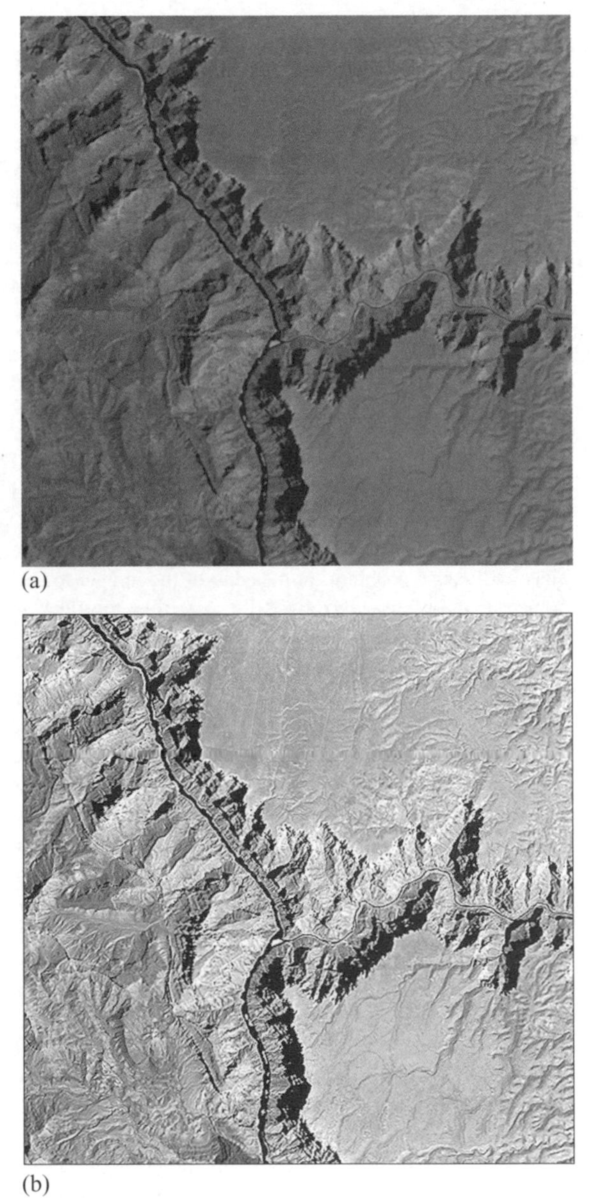

Figure 7.11 Landsat TM image of the Little Colorado River, Painted Desert, Arizona. The unenhanced image is shown in (a). The image shown in (b) has been subjected to the 'image minus Laplacian' operation followed by a 5%–95% linear contrast stretch.

Rosenfeld and Kak (1982, p. 241–244) give reasons why this reduction in haziness is be observed. If the discussion of the Point Spread Function (Chapter 2) is recalled, it will be realised that the effect of imaging through the atmosphere and the use of lenses in the optical system is to diffuse the radiance emanating from a point source so that the image of a sharp point source appears as a circular blob. Rosenfeld and Kak show that the Laplacian operator approximates in mathematical terms to the equation known as Fick's Law, which describes the two-dimensional diffusion process. Thus, subtracting the Laplacian from the original image is equivalent to removing the diffused element of the signal from a given pixel. Another possible explanation is that the value recorded at any point contains a contribution from the neighbouring pixels. This is a reasonable hypothesis for the contribution could consist of the effects of diffuse radiance, that is, radiance from other pixels that has been scattered into the field of view of the sensor. The Laplacian operator effectively subtracts this contribution.

The weight matrix to be passed across the image to compute the Laplacian is shown in Table 7.2(a), while the 'image-minus-Laplacian' operation can be performed directly using the weight matrix shown in Table 7.2(b). Other forms of the weight matrix are conceivable; for example, diagonal differences rather than vertical and horizontal differences could be used, or the diagonal differences plus the vertical/horizontal differences. A wider neighbourhood could be used, with fractions of the difference being applied. There seems to be little or no reason why such methods should be preferred to the basic model unless the user has some motive based upon the physics of the imaging process.

The basic model of a high-pass image-domain filter involves the subtraction of the pixel values within a window from a multiple of the central pixel. The size of the window is not limited to 2×2 or 3×3 which are used in the derivative-based filters described above. Generally, if the number of pixels in a window is k then the weight given to the central pixel is $(k - 1)$ while all other pixels have a weight of -1. The product of the window weights and the underlying image pixel values is subsequently divided by k. The size of the window is proportional to the wavelengths allowed through the filter. A low-pass filter will remove more of the high-frequency components as the window size increases (i.e. the degree of smoothing is proportional to the window size). A high-pass filter will allow through a broader range of wavebands as the window size increases. Unless precautions are taken, the use of very large window sizes will cause problems at the edge of the image; for instance, if the window size is 101×101 then the furthest left that the central pixel can be located is at row 51, giving a margin of 50 rows that cannot be filtered. For 3×3 filters this margin would be one pixel wide, and it could be filled with zeros. A zero margin 50 pixels wide at the top, bottom, left and right of an image might well be unacceptable. One way around this problem is to ignore those window weights that overlap the image boundary, and compute the filtered value using the weights that fall inside the image area. The value of the central weight will need to be modified according to the number of weights that lie inside the image area. This implies that the bandwidth of the filter will vary from the edge of the image until the point at which all the window weights lie inside the image area.

High-pass filters are used routinely in image processing, especially when high-frequency information is the focus of interest. For instance, Ichoku *et al.* (1996) use the 'image minus Laplacian' filter as part of a methodology to extract drainage-pattern information from satellite imagery. Krishnamurthy *et al.* (1992) and Nalbant and Alptekin (1995) demonstrate the value of high-frequency enhancement and directional filtering in geological studies. Al-Hinai *et al.* (1991) use a high-pass filter to enhance images of sand dunes in the Saudi Arabian desert.

7.4 SPATIAL DOMAIN EDGE DETECTORS

A high-pass filtered image that is added back to the original image is a high-boost filter and the result is a sharpened or de-blurred image. The high-pass filtered image can be used alone, particularly in the study of the location and geographical distribution of 'edges'. An edge is a discontinuity or sharp change in the grey scale value at a particular pixel point and it may have some interpretation in terms of cultural features, such as roads or field boundaries, or in terms of geological structure or relief. We have already noted that the first difference can be computed for the horizontal, vertical and diagonal directions, and the magnitude and direction of the maximum spatial gradient can also be used. Other methods include the subtraction of a low-pass filtered image from the original (section 7.3.1) or the use of the Roberts Gradient. A method not so far described is the Sobel non-linear edge operator (Gonzales and Woods, 1992), which is applied to a 3×3 window area. The value of this operator for the 3×3 window defined by:

A	*B*	*C*
D	*E*	*F*
G	*H*	*I*

is given for the pixel underlying the central window weight (E) by the function:

$$S = \sqrt{X^2 + Y^2}$$

where

$$X = (C + 2F + I) - (A + 2D + G)$$
$$Y = (A + 2B + C) - (G + 2H + I)$$

This operation can also be considered in terms of two sets of filter weight matrices. X is given by the following weight matrix, which determines horizontal differences in the neighbourhood of the centre pixel:

−1	0	1
−2	0	2
−1	0	1

while Y is given by a weight matrix which involves vertical differences:

−1	−2	−1
0	0	0
1	2	1

An example of the output from the Sobel filter for a Landsat MSS near-infrared image of part of the Tanzanian coast is shown in Figure 7.12.

Shaw *et al.* (1982) provide a comparative assessment of these techniques of edge detection. They conclude that first-differencing methods reveal local rather than regional boundaries, and that increasing the size of a high-pass filter window increases the amount of regional-scale information. The Roberts and Sobel techniques produced a too intense enhancement of local edges but did not remove the regional patterns.

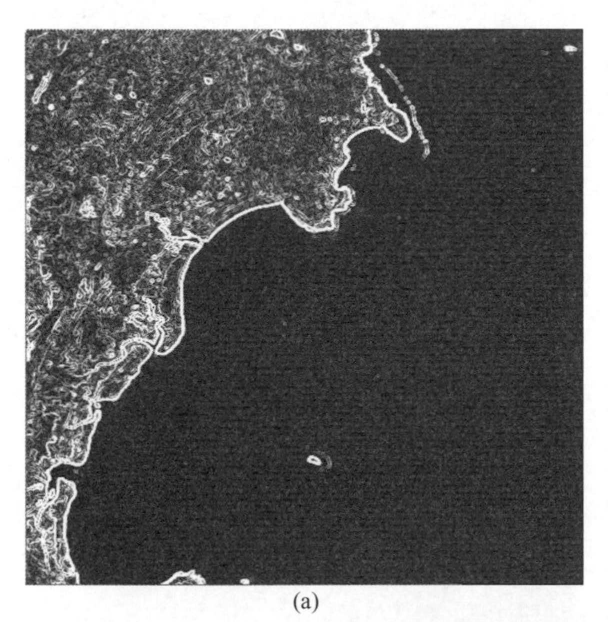

(a)

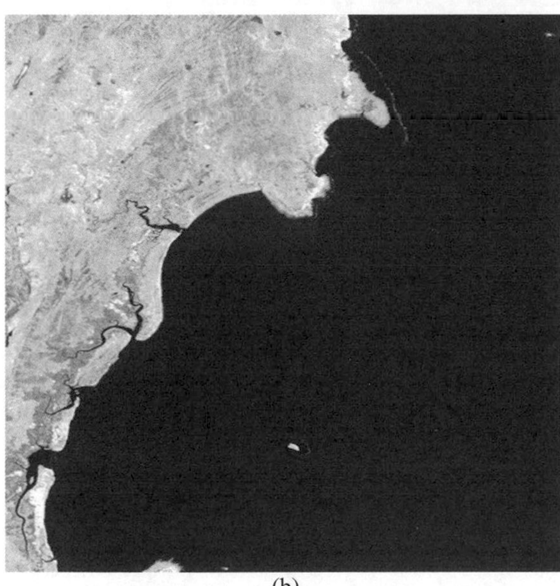

(b)

Figure 7.12 **(a)** Landsat MSS near-infrared image of the Tanzanian coast south of Dar es Salaam after the application of the Sobel filter. **(b)** Original image. Both images have been enhanced by a linear contrast stretch. Note how the strength or magnitude of the output from the Sobel filter is related to the degree of contrast between the pixels on either side of the edge in question.

One of the main uses of edge-detection techniques has been in the enhancement of images for the identification and analysis of geological lineaments, which are defined as

> 'mappable, simple or composite linear features whose parts are aligned in a rectilinear or slightly curvilinear relationship and which differ distinctly from the pattern of adjacent features and which presumably reflect a subsurface phenomenon'
>
> (O'Leary *et al.*, 1976, p. 1467).

The subsurface phenomena to which the definition refers are presumed to be fault and joint patterns in the underlying rock. However, linear features produced from remotely sensed images using the techniques described in this section should be interpreted with care. An example of the use of an edge-detection procedure to highlight linear features for geological interpretation is to be found in Moore and Waltz (1983).

Other applications of edge-detection techniques include the determination of the boundaries of homogeneous regions (segmentation) in synthetic aperture radar images in order to segment those images (Quegan and Wright, 1984). A comprehensive review by Brady (1982) considers the topic of image segmentation within the context of image understanding. Gurney (1980) discusses the problem of threshold selection for line detection algorithms. Chavez and Bauer (1982) consider the choice of filter window size for edge enhancement, and Nasrabadi *et al.* (1984) present a line detection algorithm for use with noisy images. Chittenini (1983) and Drewniok (1994) look at the problem of edge and line detection in multi-dimensional noisy images. Eberlein and Wezka (1975), Haralick (1980) and Peli and Malah (1982) consider the general problem of edge and region analysis. A widely used line detector is the one presented by Vanderbrugg (1976). Algorithms for edge detection and region segmentation are discussed by Farag (1992), Pavlidis (1982) and Pitas (1993). Reviews of edge detection and linear feature extraction methodologies are provided by Budkewitsch *et al.* (1994), Peli and Malah (1982) and Wang (1993). Riazanoff *et al.* (1990) describe ways of thinning (*skeletonising*) lines which have been identified using edge-detection techniques. Such lines are generally defined by firstly applying a high-pass filter, then thresholding the resulting image using edge magnitude or strength to give a binary image. The definitive reference is still Marr and Hildreth (1980).

7.5 FREQUENCY DOMAIN FILTERS

The Fourier transform of a two-dimensional digital image is discussed in section 6.6. The Fourier transform of an

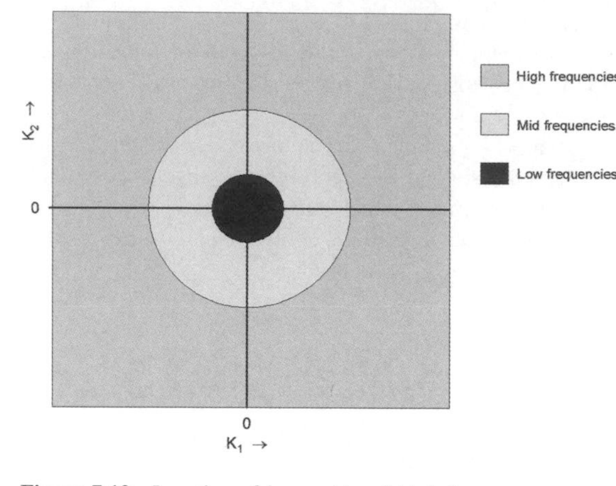

Figure 7.13 Location of low, mid and high frequency components of the two-dimensional amplitude spectrum.

image, as expressed by the amplitude spectrum, is a breakdown of the image into its frequency or scale components. Since the process of digital filtering can be viewed as a technique for separating these components, it might seem logical to consider the use of frequency-domain filters in remote sensing image processing. Such filters operate on the amplitude spectrum of an image and remove, attenuate, or amplify the amplitudes in specified wavebands. A simple filter might set the amplitudes of all frequencies less than a selected threshold to zero. If the amplitude spectrum information is converted back to the spatial domain by an inverse Fourier transform, the result is a low-pass filtered image. Any wavelength or waveband can be operated upon in the frequency domain, but three general categories of filter are considered here – low-pass, high-pass and band-pass. The terms low-pass and high-pass are defined in section 7.1. A band-pass filter removes both the high and low frequency components, but allows an intermediate range of frequencies to pass through the filter, as shown in Figure 7.13. Directional filters can also be developed, because the amplitude spectrum of an image contains information about the frequencies and orientations as well as the amplitudes of the scale components that are present in an image.

The different types of high-, low- and band-pass filters are distinguished on the basis of what are known as their 'transfer functions'. The transfer function is a graph of frequency against filter weight, though the term filter weight should, in this context, be interpreted as 'proportion of input amplitude that is passed by the filter'.

Figure 7.14(a) shows a cross-section of a transfer function that passes all frequencies up to the value f_1 without alteration. Frequencies higher in value than f_1 are subjected

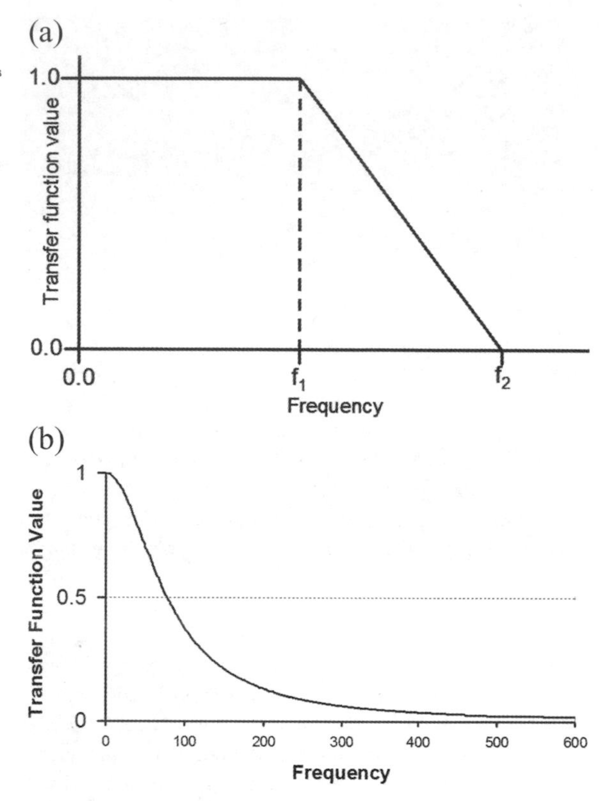

Figure 7.14 **(a)** Filter transfer function (one-dimensional slice) that passes unchanged all spatial frequencies lower than f_1, attenuates all frequencies in the range f_1–f_2 and suppresses all frequencies higher than f_2. The degree of attenuation increases linearly in the range f_1–f_2. This filter would leave low frequencies unchanged, and would suppress high frequencies. **(b)** Transfer function for a low-pass Butterworth filter with cut-off frequency D_0 equal to 50. The shape of the transfer function is smooth, which is an advantage as sharp edges in the transfer function cause 'ringing'.

to increasing attenuation until the point f_2. All frequencies with values higher than f_2 are removed completely. Figure 7.14(b) shows the transfer function of a more complex filter, a Butterworth low-pass filter, which is described in more detail below.

Care should be taken in the design of filter transfer functions. As noted earlier, the spatial domain filtered image is derived from the 2D amplitude spectrum image by multiplying the 2D amplitude spectrum by the 2D filter transfer function and then performing an inverse Fourier transform on the result of this calculation (Figure 7.15). Any sharp edges in the filtered amplitude spectrum will convert to a series of concentric circles in the spatial domain, producing a pattern of light and dark rings on the filtered image. This phenomenon is termed *ringing*, for reasons that are evident

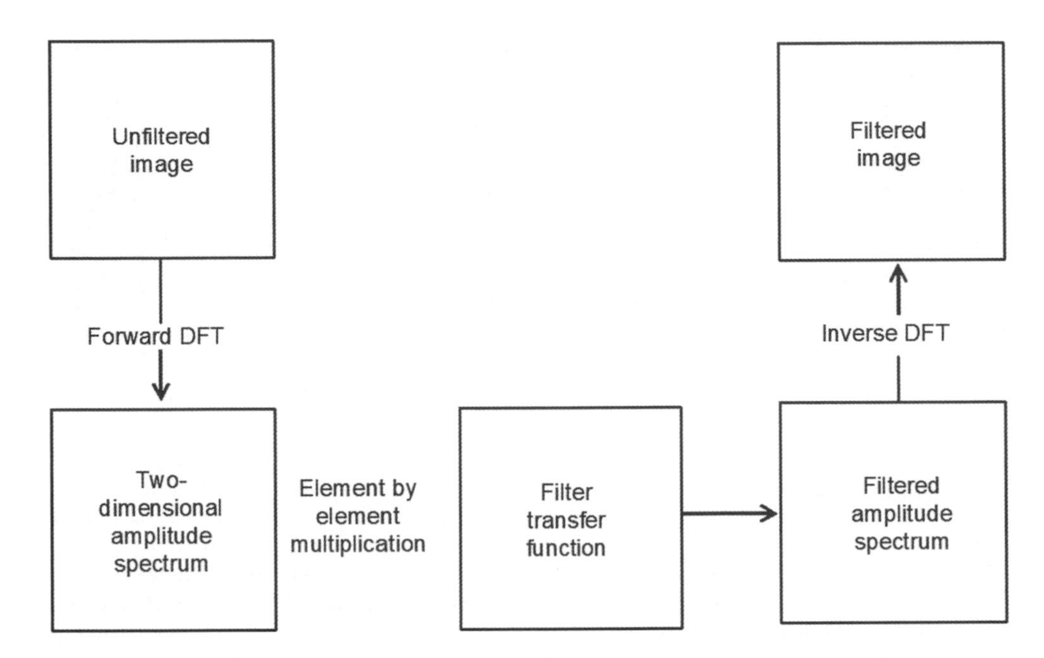

Figure 7.15 Steps in the frequency domain filtering of a digital image.

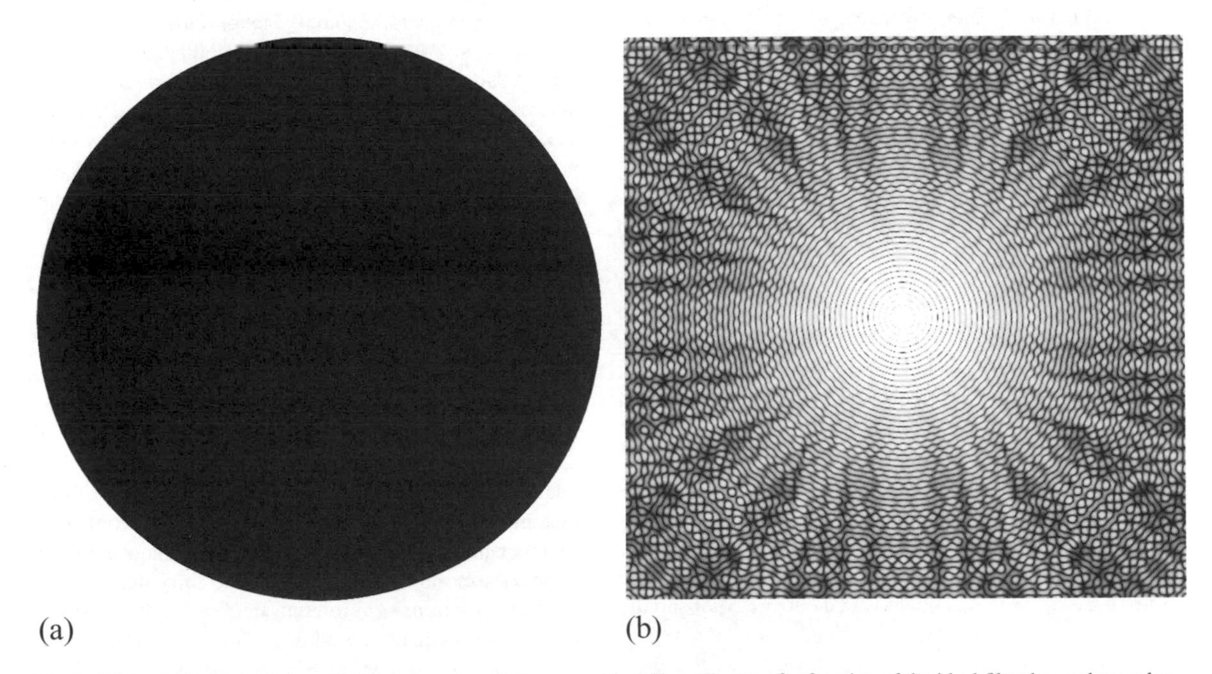

Figure 7.16 (a) Representation of a frequency-domain low-pass ideal filter. The transfer function of the ideal filter has a sharp edge, in this case at a frequency of 50 Hz. (b) The logarithm of the two-dimensional Fourier amplitude spectrum of (a). Note the concentric circles centred on the origin of the amplitude spectrum. When the inverse transform is applied, these circles are, in effect, superimposed on the forward transform of the image. The result is a pattern of ripples on the transformed image. This phenomenon is called 'ringing'.

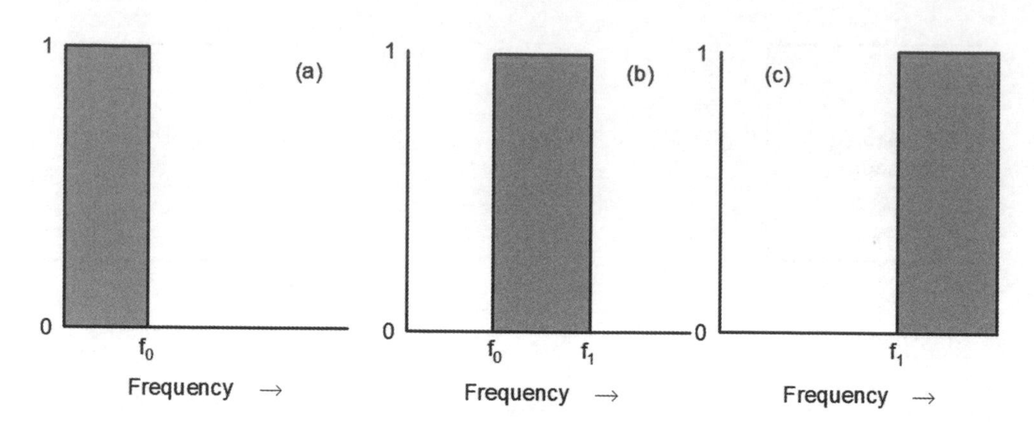

Figure 7.17 Cross sections of transfer functions for three ideal filters: (a) low-pass, (b) band-pass and (c) high-pass. The amplitude coefficients lying within the shaded area of each filter are unchanged as the transfer function value is 1.0. Those amplitude coefficients lying outside the shaded area are set to zero.

from an inspection of Figure 7.16(a) and (b). Gonzales and Woods (1992) discuss this aspect of filter design in detail.

A cross-section through the transfer function of a *low-pass ideal filter* is shown in Figure 7.17(a). The degree of smoothing achieved by the low-pass ideal filter depends on the position of the cut-off frequency, f_0. The lower the value of f_0, the greater the degree of smoothing, as more intermediate and high frequency amplitude coefficients are removed by the filter. The transfer functions for band-pass and high-pass ideal filters are also shown in Figure 7.17 Their implementation in software is not difficult, as the cut-off frequencies form circles of radii f_0 and f_1 around the centre point of the transform (also known as the DC point in the literature of image processing). Figures 7.18(a)–(c) illustrate the results of the application of increasingly severe low-pass Ideal filters to the TM band 7 image shown in Figure 1.11(b) using D_0 values of 100, 50 and 5. The degree of smoothing increases as the cut-off frequency decreases. Figure 7.18(c) shows very little real detail but is, nevertheless, one of the frequency components of the TM image.

Because of their sharp cut-off features, ideal filters tend to produce a filtered image that can be badly affected by the ringing phenomenon, as discussed earlier. Other filter transfer functions have been designed to reduce the impact of ringing by replacing the sharp edge of the ideal filter with a sloping edge or with a function that decays exponentially from the cut-off frequency. An example of this latter type is the Butterworth filter (Figure 7.14(b)), which is defined by:

$$H(u, v) = \frac{1.0}{1.0 + 0.414\,[D(u, v)/D_0]^2}$$

$H(u, v)$ is the value of the filter transfer function for frequencies u and v (remember that the origin of the coordinates u, $v$is the centre point of the frequency domain representation), $D(u, v)$ is the distance from the origin to the point on the amplitude spectrum with coordinates (u, v) and D_0 is the cut-off frequency, as shown in Figure 7.14(b), which is a plot of the value of the transfer function $H(u, v)$ against frequency. This form of the Butterworth filter ensures that $H(u, v) = 0.5$ when $D(u, v)$ equals D_0. Gonzales and Woods (1992) describe other forms of filter transfer function. The result of applying a low pass Butterworth filter with D_0 equal to 50 to the TM band 7 image (Figure 1.11(b)) is shown in Figure 7.18(d). Compare this result with the output from an Ideal low-pass filter (Figure 7.18(b)). It is clear that the Butterworth filter has retained more high frequency information.

Directional filters can be implemented by making use of the fact that the amplitude spectrum contains scale and orientation information (section 6.6). A filter such as the one illustrated in Figure 7.19 removes all those spatial frequencies corresponding to sinusoidal waves oriented in an east-west direction. Such filters have been used in the filtering of binary images of geological fault lines (McCullagh and Davis, 1972).

High-frequency enhancement is accomplished by firstly defining the region of the amplitude spectrum containing 'high' frequencies and then adding a constant, usually 1.0, to the corresponding amplitudes before carrying out an inverse Fourier transform to convert from the frequency to the spatial domain representation. The transfer function for this operation is shown in Figure 7.20. It is clear that it is simply a variant of the ideal filter approach with the transfer function taking on values of 1 and 2 rather than 0 and 1.

Filtering in the frequency domain can be seen to consist of a number of steps, as follows (see also Example 7.3):

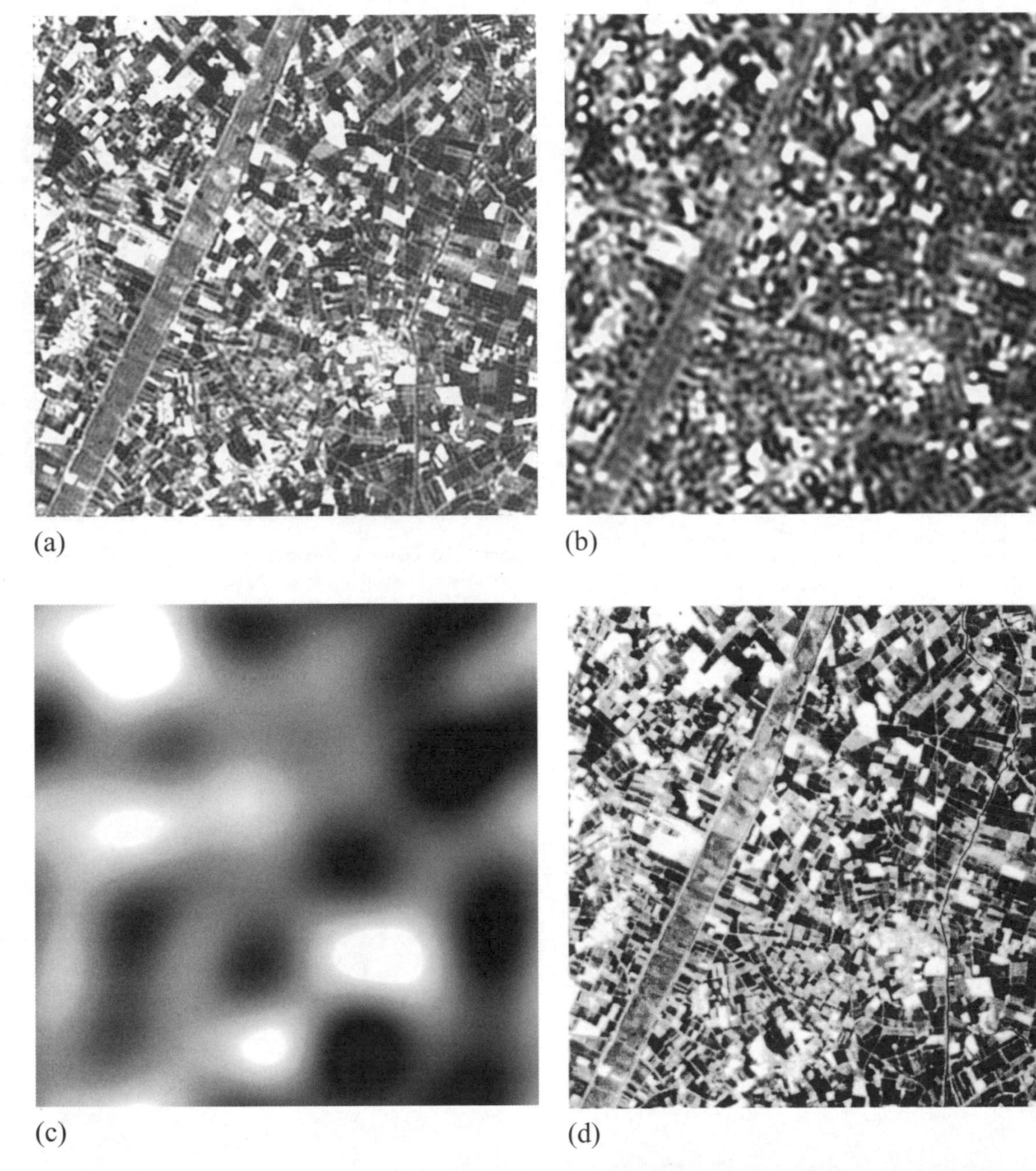

Figure 7.18 Illustrating the results of the application of increasingly severe low-pass Ideal filters to the Littleport TM band 7 image shown in Figure 1.11(b). The filter radii D_0 used in the Ideal filter are (a) 50, (b) 100 and (c) 5. Figure 7.18(d) shows the result of a low-pass Butterworth filter applied to the same source image with a cut-off frequency of 50.

EXAMPLE 7.3: FREQUENCY DOMAIN FILTERING

The example of the Fourier Transform in Chapter 6 demonstrated the use of the forward Fourier transform to generate an amplitude spectrum. The log of the amplitude spectrum of a Landsat TM image of part of the Red Sea Hills in eastern Sudan is displayed in that example. In this example,

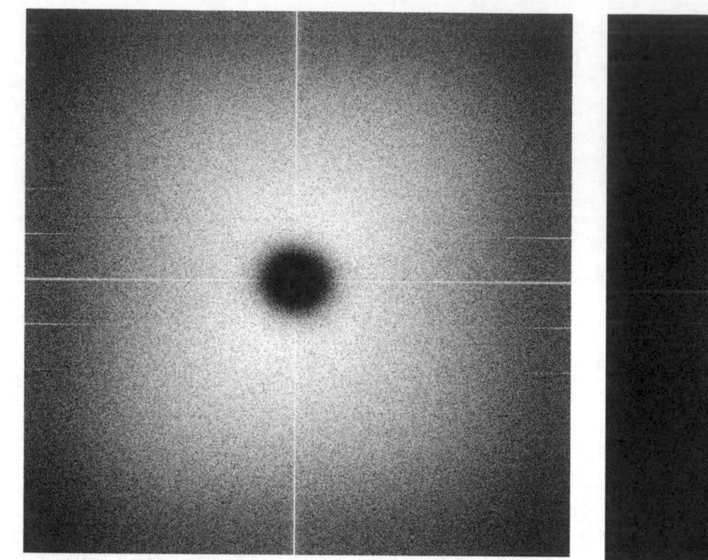

Example 7.3 Figure 1 Logarithm of the filtered amplitude spectrum of the image shown in Example 6.2 Figure 1. The full amplitude spectrum is shown in Example 6.2 Figure 2. The dark circle in the centre of the filtered amplitude spectrum shows that, in comparison with Example 6.2 Figure 2, the frequency components close to the coordinate centre have been suppressed or attenuated. This figure therefore illustrates a high-pass filter.

Example 7.3 Figure 3 The logarithm of the amplitude spectrum of the image shown in Example 6.2 Figure 1, after the application of a low-pass filter (in this example, a Butterworth low-pass filter with a cut-off frequency of 100 was used).

Example 7.3 Figure 2 The result of applying the Inverse Discrete Fourier Transform to the filtered amplitude spectrum shown in Example 7.3 Figure 1. The original image is shown in Example 7.3 Figure 1. It is clear that the high-pass filter has removed the background information, leaving behind the high-frequency information (sharp changes in greylevel and edges).

Example 7.3 Figure 4 Image recovered from the amplitude spectrum shown in Figure 3. This image is effectively the complement of the high-pass filtered image shown in Example 7.3 Figure 2. The detail has been lost (compare Example 6.2 Figure 1) but the overall background pattern of light and dark (together with some major transitions in grey level, which represent geological faults or fractures) are still apparent.

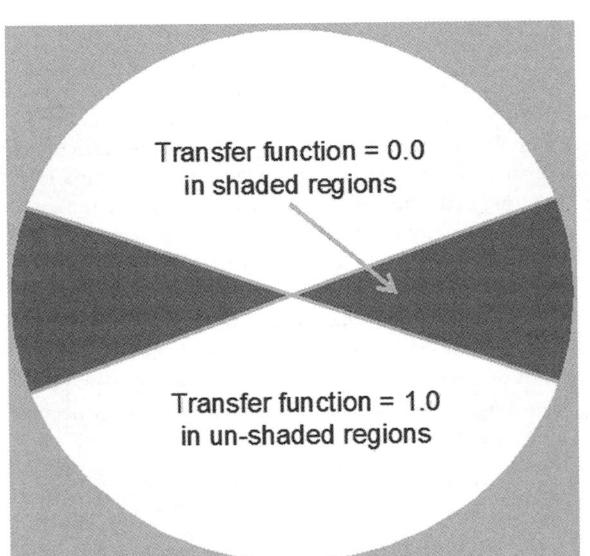

Figure 7.19 Frequency domain directional filter designed to eliminate all horizontal frequency components. Note that the origin of the (u, v) frequency domain coordinates is the centre of the circle, which has a radius equal to $n/2$, where n is the dimension of the image (assumed to be square; if the image is rectangular then the circle becomes an ellipse). Thus, for a 512×512 image the maximum frequency is 256 Hz.

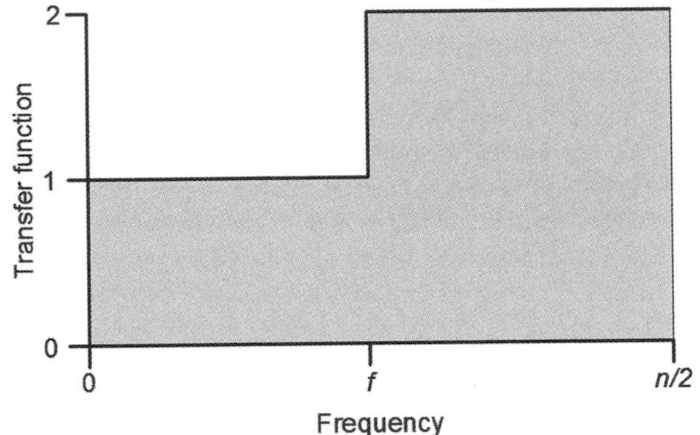

Figure 7.20 High-frequency boost filter. Spatial frequencies less than f Hz are left unchanged, as the corresponding transfer function value (y-axis) is 1.0. Frequencies higher than f Hz are doubled in magnitude. When the inverse transform is completed (Figure 7.15), the spatial domain representation of the filtered image will show enhanced.

a filter is applied to the log of the same amplitude spectrum. Two types of filtering are demonstrated – high-pass and low-pass. Example 7.3 Figure 1 shows the log of the filtered amplitude spectrum of the Sudan image. The black hole in the centre is result of applying a high-pass Butterworth filter with a cut-off frequency of 50 (pixels from centre). The magnitudes of the amplitudes within that circle have been modified using the transfer function described in section 7.5. Example 7.3 Figure 2 shows the spatial domain image that results from the application of inverse discrete Fourier transform to the filtered amplitude spectrum (Example 7.3 Figure 1). It is clear that much of the tonal information in the original image is low frequency in nature, because the removal of the central disc of the amplitude spectrum (Example 7.3 Figure 1) has eliminated most of the tonal variation, leaving an image that comprises the medium and high frequency components. The high frequency components correspond to sharp edges that may be related to the positions of linear features such as fault lines.

Example 7.3 Figure 3 shows the amplitude spectrum after the application of a Butterworth low-pass filter that suppresses frequencies beyond a cut-off point of 100 (pixels from the centre of the transform). Example 7.3 Figure 4 was reconstructed by applying an inverse Fourier transform to the filtered amplitude spectrum (Example 7.3 Figure 3). Most of the tonal variation is retained, as this varies only slowly across the image from light to dark. However, some of the detail has been removed.

These operations were performed by using the Fourier entry on the MIPS *Filter* menu, and selecting (in sequence): *Forward Transform, Filter|Butterworth High (or Low) Pass*, and *Inverse Transform*. See section 6.6 for further details of the Fourier transform and section 7.5 for a discussion of frequency domain filtering.

1. perform a forward Fourier transform of the image and compute the amplitude spectrum (section 6.6).
2. select an appropriate filter transfer function and multiply the elements of the amplitude spectrum by the appropriate transfer function.
3. apply an inverse Fourier transform to convert back to spatial domain representation.

Although frequency domain methods are far more flexible that the spatial domain filtering techniques the cost of computing the forward and inverse Fourier transforms has limited its use in the past. As noted in section 6.6, the two-dimensional Fourier transform requires the transposition of two large matrices holding the intermediate sine and cosine coefficients. This used to be a time-consuming operation when computer memory was limited, but this is no longer the case.

Examples of the use of frequency-domain filtering include de Souza Filho *et al.* (1996), who describe a method to remove noise in JERS-1 imagery. Lei *et al.* (1996) also use frequency-domain filtering methods to clean up MOMS-02 images. Aeromagnetic data is analysed using frequency-domain techniques by Hussein *et al.* (1996). Gonzales and Woods (1992) and Pitas (1993) provide more detailed accounts of frequency domain filtering than the one presented here, though the level of mathematics required to understand their presentations is rather higher than that used here.

7.6 SUMMARY

Filtering of digital images is used to remove, reduce or amplify specific frequency components of an image. The most commonly used filters operate in the spatial domain and can be divided into low-pass or smoothing filters and high-pass or sharpening filters. Uses of smoothing filters include the suppression of noise and other unwanted effects, such as the banding phenomenon, which affects some Landsat ETM+ images. Sharpening filters are used to improve the visual interpretability of the image by, for example, de-blurring the signal. Edge and line detection is seen as an extension of the technique of image sharpening. Filtering in the frequency domain is achieved via the application of the principles of the Fourier transform, discussed in section 6.6. While these methods are inherently more flexible than are spatial domain filters the computational cost of applying them is considerable, and they are often understood less intuitively. Recent developments in computer hardware, especially random access memory and processor speed, mean that frequency-domain methods may become more popular.

8

Classification

'What is or is not a cow is for the public to decide.'

(L. Wittgenstein)

8.1 INTRODUCTION

This chapter is written with two audiences in mind. The first of these consists of undergraduates following second and third year courses in remote sensing and GIS who want a gentle, non-mathematical introduction to the ideas behind pattern recognition and classification. They will find that the first few sections can accommodate their needs. The remainder of the chapter is intended for a more advanced audience. The fact that the chapter presents a progressive view of the subject should encourage the more advanced reader to 'brush up' on his or her knowledge of the basic geometrical ideas underlying the topic, while at the same time encouraging the less advanced reader to absorb some of the more intricate material that is normally presented at Masters level. Compared with other chapters, little has been added to Chapter 8 since the second edition was published, largely because this book is not intended to be a research monograph. Readers requiring a more sophisticated approach should consult Landgrebe (2003) and Tso and Mather (2001).

The process of classification consists of two stages. The first is the recognition of categories of real-world objects. In the context of remote sensing of the land surface these categories could include, for example, woodlands, water bodies, grassland and other land cover types, depending on the geographical scale and nature of the study. The second stage in the classification process is the labelling of the entities (normally pixels) to be classified. In digital image classification these labels are numerical, so that a pixel that is recognised as belonging to the class 'water' may be given the label '1', 'woodland' may be labelled '2', and so on. The process of image classification requires the user to perform the following steps:

(i) determine *a priori* the number and nature of the categories in terms of which the land cover is to be described, and
(ii) assign numerical labels to the pixels on the basis of their properties using a decision-making procedure, usually termed a classification rule or a decision rule.

Sometimes these steps are called *classification* and *identification* (or *labelling*), respectively.

In contrast, the process of *clustering* does not require the definition of a set of categories in terms of which the land surface is to be described. Clustering is a kind of exploratory procedure, the aim of which is to determine the number (but not initially the identity) of distinct land cover categories present in the area covered by the image, and to allocate pixels to these categories. Identification of the clusters or categories in terms of the nature of the land cover types is a separate stage that follows the clustering procedure. Several clusters may correspond to a single land-cover type. Methods of relating the results of clustering to real-world categories are described by Lark (1995).

These two approaches to pixel labelling are known in the remote sensing literature as *supervised* and *unsupervised* classification procedures, respectively. They can be used to segment an image into regions with similar attributes. Although land cover classification is used above as an example, similar procedures can be applied to clouds, water bodies and other objects present in the image. In all cases, however, the properties of the pixel to be classified are used to label that pixel. In the simplest case, a pixel is characterised by a vector whose elements are its grey levels in each spectral band. This vector represents the spectral properties of that pixel.

A set of greyscale values for a single pixel measured in a number of spectral bands is known as a *pattern*. The spectral bands (such as the seven Landsat ETM+ bands) or other, derived, properties of the pixel (such as context and texture, which are described in later sections of this chapter) that define the pattern are called *features*. The classification process may also include features such as land surface elevation or soil type that are not derived from the image. A pattern is thus a set of measurements on the chosen features for the individual that is to be classified. The classification process may therefore be considered as a form of pattern recognition, that is, the identification of the pattern associated with each pixel position in an image in terms of the characteristics of the objects or materials

Computer Processing of Remotely-Sensed Images: An Introduction, Third Edition. Paul M. Mather.
 ISBNs: 0-470-84918-5 (HB); 0-470-84919-3 (PB)

that are present at the corresponding point on the Earth's surface.

Pattern recognition methods have found widespread use in fields other than Earth observation by remote sensing; for example, military applications include the identification of approaching aircraft and the detection of targets for cruise missiles. Robot or computer vision involves the use of mathematical descriptions of objects 'seen' by a television camera representing the robot eye and the comparison of these mathematical descriptions with patterns describing objects in the real world. In every case, the crucial steps are: (i) selection of a set of features which best describe the pattern and (ii) choice of a suitable method for the comparison of the pattern describing the object to be classified and the target patterns. In remote sensing applications it is usual to include a third stage, that of assessing the degree of accuracy of the allocation process.

A geometrical model of the classification or pattern recognition process is often helpful in understanding the procedures involved; this topic is dealt with in section 8.2. The more common methods of unsupervised and supervised classification are covered in sections 8.3 and 8.4. Supervised methods include those based on statistical concepts and those based on artificial neural networks. The methods described in these sections have generally been used on spectral data alone (that is, on the individual vectors of pixel values). This approach is called 'per-point' or 'per-pixel' classification based on spectral data. The addition of features that are derived from the image data has been shown to improve the classification in many cases. *Texture* is a measure of the homogeneity of the neighbourhood of a pixel, and is widely used in the interpretation of aerial photographs. Objects on such photographs are recognised visually not solely by their greyscale value (tone) alone but also by the variability of the tonal patterns in the region or *neighbourhood* that surrounds them. Texture is described in section 8.7.1.

Visual analysis of a photographic image often involves assessment of the context of an object as well as its tone and texture. *Context* is the relationship of an object to other, nearby, objects. Some objects are not expected to occur in a certain context; for example, jungles are not observed in Polar regions in today's climatic conditions. Conversely glacier ice is unlikely to be widespread in southern Algeria within the next few years. In the same vein, a pixel labelled 'wheat' may be judged to be incorrectly identified if it is surrounded by pixels labelled 'snow'. The decision regarding the acceptability of the label might be made in terms of the pixel's context rather than on the basis of its spectral reflectance values alone. Contextual methods are not yet in widespread use, though they are the subject of on-going research. They are described in section 8.8.

The number of spectral bands used by satellite and airborne sensors ranges from the single band of the SPOT HRV in panchromatic to several hundred bands provided by imaging spectrometers (section 9.3). The methods considered in this chapter are, however, most effective when applied to multispectral image data in which the number of spectral bands is less than 12 or so. The addition of other 'bands' or features such as texture descriptors or external data such as land surface elevation or slope derived from a digital elevation model can increase the number of features available for classification. The effect of increasing the number of features on which a classification procedure is based is to increase the computing time requirements but not necessarily the accuracy of the classification. Some form of feature selection process to allow a trade-off between classification accuracy and the number of features is therefore desirable (section 8.9). The assessment of the accuracy of a thematic map produced from remotely-sensed data is considered in section 8.10.

8.2 GEOMETRICAL BASIS OF CLASSIFICATION

One of the easiest ways to perceive the distribution of values measured on two features is to plot one feature against the other. Figure 8.1 is a plot of catchment area against stream discharge for a hypothetical set of river basins. Visual inspection is sufficient to show that there are two basic types of river basin. The first type has a small catchment area and a low discharge whereas the second type has a large area and a high discharge. This example might appear trivial

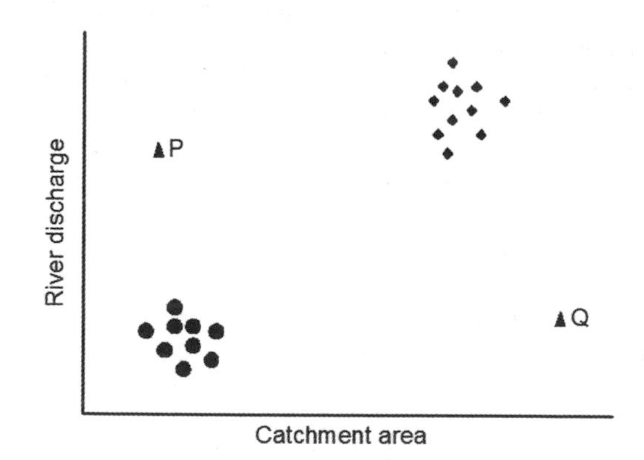

Figure 8.1 Plot of catchment (watershed) area against river discharge at mouth of catchment. Two distinct groups of river catchments can be seen – small catchments with low river discharge and large catchments with high river discharge. It is difficult to say to which of these groups the catchments represented by points P and Q belong.

but it demonstrates two fundamental ideas. The first is the representation of the selected features of the objects of interest (in this case the catchment area and discharge) by the axes of a Euclidean space (termed 'feature space'), and the second is the use of measurements of distance (or, conversely, closeness) in this Euclidean space to measure the resemblance of pairs of points (representing river basins) as the basis of decisions to classify particular river basins as large area/high discharge or small basin/low discharge. The axes of the graph in Figure 8.1 are the x, y axes of a Cartesian coordinate system. They are orthogonal (at right angles) and define a two-dimensional Euclidean space. Variations in basin area are shown by changes in position along the x-axis and variations in river discharge are shown by position along the y-axis of this space. Thus, the position of a point in this two-dimensional space is directly related to the magnitude of the values of the two features (area and discharge) measured on the particular drainage basin represented by that point.

The eye and brain combine to provide what is sometimes disparagingly called a 'visual' interpretation of a pattern of points such as that depicted in Figure 8.1. If we analyse what the eye/brain combination does when faced with a distribution such as that shown in Figure 8.1 we realise that a 'visual' interpretation is not necessarily a simple one, though it might be intuitive. The presence of two clusters of points is recognised by the existence of two regions of feature space that have a relatively dense distribution of points, with more or less empty regions between them. A point is seen as being in cluster 1 if it is closer to the centre of cluster 1 than it is to the centre of cluster 2. Distance in feature space is being used as a measure of similarity (more correctly 'dissimilarity' as the greater the inter-point distance the less the similarity). Points such as those labelled P and Q in Figure 8.1 are not allocated to either cluster, as their distance from the centres of the two clusters is too great. We can also visually recognise the compactness of a cluster using the degree of scatter of points (representing members of the cluster) around the cluster centre. We can also estimate the degree of separation of the two clusters by looking at the distance between their centres and the scatter of points around those centres. It seems as though a visual estimate of distance (closeness and separation) in a two-dimensional Euclidean space is used to make sense of the distribution of points shown in the diagram. However, we must be careful to note that the scale on which the numerical values are expressed is very important. If the values of the y-coordinates of the points in Figure 8.1 were to be multiplied or divided by a scaling factor, then our visual interpretation of the inter-point relationships would be affected. If we wished to generalise, we could draw a line in the space between the two clusters to represent the boundary between the two kinds of river basin. This line is called a *decision boundary*.

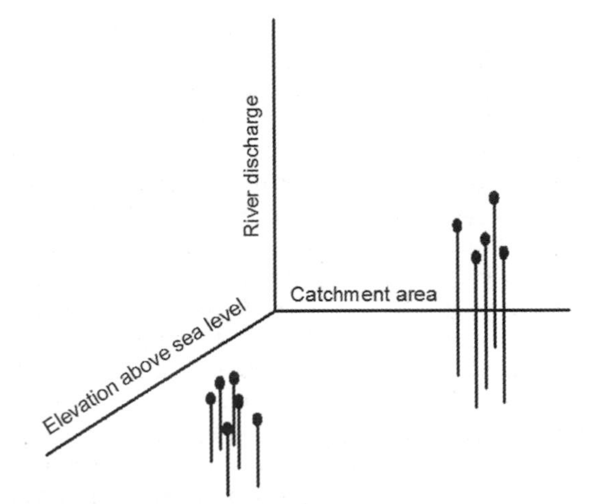

Figure 8.2 Plot of catchment (watershed) area, river discharge and elevation above sea level for a sample of drainage basins. Two groups of drainage basins are identifiable in this three-dimensional feature space.

The same concepts – the association of a feature or characteristic of an object with one axis of a Euclidean space and the use of inter-point distance as the basis of a decision rule – can easily be extended to three dimensions. Figure 8.2 shows the same hypothetical set of river basins, but this time they are represented in terms of elevation above sea level as well as area and discharge. Two groupings are evident as before, though it is now clear that the small basins with low discharge are located at higher altitudes than the large basins with high discharge. Again, the distance of each point from the centres of the two clouds can be used as the basis of an allocation or decision rule but, in this three-dimensional case, the decision boundary is a plane rather than a line.

Many people seem to find difficulty in extending the concept of inter-point distance to situations in which the objects of interest have more than three characteristics. There is no need to try to visualise what a four, five or even seven-dimensional version of Figure 8.2 would look like; just consider how straight-line distance is measured in one-, two- and three-dimensional Euclidean spaces in which x, y and z represent the axes:

$$d_{12} = \sqrt{(x_1 - x_2)^2}$$
$$d_{12} = \sqrt{(x_1 - x_2)^2 + (y_1 - y_2)^2}$$
$$d_{12} = \sqrt{(x_1 - x_2)^2 + (y_1 - y_2)^2 + (z_1 - z_2)^2}$$

The squared differences on each axis are added and the square root of the sum is the Euclidean distance from point 1

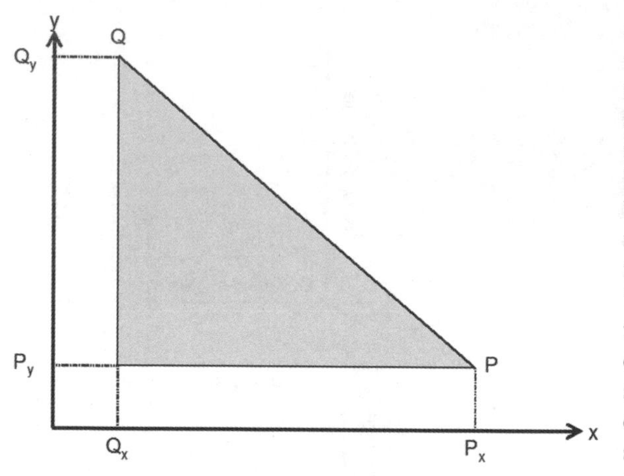

Figure 8.3 P and Q represent two objects to be compared, such as two trees. We have measurements of height and diameter of the two trees, and these features are represented by axes x and y, respectively. Pythagoras's Theorem is used to calculate the distance PQ in a two-dimensional feature space using the formula $PQ = \sqrt{(P_x - Q_x)^2 + (P_y - Q_y)^2}$.

to point 2. This is a simple application of the theorem of Pythagoras (Figure 8.3). If we replace the terms x_j, y_j and z_j (where j is an index denoting the individual point) by a single term x_{ij}, where i is the axis number and j the identification of the particular point, then the three expressions above can be seen to be particular instances of the general case, in which the distance from point a to point b is:

$$d_{ab} = \sqrt{\sum_{i=1}^{p} (x_{ia} - x_{ib})^2}$$

where d_{ab} is the Euclidean distance between point a and point b measured on p axes or features. There is no reason why p should not be any positive integer value – the algebraic formula will work equally well for $p = 4$ as for $p = 2$ despite the fact that most people cannot visualise the $p > 3$ case. The geometrical model that has been introduced in this section is thus useful for the appreciation of two of the fundamental ideas underlying the procedure of automated classification, but the algebraic equivalent is preferable in real applications because (i) it can be extended to beyond three dimensions and (ii) the algebraic formulae are capable of being used in computer programs.

It may help to make things clearer if an example relating to a remote sensing application is given at this point. The discussion of the spectral response of Earth-surface materials in section 1.3.2 shows that deep, clear water bodies have a very low reflectance in the near infrared waveband, and their reflectance in the visible red waveband is not much higher. Vigorous vegetation, on the other hand, reflects strongly in the near-infrared waveband whereas its reflectance in the visible red band is relatively low. The red and near infrared wavebands might therefore be selected as the features on which the classification is to be based. Estimates can be made of the pixel greyscale values in each spectral band for sample areas on the image that can be identified *a priori* as 'water', 'cloud top' and 'vigorous vegetation' on the basis of observations made in the field, or from maps or aerial photographs, and these estimates used to fix the mean position of the points representing these three catagories in Figure 8.4. The two axes of the figure represent near-infrared and red reflectance, respectively, and the mean position of each type is found by finding the average red reflectance (y coordinate) and near-infrared reflectance (x coordinate) of the sample values for each of the two categories. Fixing the number and position of the large circles in Figure 8.4 represents the first stage in the strategy outlined at the start of this section, namely, the building of a classification.

Step two is the labelling of unknown objects (we could use verbal labels, as we have done up to now, or we could use numerical labels such as '1', '2' and '3'. Remember that these numbers are merely labels. The points labelled *a–f* in Figure 8.4 represent unclassified pixels. We might choose a decision rule such as 'points will be labelled as members of the class whose centre is closest in feature space to the

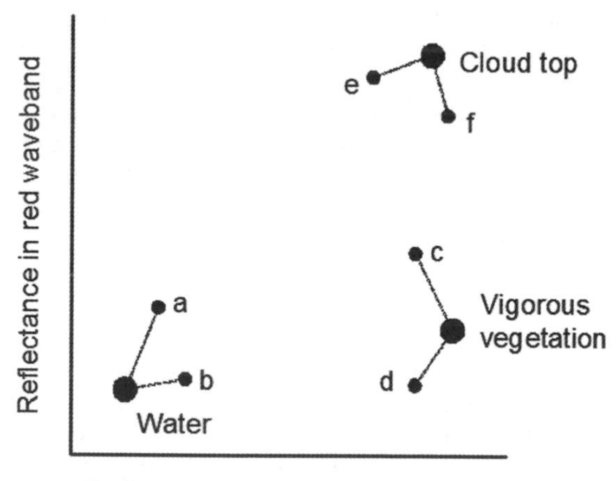

Figure 8.4 The positions of points representing the average reflectance of vigorous vegetation, water and cloud tops can be estimated from knowledge of their spectral reflectance characteristics (Chapter 1). Points a–f can be allocated to one of these three categories using the criterion of 'minimum distance', i.e. maximum similarity.

point concerned'. The distance formula given above could then be used on the points taken one at a time to give the Euclidean straight-line distance from that point (representing a pattern associated with a particular pixel) to each of the centres. Those points that are closer to the mean value for vigorous vegetation are labelled '1' while those closer to the central water point are labelled '2'. Finally, points nearest the cloud top point are labelled '3'. If this procedure is applied to a two-band image, as shown in Figure 8.4, the end product is a matrix of the same dimensions as the image being classified. The elements of this new matrix are numerical pixel labels, which in this example are either 1's, 2's or 3's. If the colour green is associated with the value '1', the colour blue with the value '2' and the colour white with the label '3' then a colour-coded thematic map of the image area would result, in which water would be blue, vigorous vegetation would be green, and cloud tops would appear as white, assuming, of course, that the classification procedure was a reliable one. The position of the decision boundaries is given by the set of points that are equidistant from all three class centres.

It will be shown later that the decision rule used in this example ('allocate an unknown pixel to the closest class centroid') is not the only one that can be applied. However, the process of image classification can be understood more clearly if the geometrical basis of the example is clearly understood.

8.3 UNSUPERVISED CLASSIFICATION

It is sometimes the case that insufficient observational or documentary evidence of the nature of the land-cover types is available for the geographical area covered by a remotely-sensed image. In these circumstances, it is not possible to estimate the mean centres of the classes, as described above. Even the number of such classes might be unknown. In this situation we can only 'fish' in the pond of data and hope to come up with a suitable catch. In effect, the automatic classification procedure is left largely, but not entirely, to its own devices – hence the term 'unsupervised clustering'. The relationship between the labels allocated by the classifier to the pixels making up the multispectral image and the land-cover types existing in the area covered by the image is determined after the unsupervised classification has been carried out. Identification of the spectral classes picked out by the unsupervised classifier in terms of information classes existing on the ground is achieved using whatever information is available to the analyst. The term 'exploratory' might be used in preference to 'unsupervised' because a second situation in which this type of analysis might be used can be envisaged. The analyst may well have considerable ground data at his or her disposal but may not be certain (a) whether the spectral classes he or she proposes to use can, in fact, be discriminated given the data available, and/or (b) whether the proposed spectral classes are 'pure' or 'mixed'. As we see in section 8.4, some methods of supervised classification require that the frequency distribution of points belonging to a single spectral class in the p-dimensional feature space has a single mode or peak. In either case, exploratory or unsupervised methods could be used to provide answers to these questions.

8.3.1 The k-means algorithm

An exploratory classification algorithm should require little, if any, user interaction. The workings of such a technique, called the k-means clustering algorithm, are now described by means of an example. Figure 8.5 shows two well-separated groups of points in a two-dimensional feature space. The members of each group are drawn from separate bivariate-normal distributions. It is assumed that we know that there are two groups of points but that we do not know the positions of the centres of the groups in the feature space. Points '1_0' and '2_0' represent a first guess at these positions. The 'shortest distance to centre' decision rule, as described earlier, is used to label each unknown

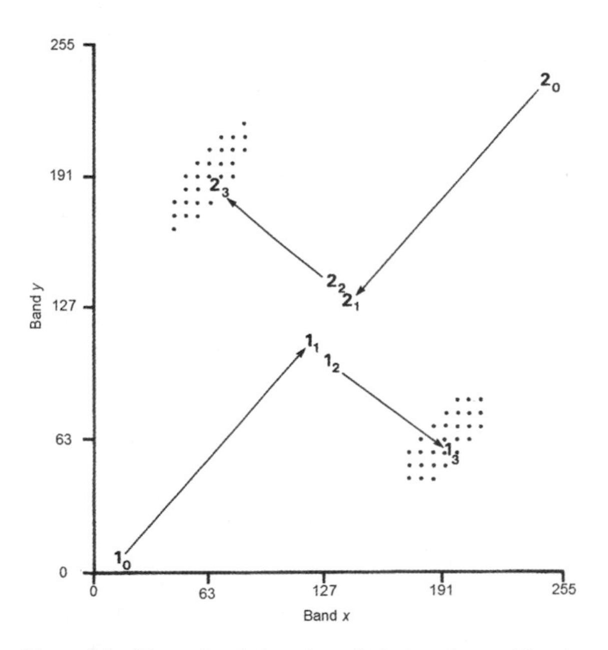

Figure 8.5 Illustrating the iterative calculation of centroid positions for two well-separated groups of points in a two-dimensional feature space defined by axes labelled Band x and Band y. Points 1_0 and 2_0 migrate in three moves from their initial random starting positions to the centres of the two clouds of points that represent the two classers of pixels.

point (represented by a dot in the figure) with a '1' or a '2' depending on the relative Euclidean distance of the point from the initial cluster centres, labelled '1_0' and '2_0'. Thus, the (squared) Euclidean distances to cluster centres 1 and 2 (d_{q1}^2 and d_{q2}^2) are computed for each point q, and q is allocated the label '1' if d_{q1}^2 is less than d_{q2}^2 or the label '2' if d_{q2}^2 is less than d_{q1}^2. If the two squared distances are equal, then the point is arbitrarily allocated the label '1'.

At the end of this labelling sequence the mean of the values of all points labelled '1' is computed for each of the axes of the feature space, and the same is done for all points labelled '2' to give the coordinates in the feature space of the centroids of the two groups of points. These new centroids are shown in the diagram as '1_1' and '2_1'. The points are re-labelled again using the shortest-distance-to-mean decision rule, based this time on the new positions of the centroids. Again, the position of the centroid of the points labelled '1' at this second iteration is computed and is shown as '1_2'. The centroid of the set of points labelled '2' is found in a similar fashion and is shown as '2_2'. Distances from all points to these new centres are calculated and another pair of new centroids are found ('1_3' and '2_3'). These centroids are now at the centres of the two groups of points that were artificially generated for this example, and re-labelling of the points does not cause any change in the position of the centroids, hence this position is taken as the final one.

To show that the technique still works even when the two groups of points are not so well separated, as in Figure 8.5, a second pair of groups of points can be generated. This time the coordinates of the points in a two-dimensional feature space are computed by adding random amounts to a pre-selected pair of centre points to give the distribution shown in Figure 8.6. The start positions of the migrating centroids are selected randomly and are shown on the figure as '1_0' and '2_0' respectively. The same re-labelling and recalculation process as that used in the previous example is carried out and the centroids again migrate towards the true centres of the point sets, as shown in Figure 8.6. However, this time the decision boundary is not so clear-cut and there may be some doubt about the class membership (label) of points that are close to the decision boundary.

Since the re-labelling procedure involves only the rank orders of the distances between point and centroids, the squared Euclidean distances can be used, for the squares of a set of distance measures have the same rank order as the original distances. Also, it follows from the fact that the squared Euclidean distances are computed algebraically that the feature space can be multi-dimensional. The procedures in the multi-dimensional case involve only the additional summations of the squared differences on the feature

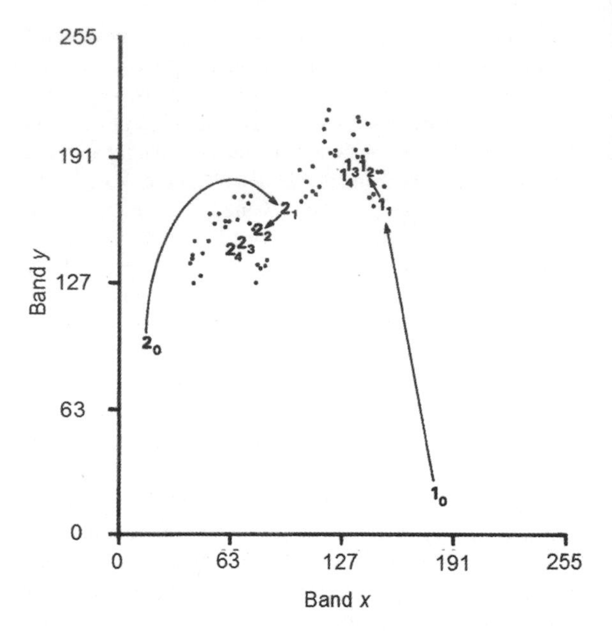

Figure 8.6 Iterative calculation of centroid positions for two diffuse clusters. See text for discussion.

axes as shown in section 8.2; no other change is needed. Also note that the user can supply the starting centroid values in the form of mean values for each cluster for each feature. If this starting procedure is used then the method can no longer be described as 'unsupervised'.

8.3.2 ISODATA

In the examples used so far it has been assumed that the number of clusters of points is known in advance. More elaborate schemes are needed if this is not the case. The basic assumption on which these schemes are based is that the clusters present in the data are 'compact' (that is, the points associated with each cluster are tightly grouped around the cluster centre, and thus occupy a spheroidal region of feature space). A measure of the compactness of a cluster can be taken as the set of standard deviations for the cluster measured separately for each feature (Figure 8.7). If any of these feature standard deviations for a particular cluster is larger than a user-specified value then that cluster is considered to be elongated in the direction of the axis representing that feature.

A second assumption is that the clusters are well separated in that their inter-centre distances are greater than a preselected threshold. If the feature-space coordinates of a trial number of cluster centres are generated randomly (call the number of centres k_0) then the closest-distance-to-centre

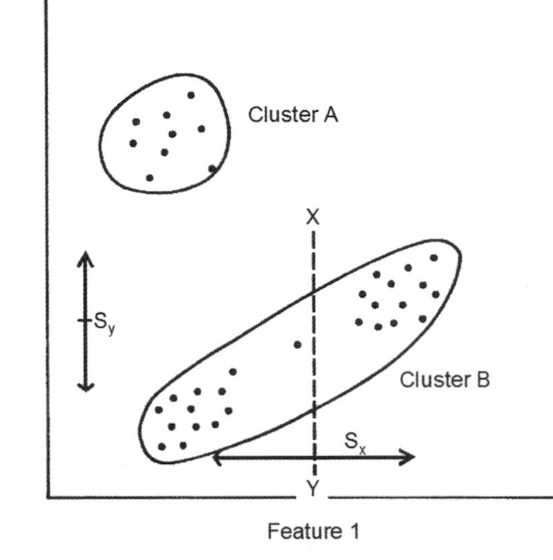

Figure 8.7 Compact (A) and elongated cluster (B) in a two-dimensional feature space. Cluster B has standard deviations s_x and s_y on features 1 and 2, respectively. Since s_x is larger than a user-specified threshold value, cluster B is split along the line XY.

decision rule can be used to label the pixels which, as before, are represented by points in feature space. Once the pixels have been labelled then (a) the standard deviation for each feature axis is computed for each of the non-null k_0 clusters and (b) the Euclidean distances between the k_0 cluster centres are found. Any cluster that has one or more 'large' standard deviations is split in half along a line perpendicular to the feature axis concerned (Figure 8.7) while any clusters that are closer together than a second user-supplied threshold (in terms of their inter-centre distance) are amalgamated. Application of this split-and-merge routine results in a number k_1 of new cluster centre positions, and the pixels are re-labelled with respect to these k_1 centroids. The split and merge function is again applied, and a new number k_2 of centres is found. This process iterates until no clusters are split or merged and no pixels change cluster. This gives the number k_p of clusters and the positions of their centroids in the feature space. At each cycle, any cluster centres that are associated with less than a pre-specified number of pixels are eliminated. The corresponding pixels are either ignored as unclassifiable in subsequent iterations or else are put back into the pool for re-labelling at the next iteration.

This split and merge procedure forms the basis of the ISODATA algorithm (ISODATA is an acronym derived from *I*terative *S*elf-*O*rganising *D*ata *A*nalysis *T*echniques, with a terminal '*A*' added for aesthetic reasons). Note that some commercial software packages use a basic k-means unsupervised clustering (section 8.3.1) but call it 'ISODATA'. The ISODATA algorithm can be surprisingly voracious in terms of computer time if the data are not cleanly structured (i.e. do not possess clearly separated and spheroidal clusters). Unless precautions are taken it can easily enter an endless loop when clusters that are split at iteration i are merged again at iteration $i + 1$, then are split at iteration $i + 2$. Little general guidance can be given on the choice of the number of initial cluster centres k_0 or on the choice of the elongation and closeness threshold values to be used. It is often sensible to experiment with small subsets of the image to be classified to get a 'feel' for the data. This algorithm has been in use for many years. A full description is given by Tou and Gonzales (1974) while the standard reference is Duda and Hart (1973). A more recent account is Bow (1992), who provides a flow chart of the procedure as well as a lengthy example.

The ISODATA procedure as described above is implemented in MIPS. The coordinates of the initial centres in feature space are generated by randomly selecting (x, y) coordinate pairs and choosing to use the vector of pixel values corresponding to the pixel in row y, column x as a starting centre. This strategy helps to prevent the generation of cluster centres that are located away from any pixel points in feature space. Input includes the 'desired' number of clusters, which is usually fewer than the initial number of clusters, and the maximum number of clusters, together with the splitting and merging threshold distances. The stopping criteria are (i) when the average inter-centre Euclidean distance falls below a user-specified threshold or (ii) when the change in average inter-centre Euclidean distance between iteration i and iteration $i-1$ is less than a user-specified amount. Alternatively, a standard set of defaults values can be selected, though it should be said that these defaults are really no more than guesses. The MIPS implementation of ISODATA outputs the classified image to the screen at each iteration, and the cluster statistics are listed on the lower toolbar, which should be kept visible.

Example 8.1 provides some insights into the workings of the ISODATA procedure as implemented in MIPS.

8.3.3 A modified k-means algorithm

A second, somewhat less complicated, method of estimating the number of separable clusters in a data set involves a modification of the k-means approach that is outlined in section 8.3.1 to allow merging of clusters. An overestimate of the expected number of cluster centres (k_{max}) is provided by the user, and pixels are labelled as described in section 8.3.1, using the closest-distance-to-centre decision rule. Once the pixels have been labelled the centroids of the clusters are calculated and the re-labelling procedure is employed to find stable positions for the centroids. Once such a position has been located, a measure of the compactness of each cluster i is found by summing the squared Euclidean distances from the pixels belonging to cluster i to the centroid of cluster i. The square root of this sum divided by the number of points in the cluster gives the root mean squared deviation for that cluster. It is now necessary to find the pair of clusters that can be combined so as (a) to reduce the number of cluster centres by one, and (b) at the same time cause the least increase in the overall root mean square deviation. This is done by computing the quantity P from:

$$P = \frac{n_i n_j}{n_i + n_j} \sum_{k=1}^{p} (y_{ik} - y_{jk})^2 \qquad (i = 2, k; j = 1, i - 1)$$

for every pair of cluster centres (y_i and y_j) where p is the number of dimensions in the feature space and n_i is the number of pixels assigned to cluster i. If clusters $i = r$ and $j = s$ give the lowest value of P then the centroids of clusters r and s are combined by a weighted average procedure, the weights being proportional to the numbers of pixels in clusters r and s. If the number of clusters is still greater than or equal to a user-supplied minimum value $k_{\min}$, then the re-labelling procedure is then employed to reallocate the pixels to the reduced number of centres and the overall root mean square deviation is computed. The procedure is repeated for every integral value of k (the number of clusters) between $k_{\max}$ and $k_{\min}$ or until the analyst terminates the procedure after visually inspecting the classified image. As with the ISODATA algorithm, empty clusters can be thrown away at each iteration. Mather (1976) provides a FORTRAN program to implement this procedure.

The result of an unsupervised classification is a set of labelled pixels, the labels being the numerical identifiers of the classes. The label values run from 1 to the number of classes (k) picked out by the procedure. The class numbered zero (or $k + 1$) can be used as the label for uncategorised pixels. The image made up of the labels of the pixels is displayed by assigning a colour or a grey tone to each label. From a study of the geographical location of the pixels in each class, an attempt is normally made to relate the spectral classes (groups of similar pixels) to corresponding information class (categories of ground cover). Alternatively, a method of hierarchical classification can be used to produce a linkage tree or dendrogram from the centroids of the unsupervised classes, and this linkage tree can be used to determine which spectral classes might best be combined (Example 8.1). The relationship between spectral classes and information classes is likely to be tenuous unless external information can be used for, as noted earlier, unsupervised techniques of classification are used when little or no detailed information exists concerning the distribution of ground cover types. An initial unsupervised classification can, however, be used as a preliminary step in refining knowledge of the spectral classes present in the image so that a subsequent supervised classification can be carried out more efficiently. The classes identified by the unsupervised analysis could, for example, form the basis for the selection of training samples for use in a supervised technique of classification (section 8.4). General references covering the material presented above are Anderberg (1973) and Hartigan (1975). Example 8.1 gives some practical advice on cluster labelling.

EXAMPLE 8.1: ISODATA UNSUPERVISED CLASSIFICATION

The aim of this example is to demonstrate the operation of the ISODATA unsupervised classifier. An outline of the workings of this algorithm is provided in the main text.

We begin by selecting *Classify|Isodata* from the main menu, and then choosing the INF file *missis.inf* (provided on the CD and, if MIPS was properly installed, copied to your hard disc). Next, specify that you wish to base the classification on bands 1–5 and 7 of this Landsat TM sub-image. The sub-image size is quite small (512 × 512 pixels) so select the whole of the sub-image for analysis. Now we have to decide whether or not to accept the default options for the user-specified parameters (Example 8.1 Table 2). As we do not know yet whether these parameter values are suitable, and given that the sub-image is not too big, select 'Use Defaults', just to see what happens. Do not save the starting centres (this option is available for users who wish to restart the procedure using the same starting cluster centres but with different parameter values, as noted below). Instead, select 'Generate Randomly' so that the initial cluster centres are selected from the image using random x and y pixel coordinates. The next dialogue box asks if you want to save the randomly generated cluster centres. Click *Don't Save*. Note that if you repeat the ISODATA classification on the same image set using the 'Generate Randomly' option, then the final results may well differ, as the final solution depends to a considerable extent on the starting configuration. This is one of the less welcome features of iterative optimisation algorithms.

The Mississippi image now appears on the screen in colour-coded form, with a colour table. Details of the number of clusters and the overall pixel-to-centre distance measure are listed on the lower toolbar. You can continue for a further iteration or quit at this point. If you continue, you will see the colours on the classified image change

Example 8.1 Figure 1 Greyscale representation of the output from the ISODATA procedure applied to the Landsat TM image set referenced by the file *missis.inf*. The sub-image size is 512 × 512 pixels. See text for elaboration.

as clusters are merged or split. Eventually, the change in the inter-cluster squared Euclidean distance will fall below the threshold, and the iterative process will terminate. Alternatively, the default number of iterations (35) will be executed and the process will again terminate. The result should be similar to the image shown in Example 8.1 Figure 1.

You are then offered the option of performing a hierarchical cluster analysis on the ISODATA results (Example 8.1 Figure 2). A dendrogram is a two-dimensional hierarchical representation of the distance (dissimilarity) relationships among a set of objects, which in this case are the cluster centres. You can cut the dendrogram at any point along the x-axis. A vertical line at the cutting point at, for example, a distance of 9 units in Example 8.1 Figure 2 represents a four-cluster solution, with IOSDATA cluster centres numbered 1, 7, 3, 11, 5, 8 and 12 forming the first composite cluster. The second cluster groups together centres 9 and 14, and the third amalgamates cluster centres 2, 15 and 10. The final grouping consists of ISODATA cluster centres 4, 13 and 6. These relationships are useful in understanding the nature of the ISODATA results, for they show the structure present at different levels of dissimilarity (x-axis). The groupings are also used in the reclassification process, which is described next.

Reclassification is simply the allocation of the same colour code to two or more ISODATA classes. When you select this option you can type the number of the class you have chosen then use the mouse to left-click on a colour in the palette. You can choose the same colour for several classes – for instance, you may decide to colour classes 4, 13 and 6 in blue. These three ISODATA classes have been grouped in the dendrogram, as explained in the preceding paragraph. If you are not satisfied with the result, you can repeat the reclassification exercise.

The final decision to be made in the ISODATA process is how to save the classified image. You can use the *Utilities|Copy to Clipboard|Copy Image* option to place the image on the Windows clipboard, or the *File|Export Image Set* to save the result as a TIF or bitmap image, or you can use the final ISODATA option, which is to save the image (together with an associated INF file) as a set of labels. These labels are simply the ISODATA class identifiers of the pixels. The same class identifiers are used in the reclassification process. You may wish to save the label image so that you can use a common colour scheme on the results of ISODATA classifications of several images, or the output from ISODATA for a single image set but using different parameters. Note that the bitmap and TIF representations save the RGB colours associated with each pixel, not the class labels.

Since the ISODATA procedure is started by picking random pixels to act as the initial centres, it is impossible to say what exactly will happen when you run the ISODATA module. The final result should look something like Example 8.1 Figure 1 and the dendogram, showing the relationships between the ISODATA classes, looks like Example 8.1 Figure 2. Remember that the ISODATA classes are identified solely on the basis of spectral characteristics (modified, perhaps, by topographic effects) and, as yet, they have no 'real-world' interpretation. Using a combination of subject-matter reasoning (i.e., inferring the nature of the spectral classes from geographical location and spectral reflectance properties), and sources of information such as existing maps, it may be possible to assign labels to some of the classes in the classified image. For example, the dark, wide linear feature running from the top right to the bottom left of the image shown in Example 8.1 Figure 1 is probably the Mississippi River.

The dendrogram (Example 8.1 Figure 2), which shows the similarities between the spectral classes at different levels of generalisation, can also be used to aid interpretation. Starting at the right-hand side of the dendrogram, we see that at a Euclidean distance of about 17.5 all of the pixels belong to a single class (the Earth's surface). At a dissimilarity level of 17.5, the ISODATA classes split into two groups, one containing six classes and the other consisting of the remaining nine classes. By locating these six and nine classes in the image, we can again use inferential reasoning to label the spectral classes.

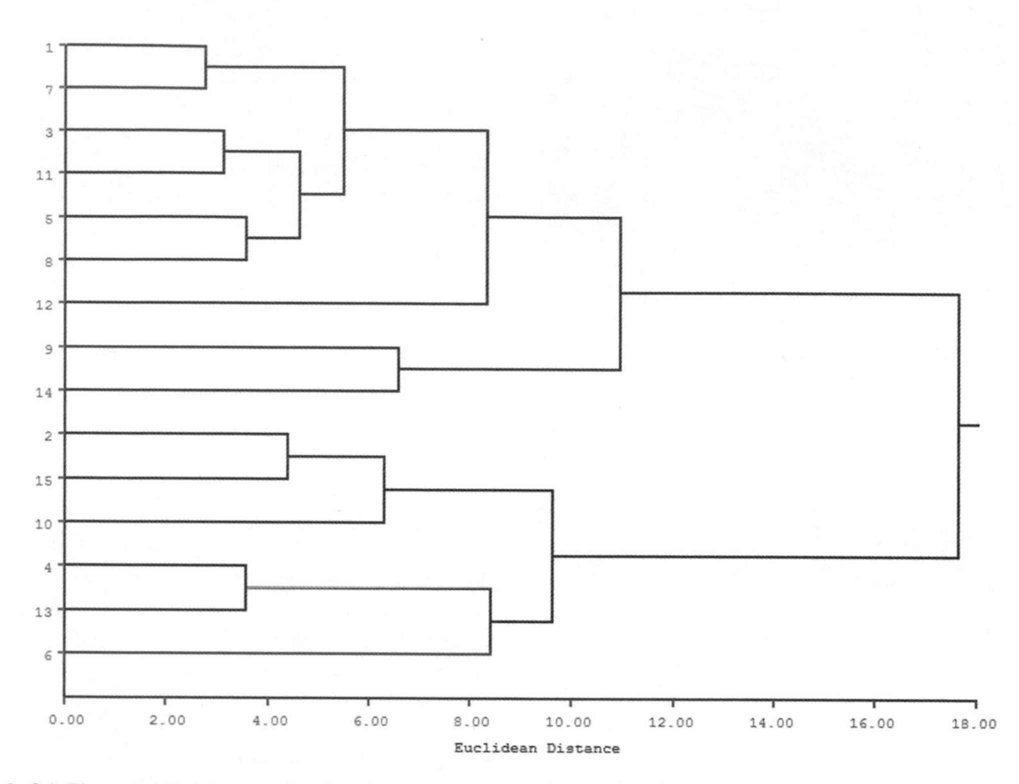

Example 8.1 Figure 1 Dendrogram showing the hierarchical dissimilarity relationships among the ISODATA cluster centres. The class labels associated with the cluster centres are shown on the y-axis. The Euclidean distance (x-axis) is a measure of dissimilarity (in that a value of 0.0 indicates perfect similarity). See text for discussion.

Example 8.1 Table 1 Summary of the output from the ISODATA unsupervised classification.

Cluster number	Number of pixels	TM band 1	TM band 2	TM band 3	TM band 4	TM band 5	TM band 7	Mean squared distance
1	15077	56.2	18.6	20.1	21.3	35.0	15.8	26.66
2	10883	56.8	19.3	20.5	16.5	18.0	8.1	32.19
3	30967	55.4	18.0	19.6	24.0	47.0	20.9	13.85
4	44676	61.8	23.9	27.1	13.9	3.1	1.5	9.54
5	7261	58.5	20.8	22.6	33.8	45.0	18.4	35.84
6	3579	70.6	29.6	36.7	22.8	5.1	2.3	56.94
7	13486	56.6	18.7	20.5	22.4	40.9	18.7	21.27
8	9547	61.8	22.1	26.0	26.9	43.9	20.4	48.45
9	23545	66.1	25.4	30.5	36.2	70.5	35.1	172.19
10	3199	64.2	24.8	29.3	22.4	20.6	9.2	78.39
11	26836	57.7	19.5	22.1	26.4	52.4	23.8	23.39
12	15853	59.6	21.9	23.0	43.1	52.4	21.1	58.89
13	6255	58.1	21.2	21.7	11.2	7.0	3.2	35.18
14	36024	62.3	22.6	26.6	29.7	58.6	28.8	44.14
15	14956	55.7	18.3	19.3	19.0	27.3	12.5	26.84

Example 8.1 Table 2 ISODATA parameters and their effects.

Parameter number	Parameter description	Default value	Action
1	Starting number of clusters.	20	Affects final number of clusters.
2	Desired number of clusters.	10	Should be half the value of parameter 1.
3	Maximum number of clusters.	50	Stops excessive splitting.
4	Minimum number of pixels in a cluster.	50	Kills off small clusters by declaring them to be 'dead'.
5	Exclusion distance.	200	Any pixels further than the exclusion distance from their nearest centre are declared to be unclassified (label 0). This parameter can be used to encourage more spherical clusters.
6	Closeness criterion.	30	Cluster centres closer than this can be merged. Decrease this value if merging is too voracious.
7	Elongation criterion.	16	Clusters that extend further than this criterion along one axis are split perpendicular to that axis. Increase this value if splitting is excessive.
8	Maximum number of iterations.	35	This is normally sufficient.
9	Maximum number of clusters that can be merged at one time.	2	Use this to increase or decrease the merging tendency.
10	Relative decrease in inter-cluster-centre distance.	1	Stops ISODATA if the decrease I the value of the inter-cluster-centre distance becomes less than this.
11	Absolute value of inter-cluster-centre distance.	5	Stops ISODATA if the value of the inter-cluster-centre distance becomes less than this.

MIPS outputs a lot of information to the log file, so that you can work out how the clusters are split and merged. The final results for this example are shown in Example 8.1 Table 1. Use the information in column two (number of pixels allocated to this class) to 'weed out' any small, inconsequential, clusters. Next, look at the shape of the spectrum of the mean or centroid of each of the remaining classes. A knowledge of the spectral reflectance characteristics of the major land surface cover types is useful. Finally, the last column gives a measure of the compactness of each cluster. A low value indicates a very homogeneous class, and a high value indicates the reverse. Look at class number four, which is the largest of the classes, with 44,676 of the 262,144 pixels comprising the image. It is very compact (the mean squared distance of 9.54 is calculated by measuring the mean Euclidean distance from each of the 44,676 pixels in the class to the centre point of the class, which is described by the values in columns 3–8. The centroid (mean) value on TM bands 1–5 and 7 shows very low values in the near-infrared region, with moderate values in the visible bands (1–3). Figure 1.1 shows a typical reflectance curve for water, which corresponds quite well to the profile of class number four. Class 13 is similar to class four, according to the dendrogram (Example 8.1 Figure 2), and its centroid profile is also typical of water. It may be concluded that classes four and 13 represent two different but related water classes, though class 13 contains only about 6,000 pixels.

Other classes are less compact than class four. Class nine in particular is very diffuse, though it contains more than 23,000 pixels. It is not easy to interpret the centroid values in columns 3–8 in terms of a specific cover type, but use of the *Reclassify* procedure in MIPS may give some indication of the spatial pattern of class nine, and the dendrogram may also provide some clues.

You can try running the program again, using default parameters. Each time, different pixels are selected to act as starting centres so the result is never the same twice in succession. Sometimes the algorithm spirals out of control and produces a single class. When you feel that you understand how the procedure works, you can experiment with the parameter values. If you do this, then store a set of starting centre coordinates so that you can re-use them each time. By doing this you can eliminate the effects of starting the clustering process from different points and you will therefore isolate the effects of the changes you have made to the default parameters. The main parameters are shown in Example 8.1 Table 2.

8.4 SUPERVISED CLASSIFICATION

Supervised classification methods are based on external knowledge of the area shown in the image. Unlike some of the unsupervised methods discussed in section 8.3 supervised methods require some input from the user before the chosen algorithm is applied. This input may be derived from fieldwork, air photo analysis, reports, or from the study of appropriate maps of the area of interest. In the main, supervised methods are implemented using either statistical or neural algorithms. Statistical algorithms use parameters derived from sample data in the form of training classes, such as the minimum and maximum values on the features, or the mean values of the individual clusters, or the mean and variance–covariance matrices for each of the classes. Neural methods do not rely on statistical information derived from the sample data but are trained on the sample data directly. This is an important characteristic of neural methods of pattern recognition, for these methods make no assumptions concerning the frequency distribution of the data. In contrast, statistical methods such as the maximum likelihood procedure are based on the assumption that the frequency distribution for each class is multivariate normal in form. Thus, statistical methods are said to be *parametric* (because they use statistical parameters derived from training data) whereas neural methods are *non-parametric*. The importance of this statement lies in the fact that additional non-remotely-sensed data such as slope angle or soil type can more easily be incorporated into a classification using a non-parametric method.

Since all methods of supervised classification use training data samples it is logical to consider the characterisation of training data in the next section.

8.4.1 Training samples

Supervised classification methods require prior knowledge of the number and, in the case of statistical classifiers, certain aspects of the statistical nature of the information classes with which the pixels making up an image are to be identified. The statistical characteristics of the classes that are to be estimated from the training sample pixels depend on which method of supervised classification is used. The simple parallelepiped method requires estimates of the extreme values on each feature for each class, while the k-means or centroid method needs estimates of the multivariate means of the classes. The most elaborate statistical method discussed in this book, the maximum likelihood algorithm, requires estimates of the mean vector and variance–covariance matrix of each class. Neural classifiers operate directly on the training data, but are strongly influenced by mis-identification of training samples as well as by the size of the training data sets. Mis-identification of an individual training sample pixel may not have much influence on a statistical classifier, but the impact on a neural classifier could be considerable. The material contained in this section must be interpreted in the light of whichever of these methods is used.

It is of crucial importance to ensure that the *a priori* knowledge of the number and statistical characteristics of the classes is reliable. The accuracy of a supervised classification analysis will depend upon two factors: (i) the representativeness of the estimates of both the number and the statistical nature of the information classes present in the image data and (ii) the degree of departure from the assumptions upon which the classification technique is based. These assumptions vary from one technique to another; in general, the most sophisticated techniques have the most stringent assumptions. These assumptions will be mentioned in the following sub-sections. In this section we concentrate on the estimation of statistical properties, in particular the mean and variance of each spectral band and the covariances of all pairs of spectral bands. These methods can be used to locate aberrant pixels that can then be eliminated or down-weighted.

The validity of statistical estimates depends upon two factors – the size and the representativeness of the sample. Sample size is not simply a matter of 'the bigger the better' for cost is, or should be, an important consideration. Sample size is related to the number of variables (spectral bands in this case) whose statistical properties are to be estimated, the number of those statistical properties, and the degree of variability present in the class. In the case of a single variable and the estimation of a single property (such as the mean or the variance) a sample size of 30 is usually held to be sufficient. For the multivariate case the size should be at least $30p$ pixels per class where p is the number of features (e.g., spectral bands), and preferably more. Dobbertin and Biging (1996) show that classification accuracy tends to improve as sample size increases. However, neural-based classifiers appear to work better than statistical classifiers for small training sets, though better results were achieved when training set size was proportional to class size and variability (Blamire, 1996; Foody *et al.*, 1995, Hepner *et al.*, 1990). Small training samples must be representative, however. If you inspect Table 1 in Example 8.1 you will see that some of these (spectral) classes are very variable and will require a greater number of training samples in order to be properly characterised. Other classes are much more compact so that a smaller number of training samples will be adequate to represent their characteristics.

Training samples are normally located by fieldwork or from air photograph or map interpretation, and their positions on the image found either by visual inspection or by carrying out a geometric correction on the image to be classified. It is not necessary to carry out the procedure of geometric transformation on the full image set to be classified, unless the resulting classified image is to be input to a GIS. All that is required is the set of transform equations that will convert a map coordinate pair to the corresponding image column and row coordinates (section 4.3). Using these equations, the location on the image of a training sample whose map coordinates are known is a relatively simple matter, provide that the geometric transform is accurate. If geometric correction is required, it is best carried out on a single-band classified image (in which the pixel 'values' are the labels of the classes to which the pixels have been allocated) rather than on the p images to be classified. Not only is this less demanding of computer resources, but it ensures that the radiometric (pixel) values are not distorted by any resampling procedure (Khan *et al.*, 1995). If external data are used in the classification (section 8.7.2) then geometric correction of image and non-image data to a common reference system is a necessary pre-requisite.

The minimum sample size specified in the preceding paragraphs is valid only if the individual members of the training sample are independent, as would be the case if balls were drawn randomly from a bag. Generally, however, adjacent pixels are not independent – if you were told that pixel a was identified as 'forest' you might be reasonably confident that its neighbour, pixel b, would also be a member of the class 'forest'. If a and b were statistically independent there would be an equal chance that b was a member of any other of the candidate classes, irrespective of the class to which a was allocated. The correlation between nearby points in an image is called *spatial autocorrelation*.

It follows that the number n of pixels in a sample is an over-estimate of the number of fully independent pieces of information in the sample if the pixels making up the training sample are autocorrelated, which may be the case if blocks of pixels are selected rather than scattered, individual pixels. The consequence of autocorrelation is that the use of the standard statistical formulae to estimate the means and variances of the features, and the correlations among the features, will give biased results. Correlations between spectral bands derived from spatially autocorrelated training data will, in fact, be underestimated and the accuracy of the classification will be reduced as a result. Campbell (1981) found that variance–covariance matrices (the unstandardised analogue of the correlation matrix) were considerably greater when computed from randomly selected pixels within a class rather than from contiguous blocks of pixels from the same class (Table 8.1). Figure 8.8 shows the ellipsoids defined by the mean vectors and variance–covariance matrices for Landsat MSS bands

Table 8.1 Variance/covariance matrices for four Landsat MSS bands obtained from random sample (upper figure) and contiguous sample (in parentheses) drawn from same data. Source: Campbell (1981), Table 7.

	MSS 1	MSS 2	MSS 3	MSS 4
MSS 1	1.09 (0.40)			
MSS 2	1.21 (0.21)	3.50 (1.01)		
MSS 3	−1.00 (−0.78)	−1.65 (−0.19)	23.15 (14.00)	
MSS 4	−0.51 (−0.43)	−1.85 (−1.10)	12.73 (9.80)	11.58 (8.92)

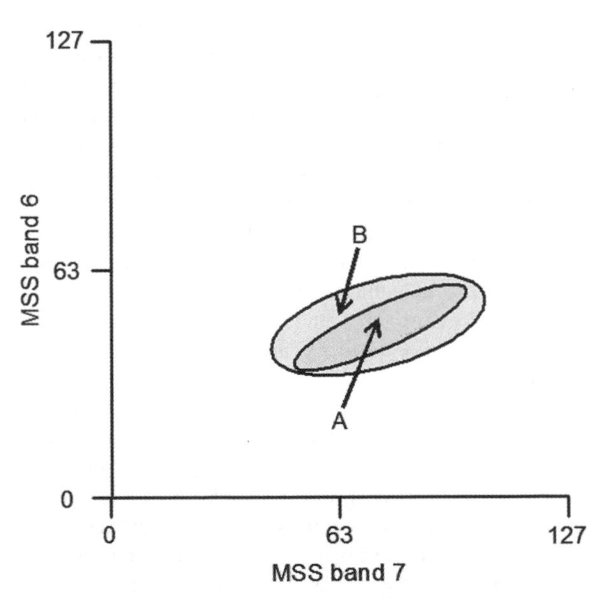

Figure 8.8 Ellipsoids derived from the variance-covariance matrices for training sets based on contiguous (A) and random (B) samples (derived from Campbell, 1981).

6 and 7 of Campbell's (1981) contiguous and random data (marked A and B, respectively). The locations of the centres of the centres of the two ellipses are not too far apart but their orientation, size and shape differ somewhat. Campbell (1981) suggests taking random pixels from within a training area rather than using contiguous blocks, while Labovitz and Matsuoko (1984) prefer a systematic sampling scheme with the spacing between the samples being determined by the degree of positive spatial autocorrelation in the data. Dobbertin and Biging (1996) report that classification accuracy is reduced when images show a high level of spatial autocorrelation. Derived features such as texture might be expected to display a higher degree of spatial autocorrelation than the individual pixel values in the raw images, because such measures are often calculated from overlapping windows. Better results were obtained from randomly selected training pixels than from contiguous blocks of training pixels, a conclusion also reached by Gong *et al.* (1996) and Wilson (1992). The variances of the training samples were also higher when individual random training pixels were used rather than contiguous pixel blocks. The method of automatically collecting training samples, described by Bolstad and Lillesand (1992) may well generate training data that are highly autocorrelated.

The degree of autocorrelation will depend upon (a) the natural association between adjacent pixels, (b) the pixel dimensions and (c) the effects of any data pre-processing. The degree of autocorrelation can be calculated by taking sequences of pixels that are spaced 1, 2, 3, ..., units apart and plotting the correlations between a set of pixels and its first, second, third and subsequent nearest neighbours in the form of a correlogram. A diagram of the kind shown in Figure 8.9 might result, and the autocorrelation distance (in terms of number of pixels) can be read directly from it. As pixel size increases so the autocorrelation distance will diminish. The problem of spatially-autocorrelated samples is considered in more detail in the papers cited above, and in Basu and Odell (1974), Craig (1979), and Labovitz *et al.* (1982). The definitive reference is still Cliff and Ord (1973). Geostatistical methods are based upon the spatial autocorrelation property, and are discussed briefly in section 8.5.

Another source of error encountered in the extraction of training samples is the presence in the sample of atypical values. For instance, one or more vectors of pixel measurements in a given training sample may be contaminated in some way; hence, the sample mean and variance–covariance matrix for that class will be in error. Campbell (1980) considers ways in which these atypical values can be detected and proposes estimators of the mean and variance–covariance matrix which are robust (that is, they are not unduly influenced by the atypical values). These estimators give full weight to observations that are assumed to come from the main body of the data but reduce the weight given to observations identified as aberrant. A measure called the Mahalanobis distance D is used to identify deviant members of the sample. Its square is defined by:

$$D^2 = (\mathbf{x_m} - \bar{\mathbf{x}})' \mathbf{S}^{-1} (\mathbf{x_m} - \bar{\mathbf{x}})$$

where m is the index counting the elements of the sample, $\mathbf{x_m}$ is the mth sample value (pixel vector). The sample mean vector is $\bar{\mathbf{x}}$ and $\mathbf{S}$ is the sample variance–covariance matrix. The transpose of vector $\mathbf{x}$ is written as $\mathbf{x}'$. The Mahalanobis distance, or some function of that distance, can be plotted against the normal probabilities and outlying elements of the sample can be visually identified (Healy, 1968; Sparks, 1985). Robust estimates (that is, estimates that are less affected by outliers) of $\bar{\mathbf{x}}$ and $\mathbf{S}$ are computed using weights which are functions of the Mahalanobis distance. The effect is to downgrade pixel values with high Mahalanobis distances (i.e., low weights) that are associated with pixels that are relatively far from (dissimilar to) the mean of the training class taking into account the shape of the probability distribution of training-class members. For uncontaminated data these robust estimates are close to those obtained from the usual estimators. The procedure for obtaining the weights is described and illustrated

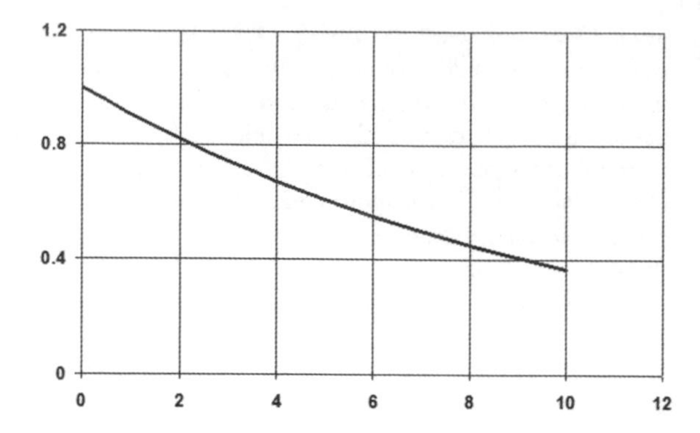

Figure 8.9 Illustrating the concept of autocorrelation. The diagram shows the correlation between the pixels in an image and their nth nearest neighbours in the x direction ($n = 1, \ldots, 10$). The correlation at distance (lag) one is computed from two sets of data – the values at pixels numbered 1,2,3, ..., $n-1$ along each scan line and the values at the pixels numbered 2,3,4, ..., n along the same scan lines. The higher the correlation the greater the resemblance between pixels spaced one unit apart on the x-axis. The same procedure is used to derive the second, third, ..., mth lag autocorrelation.

by Campbell (1980); it is summarised here for completeness.

$$\bar{x}_k = \frac{\sum_{i=1}^{n} w_i x_{ki}}{\sum_{i=1}^{n} w_i} \qquad (k = 1,2, \ldots, p)$$

$$s_{jk} = \sum_{i=1}^{n} w_i^2 (x_{ji} - \bar{x}_j)(x_{ki} - \bar{x}_k) \qquad (j = 1, 2, \ldots, p) \quad (k = j, j+1, \ldots, p)$$

where n is number of pixels in the training sample, p is number of features, w_i is weight for pixel i, x_{ki} is value for pixel i on feature k, $\bar{x}_j$ is mean of jth feature for this class and s_{jk} is element j, k of the variance–covariance matrix for this class.

The weights are found from:

$$w_i = F(d_i)/d_i$$

given

$$F(d_i = \begin{cases} d_i & d_i \leq d_0 \\ d_0\, exp[-0.5\,(d_i - d_0)^2/b_2^2 & \text{otherwise} \end{cases}$$

and

d_i is the Mahalanobis distance of pixel i for this class, d_0 is $\sqrt{p} + b_1/\sqrt{2}$, b_1 is 2, b_2 is 1.25.

The weights w_i are initially computed from the Mahalanobis distances which in turn are computed from $\bar{\mathbf{x}}_\mathbf{j}$ and $\mathbf{S_j}$ derived from the above formulae but using unit weights. The Mahalanobis distances and the weights are recalculated iteratively until successive weight vectors converge within an acceptable limit, when any aberrant pixel vectors should have been given very low weights, and will therefore contribute only negligibly to the final (robust) estimates of $\bar{\mathbf{x}}_j$ and $\mathbf{S_j}$ which are required in the maximum likelihood classification scheme, which is described later.

The reason for going to such apparently great lengths to obtain robust estimates of the mean and variance–covariance matrix for each of the training samples for use in maximum likelihood classification is that the probabilities of class membership of the individual pixels depend on these statistics. The performance of both statistical and neural classifiers depends to a considerable extent on the reliability and representativeness of the sample. It is easy to use an image processing system to extract 'training samples' from an image, but it is a lot more difficult to ensure that these training samples are not contaminated either by spatial autocorrelation effects or by the inclusion in the training sample of pixels which are not 'pure' but 'mixed' and therefore atypical of the class which they are supposed to represent.

The use of unsupervised classification techniques applied to the training classes has already been described as a method of ensuring that the classes have been well chosen to represent a single spectral class (that is one with a single mode or peak in their frequency distributions). An alternative way to provide a visual check of the distribution of the training sample values is to employ an ordination method. Ordination is the expression of a multivariate data set in terms of a few dimensions (preferably two) with the minimum loss of information. The Nonlinear Mapping procedure of Sammon (1969) projects the data in a p-dimensional space onto an m-dimensional subspace (m being less than p) whilst minimising the error introduced into the inter-point Euclidean distances. That is to say, the m-dimensional representation of the distances between data points is the best possible for that number of dimensions in terms of the maintenance of inter-point Euclidean relationships. If $m = 2$ or 3 the results can be presented graphically and the presence of multiple modes or outlying members of the sample can be picked out by eye. Figure 8.10 shows a training sample projected onto two dimensions using the MIPS *Nonlinear Mapping* module. The training data coordinates were collected using the MIPS *Classify|Collect Training Data* option and the pixel data were extracted (cut) from the log file and pasted into a new data file, which was then edited using Windows Notepad. The 'point and click' facility of the Nonlinear Mapping module allows the identification of extreme points, perhaps representing aberrant pixels.

The performance of a classifier is usually evaluated using measures of classification accuracy (section 8.10). These accuracy measures use a test set of known data that is collected using the same principles as those described above for the training data set. One could thus think of training data being used to calibrate the classifier and test data being used for validation, a point that is explored by Muchoney and Strahler (2002).

8.4.2 Statistical classifiers

Three algorithms are described in this section. All require that the number of categories (classes) be specified in advance, and that certain statistical characteristics of each class are known. The first method is called the parallelepiped or box classifier. A parallelepiped is simply a geometrical shape consisting of a body whose opposite sides are straight and parallel. A parallelogram is a two-dimensional parallelepiped. To define such a body all that is required is an estimate for each class of the values of the

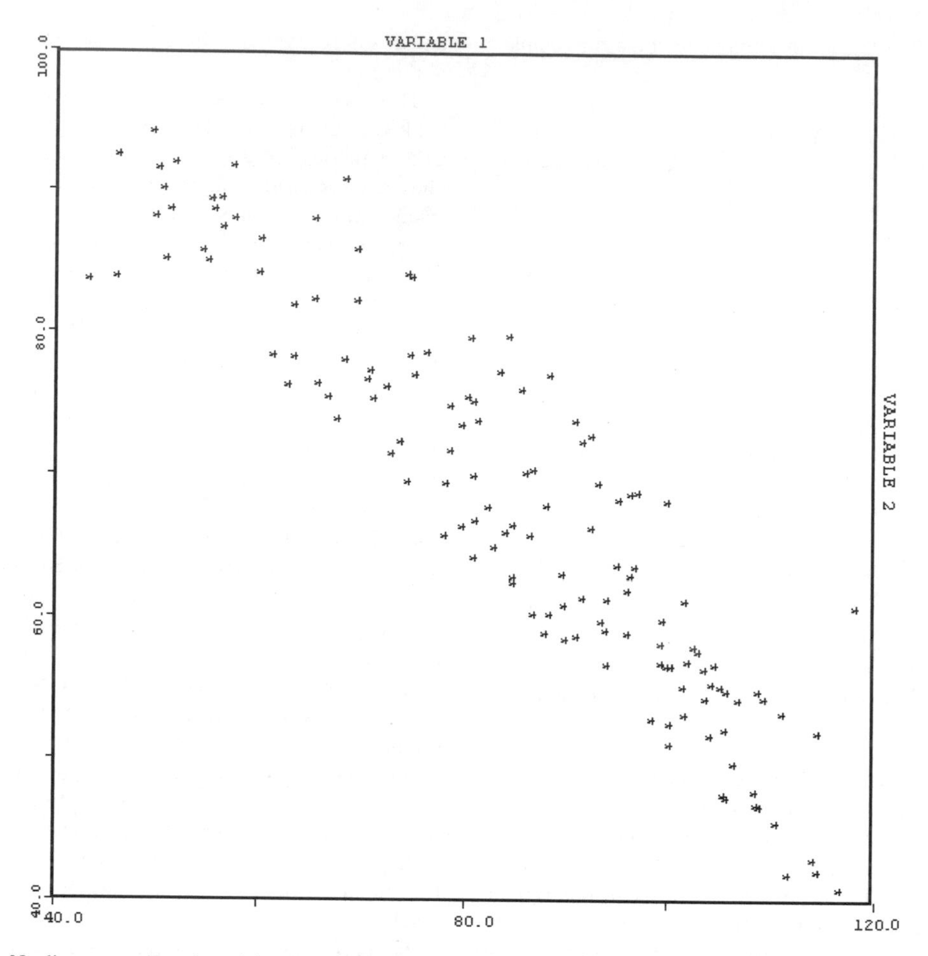

Figure 8.10 Nonlinear mapping of a training sample. The raw data are measured on six spectral bands. Nonlinear mapping has been used to project the sample data onto two dimensions. The MIPS module *Plot|Nonlinear Mapping* allows the user to 'query' the identity of any aberrant or outlying pixels. Original in colour.

lowest and highest pixel values in each band or feature used in the analysis. Pixels are labelled by determining the identifier of the box into which they fall (Figure 8.11, section 8.4.2.1). The second method, which is analogous to the k-means unsupervised technique, uses information about the location of each class in the p-dimensional Cartesian space defined by the p bands (features) to be used as the basis of the classification. The location of each class in the p-space is given by the class mean or centroid (Figure 8.12, section 8.4.2.2). This third method also uses the mean as a measure of the location of the centre of each class in the p-space and, in addition, makes use of a measure summarising the disposition or spread of values around the mean along each of the p axes of the feature space. The third method is that of maximum likelihood (section 8.4.2.3). All three methods require estimates of certain statistical characteristics of the classes to which the pixels are to be allocated. These estimates are derived from samples of pixels, called training samples, which are extracted from the image to be classified (section 8.4.1).

8.4.2.1 Parallelepiped classifier

The parallelepiped classifier requires the least information from the user of the statistical supervised classification methods described in this chapter. For each of the k classes specified, the user provides an estimate of the minimum and maximum pixel values on each of the p bands or features. Alternatively, a range, expressed in terms of a given number of standard deviation units on either side of the mean of each feature, can be used. These extreme values allow the estimation of the position of the boundaries of the

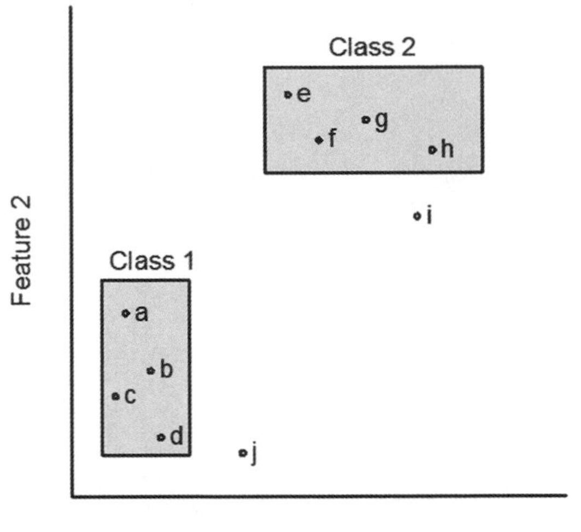

Figure 8.11 Parallelepiped classifier in two dimensions. Points a, b, c, d lie in the region bounded by parallelepiped 1 and are therefore assigned to class 1. Points e, f, g, h are similarly labelled '2'. Point i is unclassified.

parallelepipeds, which define regions of the p-dimensional feature space that are identified with particular land cover types (or information classes). Regions of the p-space lying outside the boundaries of the set of parallelepipeds form a *terra incognita* and pixels lying in these regions are usually assigned the label zero. The decision rule employed in the parallelepiped classifier is simple. Each pixel to be classified is taken in turn and its values on the p features are checked to see whether they lie inside any of the parallelepipeds. Two extreme cases might occur. In the first, the point in p-space representing a particular pixel does not lie inside any of the regions defined by the parallelepipeds. Such pixels are of an unknown type. In the second extreme case the point lies inside just one of the parallelepipeds, and the corresponding pixel is therefore labelled as a member of the class represented by that parallelepiped. However, there is the possibility that a point may lie inside two or more overlapping parallelepipeds, and the decision then becomes more complicated. The easiest way around the problem is to allocate the pixel to the first (or some other arbitrarily selected) parallelepiped inside whose boundaries it falls. The order of evaluation of the parallelepipeds then becomes of crucial importance, and there is often no sensible rule that can be employed to determine the best order.

The method can therefore be described as 'quick and dirty'. If the data are well structured (that is, there is no overlap between the classes) then the quick-and-dirty method might generate only a very few conflicts but, unfortunately, many image data sets are not well structured. A more complicated rule for the resolution of conflicts might be to calculate the Euclidean distance between the doubtful pixel and the centre point of each parallelepiped and use a 'minimum distance' rule to decide on the best classification. In effect, a boundary is drawn in the area of overlap between the parallelepipeds concerned. This boundary is equidistant from the centre points of the parallelepipeds, and pixels can be allocated on the basis of their position relative to the boundary line. On the other hand, a combination of the parallelepiped and some other, more powerful, decision rule could be used. If a pixel falls inside one single parallelepiped then it is allocated to the class that is represented by the parallelepiped. If the pixels falls inside two or more parallelepipeds, or is outside all of the parallelepiped areas, then a more sophisticated decision rule could be invoked to resolve the conflict.

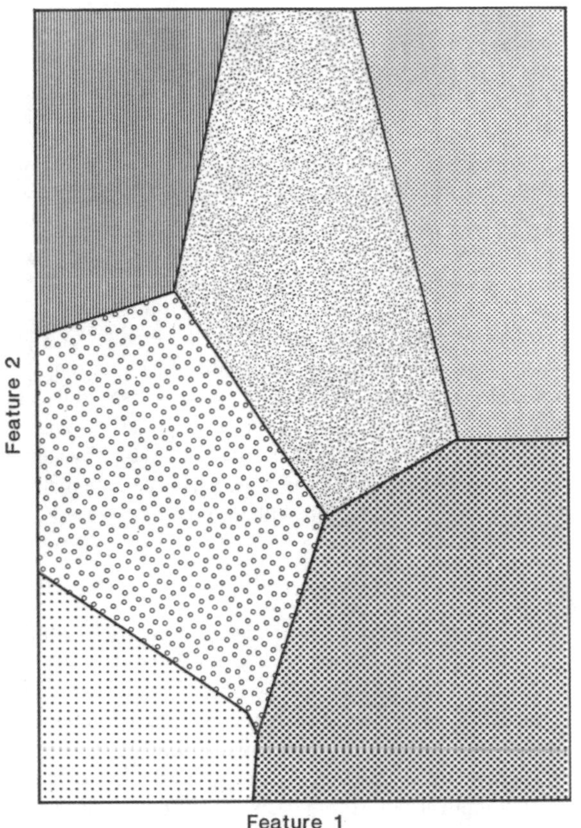

Figure 8.12 Two-dimensional feature space partitioned according to distance from the centroid of the nearest group. There are six classes.

Figure 8.11 shows a geometric representation of a simple case illustrating the parallelepiped classifier in action. Points a, b, c and d are allocated to class 1 and points e, f, g and h are allocated to class 2. Points i and j are not identified and are labelled as 'unknown'. The technique is easy to program and is relatively fast in operation. Since, however, the technique makes use only of the minimum and maximum values of each feature for each training set it should be realised that (a) these values may be unrepresentative of the actual spectral classes that they purport to represent, and (b) no information is garnered from those pixels in the training set other than the largest and the smallest in value on each band. Furthermore it is assumed that the shape of the region in p-space occupied by a particular spectral class can be enclosed by a box. This is not necessarily so. Consequently the parallelepiped method should be considered as a cheap and rapid but not particularly accurate method of associating image pixels with information classes.

8.4.2.2 *Centroid (k-means) classifier*

The centroid or k-means method does make use of all the data in each training class, for it is based upon the 'nearest centre' decision rule that is described in section 8.3. The centroid (mean centre) of each training class is computed - it is simply the vector comprising the mean of each of the p features used in the analysis, perhaps weighted to diminish the influence of extreme values as discussed in section 8.4.1. The Euclidean distance from each unknown pixel is then calculated for each centre in turn and the pixel is given the label of the centre to which its Euclidean distance is smallest. In effect, the p-space is divided up into regions by a set of rectilinear boundary lines, each boundary being equidistant from two or more centres (Figure 8.12). Every pixel is classified by this method, for each point in p-space must be closer to one of the k centres than to the rest, excluding the case in which a pixel is equidistant from two or more centres. A modification to the 'closest distance' rule could be adopted to prevent freak or outlying pixel values from being attached to one or other of the classes. This modification could take the form of a distance threshold, which could vary for each class depending upon the expected degree of compactness of that class. Compactness might be estimated from the standard deviation for each feature of the pixels making up the training sample for a given class. Any pixel that is further away from the nearest centre than the threshold distance is left unclassified. This modified rule is actually changing the geometry of the decision boundaries from that shown in Figure 8.12 to that shown in Figure 8.13. In the latter, the p-space is subdivided into k hyper-spherical regions each centred on a class mean. In the same way that the parallelepiped method gets into difficulties with overlapping boxes and has to adopt a nearest-centre rule to break the deadlock, so the k-means method can be adapted to utilise additional information in order to make it intuitively more efficient. The alteration to the decision rule involving a threshold distance is effectively acknowledging that the shape of the region in p-space that is occupied by pixels belonging to a particular class is important. The third classification technique described in this section (the maximum likelihood method) begins with this assumption and uses a rather more refined method of describing the shapes of the regions in p-space that are occupied by the members of each class.

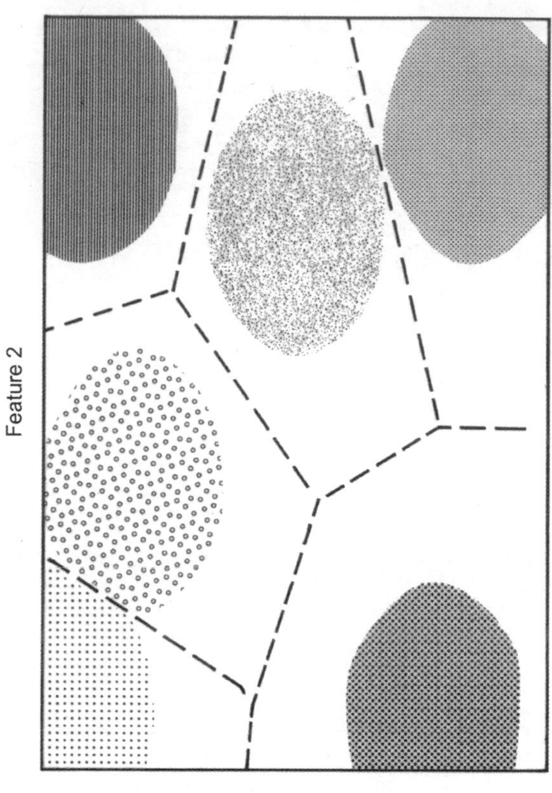

Figure 8.13 Group centroids are the same as in Figure 8.12 but a threshold has been used to limit the extent of each group. Blank areas represent regions of feature space that are not associated with a spectral class. Pixels located in these regions are not labelled.

A k-means procedure is implemented in the MIPS program using training data collected interactively from the display. The resulting classification can be based on any selection of image bands that are referenced in the chosen INF file.

8.4.2.3 *Maximum likelihood method*

The geometrical shape of a cloud of points representing a set of image pixels belonging to a class or category of interest can often be described by an ellipsoid. This knowledge is used in Chapter 6 in the discussion of the principal components technique. In that chapter it is shown that the orientation and the relative dimensions of the enclosing ellipsoid (strictly speaking, a hyper-ellipsoid if p is greater than three) depends on the degree of covariance among the p features defining the pattern space. Examples of two-dimensional ellipses are shown in Figure 8.14. A shape such as that of ellipse A (oriented with the longer axis sloping upwards to the right) implies high positive covariance between the two features. If the longer axis sloped upwards to the left the direction of covariance would be negative. The more circular shape of ellipse B implies lower covariances between the features represented by x and y. The lengths of the major and minor axes of the two ellipses projected onto the x- or y-axes are proportional to the variances of the two variables. The location, shape and size of the ellipse therefore reflects the means, variances and covariances of the two features, and the idea can easily be extended to three or more dimensions. The ellipses in Figure 8.14 do not enclose all the points that fall into a particular class; indeed, we could think of a family of concentric ellipses centred on the p-variate mean of a class, such as points '1' and '2' in Figure 8.14. A small ellipse centred on this mean point might enclose only a few per cent of the pixels, which are members of the class, and progressively larger ellipses will enclose an increasingly larger proportion of the class

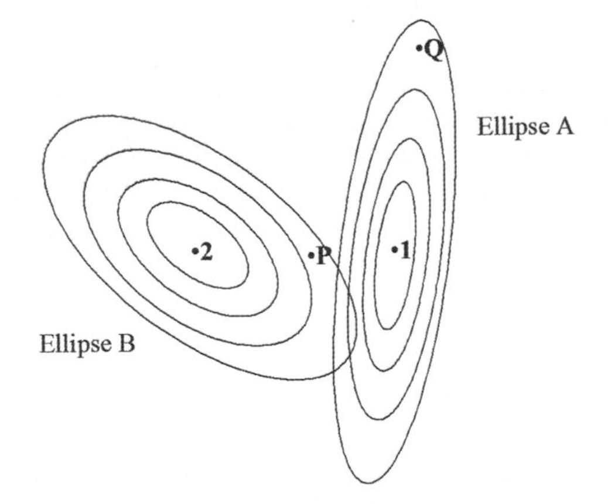

Figure 8.14 Showing the equi-probability contours for two bivariate-normal distributions with means located at points 1 and 2. Point P is closer to the mean centre of distribution 1 than it is to the centre of distribution 2 yet, because of the shapes of the two ellipses, P is more likely to be a member of class 2. Similarly, point P is closer to the centre of distribution 1 than is point Q, yet Q is more likely to be a member of class 1.

members. These concentric ellipses represent contours of probability of membership of the class, with the probability of membership declining away from the mean centre. Thus, membership probability declines more rapidly along the direction of the shorter axis than along the longer axis.

Distance from the centre of the training data is not now the only criterion for deciding whether a point belongs to one class or another, for the shape of the probability contours depends on the relative dimensions of the axes of the ellipse as well as on its orientation. In Figure 8.14 point P is closer than point Q to the centre of class 1 yet, because of the shape of the probability contours, point Q is seen to be more likely to be a member of class 1 while point P is more likely to be a member of class 2.

If equi-probability contours can be defined for all k classes of interest then the probability that a pixel shown by a point in the p-dimensional feature belongs to class i ($i = 1, 2, \ldots, k$) can be measured for each class in turn, and that pixel assigned to the class for which the probability of membership is highest. The resulting classification might be expected to be more accurate than those produced by either the parallelepiped or the k-means classifiers because the training sample data are being used to provide estimates of the shapes of the distribution of the membership of each class in the p-dimensional feature space as well as of the location of the centre point of each class. The coordinates of the centre point of each class are the mean values on each of the p features, while the shape of the frequency distribution of the class membership is defined by the covariances among the p features for that particular class, as we saw earlier.

It is important to realise that the maximum likelihood method is based on the assumption that the frequency distribution of the class membership can be approximated by the multivariate normal probability distribution. This might appear to be an undue restriction for, as an eminent statistician once remarked, there is no such thing as a normal distribution. In practice, however, it is generally accepted that the assumption of normality holds reasonably well, and that the procedure described above is not too sensitive to small departures from the assumption provided that the actual frequency distribution of each class is uni-modal (i.e., has one peak frequency). A clustering procedure (unsupervised classification) could be used to check the training sample data for each class to see if that class is multi-modal, for the clustering method is really a technique for finding multiple modes.

The probability P(**x**) that a pixel vector **x** of p elements (a pattern defined in terms of p features) is a member of class i is given by the multivariate normal density:

$$P(\mathbf{x}) = 2\pi^{-0.5p}\,|\mathbf{S}_i|^{-0.5}\exp\left[-0.5\left(\mathbf{y}'\,\mathbf{S}_i^{-1}\,\mathbf{y}\right)\right]$$

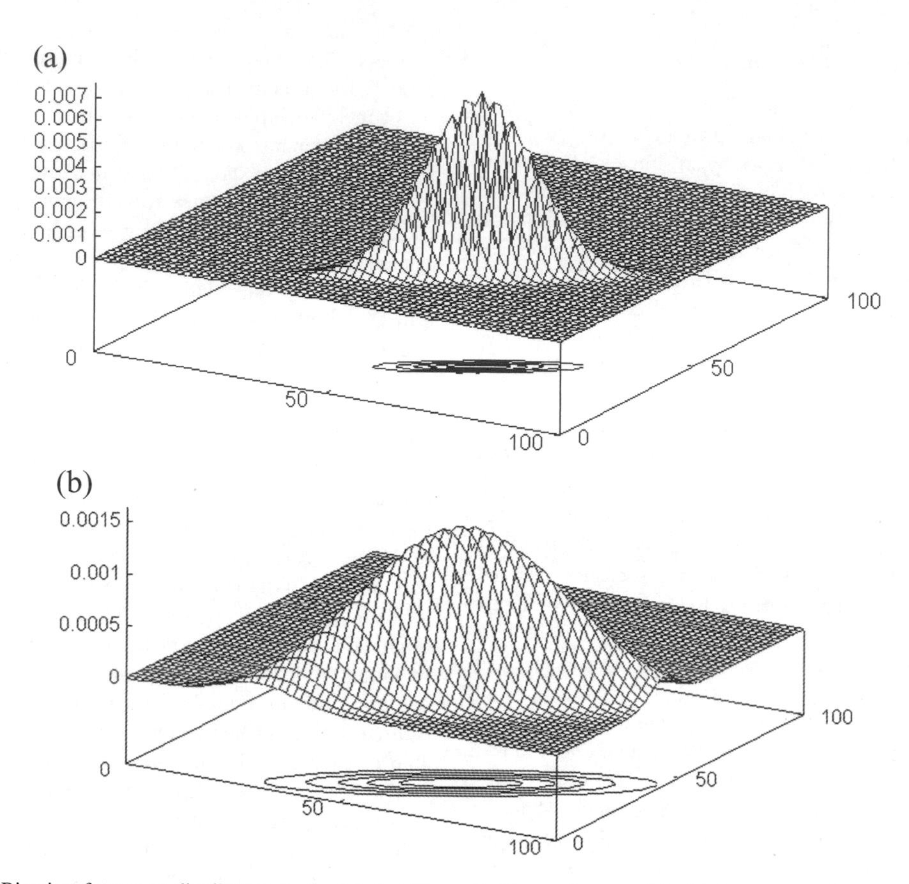

Figure 8.15 Bivariate frequency distribution for the Landsat Thematic Mapper images of the Littleport area (Figures 1.10 and 1.11). (a) Bands 1 and 2, (b) Bands 5 and 7. The mean vectors for both data sets are approximately the same, and both show a positive covariance. The variances of bands 5 and 7 are substantially larger than the variances of bands 1 and 2 (see text).

where |.| denotes the determinant of the specified matrix, $\mathbf{S}_i$ is the sample variance–covariance matrix for class i, $\mathbf{y} = (\mathbf{x} - \bar{\mathbf{x}}_i)$ and $\bar{\mathbf{x}}_i$ is the multivariate mean of class i. Note that the term $\mathbf{y}' \mathbf{S}^{-1} \mathbf{y}$ is the Mahalanobis distance, used in section 8.4.1 to measure the distance of an observation from the class mean, corrected for the variance and covariance of class i.

Understanding of the relationship between equi-probability ellipses, the algebraic formula for class probability, and the placing of decision boundaries in feature space will be enhanced by a simple example. Figure 8.15 shows the bivariate frequency distribution of two samples, drawn respectively from (a) bands 1 and 2 and (b) bands 5 and 7 of the TM images shown in Figures 1.10 and 1.11. The contours delimiting the equi-probability ellipses are projected onto the base of each of the diagrams. It is clear that the area of the two-dimensional feature space that is occupied by the band 5–band 7 combination (Figure 8.15(b)) is greater than that occupied by the band 1–band 2 combination. The orientation of the probability ellipses is similar, and the two ellipses are located at approximately the same point in the feature space. These observations can be related to the elements of the variance–covariance matrices on which the two plots are based. The following matrices were used in the derivation of the bivariate probability distributions Figure 8.15:

$$\bar{\mathbf{x}}_{12} = \begin{bmatrix} 65.812 \\ 28.033 \end{bmatrix} \quad \mathbf{S}_{12} = \begin{bmatrix} 78.669 & 45.904 \\ 45.904 & 31.945 \end{bmatrix}$$

$$\mathbf{S}_{12}^{-1} = \begin{bmatrix} 0.079 & -0.113 \\ -0.113 & 0.194 \end{bmatrix} \quad |\mathbf{S}_{12}| = 405.904$$

$$\bar{\mathbf{x}}_{57} = \begin{bmatrix} 64.258 \\ 23.895 \end{bmatrix} \quad \mathbf{S}_{57} = \begin{bmatrix} 329.336 & 181.563 \\ 181.563 & 128.299 \end{bmatrix}$$

$$\mathbf{S}_{57}^{-1} = \begin{bmatrix} 0.014 & -0.020 \\ -0.020 & 0.035 \end{bmatrix} \quad |\mathbf{S}_{57}| = 9288.356$$

The variances are the diagonal elements of the matrix **S**, and it is clear that the variances of bands 5 and 7 (329.336 and 128.299) are much larger than the variances of bands 1 and 2 (78.669 and 31.945). Thus, the 'spread' of the two ellipses in the x and y directions is substantially different. The covariance of bands 5 and 7 (181.563) and bands 1 and 2 (45.904) are both positive, so the ellipses are oriented upwards towards the $+x$ and $+y$ axes (if the plot were a two dimensional one, we could say that the ellipses sloped upwards to the right, indicating a positive correlation between x and y). The covariance of bands 5 and 7 is larger than that of bands 1 and 2, so the degree of scatter is less for bands 5 and 7, relative to the magnitude of the variances. This example illustrates the fact that the mean controls the location of the ellipse in feature space, while the variance–covariance matrix controls the 'spread' and orientation of the ellipse. It is not possible to illustrate these principles in higher-dimensional spaces, though if it were then the same conclusions would be drawn.

The function $P(\mathbf{x})$ can be used to evaluate the probability that an unknown pattern $\mathbf{x}$ is a member of class i ($i = 1, 2, \ldots, k$). The maximum value in this set can be chosen and $\mathbf{x}$ allocated to the corresponding class. However, the cost of carrying out these computations can be reduced by simplifying the expression. Savings can be made by first noting that we are only interested in the rank order of the values of $P(\mathbf{x})$. Since the logarithm to base e of a function has the same rank order as the function, the evaluation of the exponential term can be avoided by evaluating:

$$\ln(P(x) = -0.5p\ln(2\pi) - 0.5\ln|\mathbf{S}| - 0.5\left(\mathbf{y}'\,\mathbf{S}_i^{-1}\,\mathbf{y}\right)$$

The rank order is unaffected if the right-hand side of this expression is multiplied by -2 and if the constant term $p \ln(2\pi)$ is dropped. The expression also looks tidier if it is multiplied by -1 and the smallest value for all k classes chosen, rather than the largest. These modifications reduce the expression to

$$-\ln(P(\mathbf{x}) = \ln(|\mathbf{S}|) + \mathbf{y}'\,\mathbf{S_i^{-1}}\,\mathbf{y}$$

Further savings can be made if the inverse and determinant of each $\mathbf{S}_i$ (the variance–covariance matrix for class i) are computed in advance and read from a file when required, rather than calculated when required. The computations then reduce to the derivation of the Mahalanobis distance, the addition of the logarithm of the determinant of the estimated variance–covariance matrix for each of the k classes in turn, and the selection of the minimum value from among the results. Note that, because we have multiplied the original expression by -0.5 we minimise $\{-\ln P(\mathbf{x})\}$ so as to achieve the same result as maximising $P(\mathbf{x})$.

The maximum likelihood equations given above are based upon the presumption that each of the k classes is equally likely. This may be the safest assumption if we have little knowledge of the extent of each land cover type in the area covered by the image. Sometimes the proportion of the area covered by each class can be estimated from reports, land-use maps, aerial photographs or previous classified images. An unsupervised classification of the area would also provide some guide to the areal extent of each cover type. An advantage of the maximum likelihood approach to image classification is that this prior knowledge can be taken into account. *A priori* knowledge of the proportion of the area to be classified that is covered by each class can be expressed as a vector of *prior probabilities*. The probabilities are proportional to the area covered by each class, and can be thought of as weights. A high prior probability for class i in comparison with class j means that any pixel selected at random is more likely to be placed in class i than class j because class i is given greater weight. These weights are incorporated into the maximum likelihood algorithm by subtracting twice the logarithm of the prior probability for class i from the log likelihood of the class as given by the equation above. Strahler (1980) provides further details and shows how different sets of prior probabilities can be used in cases where the image area can be stratified into regions according to an external variable such as elevation (for instance, regions described as high, intermediate or low elevation might have separate sets of prior probabilities as described in section 8.7.2). Maselli *et al.* (1995) discuss a nonparametric method of estimating prior probabilities for incorporation into a maximum likelihood classifier.

In the same way that the parallelepiped classifier (section 8.4.1) allows for the occurrence of pixels that are unlike any of the training patterns by consigning such pixels to a 'reject' class, so the probability of class membership can be used in the maximum likelihood classifier to permit the rejection of pixel vectors for which the probability of membership of any of the k classes is considered to be too low. The Mahalanobis distance is distributed as chi-square, and the probability of obtaining a Mahalanobis distance as high as that observed for a given pixel can be found from tables, with degrees of freedom equal to p, the number of feature vectors used (Moik, 1980). For $p = 4$ the tabled chi-square values are 0.3 (99%), 1.06 (90%), 3.65 (50%), 7.78 (10%) and 15.08 (1%). The figures in brackets are probabilities, expressed in per cent form. They can be interpreted as follows: a Mahalanobis distance as high as (or higher than) 15.08 would, on average, be met in only 1% of cases in a long sequence of observations of four-band pixel vectors drawn from a multivariate normal population whose true mean and variance–covariance matrix are estimated by the mean and variance–covariance matrix on which the calculation of the Mahalanobis distance is based. A Mahalanobis distance of 1.06 or more would be observed in 90% of all such observations. It is self-evident that 100% of all observations drawn from the given population will have Mahalanobis distances of

0 or more. A suitable threshold probability (which need not be the same for each class) can be specified. Once the pixel has been tentatively allocated to a class using the maximum likelihood decision rule, the value of the Mahalanobis distance (which is used in the maximum likelihood calculation) can be tested against a threshold chi-square value. If this chi-square value is exceeded then the corresponding pixel is placed in the 'reject' class, conventionally labelled '0'. The use of the threshold probability helps to weed out atypical pixel vectors. It can also serve another function – to indicate the existence of spectral classes which may not have been recognised by the analyst and which, in the absence of the probability threshold, would have been allocated to the most similar (but incorrect) spectral class.

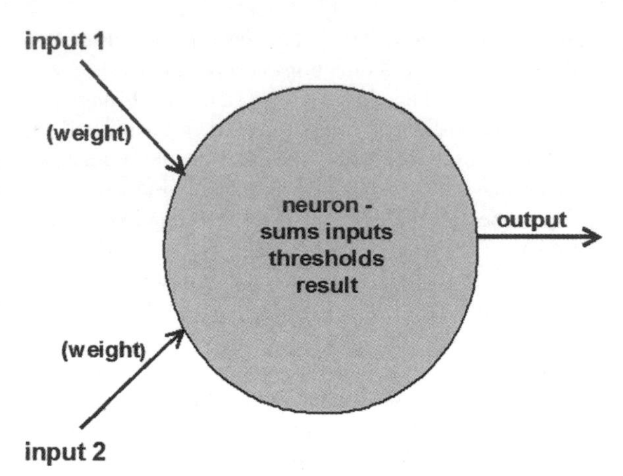

Figure 8.16 A basic neuron receives two weighted inputs, sums them, applies a threshold and outputs the result.

8.4.3 Neural classifiers

The best image-interpretation system that we possess is the combination of our eyes and our brain (Gregory, 1998). Signals received by two sensors (our eyes) are converted into electrical impulses and transmitted to the brain, which interprets them in real time, producing labelled (in the sense of recognised) three-dimensional images of our field of view (section 5.2). Operationally speaking, the brain is thought to be composed of a very large number of simple processing units called neurons. Greenfield (1997, p. 79) estimates the number of neurons as being of the order of a hundred billion, a number that is of the same order of magnitude as the number of trees in the Amazon rain forest. Each neuron is connected to perhaps 10 000 other neurons (Beale and Jackson, 1990). Of course, brain size varies – for example, men have – on average – larger brains than women, though most women wonder why they do not use them. Many neurons are dedicated to image processing, which takes place in a parallel fashion (Gregory, 1999; Greenfield, 1997, p. 50). The brain's neurons are connected together in complex ways so that each neuron receives as input the results produced by other neurons, and it in turn outputs its signals to other neurons. It is not possible at the moment to specify how the brain actually works, or even whether the connectionist model really represents what is going on in the brain. It has been suggested that if there were fewer neurons in the brain then we might have a chance of understanding how they interact but, unfortunately, if our brains possessed fewer neurons we would probably be too stupid to understand the implications of the present discussion.

One model of the brain is that it is composed of sets of neural networks that perform specific functions such as vision or hearing. Artificial neural networks (ANN) attempt, in a very simple way, to use this model of the brain by building sets of linked processing units (by analogy with the neurons of the brain) and using these to solve problems. Each neuron is a simple processing unit which receives weighted inputs from other neurons, sums these weighted inputs, performs a simple calculation on this sum such as thresholding, and then sends this output to other neurons (Figure 8.16).

The two functions of the artificial neuron are to sum the weighted inputs and to apply a thresholding function to this sum. The summation procedure can be expressed by:

$$S = \sum_{i=1}^{n} w_i x_i$$

where S represents the sum of the n weighted inputs, w_i is the weight associated with the ith input and x_i is the value of the ith input (which is an output from some other neuron). The thresholding procedure, at its simplest, a comparison between S and some pre-set value, say T. If S is greater than T then the neuron responds by sending an output to other neurons to which it is connected further 'down the line'. The term *feed-forward* is used to describe this kind of neural network model because information progresses from the initial inputs to the final outputs.

A simple ANN such as the one presented above lacks a vital component - the ability to learn. Some training is necessary before the connected set of neurons can perform a useful task. Learning is accomplished by providing training samples and comparing the actual output of the ANN with the expected output. If there is a difference between the two then the weights associated with the connections between the neurons forming the ANN are adjusted so as to improve the chances of a correct decision and diminish the chances of the wrong choice being made, and the training step is repeated. The weights are initially set to random values. This 'supervised learning' procedure is followed until the ANN gets the correct answer. To a parent, this is perhaps reminiscent of teaching a child to read; repeated correction of mistakes in identifying the letters of the alphabet and the sounds associated with them eventually results in the development of an ability to read. The method is called *Hebbian learning* after its developer, D.O. Hebb.

This simple model is called the *single layer perceptron* and it can solve only those classification problems in which the classes can be separated by a straight line (in other words, the decision boundary between classes is a straight line as in the simple example given in section 8.2 and shown in Figure 8.1). Such problems are relatively trivial and the inability of the perceptron to solve more difficult problems led to a lack of interest in ANN on the part of computer scientists until the 1980s when a more complex model, the *multi-layer perceptron*, was proposed. Firstly, this model uses a more complex thresholding function rather than a step function, in which the output from the neuron is 1 if the threshold is exceeded and 0 otherwise. A sigmoid function is often used and the output from the neuron is a value somewhere between 0 and 1. Secondly, the neurons forming the ANN are arranged in layers as shown in Figure 8.17. There is a layer of input neurons which provide the link between the ANN and the input data, and a layer of output neurons which provide information on the category to which the input pixel vector belongs (for example, if output neuron number 1 has a value near to 1 and the remaining output neurons have values near zero then the input pixel will be allocated to class 1).

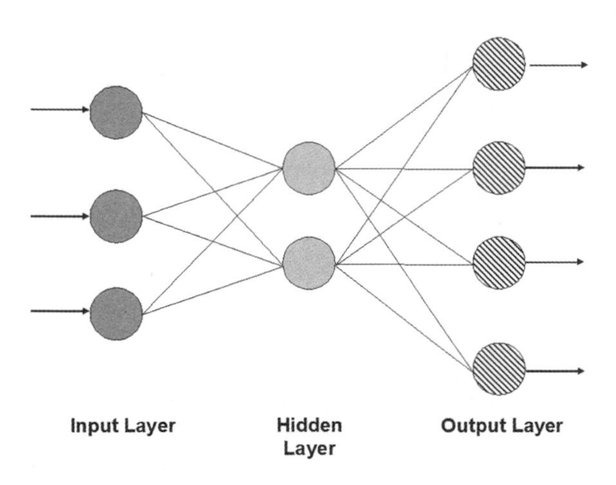

Figure 8.17 Multi-layer perceptron. The input layer (left side, dark shading) connects to the hidden layer (centre, light shading) which in turn connects to the output layer (right, hatched).

The multi-layer perceptron shown in Figure 8.17 could, for example, be used to classify an image obtained from the SPOT HRV sensor. The three neurons of the input layer represent the values for the three HRV spectral bands for a specific pixel. These inputs might be normalised, for example to the 0–1 range, as this procedure has been shown to improve the network's performance. The four neurons in the output layer provide outputs (or *activations*) which form the basis for the decision concerning the pixel's class membership, and the two neurons in the centre (hidden) layer perform summation and thresholding, possibly using a sigmoidal thresholding function. All the internal links (input to hidden and hidden to output) have associated weights. The input neurons do not perform any summation or thresholding – they simply pass on the pixel's value to the hidden layer neurons (so that input layer neuron 1 transmits the pixel value in band HRV-1, neuron 2 in the input layer passes the pixel value for band HRV-2 to the hidden layer neurons, and the third input neuron provides the pixel value for band HRV-3). The multi-layer perceptron is trained using a slightly more complex learning rule than the one described above; the new rule is called the *generalised delta function rule* or the *back-propagation rule*. Hence the network is a feed-forward multi-layer perceptron using the back-propagation learning rule, which is described by Paola and Schowengerdt (1995a).

Assume that a certain training data pixel vector, called a *training pattern,* is fed into the network and it is known that this training pattern is a member of class i. The output from the network consists of one value for each neuron in the output layer. If there are k possible classes then the expected output vector $\mathbf{o}$ should have elements equal to zero except for the ith element, which should be equal to one. The actual output vector, $\mathbf{a}$, differs from $\mathbf{o}$ by an amount called the error, e.Thus:

$$e = \frac{1}{2}\sum_{j=1}^{k}(o_j - a_j)^2$$

Multiplication by ½ is performed for arcane reasons of mathematics. The error is used to adjust the weights by a procedure which (in effect) maps the isolines or contours of the distribution of the error against the values of the weights, and then uses these isolines to determine the direction to move in order to find the minimum point in this map. The idea is known as the method of *steepest descent*. It requires (i) the direction of the steepest downhill slope and (ii) the length of the step that should be taken. Imagine that you have reached the summit of a hill when a thick mist descends. To get down safely you need to move downhill, but you may not be confident enough to take long strides downhill for fear of falling over a cliff. Instead, you may prefer to take short steps when the slope is steep and longer steps on gentle slopes. The steepest descent method uses the same approach. The gradient is measured by the first derivative of the error in terms of the weight; this gives both the magnitude and direction of the gradient. The step length is fixed and, in the terminology of ANNs, it is called the *learning rate*. A step is taken from the current position in the direction of maximum gradient and new values for the weights are determined. The error is propagated backwards through the net from the output layer to the input data, hence the term back-propagation.

In your descent down a fog-bound hill you may come to an enclosed hollow. No matter which way you move, you will go uphill. You might think that you were at the bottom of the hill. In mathematical terms this is equivalent to finding a local minimum of the function relating weights to error, rather than a global minimum. The method of steepest descent thus may not give you the right answer; it may converge at a point that is far-removed from the real (global) minimum. Another problem may arise if you take steps of a fixed length. During the descent you may encounter a small valley. As you reach the bottom of this valley you take a step across the valley floor and immediately start to go uphill. So you step back, then forward, and so on in an endless dance rhythm. But you never get across the valley and continue your downhill march. This problem is similar to that of the local minimum and both may result in oscillations. Some ANN software allow you to do the equivalent of taking a long jump across the surface when such oscillations occur, and continuing the downhill search from the landing point. There are other, more powerful methods of locating the minimum of a function that can be used in ANN training, but these are beyond the scope of this book.

The advantages of the feed-forward multi-layer ANN using the back-propagation training method are:

(i) it can accept all kinds of numerical inputs whether or not these conform to a statistical distribution or not. So non-remotely-sensed data can be added as additional inputs, and the user need not be concerned about multivariate normal distributions or multi-modality. This feature is useful when remote sensing data are used within a GIS, for different types of spatial data can be easily registered and used together in order to improve classifier performance.

(ii) ANNs can generalise. That is, they can recognise inputs that are similar to those which have been used to train them. They can generalise more successfully when the unknown patterns are intermediate between two known patterns, but they are less good at extending their ability to new patterns that exist beyond the realms of the training patterns. Bigger networks (with more neurons) tend to have a poorer generalisation capability than small networks.

(iii) because ANNs consist of a number of layers of neurons, connected by weighted links, they are tolerant to noise present in the training patterns. The overall result may not be significantly affected by the loss of one or two neurons as a result of noisy training data.

Disadvantages associated with the use of ANN in pattern recognition are:

(i) problems in designing the network. Usually one or two hidden layers suffice for most problems, but how many hidden layers should be used in a given case? How many neurons are required on each hidden layer? (A generally used but purely empirical rule is that the number of hidden layer neurons should equal twice the number of input neurons.) Are all of the inter-neuron connections required? What is the best value for the learning rate parameter? Many authors do not say why they select a particular network architecture. For example, Kanellopoulos *et al.* (1992) use a four-layer network (one input, one output and two hidden layers). The two hidden layers contain, respectively, 18 and 54 neurons. Foody (1995), on the other hand, uses a three-layer architecture with the hidden layer containing three neurons. Ardö *et al.* (1997) conclude that

> '. . . no significant difference was found between networks with different numbers of hidden nodes, or between networks with different numbers of hidden layers'

(a conclusion also reached by Gong *et al.*, 1996, and Paola and Schowengerdt, 1997), though it has already been noted that network size and generalisation capability appear to be inversely related. This factor is the motivation for network pruning (see below, point (v)).

(ii) training times may be long, possibly of the order of several hours. In comparison, the ML statistical classifier requires collection of training data and calculation of the mean vector and variance–covariance matrix for each training class. The calculations are then straightforward rather than iterative. The ML algorithm can classify a 512 × 512 image on a 486-based PC in less than a minute, while an ANN may require several hours of training in order to achieve the same level of accuracy.

(iii) the steepest-descent algorithm may reach a local rather than a global minimum, or may oscillate.

(iv) Results achieved by an ANN depend on the initial values given to the inter-neuron weights, which are usually set to small, random values. Differences in the initial weights may cause the network to converge on a different (local) minimum and thus different classification accuracies can be expected. Ardö *et al.* (1997), Blamire (1996) and Skidmore *et al.* (1997) note differences in classification accuracy resulting from different initial (but still randomised) choice of weights (around 6% in Blamire's experiments, but up to 11% in the results reported by Ardö *et al.* (1997)). Skidmore *et al.* (1997, p. 511) remark that

> '...the oft-quoted advantages of neural networks...were negated by the variable and unpredictable results generated'.

Paola and Schowengerdt (1997) also find that where the number of neurons in the hidden layer is low then the effects of changes in the initial weights may be considerable.

(v) The generalising ability of ANNs is dependent on a complex fashion on the numbers of neurons included in the hidden layers and on the number of iterations achieved during training. 'Over-training' may result in the ANN becoming too closely adjusted to the characteristics of the training data and losing its ability to identify patterns that are not present in the training data. Pruning methods, which aim to remove inter-neuron links without reducing the classifier's performance, have not been widely used in remote sensing image classification, but appear to have some potential in producing smaller networks that can generalise better and run more quickly.

Further details of applications of ANNs in pattern recognition are described by Aleksander and Morton (1990), Bishoff *et al.* (1992), Cappellini *et al.* (1995) and Gopal and Woodcock (1996). Austin *et al.* (1997) and Kanellopoulos *et al.* (eds.) (1997) provide an excellent summary of research problems in the use of ANNs in remote sensing. The topic of pruning ANNs in order to improve their generalisation capability is discussed by Kavzoglu and Mather (1999), Le Cun *et al.* (1990), Sietsma and Dow (1988) and Tidemann and Nielsen (1997). Jarvis and Stuart (1996) summarise the factors that affect the sensitivity of neural nets for classifying remotely-sensed data. A good general textbook is Bishop (1995). Hepner *et al.* (1990) compare the performance of an ANN classifier with that of ML, and find that (with a small training data set) the ANN gives superior results, a finding that is repeated by Foody *et al.* (1995). Kanellopoulos *et al.* (1992) compare the ML classifier's performance with that of a neural network with two hidden layers (with 18 and 54 neurons, respectively), and report that the classification accuracy (section 8.10) rises from 51% (ML) to 81% (ANN), though this improvement is very much greater than that reported by other researchers. Paola and Schowengerdt (1995b) report on a detailed comparison of the performance of a standard ANN and the ML statistical technique for classifying urban areas. Their paper includes a careful analysis of decision boundary positions in feature space. Although the ANN slightly out-performed the ML classifier in terms of percentage correctly classified test pixels, only 62% or so of the pixels in the two classified images were in agreement, thus emphasising the point that measures of classification accuracy based upon error matrices (section 8.10) do not take the spatial distribution of the classified pixels into account.

The feed-forward multi-layer perceptron is not the only form of artificial neural net that has been used in remote sensing nor, indeed, is it necessarily true that ANNs are always used in supervised mode. Chiuderi and Cappellini (1996) describe a quite different network architecture, the Kohonen Self-Organising Map (SOM). This is an unsupervised form of ANN. It is first trained to learn to distinguish between patterns in the input data ('clusters') rather than to allocate pixels to pre-defined categories. The clusters identified by the SOM are grouped on the basis of their mutual similarities, and then identified by reference to training data. The architecture differs from the conventional perceptron in that there are only two layers. The first is the input layer, which – as in the case of the perceptron – has one input neuron per feature. The input neurons are connected to all of the neurons in the output layer, and input pixels are allocated to a neighbourhood in the output layer, which is arranged in the form of a grid. Chiuderi and Cappellini (1996) use a 6 by 6 output layer and report a classification accuracy in excess of 85% in an application to land cover classification using airborne thematic mapper data. Schaale and Furrer (1995) successfully use a SOM network, also to classify land cover, while Hung (1993) gives a description of the learning mechanism used in SOM. Carpenter *et al.* (1997) report on the use of another type of neural network, the ART network, to classify vegetation. A special issue of *International Journal of Remote Sensing* (volume 18, number 4, 1997) is devoted to 'Neural Networks in Remote Sensing'. See also Tso and Mather (2001).

8.5 FUZZY CLASSIFICATION AND LINEAR SPECTRAL UNMIXING

The techniques described in the first part of Chapter 8 are concerned with 'hard' pixel labelling. All of the different classification schemes require that each individual pixel is given a single, unambiguous, label. This objective is a justifiable one whenever regions of relatively homogeneous land cover occur in the image area. These regions should be large relative to the instantaneous field of view of the sensor, and may consist of fields of agricultural crops or deep, clear water bodies that are tens of pixels in each dimension. In other instances, though, the instantaneous field of view of the sensor may be too large for it to be safely assumed that a single pixel contains just a single land cover type. In many cases, a 1 km $\times$ 1 km pixel of an AVHRR or ATSR image is unlikely to contain just one single cover type. In areas covered by semi-natural vegetation, natural variability will be such as to ensure that, even in a 20 or 30 m square pixel, there will be a range of different cover types such as herbs, bare soil, bushes, trees and water. The question of scale is one that bedevils all spatial analyses. The resolution of the sensor is not the only factor that relates to homogeneity, for much depends what is being sought. If generalised classes such as wheat, barley or rice are the targets then a resolution of 30 m rather than 1 km may be appropriate. A 30 m resolution would be quite inappropriate, however, if the investigator wished to classify individual plants. Fuzziness and hardness, heterogeneity and homogeneity are properties of the landscape at a particular geographical scale of observation that is related to the aims of the investigator. Questions of scale are considered by de Cola (1994), Levin (1991), and Ustin *et al.* (1996). The use of geostatistics in estimating spatial scales of variation is summarised by Curran and Atkinson (1998). Other references relevant to the use of geostatistics are Hyppänen (1996), Jupp *et al.* (1988, 1989), van Gardingen *et al.* (1997), Woodcock *et al.* (1988 a, b) and Woodcock and Strahler (1987).

To the investigator whose concern is to label each pixel unambiguously, the presence of large heterogeneous pixels or smaller pixels containing several cover types is problem, since they do not fall clearly within one or other of the available classes. If a conventional 'hard' classifier is used, the result will be low classification accuracy. 'Mixed pixels' represent a significant problem in the description of the Earth's terrestrial surface where that surface is imaged by an instrument with a large (1 km or more) instantaneous field of view, when natural variability occurs over a small area, or where the scale of variability of the target of interest is less than the size of the observation unit, the pixel.

Several alternatives to the standard 'hard' classifier have been proposed. The method of mixture modelling starts from the explicit assumption that the characteristics of the observed pixels constitute mixtures of the characteristics of a small number of basic cover types, or end members. Alternatively, the investigator can use a 'soft' or 'fuzzy' classifier, which does not reach a definite conclusion in favour of one class or another. Instead, these soft classifiers present the user with a measure of the degree (termed membership grade) to which the given pixel belongs to some or all of the candidate classes, and leaves to the investigator the decision as to the category into which category the pixel should be placed. In this section, the use of linear mixture modelling is described, and the use of the maximum likelihood and artificial neural net classifiers to provide 'soft' output is considered.

8.5.1 The linear mixture model

If it can be assumed that a single photon impinging upon a target on the Earth's surface is reflected into the field of view of the sensor without interacting with any other ground surface object, then the total number of photons reflected from a single pixel area on the ground and intercepted by a sensor can be described in terms of a simple linear model, as follows:

$$r_i = \sum_{j=1}^{m} a_{ij} f_j + e_i$$

in which r_i is the reflectance of a given pixel in the ith of m spectral bands. The number of mixture components is n, f_j is the jth fractional component (proportion of end member j) in the makeup of r_i, and a_{ij} is the reflectance of mixture component j in spectral band i. The term e_i expresses the difference between the observed pixel reflectance r_i and the reflectance for that pixel computed from the model. In order for the components of $\mathbf{r}$ $(= r_i)$ to be computable, the number of mixture components n must be less than the number of spectral bands, m. This model is simply expressing the fact that if there are n land cover types present in the area on the land surface that is covered by a single pixel, and if each photon reflected from the pixel area interacts with only one of these n cover types, then the integrated signal received at the sensor in a given band (r_I) will be the linear sum of the n individual interactions. This model is quite well known; for example, Figure 8.18 shows how an object (P) can be described in terms of three components (A, B and C). The object may be a soil sample, and A, B and C could be the percentage of sand, silt and clay in the sample.

A simple example shows how the process works. Assume that we have a pixel of which 60% is covered by material with a spectral reflectance curve given by the lower curve in Figure 8.19 and 40% is covered by material with a spectral reflectance curve like the upper curve in Figure 8.19. The values 0.6 and 0.4 are the proportions of these two "end members" contributing to the pixel reflectance. The values of these mixture components are shown in the first two columns of Table 8.2, labelled C1 and C2. A 60:40 ratio mixture of the two mixture components is shown as the

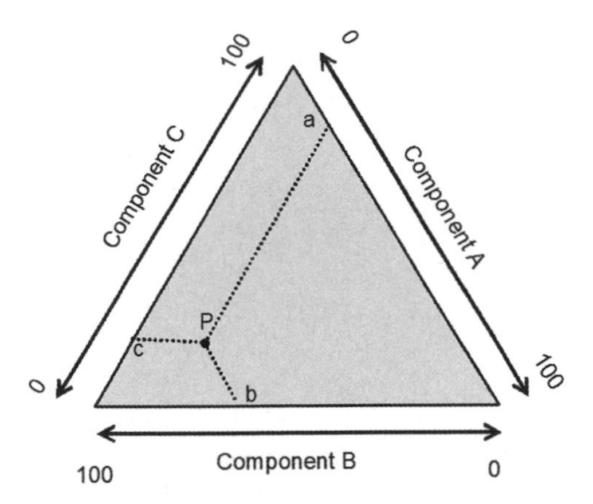

Figure 8.18 Point P represents an object, such as a soil sample, which is composed of a mixture of three components, a, b and c. The proportions of the components (in percent) are a, b and c. The components could, for example, represent the proportions of sand, silt and clay making up the soil sample. These proportions could be determined by sieving. In a remote sensing context, components a, b and c represent three land cover types. The objective of linear spectral unmixing is to determine statistically the proportions of the different land cover types within a single pixel, given the pixel's spectral reflectance curve and the spectral reflectance curves for components a, b and c.

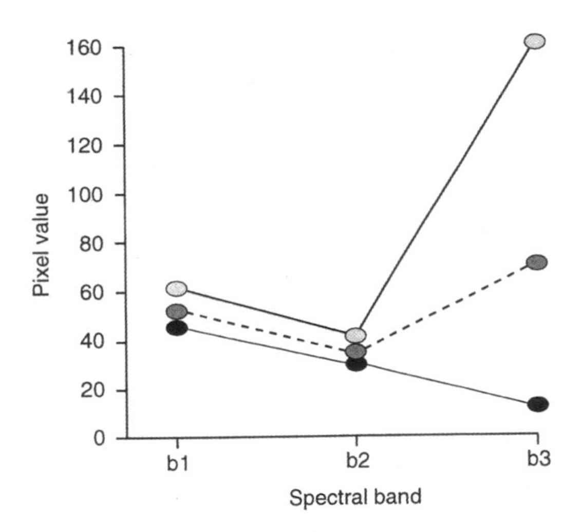

Figure 8.19 Top and bottom curves represent end member spectra. The centre curve is the reflectance spectrum of an observed pixel, and is formed by taking 60% of the value of the lower curve and 40% of the value of the upper (dashed line).

middle (dashed) curve in Figure 8.19 and in column M of Table 8.2. The rows of the table (b1, b2 and b3) represent three spectral bands in the green, red and near infrared respectively, so C1 may be turbid water and C2 may be

Table 8.2 Columns $C1$ and $C2$ show the reflectance spectra for two pure types. Column M shows a 60:40 ratio mixture of $C1$ and C2. See text for discussion.

	C1	C2	M
b1	46	62	52.4
b2	31	42	35.1
b3	12	160	68.8

vigorous vegetation of some kind. The data in Table 8.2 therefore describe two end members, with contributions shown by the proportions in columns C1 and C2, from which a mixed pixel vector, M, is derived. We will now try to recover the values of the mixture proportions f_1 and f_2 from these data, knowing in advance that the correct answer is 0.6 and 0.4.

Firstly, define a matrix **A** with three rows (the spectral bands) and two columns (the end member proportions, represented by columns C1 and C2 in Table 8.2). Vector **b** holds the measurements for each spectral band of the mixed pixel (column M of Table 8.2). Finally **f** is an unknown vector, which will contain the proportions f_1 and f_2 as the elements of its two rows. Assume that the relationship between **A** and **b** is of the form: $\mathbf{Af} = \mathbf{b}$, which is equivalent to the following set of simultaneous linear equations:

$$52.4 = 46f_1 + 62f_2$$
$$35.4 = 31f_1 + 42f_2$$
$$71.2 = 12f_1 + 160f_2$$

Because we have made up the columns of **A** and **b** we know that f_1 and f_2 must equal 0.6 and 0.4, respectively, but readers should check that this is in fact the case by, for example, entering the values of **A** and **b** into a spreadsheet and using the Multiple Regression option, remembering to set the intercept to 'zero' rather than 'computed'. This should give values of f_1 and f_2 equal to 0.6 and 0.4, respectively. If linear mixture modelling were being applied to an image then the values f_1 and f_2 would be scaled onto the range 0–255 and written to file as the two output fraction images, thus generating an output fraction image for each mixture component.

The example given above assumes that the spectral reflectance curve that is derived as a mixture of two other spectral reflectance curves is unique and is different from other spectral reflectance curves. In reality, this may not be the case, as Price (1994), shows. For instance, the spectral reflectance curve for corn is intermediate between the spectral reflectance curves of soya beans and winter wheat. Hence, the procedure outlined in the example above would be incapable of distinguishing a pixel that was split between two agricultural crops (soya beans and winter wheat) and a pure pixel covered by corn. Applications of spectral unmixing with multispectral data use general classes such as 'green vegetation' and 'soil', so this confusion may not be too serious. However, hyperspectral data may be used to identify specific mineral/rock types, and Price's point is relevant to such studies. Sohn and McCoy (1997) also consider problems in end member selection.

In a real application, rather than a fictitious example, it is necessary to go through a few more logical steps before determining the values of the fractional (end member) components, otherwise serious errors could result. First of all, in our example we knew that the mixed pixel values are composed of two components, C1 and C2 in Table 8.2. In reality, the number of end members is not known. Secondly, the values of the fractions (the proportions of the mixture components used to derive column M from columns C1 and C2) were known in advance to satisfy two logical requirements, namely:

$$0.0 \leq f_i \leq 1.0$$

$$\sum_{j=1}^{n} f_j \leq 1.0$$

These constraints specify that the individual fractions f_i must take values between 0% and 100%, and that the fractions for any given mixed pixel must sum to 100% or less. These two statements are a logical part of the specification of the mixture model, which, in effect, assumes that the reflectance of a mixed pixel is composed of a linear weighted sum of a set of end member reflectances, with no individual proportion exceeding the range 0–1 and the sum of the proportions being 1.0 at most. This statement specifies the linear mixture model.

There are several possible ways of proceeding from this point. One approach, called 'unconstrained', is to solve the mixture model equation without considering the constraints at all. This could result in values of f_i that lie outside the 0–1 range (such illogical values are called under-shoots or over-shoots, depending whether they are less than 0 or greater than 1). However, the value of the unconstrained approach is that it allows the user to estimate how well the linear mixture model describes the data. The following criteria may be used to evaluate the goodness of fit of the model:

1. *The size of the residual term, e_i, in the mixture equation.* There is one residual value in each spectral band, representing the difference between the observed pixel value and the value computed from the linear mixture model equation. In the simple example given above, the three residuals are all zero. It is normal to take the square root of the sum of squares of all the residuals for a given pixel divided by the number of spectral bands, m, to give the root mean squared (RMS) error for that pixel:

$$\text{RMS} = \sqrt{\frac{\sum_{b=1}^{m} e_b^2}{m}}$$

The RMS error is calculated for all image pixels and scaled to the 0–255 range in order to create an RMS error image. The larger the RMS error, the worse the fit of the model. Since the residuals are assumed to be random, then any spatial pattern that is visible in the RMS image can be taken as evidence that the model has not fully accounted for the systematic variation in the image data, which in turn implies that potential end members have been omitted from the model, or that the selected end members are deficient, or that the assumed linear relationship $\mathbf{Af} = \mathbf{b}$ does not describe the relationship between the pixel spectrum and the end member spectra.

2. *The number of pixels that have proportions f_i that lie outside the logical range of 0–1.* These under-shoots (f_i less than 0.0) and over-shoots (f_i greater than 1.0) indicate that the model does not fit. If there are only a few per cent of pixels showing under- and over-shoots then the result can be accepted, but if large numbers (say greater than 5%) of pixels under- or over-shoot then the model does not fit well.

Under-shoots and over-shoots can be coded by a method which maps the legitimate range of the fractions f_i (i.e., 0–1) to the range 100–200 by multiplying each f_i by 100 and adding 100 to the result, which is constrained to lie in the range 0–255. This ensures that under-shoots and over-shoots are mapped onto the range 0–99 and 201–255, respectively. Any extremely large under- and over-shoots map to 0 and 255, respectively. Each fraction image can then be inspected by, for example, applying a pseudo-colour transform to the range 100–200. All of the legitimate fractions will then appear in colour and the out of range fractions will appear as shades of grey. Alternatively, the ranges 0–99 and 201–255 can be pseudo-coloured (section 5.4).

If unconstrained mixture modelling is used then the number of under- and over-shoots, and the overall RMS error, should be tabulated in order to allow others to evaluate the fit of the model. Some software packages simply map the range of the fractions for a given mixture component, including under- and over-shoots, onto a 0–255 scale without reporting the presence of these under- and over-shoots. Whilst this approach may produce an interpretable image for a naive user, it fails to recognise a basic scientific principle, which holds that sufficient information should be provided to permit others independently to replicate your experiments and test your results.

The model may not fit well for one or more of several reasons. Firstly, the pixels representing specific mixture components may be badly chosen. They should represent 'pure' pixels, composed of a single land-cover type, that determine the characteristics of all of other image pixels. It is easy to think of simple situations where mixture component selection is straightforward; for example, a flat, shadow-free landscape composed of dense forest, bare soil and deep, clear water. These three types can be thought of as the vertices of a triangle within which all image pixels are located, by analogy with the triangular sand-silt-clay diagram used in sedimentology (Figure 8.18). If the forest end member is badly chosen and, in reality, represents a ground area that is only 80% forest then any pixel with a forest cover of more than 80% will be an over-shoot. There may be cases in which no single pixel is completely forest-covered, so the forest end member will not represent the pure case.

Secondly, a mixture component may have been omitted. This is a problem where the landscape is complex and the number of image bands is small because, as noted earlier, the number of mixture components cannot exceed the number of bands. Thirdly, the data distribution may not be amenable to description in terms of the given number of mixture components. Consider the triangle example again. The data distribution may be circular or elliptical, and thus all of the pixels in the image may not fit within the confines of the triangle defined by three mixture components. Fourthly, the assumption that each photon reaching the sensor has interacted with only a single object on the ground may not be satisfied, and a nonlinear mixing model may be required (Ray and Murray, 1996). The topic of choice of mixture components is considered further in the following paragraphs.

Some software uses a simple procedure to make the model appear to fit. Negative fractions (under-shoots) are set to zero, and the remaining fractions are scaled so that they lie in the range 0–1 and add to 1. This might seem like cheating, and probably is. An alternative is to use an algorithm that allows the solution of the linear mixture model equation subject to the constraints. This is equivalent to searching for a solution within a specified range, and is no different from the use of the square root key on a calculator. Generally speaking, when we enter a number and press the square root key we want to determine the positive square root, and so we ignore the fact that a negative square root also exists. If anyone were asked 'What is the square root of four?' he or she would be unlikely to answer 'Minus two', though this is a correct answer. Similarly, the solution of a quadratic equation may result in an answer that lies in the domain of complex numbers. That solution, in some instances, would be unacceptable and the alternative solution is therefore selected.

Lawson and Hansen (1995) provide a Fortran-90 subroutine, *BVLS*, which solves the equations $\mathbf{Af} = \mathbf{b}$ subject to constraints on the elements of **f**. This routine, which is available on the World Wide Web via the Netlib library, does not allow for a constraint on the sum of the f_i. An alternative is a Fortran-90 routine from the IMSL mathematical library provided with Microsoft Powerstation Fortran v4.0 Professional Edition. This routine, *LCLSQ*, allows the individual f_i to be constrained and it also allows the sum of the f_i to be specified.

The solution of the mixture model equation in the unconstrained mixture model requires some thought. The standard solution of the matrix equation $\mathbf{Af} = \mathbf{b}$ generally involves the calculation of the inverse of **A** (represented as $\mathbf{A}^{-1}$) from which **f** is found from the expression $\mathbf{f} = \mathbf{A}^{-1}\mathbf{b}$. A number of pitfalls are likely to be encountered in evaluating this innocent-looking expression. If **A** is an orthogonal matrix (that is, its columns are uncorrelated) then the derivation of the inverse of **A** is easy (it is the transpose of **A**). If the columns of **A** (which contain the reflectances of the end members) are correlated then error enters the calculations. The greater the degree of inter-dependence between the columns the more likely it is that error may become significant. When the columns of **A** are highly correlated (lin early dependent) then matrix **A** is said to be *near-singular*. If one column of **A** can be calculated from the values in the other columns then *A* is *singular*, which means that it does not possess an inverse (just has the value 0.0 does not possess a reciprocal). Consequently, the least-squares equations cannot be solved. However, it is the problem of near-singularity of **A** that should concern users of linear mixture modelling because it is likely that the end member spectra are similar. The solution (the elements of vector **f**) may, in such cases, be significantly in error, especially in the case of multispectral (as opposed to hyperspectral) data when the least-squares equations are being solved with relatively small numbers of observations.

Boardman (1989) suggests that a procedure based on the singular value decomposition (SVD) handles the problem of near-singularity of *A* more effectively. The inverse of matrix **A** (i.e., $\mathbf{A}^{-1}$) is found using the SVD:

$$\mathbf{A} = \mathbf{UWV}'$$

where **U** is an $m \times n$ column-orthogonal matrix, **W** is an $n \times n$ matrix of singular values and **V** is an $n \times n$ matrix of orthogonal columns. $\mathbf{V}'$ is the matrix transpose of **V**. The inverse of **A** is found from the following expression:

$$\mathbf{A}^{-1} = \mathbf{VW}^{-1}\mathbf{U}'$$

The singular values, contained in the diagonal elements of **W**, give an indication of the dimensionality of the space containing the spectral end members. They can be thought of as analogous to the principal components of a correlation or covariance matrix. If the spectral end members are completely independent then the information in each dimension of the space defined by the spectral bands is equal and the singular values $\mathbf{W}_{ii}$ are all equal. Where one spectral end member is a linearly combination of the remaining end members then one of the diagonal elements of **W** will be zero. Usually neither of these extreme cases is met with in practice, and the singular values of **A** (like the eigenvalues of a correlation matrix, used in principal components analysis) take different magnitudes. An estimate of the true number of end members can therefore be obtained by observing the magnitudes of the singular values, and eliminating any singular values that are close to zero. If this is done then the inverse of **A** can still be found, whereas if matrix inversion methods are used then the numerical procedure either fails or produces a result which is incorrect.

If the landscape is composed of a continuously varying mixture of idealised or pure types, it might appear to be illogical to search within that landscape for instances of these pure types. Hence, in a number of studies, laboratory spectra have been used to characterise the mixture components. Adams *et al.* (1993) term these spectra *reference end members*. They are most frequently used in geological studies of arid and semi-arid areas. Since laboratory spectra are recorded in reflectance or radiance units, the image data must be calibrated and atmospherically corrected before use, as described in Chapter 4.

In other cases, empirical methods are used to determine which of the pixels present in the image set can be considered to represent the mixture components. Murphy and Wadge (1994) use principal components analysis (PCA, Chapter 6) to identify candidate image end members. A graphical display of the first two principal components of the image set is inspected visually and

> '...the pixels representing the end member for each cover type should be located at the vertices of the polygon that bounds the data space of the principal components which contain information'
>
> (Murphy and Wadge, 1994, p. 73).

This approach makes the assumption that the m dimensional space defined by the spectral bands can be collapsed onto two dimensions without significant loss of proximity information; in other words, it is assumed that pixels that lie close together in the PC1–PC2 plot are actually close together in the m-dimensional space, and vice versa. Principal components analysis is discussed in section 6.4, where it is shown that the technique is based on partitioning the total variance of the image data set in such a way that the first principal component accounts for the maximum variance of a linear combination of the spectral bands, principal component two accounts for a maximum of the remaining variance and so on, with the restriction that the principal components are orthogonal. Since PCA has the aim of identifying dimensions of variance (in that the principal components are ordered on the basis of their variance), principal components 1 and 2 will, inevitably, contain much of the information present in the image data set. However, significant variability may remain in the lower-order principal components. Furthermore, if the PCA is based on covariances, then the spectral bands may contribute unequally to the total variance analysed. Tompkins *et al.* (1997) present an elaborate methodology for end member selection; their procedure estimates both the end member proportions and the end member reflectances simultaneously.

An empirical method for selecting mixture components is Sammon's (1969) Nonlinear Mapping. The method uses a measure of goodness of fit between the inter-pixel distances measured in the original m-dimensional space and those measured in the reduced-dimensionality space, usually of two or three dimensions (Figure 8.10). The two- or three-dimensional representation is updated so as to reduce the error in the inter-pixel distances measured in the two- or three-dimensional space compared to the equivalent distance measured in the full space of m spectral bands. A minimum of the error function is sought using the method of steepest descent, similar to that used in the back-propagating artificial neural net. Study of the plots of pairs of dimensions provides some interesting insights into the nature of the materials in the image. Nonlinear Mapping analysis appears to be a better way of analysing image information content than the principal components approach, as the analysis is based on inter-pixel distances. Bateson and Curtiss (1996) discuss another multi-dimensional visualisation method of projecting points (representing pixels) onto two dimensions. Their method is based on a procedure termed parallel coordinate representation (Wegman, 1990) which allows the derivation of 'synthetic' end member spectra that do not coincide with image spectra. The method is described in some detail by Bateson and Curtiss (1996).

A simple but effective alternative to linear spectral unmixing that does not have any statistical overtones is provided by the method of Spectral Angle Mapping (SAM), which is based on the well-known coefficient of proportional similarity, or cosine theta ($\cos\theta$). This coefficient measures the difference in the shapes of the spectral curves (Imbrie, 1963; Weinand, 1974). It is insensitive to the magnitudes of the spectral curves, so two curves of the same shape are considered to be identical. If two spectral reflectance curves have the same shape but differ in magnitude, one might conclude that the difference is due to changes in illumination conditions. This is a reasonable claim, but the method does not *correct* for illumination variations.

In the context of image processing, a small number of pixels is selected as the reference set **r** and the remaining image pixels (represented by vectors **t**) are compared to these reference pixels. Each pixel is considered as a geometric vector. For simplicity, let us use a feature space defined by two spectral bands, as shown in Figure 8.20(a). If the pixel values stored in **r** and **t** are measured on an 8-bit, 16 bit or 32-bit unsigned integer scale then the points representing pixels will lie in the first quadrant, as they are all non-negative. Figure 8.20(a) shows a reference vector **r** and an image pixel vector **t**. Both vectors are measured on a scale from 0 to 255, and $0 \leq \theta \leq 90°$.

If the line (vector) joining the pixel of interest **t** to the origin is coincident with the reference pixel vector **r** then the angle between the two vectors (pixel and reference) is zero. The cosine of zero is one, so $\cos(\theta) = 1$ means complete similarity. The maximum possible angle between a reference pixel vector and an image pixel vector is 90°, and $\cos(90°) = 0$, which implies complete dissimilarity.

For image data measured on a 0–255 scale, the output image can either be the cosine of the angle θ (which lies on a 0–1 scale with 1 meaning similar and 0 meaning dissimilar, so that similar pixels will appear in light shades of grey) or the value of the angle θ (on a scale of 0–90° or 0–$\pi/2$ radians). Both of these representations would require that the output image be written in 32-bit real format (section 3.2). If the angular representation is used then you must remember that pixels with a value of θ equal to 0° (i.e., complete similarity) will appear as black, while dissimilar pixels ($\theta = 90°$ or $\pi/2$ radians) will appear in white. Of course, the real number scales 0.0–1.0, 0.0–90.0 or 0.0–$\pi/2$) can be mapped onto a 0–255 scale using one of the methods described in section 3.2.1.

If the input data are represented in 32-bit real form, then it is possible that some image pixel values will be negative (for example, principal component images can have negative pixel values). In this case, additional possibilities exist. A reference vector in two-dimensional feature space may take the values (90, 0) so that its corresponding vector joins the origin to the point (90, 0). Imagine a pixel vector with the values (−90, 0). This is equivalent to a vector joining the origin to the point (-90, 0), so that the angle between the reference and pixel vectors is 180°. The cosine of 180°is −1.0. This value does not mean that the two vectors have no relationship – they do, but it is an inverse one, so that the

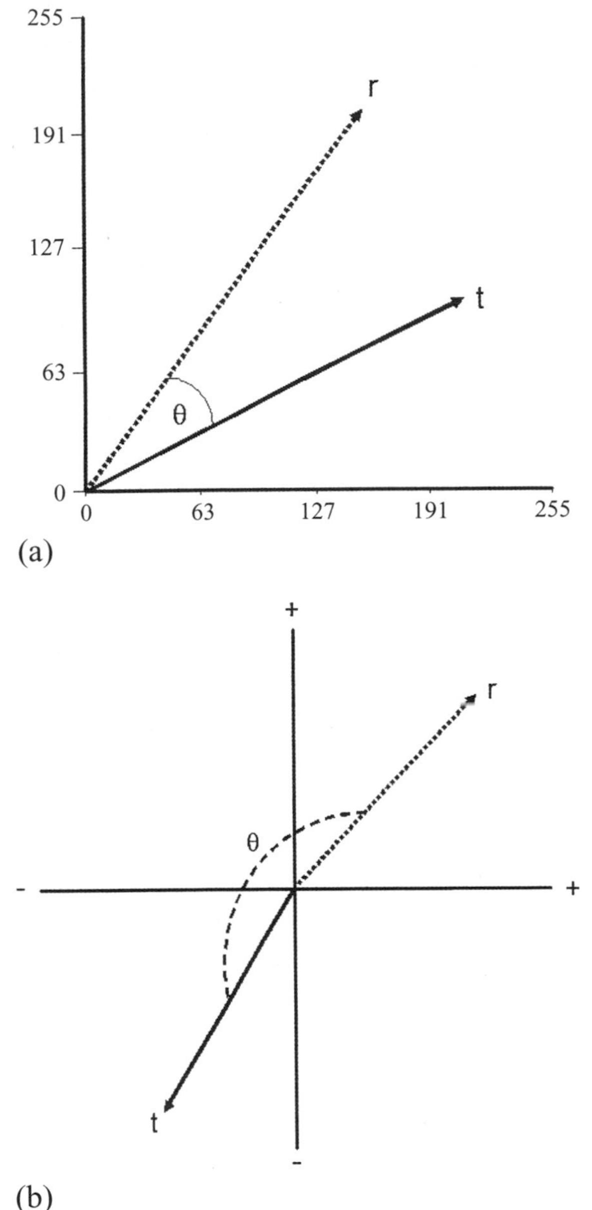

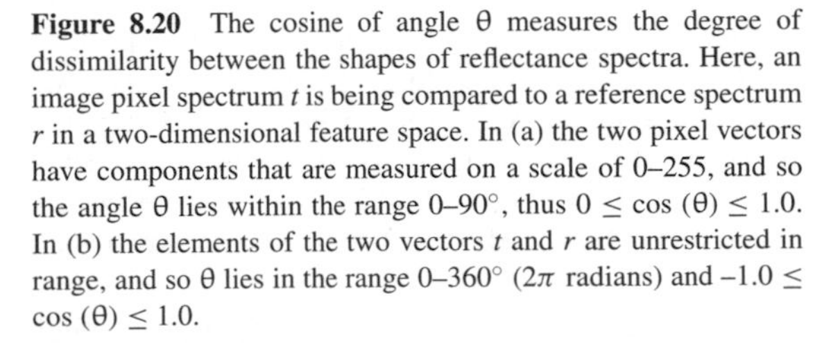

Figure 8.20 The cosine of angle θ measures the degree of dissimilarity between the shapes of reflectance spectra. Here, an image pixel spectrum t is being compared to a reference spectrum r in a two-dimensional feature space. In (a) the two pixel vectors have components that are measured on a scale of 0–255, and so the angle θ lies within the range 0–90°, thus $0 \leq \cos(\theta) \leq 1.0$. In (b) the elements of the two vectors t and r are unrestricted in range, and so θ lies in the range 0–360° (2π radians) and $-1.0 \leq \cos(\theta) \leq 1.0$.

pixel vector takes the same values as the reference vector but with the opposite sign. The output pixel values will thus lie within the range (−1.0, 1.0). When the output image is displayed then pixel vectors that are identical to a reference vector will appear white (1.0), whereas pixels that are not related to a reference vector will take the value 127 (if a linear mapping is performed from (−1.0, 1.0) to (0, 255)). Unrelated pixels will appear as mid-grey, whereas pixels whose vectors are the reciprocals of the reference vector will appear black (−1.0). Figure 8.20(b) shows a reference pixel **r** and an image pixel **t**. Both **r** and **t** are 32-bit real quantities, and the angle θ between them is almost 180°, implying that **r** is the mirror image of **t** (or vice versa).

If the ith element of the reference pixel vector is represented by r_i and any other image pixel vector is written as t_i then the value of cosine theta is calculated from:

$$\cos(\theta) = \frac{\sum_{i=1}^{N} r_i t_i}{\left(\sum_{i=1}^{N} r_i^2\right)^{0.5} \left(\sum_{i=1}^{N} t_i^2\right)^{0.5}}$$

given N bands of image data. One image of cosine theta values is produced for each reference vector **r**. There are no statistical assumptions or other limitations on the number or nature of the reference vectors used in these calculations. Kruse *et al.* (1993), Ben-Dor and Kruse (1995) and van der Meer (1996 c) provide examples of the use of the method, in comparison with the linear spectral unmixing technique.

Applications of linear mixture modelling in remote sensing are numerous, though some authors do not give enough details for their readers to discover whether or not the linear mixture model is a good fit. If the data do not fit the model then, at best, the status of the results is questionable. Good surveys of the method are provided by Adams *et al.* (1989, 1993), Ichoku and Karnieli (1996), Mustard and Sunshine (1999), Settle and Drake (1993) and Settle and Campbell (1998). A selection of typical applications is: Cross *et al.* (1991), who use AVHRR data to provide sub-pixel estimates of tropical forest cover, and Gamon *et al.* (1993), who use mixture modelling to relate AVIRIS (imaging spectrometer) data to ground measurements of vegetation distribution. Hill (1993–4) describes the technique in the context of land degradation studies in the European Mediterranean basin, while Shipman and Adams (1987) assess the detectability of minerals on desert alluvial fans. Smith *et al.* (1990) provide a critical appraisal of the use of mixture modelling in determining vegetation abundance in semi-arid areas. Other useful references are Bateson and Curtiss (1996), Bryant (1996), García-Haro *et al.* (1996), Hill and Horstert (1996), Kerdiles and Grondona (1995), Roberts *et al.* (1993), Shimabukuro and Smith (1991), Thomas *et al.* (1996), Ustin *et al.* (1996) and Wu and Schowengerdt (1993). Roberts *et al.* (1993) consider an interesting variant on the standard mixture modelling technique which, in a sense, attempts a global fit in that each pixel in the image is assumed to be the sum or mixture of the same set of end members. Roberts *et al.* (1993) suggest that the number and nature of the end members may vary over the image. This opens the possibility of using an approach similar to stepwise multiple regression (Grossman *et al.*, 1996) in which the best k from a pool of m possible end members are selected for each pixel in turn. Foody *et al.* (1996, 1997) describe an approach to mixture modelling using artificial neural networks (section 8.4.3).

8.5.2 Fuzzy classifiers

The distinction between 'hard' and 'soft' classifiers is discussed in the opening paragraphs of the introduction to section 8.5. The 'soft' or fuzzy classifier does not assign each image pixel to a single class in an unambiguous fashion. Instead, each pixel is given a 'membership grade' for each class. Membership grades range in value from 0 to 1, and provide a measure of the degree to which the pixel belongs to or resembles the specified class, just as the fractions or proportions used in linear mixture modelling (section 8.5.1) represent the composition of the pixel in terms of a set of end members. It might appear that membership grade is equivalent to probability, and the use of probabilities in the maximum likelihood classification rule might lend support to this view. Bezdek (1993) differentiates between precise and fuzzy data, vague rules and imprecise information. Crisp sets contain objects that satisfy unambiguous membership requirements. He notes that $H = \{r \in \Re | 6 \leq 8\}$ precisely represents the crisp or hard set of real numbers H from 6 to 8. Either a number is a member of the set H or it is not. If a set F is defined by a rule such as 'numbers that are close to 7' then a given number such as 7.2 does not have a precise membership grade for F, whereas its membership grade for H is 1.0. Bezdek (1993) also notes that

> '... the modeler must decide, based on the potential applications and properties desired for F, what m_F should be' (p. 1).

The membership function m_F for set F can take any value between 0 and 1, so that numbers far away from 7 still have a membership grade. This, as Bezdek notes, is simultaneously both a strength and a weakness. Bezdek (1994) points out that fuzzy membership represents the similarity of an object to imprecisely defined properties, while probabilities convey information about relative frequencies.

Fuzzy information may not be of much value when decisions need to be taken. For example, a jury must decide, often in the basis of fuzzy information, whether or not the defendant is guilty. A fuzzy process may thus have a crisp outcome. The process of arriving at a crisp or hard conclusion from a fuzzy process is called defuzzification. In some cases, and the mixture model discussed in the previous section is an example, we may wish to retain some degree of flexibility in presenting our results. The membership grades themselves may be of interest as they may relate to the proportions of the pixel area that is represented by each of the end members. In semi-natural landscapes one may accept that different land cover types merge in a transition zone. For instance, a forested area may merge gradually into grassland, and grassland in turn may merge imperceptibly into desert. In such cases, trying to draw hard and fast boundaries around land cover types may be a meaningless operation, akin to Kimble's (1951) definition of regional geography: putting '*...boundaries that do not exist around areas that do not matter*' (Kimble, 1951, p. 159). On the other hand, woodland patches, lakes, and agricultural fields have sharp boundaries, though of course the meaning of 'sharp' depends upon the spatial scale of observation, as noted previously. A lake may shrink or dry up during a drought, or the transition between land cover types may be insignificant at the scale of the study. At a generalised scale it may be perfectly acceptable to draw sharp boundaries between land cover types. Wang (1990) discusses these points in the context of the maximum likelihood decision rule, and proposes a fuzzy version of maximum likelihood classification using weighted means and variance–covariance matrices similar to those described above (section 8.4.1) in the context of deriving robust estimates from training samples. Gong *et al.* (1996) use the idea of membership functions to estimate classification uncertainty. If the highest membership function value for a given pixel is considerably greater than the runner-up, then the classification output is reasonably certain, but where two or more membership function values are close together then the output is less certain.

One of the most widely used unsupervised fuzzy classifiers is the fuzzy c-means clustering algorithm. Bezdek *et al.* (1984) describe a FORTRAN program. Clustering is based on the distance (dissimilarity) between a set of cluster centres and each pixel. Either the Euclidean or the Mahalanobis distance can be used. These distances are weighted by a factor m which the user must select. A value of m equal to 1 indicates that cluster membership is 'hard' while all membership function values approach equality as m gets very large. Bezdek *et al.* (1984) suggest that m should lie in the range $1 - 30$, though other users appear to choose a value of m of less than 1.5. Applications of this method are reported by Bastin (1997), Cannon *et al.* (1986), Du and Lee (1996), Foody (1996), and Key *et al.* (1989). Other uses of fuzzy classification procedures are reported by Blonda and Pasquariello (1991), and Maselli *et al.* (1995, 1996).

The basic ideas of the fuzzy c-means classification algorithm can be expressed as follows. **U** is the *membership grade matrix* with n columns (one per pixel) and p rows (one per cluster):

$$\mathbf{U} = \begin{bmatrix} u_{11} & \cdots & u_{1n} \\ \vdots & & \vdots \\ u_{p1} & \cdots & u_{pn} \end{bmatrix}$$

The sum of the membership grades for a given class (row of **U**) must be non-zero. The sum of the membership grades for a given pixel (column of **U**) must add to one, and the individual elements of the matrix, u_{ij}, must lie in the range 0–1 inclusive. The number of clusters p is specified by the user, and the initial locations of the cluster centres are either generated randomly or supplied by the user. The Euclidean distance from pixel i to cluster centre j is calculated as usual, that is:

$$d_{ij} = \sqrt{\sum_{l=1}^{k} (x_{il} - c_{jl})^2}$$

and the centroids c_j are computed from:

$$c_{jl} = \sum_{i=1}^{n} u_{ji}^m x_{il} \Big/ \sum_{i=1}^{n} u_{ji}^m$$

This is simply a weighted average (with the elements u being the weights) of all pixels with respect to centre $j (1 \leq j \leq p)$. The term x_{il} is the measurement of the ith pixel $(1 \leq i \leq n)$ on the lth spectral band or feature. The exponent m is discussed above.

Each of the membership grade values u_{ij} is updated according to its Euclidean distance from each of the cluster centres:

$$u_{ij} = \frac{1}{\sum_{c=1}^{p} \left(\frac{d_{ij}}{d_{cj}} \right)^{\frac{2}{(m-1)}}}$$

where $1 \leq i \leq p$ and $1 \leq j \leq n$ (Bezdek *et al.*, 1984). The procedure converges when the elements of **U** differ by no more than a small amount between iterations.

The columns of **U** represent the membership grades for the pixels on the fuzzy clusters (rows of **U**). A process of 'defuzzification' can be used to determine the cluster membership for pixel i by choosing the element in column i of **U** that contains the largest value. Alternatively, and perhaps more informatively, a set of classified images could be produced, one per class, with the class membership grades (measured on a scale from 0 to 1) expressed on a 0–255 range. If this were done, then the pixels with membership grades close to 1.0 would appear white, and pixels whose membership grade was close to zero would appear black. Colour composites of the classified images, taken three at a time, would be interpretable in terms of class membership. Thus, if class 1 were shown in red and class 2 in green, then a yellow pixel would be equally likely to belong to class 1 as to class 2.

The activations of the output neurons of an artificial neural network, or the class membership probabilities derived from a standard maximum likelihood classifier, can also be used as approximations of the values of the membership function. In an ideal world, the output from a neural network for a given pixel will be represented by a single activation value of 1.0 in the output layer of the network, with all other activation levels being set to 0.0. Similarly, an ideal classification using the maximum likelihood method would be achieved when the class membership probability for class i is 1.0 and the remaining class membership probabilities are zero. These ideal states are never achieved in practice. The use of a 'winner takes all' rule is normally used, so that the pixel to be classified is allocated to the class associated with the highest activation in the output layer of the network, or the highest class-membership probability, irrespective of the activation level or the magnitude of the maximum probability. It is, however, possible to generate one output image per class, rather than a single image representing all the classes of a 'hard' representation. The first image will show the activation for output neuron number one (or the class membership probability for that class, if the maximum likelihood method is used) scaled to a 0–255 range. Pixels with membership function values close to 1.0 are bright, while pixels with a low membership function value are dark. This approach allows the evaluation of the degree to which each pixel belongs to a given class. Such information will be useful in determining the level of confidence that the user can place in the results of his or her classification procedure.

8.6 OTHER APPROACHES TO IMAGE CLASSIFICATION

It is sometimes the case that different patterns can be distinguished on the basis of one, or a few, features. The one (or few) features may not be the same for each pair of patterns. On a casual basis, one might separate a vulture from a dog on the basis that one has wings and the other does not, while a dog and a sheep would be distinguished by the fact that the dog barks and the sheep baas. The feature 'possession of wings' is thus quite irrelevant when the decision to be made is 'is this a dog or a sheep?'. The decision-tree classifier is an attempt to use this 'stratified' or 'layered' approach to the problem of discriminating between spectral classes. The design of decision trees is a research topic of considerable interest (Wang 1986a and b; Lee and Richards, 1985; Quinlan, 1993); the latter authors found that the computer time for the decision-tree approach increased only linearly with the number of features while the time for the maximum likelihood classification increased as the square of the number of features, while classification accuracy for both approaches was not dissimilar. Descriptions of applications of decision tree techniques to the classification of remotely-sensed data are provided by Friedl *et al.* (1999), Gahegan and West (1998), and Muchoney *et al.* (2000). Pal and Mather (2003) provide a comparison of the effectiveness of decision tree, ANN and maximum likelihood methods in land cover classification.

Hansen *et al.* (1996) discuss a method based on the use of a 'classification tree' which, in principle, is similar to the decision tree approach mentioned in the preceding paragraph. Rule-based classifications are also related to the decision-tree approach, and represent an attempt to apply artificial intelligence methods to the problem of determining the class to which a pixel belongs (Srinivasana and Richards, 1990). The number of rules can become large, and may require the use of advanced search algorithms to perform an exhaustive analysis. The *genetic algorithm*, which uses a randomised approach based on the processes of mutation and crossover, has become popular in recent years as a robust search procedure. A brief introduction is provided by Tso and Mather (2001). Seftor and Larch (1995) illustrate its use in optimising a rule-based classifier, and Clark and Cañas (1995) compare the performance of a neural network and a genetic algorithm for matching reflectance spectra. The genetic algorithm is also described by Zhou and Civco (1996) as a tool for spatial decision making. Stolz and Mauser (1996) combine fuzzy logic (based on maximum likelihood probabilities) and a rule base to classify agricultural land use in a mountainous area of S. Germany and report a substantial improvement in accuracy.

If a single image set is to be classified using spectral data alone then the 'reliability' of the respective features is not usually taken into consideration. All bands are considered to be equally reliable, or important. Where multisource data are used, the question of reliability needs to be considered. For example, one might use data derived from maps, such as elevation, and may even use derivatives of these features, such as slope and aspect, as well as data sets acquired by different sensor systems (section 8.7). The Dempster-Shafer theory of evidence has been used by some researchers to develop the method of *evidential reasoning*, which is a formal procedure which weights individual data sources according to their reliability or importance. Duguay and Peddle (1996), Kim and Swain (1995), Lee *et al.* (1987), Peddle (1993, 1995a, 1995b), Peddle *et al.* (1994), Srinivasana and Richards (1990), and Wilkinson and Mégier (1990) provide a survey of the uses of the evidential reasoning approach in land cover classification, and compare its effectiveness with that of other classifiers. They note that, in comparison to the artificial neural network, the evidential reasoning approach is faster, is also non-parametric, is independent of scales of measurement, and can handle uncertainty about label assignments. Another approach to multi-source classification is based on Bayesian theory (Benediktsson *et al.*, 1990).

Object-oriented methods of image classification have become more popular in recent years due to the availability of software (E-Cognition) developed by the German company Definiens Imaging. This software uses a region-growing approach at different scale levels from coarse to fine, using both spectral properties and geometric attributes of the regions, such as shape. A hierarchy of regions at different scale levels is thus developed. Esch *et al.* (2003) provide references to the technical details, and compare the performance of the E-Cognition approach with that of the standard maximum likelihood method. The former proves to be about 8% more accurate than the latter, though the overall accuracy (section 8.10) of the maximum likelihood result is surprisingly high at 82%.

It can be beneficial to think of classification methods as complementary rather than as alternatives to each other. Procedures known as 'bagging' and 'boosting' can be used to improve the performance of weak classifiers. Bagging operates by generating a series of training data subsets by random sampling, and using each of these subsets to produce a classification. A voting procedure is then used to determine the class of a given pixel. The simplest such method is majority voting; if the majority of the p classifiers place pixel j into class k then pixel j is given the label k (Breiman, 1996). Boosting involves the repeated presentation of the training data set to a classifier, with each element of the training set being given a weight that is proportional to the difficulty of classifying that element correctly. The initial weights given to the pixels are unity. The Adaboost method of Freund and Schapire (1996) is probably the best known of these methods. Chan *et al.* (2001) provide a short example of the use of these procedures in land cover classification. Other interesting papers on these topics are Drucker *et al.* (1993) and Bauer and Kohavi (1999).

The image classification techniques discussed thus far are based on the labelling of individual pixels in the expectation that groups of neighbouring pixels will form regions or patches with some spatial coherence. This method is usually called the *per-pixel* approach to classification. An alternative approach is to use a process termed *segmentation*, which – as its name implies – involves the search for homogeneous areas in an image set and the identification of these homogeneous areas with information classes. A hierarchical approach to segmentation, using the concept of geographical scale, is used by Woodcock and Harward (1992) to delineate forest stands. This paper also contains an interesting discussion of the pros and cons of segmentation versus the per-pixel approach. Shandley *et al.* (1996) test the Woodcock-Harward image segmentation algorithm in a study of chaparral and woodland vegetation in Southern California.

Yet another alternative to the per-pixel approach requires *a priori* information about the boundaries of objects in the image, for example, agricultural fields. If the boundaries of these fields are digitised and registered to the image, then some property or properties of the pixels lying within the boundaries of the field can be used to characterise that field. For instance, the means and standard deviations in the six non-thermal TM bands of pixels lying within agricultural fields could be used as features defining the spectral reflectance properties of the fields. The fields, rather than the pixels, are then classified. This method thus uses a *per-field* approach. If field boundaries digitised from a suitable map are not available, it may be possible to use edge-detection and line-following techniques to derive boundary information from the image (section 7.4), though it is unlikely that a full set of boundaries will be extractable. Normally, the use of map and image data would take place within a Geographical Information System (GIS), which provides facilities for manipulating digitised boundary lines (for example, checking the set of lines to eliminate duplicated boundaries, ensuring that lines 'snap on' to nodes, and identifying illogical lines that end unexpectedly). One useful feature of most GIS is their ability to create buffer zones on either side of a boundary line. If a per-field approach is to be used then it would be sensible to create a buffer around the boundaries of the objects (agricultural fields) to be classified in order to remove pixels which are likely to be mixed. Such pixels may represent tractor-turning zones (headlands) as well as field boundary vegetation. The per-field approach is often used with SAR imagery because the individual pixels contain speckle noise, which leads to an unsatisfactory per-pixel classification. Averaging of the pixels within a defined geographical area such as a field generally gives to better results (Schotten *et al.* (1995), Wooding *et al.* (1993)). Lobo *et al.* (1996) discuss the per-pixel and per-field approaches in the context of Mediterranean agriculture.

8.7 INCORPORATION OF NON-SPECTRAL FEATURES

Two types of feature in addition to spectral values can be included in a classification procedure. The first kind are measures of the texture of the neighbourhood of a pixel, while the second kind represent external (i.e., non-remotely-sensed) information such as terrain elevation values or information derived from soil or geology maps. Use of textural information has been limited in passive remote sensing, largely because of two difficulties. The first is the operational definition of texture in terms of its derivation from the image data, and the second is the computational cost of carrying out the texture calculations relative to the increase in classification accuracy, if any. External data have not been widely used either, though digital cartographic data have become much more readily available in recent years. A brief review of both these topics is provided in this section.

8.7.1 Texture

Getting a good definition of texture is almost as difficult as measuring it. While the grey level of a single pixel in a greyscale image can be said to represent 'tone', the texture of the neighbourhood in which that pixel lies is a more elusive property, for several reasons. At a simple level, texture can be thought of as the variability in tone within a neighbourhood, or the pattern of spatial relationships among the grey-levels of neighbouring pixels, and which is usually described in terms such as 'rough' or 'smooth'. Variability is a variable property, however; it is not necessarily random – indeed, it may be structured with respect to direction as, for instance, a drainage pattern on an area underlain by dipping beds of sandstone. The observation of texture depends on two factors. One is the scale of the variation that we are willing to call 'texture' – it might be local or regional. The second is the scale of observation. Micro-scale textures that might be detected by the panchromatic band of the SPOT HRV would not be detected by the NOAA AVHRR due to the different spatial resolutions of the two sensor systems (10 m and 1.1 km, respectively). We must also be careful to distinguish between the real-world texture present, for example, in a field of potatoes (which are generally planted in parallel rows) and the texture that is measurable from an image of that field at a given spatial resolution.

The fact that texture is difficult to measure is no reason to ignore it. It has been found to be an important contributor to the ability to discriminate between targets of interest where the spatial resolution of the image is sufficient to make the concept a meaningful and useful one, for example in manual photo interpretation. On the other hand, an investigation of a number of texture measurements for satellite remotely-sensed data led Irons and Petersen (1981) to conclude that none of these measures was useful in thematic mapping. This divergence of opinion demonstrates the difficulty in defining the property known as texture, the dependence of this property on the scale of the image and the variation in scale of 'texture'.

The earliest application of texture measurements to digital remotely-sensed image data was published by Haralick *et al.* (1973). These authors proposed what has become known as the Grey Level Co-occurrence Matrix (GLCM) which represents the distance and angular spatial relationships over an image sub-region of specified size. Each element of the GLCM is a measure of the probability of occurrence of two greyscale values separated by a given distance in a given direction. The concept is more easily appreciated via a simple numerical example. Table 8.3(a) shows a small segment of a digital image quantised to four grey levels (0–3). The number of adjacent pixels with grey levels i and j is counted and placed in element (i, j) of the GLCM $\mathbf{P}$. Four definitions of adjacency are used; horizontal (0°), vertical (90°), diagonal (bottom left to top right – 45°) and diagonal (top left to bottom right – 135°). The inter-pixel distance used in these calculations is one pixel. Thus, four GLCM are calculated, denoted $\mathbf{P}_0$, $\mathbf{P}_{90}$, $\mathbf{P}_{45}$ and $\mathbf{P}_{135}$, respectively. For example, the element $\mathbf{P}_0(0, 0)$ is the number of times a pixel with greyscale value 0 is horizontally adjacent to a pixel which also has the greyscale value 0, scanning from left to right as well as right to left. Element $\mathbf{P}_0(1, 0)$ is the number of pixels with value 1 that are followed by pixels with value 0, while $\mathbf{P}_0(0, 1)$ is the number of pixels with value 0 that are followed by pixels with value 1, again looking in both the left-right and right-left directions. The four GLCM are shown in Table 8.3(b)–(e).

Haralick *et al.* (1973) originally proposed 32 textural features to be derived from each of the four GLCM. Few instances of the use of all these features can be cited; Jensen and Toll (1982) use only one, derived from the Landsat-1–3 MSS band 5 image. The first two of the Haralick measures will be described here to illustrate the general approach. The first measure (f_1) is termed the *angular second moment*, and is a measure of homogeneity. It effectively measures the number of transitions from one grey level to another and is high for few transitions. Thus, low values indicate heterogeneity. The second Haralick texture feature, *contrast* (f_2), gives non-linearly increasing weight to transitions from low to high greyscale values. The weight is the square of the difference in grey level. Its value is a function of the number of high/low or low/high transitions

Table 8.2 Example data and derived grey-tone spatial dependency matrices. (a): test data set; (b)–(e): grey-tone spatial dependency matrices for angles of 0, 45, 90 and 135°, respectively.

0	0	0	2	1
0	1	1	2	2
0	1	2	2	3
1	1	2	3	3

(a)

	0	2	3	
0	4	2	1	0
1	2	4	4	0
2	1	4	4	2
3	0	0	2	2

(b)

	0	1	2	3
0	2	2	0	0
1	1	4	3	0
2	0	3	6	0
3	0	0	0	2

(c)

	0	1	2	3
0	4	3	0	0
1	3	4	2	0
2	0	2	6	2
3	0	0	2	2

(d)

	0	1	2	3
0	0	4	1	0
1	4	0	3	0
2	1	3	2	3
3	0	0	3	0

(e)

in grey level. The two features are formally defined by:

$$f_1 = \sum_{i=1}^{N}\sum_{j=1}^{N}\left\{\frac{P(i,j)}{R}\right\}^2$$

$$f_2 = \sum_{n=0}^{N-1}\sum_{i=1}^{N}\sum_{j=1}^{N}\frac{P(i,j)}{R} \qquad |i-j| = n$$

where N is the number of grey levels, $P(i, j)$ is an element of one of the four GLCM listed above and R is the number of pairs of pixels used in the computation of the corresponding **P**. For the horizontal and vertical directions R is equal to $2N^2$ while in the diagonal direction R equals $2(N-1)^2$. Haralick *et al.* (1973) and Haralick and Shanmugam (1974) give examples of the images and corresponding values of f_1 and f_2. A grassland area gave low (0.064 to 0.128) values of f_1 indicating low homogeneity and high contrast whereas a predominantly water area has values of f_1 ranging from 0.0741 to 0.1016 and of f_2 between 2.153 and 3.129 (higher homogeneity, lower contrast). These values are averages of the values of f_1 and f_2 for all four angular grey-tone spatial dependency matrices. The values of f_1 for the example data in Table 8.3 are (for angles of 0, 45, 90 and 135 degrees): 0.074, 0.247, 0.104 and 0.216 while the values of f_2 for the same data are 0.688, 0.444, 0.438 and 1.555.

Rather than compute the values of these texture features for windows surrounding a central pixel, Haralick and Shanmugam (1974) derive them for 64 × 64 pixel sub-images. This shortcut is unnecessary nowadays, as sufficient computer power is available to compute local texture measures for individual pixels. They also use 16 rather than 256 quantisation levels in order to reduce the size of the matrices **P**. If all 256 quantisation levels of the Landsat TM were to be used, for example, then **P** would become very large. The reduction in the number of levels from 256 to 16 or 32 might be seen as an unacceptable price to pay, though if the levels are chosen after a histogram equalisation enhancement (section 5.3.2) to ensure equal probability for each level then a reduction from 256 to 64 grey levels will give acceptable results (Tso, 1997; Tso and Mather, 2001).

Improvements in computing power since the 1980s have led to an increased interest in the use of texture measures. Paola and Schowengerdt (1997) incorporate texture features into a neural network-based classification by including as network inputs the greyscale values of the eight neighbours of the pixel to be classified. The central pixel is thus classified on the basis of its spectral reflectance properties plus the spectral reflectance properties of the neighbourhood. Although the number of input features is considerably greater than would be the case if the central pixel alone were to be input, Paola and Schowengerdt report that the extra size of the network is compensated by faster convergence during training.

Mather *et al.* (1998) use a number of texture measures in a study of lithological mapping using SIR-C SAR and Landsat TM data. The first approach described is uses the GLCM. A second approach to texture quantisation uses filters in the frequency domain to measure the proportion of high frequency information in each of a number of moving windows for which texture is measured (section 7.5; see also Riou and Seyler, 1997). A third approach is based on the calculation of the fractal dimension of the region surrounding the pixel of interest, and a fourth method uses a model of the spatial autocorrelation properties of the same region as texture features. The conclusions drawn from the study are: firstly, classification accuracy, in terms of lithological mapping, is poor when Landsat TM spectral information is used alone. Values of accuracy (section 8.10) in the range 47%–57% were achieved, with the feed-forward neural network performing best. Classification accuracy rose by 12% at best when SAR-based texture measures were added. Secondly, the two best-performing texture descriptors were produced by the GLCM and autocorrelation modelling approaches, which raised classification accuracy to 69.5% and 68.8%, respectively. Results from the GLCM approach using 256 and 64 grey levels are compared by Tso (1997), who shows that accuracy is not significantly affected by the reduction in the number of grey levels, though the computational requirements are significantly reduced. The size of the moving windows was estimated by calculating a geostatistical function, the semi-variogram, for each lithological unit. Geostatistical methods are summarised above (section 8.5, introduction). Further reading on alternative approaches to and applications of texture measures in remote sensing image classification is provided by Bruzzone *et al.* (1997), Carlson and Ebell (1995), de Jong and Burrough (1995), Dikshit (1996), Fioravanti (1994), Franklin and Peddle (1987), Frankot and Chellappa (1987), Keller *et al.* (1989), Lark (1996), Marceau *et al.* (1990), Sarkar and Chaudhuri (1994), Schistad and Jain (1992), Stromberg and Farr (1986), Soares *et al.* (1997), Tso (1997), Tso and Mather (2001) and Wezka *et al.* (1976).

With few exceptions, texture measures have not been found to be cost-effective in terms of the improvement in classification accuracy resulting from their use. Two reasons could be proposed to account for this: (i) the difficulty of establishing the relationship between land-surface texture and scale in terms of the spatial resolution of the image and the scale of the textural feature on the ground relative to pixel size, and (ii) the cost of calculating texture features. Nevertheless, the value of texture as one of the fundamental pattern elements used in manual photo interpretation guarantees a continuing interest in the topic.

8.7.2 Use of external data

The term external (or ancillary) is used to describe any data other than the original image data or measures derived from these data. Examples include elevation and soil type information or the results of a classification of another image of the same spatial area. Some such data are not measured on a continuous (ratio or interval) scale and it is therefore difficult to justify their inclusion as additional feature vectors. Soil type or previous classification results are examples, both being categorical variables. Where a continuous variable, such as elevation, is used difficulties are encountered in deriving training samples. Some classes (such as water) may have little relationship with land surface height and the incorporation of elevation information in the training class may well reduce rather than enhance the efficiency of the classifier in recognising those categories.

An external variable may be used to stratify the image data into a number of categories. If the external variable is land elevation then, for example, the image may be stratified in terms of land below 500 m, between 500 and 800 m and above 800 m. For each stratum of the data an estimate of the frequency of occurrence of each class must be provided by the user. This estimate might be derived from field observation, sampling of a previously classified image or from sample estimates obtained from air photographs or maps. The relative frequencies of each class are then used as estimates of the prior probabilities of a pixel belonging to each of the k classes and the maximum likelihood algorithm used to take account of these prior probabilities (section 8.4.2.3). The category or level of the external variable is used to point to a set of prior probabilities, which are then used in the estimation of probabilities of class membership. This would assist in the distinction between classes which are spectrally similar but which have different relationships with the external variable. Strahler, Logan and Bryant (1978) used elevation and aspect as external variables; both were separated into three categories and used as pointers to sets of prior probabilities. They found that the elevation information contributed considerably to the improvement in the accuracy of forest cover classification. Whereas the spectral features alone produced a classification with an accuracy estimated as 57%, the addition of terrain information and the introduction of prior probability estimates raised this accuracy level to a more acceptable 71%. The use of elevation and aspect to point to a set of prior probabilities raised the accuracy further to 77%. Strahler (1980) provides an excellent review of the use of external categorical variables and associated sets of prior probabilities in maximum likelihood classification. He concludes that the method '*can be a powerful and effective aid to improving classification accuracy*'. Another accessible reference is Hutchinson (1982), while Maselli *et al.* (1995) discuss integration of ancillary data using a non-parametric method of estimating prior probabilities.

Elevation data sets are now available for many parts of the world, generally at a scale of 1:50 000 or coarser. Digital Elevation Models (DEM) can also be derived from stereo SPOT and ASTER images as well as from interferometric data from SAR sensors. The widespread availability of GIS means that many users of remotely-sensed data can now derive DEM by digitising published maps, or by using photogrammetric software to generate a DEM from stereoscopic images such as SPOT HRV, ASTER or IRS-1 LISS. Care should be taken to ensure that the scale of the DEM matches that of the image. GIS technology allows the user to alter the scale of a data set, and if this operation is performed thoughtlessly then error is inevitable. Users of DEM derived from digitised contours should refer to one of the many GIS textbooks now available (for example, Bonham-Carter, 1994) to ensure that correct procedures are followed.

Rather than use an external variable to stratify the image data for improved classifier performance, users may prefer to use what has become known as the *stacked vector* approach, in which each feature (spectral, textural, external) is presented to the classifier as an independent input. Where a statistical classifier, such as maximum likelihood, is used then this approach may well not be satisfactory. Some external variables, such as elevation, may be measured on a continuous scale but may not be normally distributed or even uni-modal for a given class. Other variables, such as lithology or soil type, may be represented by a categorical label that the maximum likelihood classifier cannot handle. The value of artificial neural network and decision tree classifiers is that they are non-parametric, meaning that the frequency distribution and scale of measurement of the individual input feature is not restricted. Thus, the ANN-based classifier can accept all kinds of input features without any assumption concerning the normality or otherwise of the associated frequency distribution and without consideration of whether the feature is measured on a continuous, ordinal or categorical scale. One problem with an indiscriminate approach, however, is that all features may not have equal influence on the outcome of the classification process. If one is trying to distinguish between vultures and dogs, then 'possession of wings' is a more significant discriminating feature than 'colour of eyes', though the latter may have some value. Evidential reasoning (section 8.6) offers a more satisfactory approach.

8.8 CONTEXTUAL INFORMATION

Geographical phenomena generally display order or structure, as shown by the observation that landscapes are not, in general, randomly organised. Thus, trees grow together in forests and groups of buildings form towns and villages. The relationship between one element of a landscape and the whole defines the context of that element. So too the relationship between one pixel and the pixels in the remainder of the image is the context of that pixel. Contextual information is often taken into account after a preliminary classification has been produced, though at the research level investigations are proceeding into algorithms which can incorporate both contextual and spectral information simultaneously (Kittler and Föglein, 1984). The simplest methods are those which are applied following the classification of the pixels in an image using one of the methods described in section 8.4. These methods are similar in operation to the spatial filtering techniques described in Chapter 7 for they use a moving window algorithm.

The first of these methods is called a 'majority filter'. It is a logical rather than numerical filter since a classified image consists of labels rather than quantised counts. The simplest form of the majority filter involves the use of a filter window, usually measuring 3 rows by 3 columns, is centred on the pixel of interest. The number of pixels allocated to each of the k classes is counted. If the centre pixel is not a member of the majority class (containing five or more pixels within the window) it is given the label of the majority class. A threshold other than five (the absolute majority) can be applied – for example, if the centre pixel has fewer than n neighbours (in the window) that are not of the same class then re-label that pixel as a member of the majority class. The effect of this algorithm is to smooth the classified image by weeding-out isolated pixels, which were initially given labels that were dissimilar to the labels assigned to the surrounding pixels. These initial dissimilar labels might be thought of as noise or they may be realistic. If the latter is the case then the effect of the majority filter is to treat them as detail of no interest at the scale of the study, just as contours on a 1:50 000-scale map are generalised (smoothed) in comparison with those on a map of the same area at a 1:25 000 scale. A modification of the algorithm just described is to disallow changes in pixel labelling if the centre pixel in the window is adjacent to a pixel with an identical label. In this context adjacent can mean having a common boundary (i.e. to the left or right, above or below) or having a common corner. The former definition allows four pixels to be adjacent to the centre pixel, the latter eight. Thomas (1980) reports a method based on what he terms a 'proximity function'. Again, it uses a moving 3 × 3 window with the pixels numbered as shown in Figure 8.21. The function is defined by:

$$F_j = \sum_i \frac{q_i q_5}{d_{i5}^2} \qquad (j = 1, \ldots, k)(i = 2, 4, 6, 8)$$

In this expression the q_i are weights. If pixel i has been placed in class j then $q_i = 2$ otherwise $q_i = 0$ for $i = 2$, 4, 6 and 8. If pixel 5 has been placed in the jth class then $q_5 = 2$ otherwise $q_5 = 1$. The d_{i5} are distances (in metres) from the centre pixel to its neighbours. The function is evaluated for all classes j and the centre pixel is reallocated

1	2	3
4	5	6
7	8	9

Figure 8.21 Pixel labelling scheme for Thomas (1980) filter window. See text for explanation.

if the maximum value of F_j exceeds a threshold. Thomas (1980) suggests a value of 12×10^{-4} m for this threshold, and illustrates the use of the method with reference to classified Landsat MSS imagery. Like the majority filter, its effect is to remove isolated pixels and re-label them with the most frequently occurring label considering the vertical and horizontal neighbours. It might also reallocate a previously unclassified pixel that had been placed in the 'reject' class by the classification algorithm. Another example of a post-classification context algorithm is given by Wharton (1980).

Harris (1981, 1985) describes a method of post-classification processing which uses a probabilistic relaxation model. An estimate of the probability that a given pixel will be labelled li ($i = 1, 2, \ldots, k$) is required. Examination of the pixels surrounding the pixel under consideration is then undertaken to attempt to reduce the uncertainty in the pixel labelling by ensuring that pixel labels are locally consistent. The procedure is both iterative and rather complicated. The results reported by Harris (1985) showed the ability of the technique to clean up a classified image by eliminating improbable occurrences (such as isolated urban pixels in a desert area) while at the same time avoiding smoothing-out significant and probably correct classifications. However the computer time requirements are considerable. Further discussion of the probabilistic relaxation model is given in Rosenfeld (1976), Peleg (1980), Kittler (1983) and Kontoes and Rokos (1996), while an alternative approach to the problem of specifying an efficient spectral-spatial classifier is discussed by Landgrebe (1980). An excellent general survey is Gurney and Townshend (1983).

More recently, attention has been given to the use of geostatistical methods of characterising the spatial context of a pixel that is to be classified. Geostatistical methods are summarised in section 8.5. Image data are used to characterise the spectral properties of the candidate pixel, and geostatistical methods provide a summary of its spatial context, so that both are simultaneously considered in the decision-making process. See Lark (1998) and van der Meer (1994, 1996 a, b) for a fuller exposition. Flygare (1997) gives a review of advanced statistical methods of characterising context. Wilson (1992) uses a modified maximum likelihood approach to include neighbourhood information by the use of a penalty function, which increases the 'cost' of labelling a pixel as being different from its neighbours. Sharma and Sarkar (1998) review a number of approaches to the inclusion of contextual information in image classification, as do Tso and Mather (2001).

8.9 FEATURE SELECTION

Developments in remote sensing instruments over the last ten years have resulted in image data of increasingly higher resolution becoming available in more spectral channels. Thus, the volume and dimensionality of data sets being used in image classification is exceeding the ability of both available software systems and computer hardware to deal with it. However, as shown in the discussion of the decision-tree approach to classification (section 8.4.6), it is possible to base a classification on the consideration of the values measured on one spectral band at a time. In this section the idea will be extended, so that we will ask: can the dimensions of the data set be reduced (in order to save computer time) without losing too much of the information present in the data? If a subset of the available spectral bands (and other features such as textural and ancillary data) will provide almost as good a classification as the full set then there are very strong arguments for using the subset. We will consider what 'almost as good' means in this context in the following paragraphs.

Reduction of the dimensionality of a data set is the aim of principal components analysis (section 6.4). An obvious way of performing the feature selection procedure would be to use the first m principal components in place of the original p features (m being smaller than p). This does not, however, provide a measure of the relative performance of the two classifications – one based on all p features, the other on m principal components. Methods of accuracy assessment (section 8.10) might be used on training and test sites to evaluate the performance directly. Information, in terms of principal components, is directly related to variance or scatter and is not necessarily a function of inter-class differences. Thus, the information contained in the last $(p - m)$ components might represent the vital piece of information needed to discriminate between class x and class y, as shown in the example of principal components analysis in section 6.4. Principal components analysis might therefore be seen as a crude method of feature selection if it is employed without due care. It could be used in conjunction with a suitable method for determining which of the possible p components should be selected in order to maximise inter-class differences, as discussed below. Spanner *et al.* (1984) give an example.

Two methods of feature selection are discussed in this section. The first is based on the derivation of a measure of the difference between all pairs from the k groups. It is called *divergence*. The second is more empirical. It evaluates the performance of a classifier in terms of a set of test data for which the correct class assignments have been established by ground observations or by the study of air photographs or maps. The classifier is applied to subsets of the p features and classification accuracy measured for each subset using the techniques described in section 8.10. A subset is selected which gives a sufficiently high accuracy for a specific problem.

The technique based on the divergence measure requires that the measurements on the members of the k classes are distributed in multivariate normal form. The effect of departures from this assumption is not known, but one can be certain that the results of the analysis would be less reliable as the departures from normality increased. If the departures are severe then the results could well be misleading. Hence, the divergence method is only to be recommended for use in conjunction with statistical (rather than neural) classifiers. The divergence measure J based on a subset m of the p features is computed for classes i and j as follows (Singh, 1984) with a zero value indicating that the classes are identical. The greater the value of $J(i, j)$ the greater is the class separability based on the m selected features:

$$J(i,j) = 0.5 \operatorname{tr}\left\{(\mathbf{S}_i - \mathbf{S}_j)\left(\mathbf{S}_j^{-1} - \mathbf{S}_\text{i}^{-1}\right)\right\}$$
$$+0.5 \operatorname{tr}\left\{\left(\mathbf{S}_i^{-1} + \mathbf{S}_j^{-1}\right)(\bar{\mathbf{x}}_i - \bar{\mathbf{x}}_j)(\bar{\mathbf{x}}_i - \bar{\mathbf{x}}_j)'\right\}$$

The symbol tr(.) means the trace or the sum of the diagonal elements of the indicated matrix. $\mathbf{S}_i$ and $\mathbf{S}_j$ are the $m \times m$ sample variance–covariance matrices for classes i and j, computed for the m selected features, and $\bar{\mathbf{x}}_\text{i}$ and $\bar{\mathbf{x}}_\text{j}$ are the corresponding sample mean vectors. For $m = 1$ (a single feature) the divergence measure for classes i and j is:

$$J(i,j) = 0.5\left(\frac{s_i^2}{s_j^2} + \frac{s_j^2}{s_i^2} - 2\right)$$

where s_i^2 and s_j^2 are the variances of the single feature calculated separately for classes i and j. Since the divergence measure takes into account both the mean vectors and the variance–covariance matrices for the two classes being compared, it is clear that the inter-class difference is being assessed in terms of (i) the shape of the frequency distribution and (ii) the location of the centre of the distribution. The divergence will therefore be zero only when the variance–covariance matrices and the mean vectors of the two classes being compared are identical.

The distribution of $J(i, j)$ is not well known so a measure called the *transformed divergence* is used instead. This has the effect of reducing the range of the statistic, the effect increasing with the magnitude of the divergence. Thus, when averages are taken, the influence of one or more pairs of widely separated classes will be reduced. The transformed divergence is obtained from:

$$JT(i,j) = c(1 - \exp\left[-J(i,j)/8\right])$$

with c being a constant used to scale the values of J_T onto a desired range. Sometimes the value 2000 is used as a scaling factor, but a value of 100 seems to be equally reasonable as the values of J_T can then be interpreted in the same way as percentages. A value of JT of 80 or more indicates good separability of the corresponding classes i and j. The values of $J_\text{T}(i, j)$ are averaged for all possible mutually exclusive pairs of classes i and j and the average pair-wise divergence is denoted by J_Tav:

$$J_\text{Tav} = \frac{2}{k\,(k-1)} \sum_{i=1}^{k-1} \sum_{j=1}^{i} J_T(i,j)$$

Study of the individual $J_\text{T}(i, j)$ might show that some pairs of classes are not statistically separable on the basis of any subset of the available features. The feature selection process might then also include a class amalgamation component. It might be worth following another line of thought. If the aim of feature selection is to produce the subset of m features that best combines classification accuracy and computational economy then, instead of considering the average separability of all pairs of classes, why not try to find that set of m features that maximises the minimum pair-wise divergence? In effect, this is trying to find the subset of m features that best performs the most difficult classification task. The minimum pairwise divergence is:

$$J_\text{min}(i,j) = \min J(i,j) \quad i < j$$

A measure called the *Bhattacharyya distance* is sometimes used in place of the divergence to measure the statistical separability (or, more correctly, the probability of correct classification) of a pair of spectral classes. It is computed from the expression:

$$B_{12} = \frac{1}{8}(\bar{\mathbf{x}}_1 - \bar{\mathbf{x}}_2)'\frac{\mathbf{S}_1 + \mathbf{S}_2}{2}(\bar{\mathbf{x}}_1 + \bar{\mathbf{x}}_2) + \frac{1}{2}\ln\frac{(\mathbf{s}_1 - \mathbf{s}_2)/2}{|\mathbf{S}_1|^{0.5}\,|\mathbf{S}_2|^{0.5}}$$

(Haralick and Fu, 1983). The quantity B_{ij} is computed for every pair of classes given m features. The sum of B_{ij} for all $k(k-1)/2$ classes is obtained and is a measure of the overall separability of the k classes using m features. All possible combinations of m out of p features are used to decide the best combination. Again, selection algorithms such as those described above for the transformed divergence could be used to improve the efficiency of the method. Like the divergence measure the Bhattacharyya distance is based on the assumption of multivariate normality.

Given that the *raison d'etre* of feature selection is the availability of several (more than four) features, the selection of combinations of m from p features is a problem. The number of subsets of size m that can be drawn from a set with p elements is:

$$\binom{p}{m} = \frac{p!}{m!(p-m)!}$$

The symbol '!' indicates 'factorial'; for example, 3! is $3 \times 2 \times 1 = 6$. If p is large then the number of subsets soon becomes very considerable. Take the Daedalus airborne scanner as an example. This instrument generates 12 channels of spectral data. If we assume that no texture features or ancillary data are added, then the number of subsets of size $m = 4$ is 495. If subsets of size $m = 12$ are to be drawn from a data set with $p = 24$ features then the number of subsets is 2 704 156. Clearly any brute-force method involving the computation of the average pairwise divergence for such a large number of subsets is out of the question. The problem of selection of optimal subsets is not dissimilar to the problem of determining the best subset of independent variables in multiple linear regression. Any of three main approaches can be used – these are the forward selection, backward elimination and stepwise procedures. The forward selection method starts with the best subset of size $m = 1$. Call this feature f_1. Now find the best subset of size $m = 2$ including f_1, that is, f_1 plus one other feature. The best subset at the end of the second cycle will be $\{f_1, f_2\}$. The procedure continues to determine subsets $\{f_1, f_2, f_3\}$ and so on, until all features are included. The user can then evaluate the list of features included and corresponding divergence value, and must weigh up the advantages of using fewer features against the cost of lower classification accuracy.

The backward elimination method works the opposite way round. Starting with the complete set $\{f_1, f_2, \ldots, f_p\}$, remove that feature which contributes least to the average pair-wise divergence. This is done by computing the average pairwise divergence for all subsets of size $p - 1$. Repeat until $m = 1$. Neither procedure is guaranteed to produce the optimal subset; indeed, both may produce differing results unless the data set is so clearly structured that no selection procedure is needed.

Stepwise methods incorporate both the addition of features to the selected set, as in forward selection, and their removal, as in backward elimination. The single best feature is selected first, with 'best' being defined as 'generating the largest classification accuracy'. Call this feature f_1. Now add that feature drawn from the set of remaining features that, together with f_1, produces the highest classification accuracy for all pairs of features that include f_1 So now the best subset is $\{f_1, f_2\}$. The increase in classification accuracy resulting from the addition of f_2 to the best subset can be tested statistically; if the increase is not statistically significant then f_2 is eliminated and the procedure terminates. If the increase is acceptably large, then a third feature is added, and the testing procedure is applied again. A second statistical test is also used. It is concerned with the question of whether any of the features included in the best subset can be eliminated without any significant loss of classification accuracy. Features that are included at an early stage in the selection process can be eliminated later. Interaction (shown by high correlations) between variables is responsible for these apparent anomalies. The process terminates when no excluded features can be added and no included features can be eliminated.

More recent methods include the use of the genetic algorithm (GA) as a search procedure. The use of the GA in this role is noted in section 8.6. It is also applicable to the feature selection problem. A good introduction to the workings of the GA is provided by Holland (1992). Other references are Coley (1999), Man *et al.* (1999) and Mitchell (1996).

Kumar (1979) describes an experiment in which the exhaustive search, forward and backward selection algorithms were employed. he found that the forward selection method produced results that were almost as good as exhaustive search and which were better than those produced by the backward elimination method. Goodenough, Narendra and O'Neill (1978) report an implementation of a branch-and-bound algorithm based on a method published by Narendra and Fukunaga (1977). This is claimed to give a globally optimum solution. Their method is based on the assumption that given a certain criterion B then, if a subset of size greater than m has a value of average pair-wise divergence that is less than B, all sub-subsets of this subset will have values which are also less than B. The advantage of the branch-and-bound approach is that many subsets are discarded without the need to evaluate them.

Swain and King (1973) report a comparative study of feature selection using divergence, transformed divergence and Bhattacharyya distance. Their results indicate that the transformed divergence and Bhattacharyya distance performed best, though it should be noted that they used artificial data that was made to be normally distributed. Other studies of feature selection are provided by Aha and Bankert (1996), Decell and Guseman

(1979), Kailath (1967), Kanal (1974), Kavzoglu and Mather (2002), Muasher and Landgrebe (1984) and Ormsby (1992). Yool *et al.* (1986) compare the use of transformed divergence and empirical approaches to the assessment of classification accuracy (section 8.10). They found no clear agreement between the results from the two alternative approaches, and attributed the differences – which in some instances were considerable – to departures from normality and conclude that

> '. . . a divergence algorithm requiring normally-distributed data may not be a reliable indicator of performance'
>
> (Yool *et al.*, 1986, p. 689).

However, if the empirical classification accuracy approach is used then a classification analysis must be carried out on test samples for each subset of m features.

The availability of high-dimensional multispectral image data is thus seen to be a mixed blessing. Additional spectral channels provide more detailed or more extensive information on the spectral response of the ground-cover targets, though their use requires additional computer time. Classification accuracy is dependent on feature-set size, yet no clear and recommendable algorithm is available to determine the subset that will produce the best compromise between accuracy and cost. Factors other than dimensionality will affect the choice of subset; the number of classes and their relative separability will have some influence on the number and choice of features needed to discriminate between them. Studies that have been carried out to date indicate that statistical methods (based on the assumption of normal distributions) should be used with caution.

Non-parametric feature selection methods do not rely on assumptions concerning the frequency distribution of the features. One such method, which has not been widely used, is proposed by Lee and Landgrebe (1993). Benediktsson and Sveinsson (1997) demonstrate its application.

8.10 CLASSIFICATION ACCURACY

The methods discussed in section 8.9 have as their aim the establishment of the degree of separability of the k spectral classes to which the image pixels are to be allocated (though the Bhattacharyya distance is more like a measure of the probability of mis-classification). Once a classification exercise has been carried out there is a need to determine the degree of error in the end product. These errors could be thought of as being due to incorrect labelling of the pixels. Conversely, the degree of accuracy could be sought. First of all, if a method allowing a 'reject' class has been used then the number of pixels assigned to this class (which is conventionally labelled '0') will be an indication of the overall representativeness of the training classes. If large numbers of pixels are labelled '0' then the representativeness of the training data sets is called into question - do they adequately sample the feature space? The most commonly used method of representing the degree of accuracy of a classification is to build a $k \times k$ *confusion* (or *error*) *matrix*. The elements of the rows i of this matrix give the number of pixels which the operator has identified as being members of class i that have been allocated to classes 1 to k by the classification procedure (see Table 8.4). Element i of row i (the ith diagonal element) contains the number of pixels identified by the operator as belonging to class i that have been correctly labelled by the classifier. The

Table 8.4 Confusion or error matrix for six classes. The row labels are those given by an operator using ground reference data. The column labels are those generated by the classification procedure. See text for explanation. (i) number of pixels in class from ground reference data. (ii) estimated classification accuracy (percent). (iii) class i pixels in reference data but not given label by classifier. pixels given label i by classifier but not class i in reference data. The sum of the diagonal elements of the confusion matrix is 350, and the overall accuracy is therefore $(350/410) \times 100 = 85.4\%$.

	Class									
Ref	1	2	3	4	5	6	(i)	(ii)	(iii)	(iv)
1	50	3	0	0	2	5	60	83.3	10	21
2	4	62	3	0	0	1	70	88.5	8	10
3	4	4	70	0	8	3	89	81.4	19	6
4	0	0	0	64	0	0	64	100.0	0	3
5	3	0	2	0	71	1	77	92.2	6	10
6	10	3	1	3	0	33	50	66.0	17	10
Col. sums	71	72	76	67	81	43	410		60	60

other elements of row i give the number and distribution of pixels that have been incorrectly labelled. The classification accuracy for class i is therefore the number of pixels in cell i divided by the total number of pixels identified by the operator from ground data as being class i pixels. The overall classification accuracy is the average of the individual class accuracies, which are usually expressed in percentage terms.

Some analysts use a statistical measure, the kappa coefficient, to summarise the information provided by the contingency matrix (Bishop *et al.*, 1975). Kappa is computed from:

$$\kappa = \frac{N \sum_{i=1}^{r} x_{ii} - \sum_{i=1}^{r} x_{i+}x_{+i}}{N^2 - \sum_{i=1}^{r} x_{i+}x_{+i}}$$

The x_{ii} are the diagonal entries of the confusion matrix. The notation x_{i+} and x_{+i} indicates, respectively, the sum of row i and the sum of column i of the confusion matrix. N is the number of elements in the confusion matrix. Row totals (x_{i+}) for the confusion matrix shown in Table 8.4 are listed in the column headed (i) and column totals are given in the last row. The sum of the diagonal elements (x_{ii}) is 350 ($\sum_{i=1}^{r} x_{ii}$for $r = 6$), and the sum of the products of the row and column marginal totals ($\sum_{i=1}^{r} x_{i+}x_{+i}$) is 28820, and the value of kappa is:

$$\kappa = \frac{410 \times 350 - 28820}{168100 - 28820} = \frac{114680}{139280} = 0.82$$

A value of zero indicates no agreement, while a value of 1.0 shows perfect agreement between the classifier output and the reference data. Montserud and Leamans (1992) suggest that a value of kappa of 0.75 or greater shows a 'very good to excellent' classifier performance, while a value of less than 0.4 is 'poor'. However, these guidelines are only valid when the assumption that the data are randomly sampled from a multinomial distribution, with a large sample size, is met.

Values of kappa are often cited when classifications are compared. If these classifications result from the use of different classifiers (such as maximum likelihood and artificial neural networks) applied to the same data set then comparisons of kappa values are acceptable, though the per cent accuracy (overall and for each class) provides as much, if not more, information. If the two classifications have different numbers of categories then it is not clear whether a straightforward logical comparison is valid. It is hard to see what additional information is provided by kappa over and above that given by a straightforward calculation of per cent accuracy. See Congalton (1991), Kalkhan *et al.* (1997), Stehman (1997) and Zhuang *et al.* (1995).

The confusion matrix procedure stands or falls by the availability of a test sample of pixels for each of the k classes. The use of training-class pixels for this purpose is dubious and is not recommended – one cannot logically calibrate and evaluate a procedure using the same data set. A separate set of test pixels should be used for the calculation of classification accuracy. Users of the method should be cautious in interpreting the results if the ground data from which the test pixels were identified was not collected on the same date as the remotely-sensed image, for crops can be harvested or forests cleared. Other problems may arise as a result of differences in scale between test and training data and the image pixels being classified.

The literal interpretation of accuracy measures derived from a confusion matrix can lead to error. Would the same level of accuracy have been achieved if a different test sample of pixels had been used? Figure 8.22 shows an extract from a hypothetical classified image and the corresponding ground reference data. If the section outlined in the solid line in Figure 8.22(a) had been selected as test data

A	A	A	A	F	A	A
A	A	A	F	F	U	U
A	A	A	F	A	A	A
A	A	A	A	A	A	A
U	A	F	A	U	U	U
A	A	A	A	U	U	A
A	A	A	A	A	A	A

(a)

A	A	A	A	A	A	A
A	A	A	A	A	A	A
A	A	A	A	A	A	A
A	A	A	A	A	A	A
A	A	A	A	A	A	A
A	A	A	A	A	A	A
A	A	A	A	A	A	A

(b)

Figure 8.22 Cover type categories derived from ground data (a) and image classifier (b). The choice of sample locations (solid lines, bottom left and bottom right corners in (a)) will influence the outcome of accuracy assessment measures.

the user would infer that the classification accuracy was 100% whereas if the area outlined by the dashed line had been selected then the accuracy would appear to be 75%. For a given spectral class there are a very large number of possible configurations of test data and each might give a different accuracy statistic. It is likely that the distribution of accuracy values could be summarised by a conventional probability distribution, for example the hypergeometric distribution, which describes a situation in which there are two outcomes to an experiment, labelled P (successful) and Q (failure), and where samples are drawn from a population of finite size. If the population being sampled is large the binomial distribution (which is easier to calculate) can be used in place of the hypergeometric distribution. These statistical distributions allow the evaluation of confidence limits which can be interpreted as follows: if a very large number of samples of size N are taken and if the true proportion P of successful outcomes is P_T then 95% of all the sample values will lie between P_L and P_U (the lower and upper 95% confidence limits around P_T). The values of the upper and lower confidence limits depend on (i) the level of probability employed and (ii) the sample size N. The confidence limits get wider as the probability level increases towards 100% so that we can always say that the 100% confidence limits range from minus infinity to plus infinity. Confidence limits also get wider as the sample size N becomes smaller, which is self-evident.

Jensen (1986, p.228) provides a formula for the calculation of the lower confidence limit associated with a classification accuracy value obtained from a training sample of N pixels. The formula is used to determine the required r% lower confidence limits given the values of P, Q and N is:

$$s = P - \left[z\sqrt{PQ/N} + \frac{50}{N} \right]$$

where z is the $(100 - r)/100$th point of the standard normal distribution. Thus, if r equals 95% then the z value required will be that having a probability of (100 – 95)/100 or 0.05 under the standard normal curve. The tabled z value for this point is $z = 1.645$. If r were 99% then z would be 2.05. To illustrate the procedure assume that, of 480 test pixels, 381 were correctly classified, giving an apparent classification accuracy (P) of 79.375%. Q is therefore (100 – 79.375) = 20.625. If the lower 95% confidence limit was required then z would equal 1.645 and:

$$\begin{aligned} s &= 79.375 - [1.645\surd(79.375 \times 20.625/480) + 50/480 \\ &= 79.375 - [1.645 \times 1.847 + 0.104] \\ &= 76.223\% \end{aligned}$$

This result indicates that, in the long run, 95% of training samples with observed accuracies of 79.375% will have true accuracies of 76.223% or greater. As mentioned earlier, the size of the training sample influences the confidence level. If the training sample in the above example had been composed of 80 rather than 480 pixels then the lower 95% confidence level would be:

$$\begin{aligned} s &= 79.375 - [1.645\surd(79.375 \times 20.625/80) + 50/80 \\ &= 71.308\% \end{aligned}$$

This procedure can also be applied to individual classes in the same way as described above with the exception that P is the number of pixels correctly assigned to class j from a test sample of N_j pixels.

The confusion matrix can be used to assess the nature of erroneous labels besides allowing the calculation of classification accuracy. Errors of omission are committed when patterns that are really class i become labelled as members of some other class, whereas errors of commission occur when pixels that are really members of some other class are labelled as members of class i. Table 8.4 shows how these error rates are calculated. From these error rates the user may be able to identify the main sources of classification accuracy and alter his or her strategy appropriately. Congalton, Oderwald and Mead (1983), Congalton (1991) and Story and Congalton (1986) give more advanced reviews of this topic.

How to calculate the accuracy of a fuzzy classification might appear to be a difficult topic; refer to Gopal and Woodcock (1994) and Foody and Arora (1996). Burrough and Frank (1996) consider the more general problem of fuzzy geographical boundaries. The question of estimating area from classified remotely-sensed images is discussed by Canters (1997) with reference to fuzzy methods. Dymond (1992) provides a formula to calculate the root mean square error of this area estimate for 'hard' classifications (see also Lawrence and Ripple, 1996). Czaplewski (1992) discusses the effect of mis-classification on areal estimates derived from remotely-sensed data.

The use of single summary statistics to describe the degree of association between the spatial distribution of class labels generated by a classification algorithm and the corresponding distribution of the true (but unknown) ground cover types is rather simplistic. Firstly, these statistics tell us nothing about the spatial pattern of agreement or disagreement. An accuracy level of 50% for a particular class would be achieved if all the test pixels in the upper half of the image were correctly classified and those in the lower half of the image were incorrectly classified, assuming an equal number of test pixels in both halves of the image. The same degree of accuracy would be computed if the pixels in agreement (and disagreement) were randomly distributed over the image area. Secondly, statements of 'overall accuracy' levels can hide a multitude of sins. For example, a small number of generalised classes will usually be identified more accurately than would a larger number of more specific classes, especially if one of the general classes is 'water'. Thirdly, a number of researchers appear to use the same pixels to train and to test a supervised classification. This practice is illogical and cannot provide much information other than a measure of the 'purity' of the training classes. More thought should perhaps be given to the use of measures of confidence in pixel labelling. It is more useful and interesting to state that the analyst assigns label x to a pixel, with the probability of correct labelling being y, especially if this information can be presented in quasi-map form. A possible measure might be the relationship between the first and second highest membership grades output by a soft classifier. The use of ground data to test the output from a classifier is, of course, necessary. It is not always sufficient, however, as a description or summary of the value or validity of the classification output.

Foody (2002) provides a useful summary of the state of the art in land cover classification accuracy assessment. He describes a number of problems associated with procedures based on the confusion matrix, including misregistration, sampling issues and accuracy of the reference data, and concludes that 'it is unlikely that a single standardised method of accuracy assessment and reporting can be identified'. Vieira and Mather (2000) consider in more detail one of the points raised by Foody (2002), namely, the spatial distribution of error and its visualisation.

8.11 SUMMARY

Relative to other chapters of this book, the 1999 (second edition) version of this chapter showed the greatest increase in size relative to the first (1987) edition. No comparable increase has occurred between the second and third editions. This is not to imply that no interesting developments in classification methodology have occurred in recent years. However, the level of mathematical and statistical sophistication required to understand and implement the newer methods such as the use of support vector machines, simulated annealing, Markov random fields, and multifractals is well beyond the scope of this book. Readers requiring more detail on these topics should refer to advanced texts such as Landgrebe (2003) or Tso and Mather (2001). Other questions that are important in an operational as well as an academic context have not been described in any detail, such as the effects of mis-registration of image sets or the need for atmospheric correction when changes in classification labelling over time are used in change detection studies (this latter point is considered by Song *et al.* (2001).

The use of artificial neural net classifiers, fuzzy methods, new techniques for computing texture features, and new models of spatial context, which were introduced into remote sensing during the 1990s, are now better known. This chapter has hardly scratched the surface, and readers are encouraged to follow up the references provided at various points. I have deliberately avoided providing potted summaries of each paper or book to which reference is made in order to encourage readers to spend some of their time in the library. However, 'learning by doing' is always to be encouraged. The MIPS software supplied with this book contains some modules for image classification. These programs are intended to provide the reader with an easy way into image classification. More elaborate software is required if methods such as artificial neural networks, evidential reasoning and fuzzy classification procedures are to be used. It is important, however, to acquire familiarity with the established methods of image classification before becoming involved in advanced methods and applications.

Despite the efforts of geographers following in the footsteps of Alexander von Humboldt over the last 150 years, we are still a long way from being able to state with any acceptable degree of accuracy the proportion of the Earth's land surface that is occupied by different cover types. At a regional scale, there is a continuing need to observe deforestation and other types of land cover change, and to monitor the extent and productivity of agricultural crops. More reliable, automatic, methods of image classification are needed if answers to these problems are to be provided in an efficient manner. New sources of data, at both coarse and fine resolution, are becoming available. The early years of the new millennium will see a very considerable increase in the volumes of Earth observation data being collected from space platforms, and much greater computer power (with intelligent software) will be needed if the maximum value is to be obtained from these data. An integrated approach to geographical data analysis is now being adopted, and this having a significant effect on the way image classification is performed. The use of non-remotely-sensed data in image classification process is providing the possibility of greater accuracy, while – in turn – the greater reliability of image-based products is improving the capabilities of environmental GIS, particularly in respect to studies of temporal change.

All of these factors present challenges to the remote sensing and GIS communities. The focus of research will move away from specialised algorithm development to the search for methods that satisfy user needs, and which are broader in scope than the statistically based methods of the 1980s, which are still widely used in commercial GIS and image processing packages. If progress is to be made then high quality inter-disciplinary work is needed, involving mathematicians, statisticians, computer scientists and engineers as well as Earth scientists and geographers. The future has never looked brighter for researchers in this fascinating and challenging area.

9

Advanced Topics

9.1 INTRODUCTION

This chapter deals with three topics (interferometric SAR, imaging spectrometry and lidar) that are unlikely to be covered in an undergraduate course but which will be of interest to students following Masters courses or undertaking preliminary reading for postgraduate research degrees. In section 9.2 the topic of Synthetic Aperture Radar interferometry (SAR interferometry or InSAR) is introduced. InSAR is primarily used to acquire data that can be processed and calibrated to produce digital elevation models (DEM) of a target area. Differential InSAR (DInSAR) uses a time sequence of interferometric observations and can detect movements of the order of centimetres. It has been used in fields such as glaciology, volcanology and tectonics to measure small changes in the height of the Earth's surface or to measure movement of, for example, glaciers or ice sheets.

Imaging spectroscopy has been a topic that has attracted interest from researchers over the past 15 years or so. The launch of Earth Explorer-1, which carried an imaging spectrometer (Hyperion) into orbit for the first time, plus the more widespread availability of data from airborne sensors such as Hymap, DAIS and AVIRIS, means that access to such data is becoming easier. Section 9.3 provides an introductory description of methods of processing high-dimensional data in both one and two dimensions (i.e., spectrally and spatially). Methods of analysing spectra used in chemometric analysis are described in the context of remote sensing, and examples are provided to lead the reader through some of the more difficult material.

The third topic considered in this chapter is that of the interpretation and use of lidar data. ICESat, launched in late 2002, carries the first of what should be several space-borne lidars. As in the case of imaging spectrometry, lidar data collected by aircraft-mounted sensors are becoming more widely available, and so these data are becoming more familiar to students and researchers. Lidar, like InSAR, is a ranging technique that measures distance from the instrument to a point on or above the ground surface, such as the roof of a building or a tree canopy. Thus, lidar data can be used to generate a Digital Surface Model (DSM) of ground and above ground elevations. Lidar can penetrate vegetation canopies to a greater or lesser extent, and more modern lidar sensors can collect two or more returns from each ground point, making possible the study of the relationships between lidar penetration distances and the biophysical properties of vegetation canopies.

9.2 SAR INTERFEROMETRY

9.2.1 Basic principles

A SAR image is generated by processing of millions of pulses of microwave energy that are transmitted and received by air-borne or satellite-borne antennae (section 2.4). The transmitted pulses are scattered by a target, and the same antenna receives the return pulse of energy that is backscattered by the target on the ground. Distance to the target (in the range direction) can be computed from the time taken between transmission and reception of the pulse, as the microwave radiation travels at the speed of light, while the direction of movement of the platform defines the azimuth direction. When the magnitudes of the processed pulses, which relate to the strength of the backscattered signal, are displayed in both azimuth and range, then we see a radar image.

SAR instruments are described as *coherent* because they record information about the phase as well as the magnitude of the return pulses. Phase is measured as an angle. Figure 9.1 shows two curves. One is a plot of $\sin(x)$ and the second is a plot of $\sin(x + \pi/2)$. The first curve has a phase of zero (i.e., the value of the sine wave at $x = 0$ is zero). The second curve lags the first by $\pi/2$ radians. This is seen clearly around the point $x = 6.28$ radians. Imagine that the curve of $\sin(x)$ is typical of the microwave energy transmitted by a SAR, and $\sin(x + \pi/2)$ represents the return signal. The difference in phase between the transmitted and returned pulse is $\pi/2$ radians. The phase difference between transmitted and received energy as well

Computer Processing of Remotely-Sensed Images: An Introduction, Third Edition. Paul M. Mather.
 ISBNs: 0-470-84918-5 (HB); 0-470-84919-3 (PB)

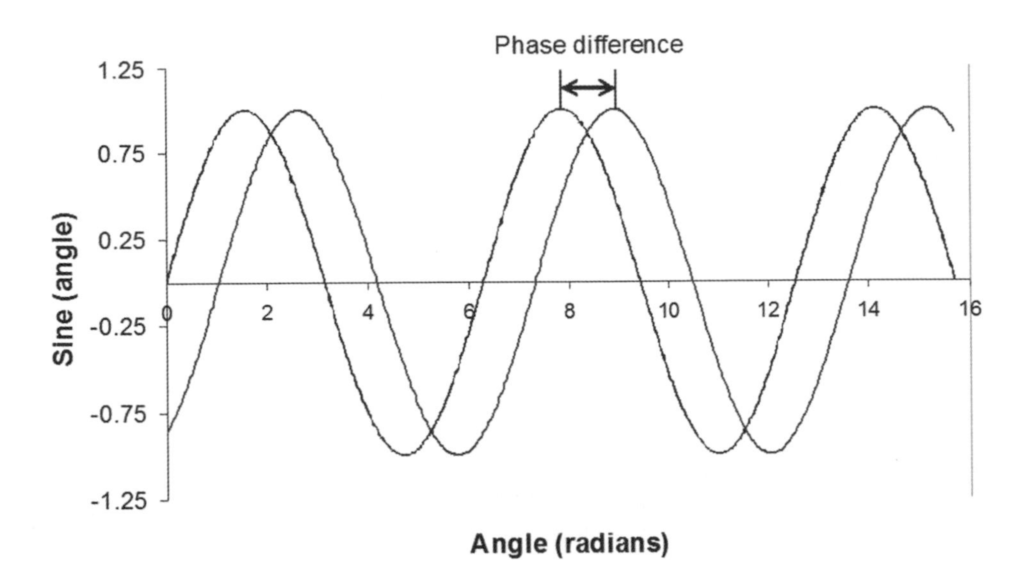

Figure 9.1 Phase difference between two sine waves. The x-axis is graduated in radians (2π radians = 360°). The shift of the peak from the first sine wave to the second is equal to $\pi/3$ radians, and this shift is the phase angle. Note that phase angle cannot be equal to or greater than 2π radians.

as the magnitude of the return signal is recorded for each pixel in a SAR image, and data sets consisting of phase and magnitude information can be acquired in the form of single-look complex (SLC) images. The phase information for a single, independent image is of no practical value. However, the technique of interferometry makes use of two or more SLC images of the same area in order to recover information about the phase differences between them. The elevations of all pixels above some datum, such as WGS84, can be computed from this phase difference information.

The mathematical and algorithmic details of interferometric SAR (InSAR) processing are well beyond the scope of this book. However, the underlying principles are reasonably straightforward, and these are presented in the following paragraphs. Readers requiring a more technical account should refer to Armour *et al.* (1998), Evans *et al.* (1992), Gens and van Genderen (1996 a), Hanssen (2001), Massonet (2000), Massonet and Feigl (1998), Rosen *et al.* (2000), and Zebker and Goldstein (1986). InSAR uses the differences in phase between the signals received by two separate SAR antennae to construct a pixel-by-pixel map of ground surface elevations. The height h in Figure 9.2 is calculated from the difference in the phase (or path difference) of the signals received by antennae A1 and A2, the length of the baseline B that separates the antennae, and the look angle of the radar. Figure 9.2 is idealised in the sense that any vegetation cover that is present may intercept the microwave radiation before it reaches the ground. The measurement h may therefore include the height of forest trees, buildings and other *person-made* structures and may therefore be described as a digital surface model (DSM) rather than as a digital elevation model (DEM).

The configuration of the two antennae A1 and A2 shown in Figure 9.2 can be achieved using one of two strategies. In *single-pass interferometry* the two antennae are carried by a single platform whereas in *repeat-pass interferometry* the signal is measured at position A1 on one orbit and at A2 on a later orbit. The US Shuttle Radar Topography Mission (SRTM), which took place during February 2000, used the single-pass approach. Two radars were carried during the mission. NASA re-used the C-band system from its SIR-C experiment of 1994, and the German Space Agency (DLR) contributed an X-band radar (Rabus *et al.*, 2003). For each instrument, one antenna was placed in the Shuttle's cargo hold and the other was located at the end of a 60 m mast that was deployed once the Shuttle reached its orbital altitude of 233 km. Single-pass systems need only one active antenna, so the backscatter from the signal transmitted from antenna A1 in Figure 9.2 is received by both antenna A1 and A2. Figure 9.3 shows a visualisation of the Wasach Mountains, Utah, produced from C-band interferometric data collected during the SRTM. The grid of interferometrically-derived elevation data is overlaid by an optical image collected by the Landsat TM sensor. Figure 9.4 is a digital elevation model of the Lac de Neuchâtel area, Switzerland, generated from SRTM X-band data by the German Space Agency, DLR.

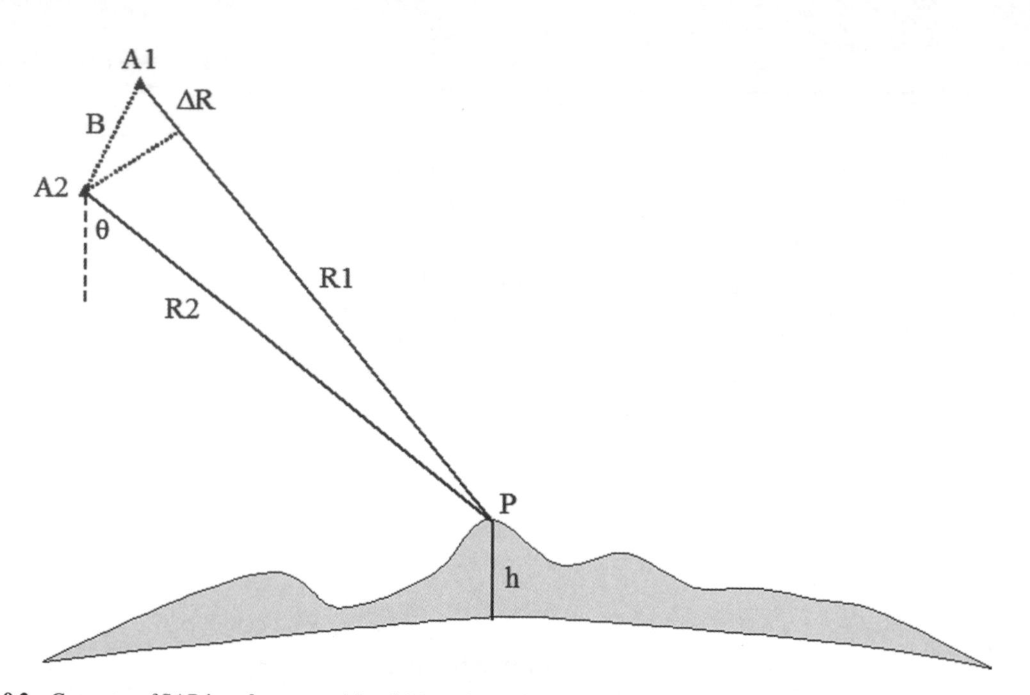

Figure 9.2 Geometry of SAR interferometry. A1 and A2 are the positions of two SAR antennae. B is the baseline (the distance between A1 and A2). R1 and R2 are the distances from A1 and A2 to the point P that has an elevation of h above a specified datum. Angle θ is the look angle of the radar.

Figure 9.3 The Wasach Mountains near Salt Lake City, Utah. This visualisation uses ground elevations derived from SAR interferometry. The data used to generate this image were collected by the joint NASA/DARA/ASI Shuttle Radar Topographic Mission (SRTM) of February 2000. Image courtesy of the U.S. Geological Survey. Original in colour.

Figure 9.4 Digital elevation model of the Lac de Neuchâtel area, Switzerland, derived from the X-band interferometric SAR carried on the SRTM, February 2000. The X-SAR system used in the SRTM was built by the German Space Agency, DLR. **Source:** http://www.dfd.dlr.de/srtm/schatztruhe/kontinente/europa/neufchatel_en.htm. Original in colour.

Repeat-pass interferometric data are collected, as the name implies, at different times. SAR data acquired by the ERS, ASAR, JERS-1 and RADARSAT radars have been used to derive interferometric DEMs. The first data take occurs when the satellite is at position A1 in Figure 9.2. At a later date, when the orbital track is different, a second set of data is acquired from antenna position A2. The length of time between successive and suitable ERS orbits varied from one day (when both ERS-1 and ERS-2 were operating in *tandem mode*) to as long as 35 days.

A single-pass interferometric configuration such as that used in the SRTM (and by aircraft-borne interferometric systems) has a number of advantages over a repeat-pass system. Firstly, the target area is imaged under virtually identical conditions, so that backscattering from the target to the two antennae is effectively the same. The SAR images collected by the antennae A1 and A2 (Figure 9.2) are thus highly correlated. The correlation between two complex-valued (SLC) SAR images is termed *coherence* (Wegmüller *et al.*, 1997; see below). High coherence is necessary for successful interferometry, in that the assumption is made that the characteristics of the backscattered energy from each pair of corresponding points in the two images is the same. If a repeat-pass system is used then the backscattering characteristics of the target may have changed between the dates that the two SAR images were collected, and the degree of correlation between the two images would therefore be reduced, causing what is known as temporal decorrelation. Temporal decorrelation leads to a reduced level of accuracy in height determination and in extreme cases makes interferometry impossible. Some targets decorrelate very quickly. For example, Askne *et al.* (1997) show that decorrelation can occur within a few minutes for forest targets, as the individual leaf orientations change quickly relative to the SAR illumination direction as a result of wind action. Water bodies also decorrelate very rapidly, whereas agricultural areas show a moderate reduction of coherence as the time between the two successive SAR image collection dates increases. Urban areas show the lowest temporal decorrelation. In vegetated areas, the impact of temporal decorrelation depends on wavelength to some extent, as shorter wavelengths are scattered by the outer leaves whereas longer wavelengths penetrate more deeply into the canopy, as noted above. The rate at which decorrelation occurs can be used to distinguish between static and dynamic targets, for example urban areas an growing crops.

A second advantage of the single-pass approach is that atmospheric conditions are similar for the SAR images collected at antennae positions A1 and A2 in Figure 9.2. If the dates of image acquisition differ then atmospheric effects may result in errors in phase angle determination. The propagation of microwave energy through the atmosphere is affected both by the presence of water vapour and by trophospheric effects that are not well understood, but which can cause significant errors in phase determination from SAR images.

A disadvantage of single-pass interferometry is the limited baseline length that can be achieved (B in Figure 9.2). The SRTM used a 60 m mast to serve as the mounting point for the second antenna, whereas a repeat-pass configuration could involve a separation between the two orbits of several hundred metres. Baseline length is important as it has an effect on the sensitivity of the relationship between height and phase. The rate of change of height with phase difference (i.e., height sensitivity) is directly proportional to baseline length. Thus, if the baseline is short then – all other things being constant – small changes in phase angle produce relatively large changes in computed ground elevations. The opposite is the case for longer baselines. However, if the baseline becomes too long then the two SAR 'views' of the target become decorrelated and the phase differences that are used to calculate terrain elevation cannot be measured. This *critical baseline length* for repeat-pass interferometry using ERS-1 and -2 is of the order of 1 km, but best results for DEM generation are obtained with a baseline length of around 200–300 m. The baseline length must be known accurately. Reigber *et al.* (1996) state that, for ERS interferometry to be successful, the baseline length must be known to an accuracy of less than 5 cm. During the SRTM the length of the mast on which the C- and X-band receive-only antennae were placed was monitored closely, for – as Rabus *et al.* (2003) note – an error of 1 mm in measuring its length would lead to an elevation error of 0.5 m on the ground. Bending of the mast tip was also a problem. A star-tracking system and an electronic distance-measuring device were employed to ensure that the position of the mast tip was known to a sufficient degree of accuracy. More details of the SRTM mission are provided by Farr *et al.* (2000).

Thus far, the question 'how do we know exactly where the antennae are?' has not been considered, yet it is a fundamental one. The orbits of the ERS-1 and ERS-2 satellites are determined using the PRARE instrument (Precise Range and Range Rate Equipment), from laser retro-reflector readings, and from the use of orbital models, which are described in section 4.3. The use of lasers to determine the locations of satellites is called Satellite Laser Ranging. Laser beams are directed from a ground station towards the satellite of interest, which carries efficient reflectors called retro-reflectors that return the pulse back to the ground. The round-trip time is measured to a high accuracy and the distance is calculated from knowledge of the speed of light. The position of the satellite is then computed from this distance plus the angle of beam projection. Using this procedure, the position of the satellite can be fixed to within one metre. Both ERS and Envisat are equipped with retro-reflectors. Reigber *et al.* (1999) discuss the effect of the accuracy of orbit determination on interferometry.

For airborne InSAR, aircraft platform altitude and attitude are monitored using GPS and an inertial navigation system (mentioned in section 9.4.2 in relation to lidar data collection from aircraft). During the SRTM the Shuttle's location in space was determined to an accuracy of about 1 m by GPS. An inertial navigation package also provided data on the Shuttle's attitude parameters (pitch, roll and yaw).

Both RADARSAT and the ASAR carried by Envisat can operate in a mode known as ScanSAR mode (also known as Wide-Swath or Global Monitoring modes in the case of ASAR). The radar antenna is capable of scanning several sub-swaths simultaneously, as illustrated in Figure 9.5. The penalty is a reduction in spatial resolution. For example, the ASAR onboard Envisat produces SAR imagery with a spatial resolution of 30 m for any single sub-swath in 'image' mode, whereas in 'global monitoring' mode all the sub-swaths are scanned, but at a resolution of 1 km. The derivation of interferograms from ScanSAR imagery presents a number of additional problems (Hellwich, 1999a and b; Monte-Guarini *et al.*, 1998). First, the sub-swaths have to be scanned in an identical fashion on the two orbits required for repeat-pass interferometry and, secondly, the critical baseline length becomes shorter, at around 400 m. However, the advantage is a greater frequency of coverage and the possibility of building up a global database of interferometric measurements that could be of value in studies of surface motion.

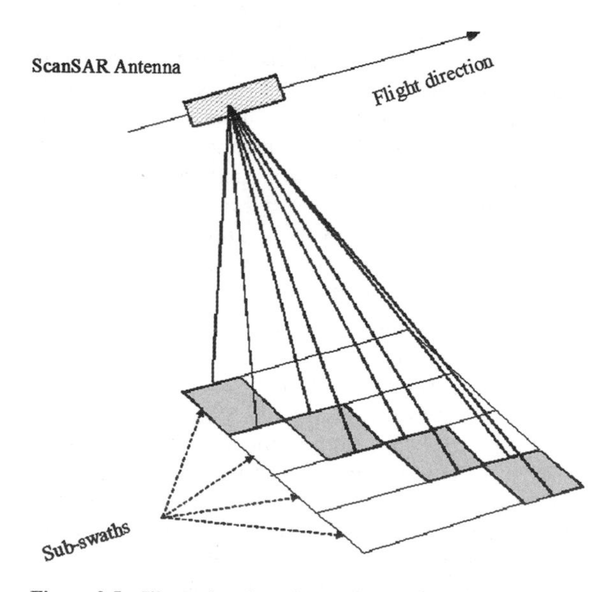

Figure 9.5 Illustrating the principle of ScanSAR. The antenna directs separate bursts towards each sub-swath to build up a large image. The ASAR onboard Envisat can produce a swath of 400 km in ScanSAR mode, but the penalty is decreased spatial resolution of either 150 or 1000 m, depending on requirements.

A further possibility is the use of polarimetric SAR in interferometry. The background to polarimetric SAR is provided in section 2.4. The SIR-C experiment in 1994 provided polarimetric SAR from space for the first time. Today, the Envisat ASAR can provide polarimetric SAR imagery from which both fine (30 m) and coarse (150/1000 m) resolution interferograms can be derived. Hellwich (1999a and b) notes that interferograms computed from different polarisation modes can be used in a number of applications, including (i) combining the polarimetric interferograms in order to create data sets for land cover classification (Chapter 8), (ii) improving interferometric DEM accuracy by inter-comparison between coherence maps produced by polarimetric SAR and (iii) deriving information on vegetation height and structure from the differential interactions between differently polarised microwaves and the components of the surface vegetation cover. Multiple-waveband SAR can also be used to infer the characteristics of the vegetation canopy. See Cloude and Papathanassiou (1998) for more details of polarimetric SAR interferometry. Applications of interferometry are summarised in section 9.2.4.

Differential SAR interferometry (DInSAR) is the estimation of differences in surface location (both plan position and height) from interferometry. The principle is the same as the use of ground surveying to collect data from which contour maps are made. Assume that we have a raster (digital) contour map of a survey conducted five years ago and a second raster contour map that was completed last week. (*Raster* means made up of pixels; in this case, the pixel values represent elevation.) If the two maps are exactly the same then subtracting one from the other on a pixel-by-pixel basis will produce an array of zeros. If any pixel in the difference map has a non-zero value then change has occurred, and the magnitude of that change is proportional to the pixel value in the difference image. In effect, we have used the first (old) raster contour map to 'remove the topography' from the newer map. The same could be done with two interferometric DEMs derived from SAR images collected before and after a significant event such as an earthquake. The difference between the two interferometric DEMs shows the change in surface geometry resulting from the earthquake. Four SLC SAR images would be required to produce two DEMs. Since the first interferometric DEM is intended to provide a good approximation to the land surface elevation, then it should be generated from a long-baseline interferometric pair, while the second interferometric DEM should show as much surface detail as possible and ideally would be derived from a short baseline interferometric pair. Where the target (such as a glacier or an ice-stream) is in motion when the second pair of SLC SAR images is acquired, the baseline should be even shorter. Critical baseline lengths should be of the order of 300 m, 20 m and 5 m for DEM generation, ground displacement, and motion analysis applications.

For DInSAR to be successful, the degree of decorrelation (as measured by loss of coherence) between the two interferograms should be as small as possible. Over water areas, or areas covered by forest, decorrelation occurs rapidly and it soon becomes impossible to separate the effects of ground deformation, land subsidence, or ice-stream motion from the effects of decorrelation. Long wavelengths tend to decorrelate less rapidly than short wavelengths, as they penetrate vegetation canopies more deeply and are less likely to be affected by the geometry of the canopy surface. The ERS and ASAR radars both operate in C band, which makes them less useful for DInSAR in vegetated areas. A second disadvantage is that displacements are measured only along the line of sight of the SAR. Atmospheric effects may also produce spurious fringes in the interferograms, especially if the gap between the dates of the two acquisitions is relatively long. However, DInSAR is capable, in theory, of measuring displacements at the millimetre scale. Gabriel *et al.* (1989) and Massonet and Feigl (1998) are the definitive references on DInSAR. Applications of DInSAR are discussed in section 9.2.4.

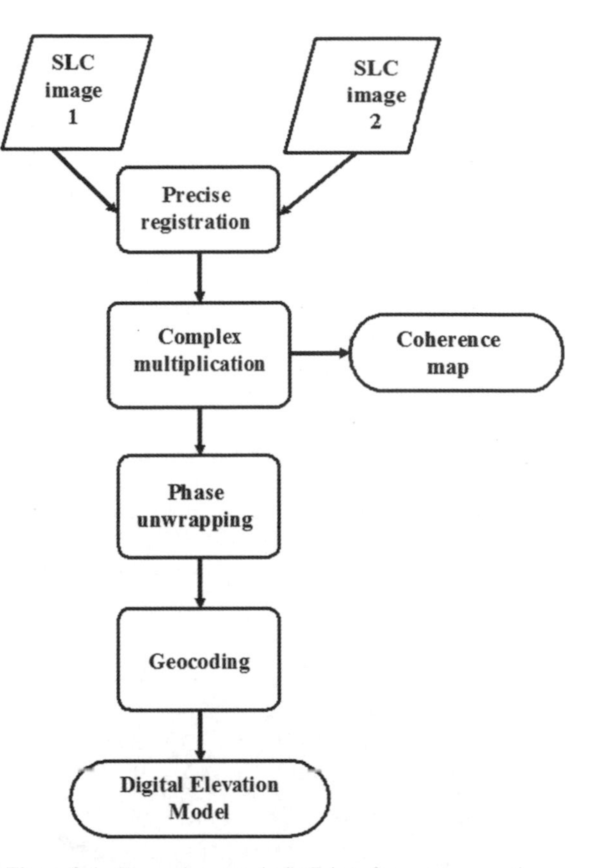

Figure 9.6 Processing steps in SAR interferogram generation.

9.2.2 Interferometric processing

The requirements for successful SAR interferometry are discussed in section 9.2.1. Given a suitable pair of SLC images, with sufficiently high coherence, known and accurate orbital parameters, and suitable baseline length, then processing can proceed. The main processing steps are: co-registering the two SLC images, complex multiplication of the two registered SLC images to generate the interferogram, computation of the coherence map, removal of the fringe pattern caused by Earth curvature, phase unwrapping and georeferencing (Figure 9.6).

In mathematical terms, multiplying one SLC SAR image by the complex conjugate of the other SLC SAR image generates an interferogram. Recall that in SLC images the backscatter characteristics of each pixel are represented by a pair of 32-bit real numbers, which form a single complex number. A complex number has two components, termed the real (a) and imaginary (b) parts in the expression $a + ib$. For example, the two elements of the complex number (4.5, 2.7) represent the terms a and b. This complex number could be written as $4.5 + i \times 2.7$, where $i = \sqrt{-1}$. The *complex conjugate* of a complex number $a + ib$ is simply $a - ib$. The SAR amplitude image is computed on a pixel-by-pixel basis from the expression amplitude $= \sqrt{a^2 + b^2}$ while the phase value is given by the expression phase $= \tan^{-1} a/b$. If the complex number (4.5, 2.7) were the value recorded for a given pixel in an SLC image then the corresponding amplitude value could be calculated as $\sqrt{4.5^2 + 2.7^2} = \sqrt{20.25 + 7.29} = \sqrt{27.54} = 5.25$, while the phase value for the same pixel would be $\tan^{-1} 2.7/4.5 = \tan^{-1}(0.6) = 0.54$ radians, or about 31°.

The complex multiplication operation is a fairly simple one, but it assumes that matching pixels in the two SLC images are precisely identifiable in the sense that if we are given the row and column coordinates of a pixel in the first SLC image then we can find the pixel covering exactly the same ground area in the second SLC image. The process of aligning the coordinate systems of two images is called *image registration*, and it is described in section 4.3.4. As noted in the preceding section, the orbital characteristics of the platform (for single-pass interferometry) or platforms (for repeat-pass interferometry) must be known accurately, and so the principles of orbital geometry can be used to register Image 1 and Image 2. Correlation methods (section 4.3) are also used to co-register the SLC images. The co-registration must be accurate to within 0.1 of a pixel. Once the two SLC images are co-registered the complex multiplication procedure is carried out to derive an interferogram and a coherence image.

The raw interferogram is represented as an image containing a repeating set of fringe patterns (Figure 9.7). Each fringe represents a single phase-difference cycle of 2π radians. The elevation difference corresponding to a single phase difference cycle can be calculated. Conventionally, each individual fringe is displayed a complete colour cycle from blue (0 radians) to red (2π radians) via cyan, green, and yellow. The raw interferogram must be corrected further before surface elevations can be derived. These corrections are discussed below. The coherence image measures the correlation between the two complex SAR images over a number of small, overlapping rectangular windows (rather like the filter windows described in section 7.2.1). The window is placed over the top left $n \times m$ area of the two registered SLC images, and a coherence (correlation) value is calculated for the $n \times m$ pixels in each SLC image lying beneath the window. This value is placed in the output image at the point corresponding to the centre of the window. The window is then moved right by one pixel, and the process repeated. When the window abuts against the right edge of the two images it is moved down by one line and back to the left-hand side of the images, and the process repeated. The result is an output image that contains the coherence values for all possible positions of the $n \times m$ window. The first position of the rectangular moving window is shown in Figure 9.8. One problem is that the choice of window size will have an effect on the resulting coherence values. If the window is too small then the coherence values will vary considerably from pixel to pixel, while an over-large window will produce a generalised result. The coherence map indicates the degree of correspondence between the backscattered signals from equivalent pixels in the two SLC images on which the interferogram is based. High values (close to 1.0) indicate close correspondence and thus high reliability. In the case of repeat-pass interferometry, low values show that the one or both of the phase and amplitude of the backscatter from the two pixels has changed, and that temporal

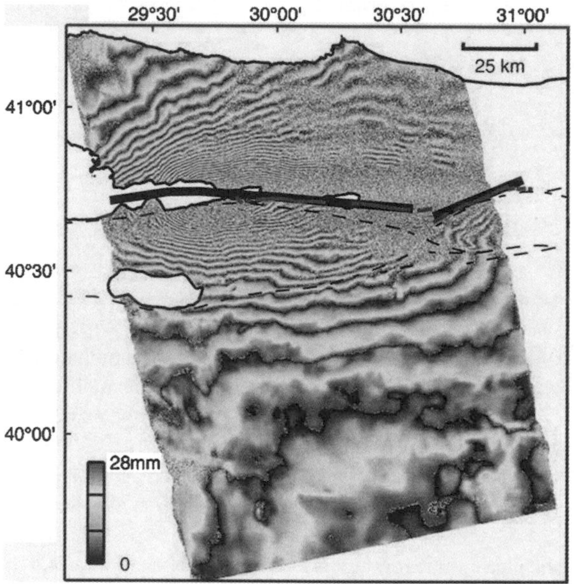

Figure 9.7 Interferogram fringe pattern recorded following the Izmit (Turkey) earthquake of 17 August 1999. Two SAR images collected by ERS-2 on 13 August 1999 and 17 September 1999 were used to derive the interferogram, which shows apparent surface deformation resulting from the earthquake. Each interferogram fringe represents 28 mm of motion towards the satellite, or about 70 mm of horizontal motion. Figure 2(a) from Wright, T., Fielding, E. and Parsons, B., 2001, Triggered slip: observations of the 17 August 1999 Izmit (Turkey) earthquake using radar interferometry. Geophysical Research Letters, **28**, page 1080. Reproduced by permission of the American Geophysical Union.

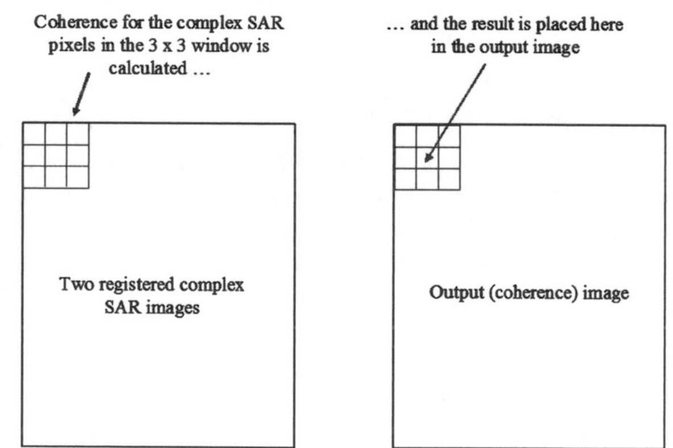

Figure 9.8 Calculation of the coherence image. The initial position of the moving window (dimension 3×3 in this example) is shown on the left, where it is placed over the co-registered SLC images. The coherence for the two sets of nine pixels underlying the window is computed and the result placed in the output coherence image (right) at the point corresponding to the central window pixel. The moving window steps right by one pixel and the process is repeated. When the moving window reaches the right side of the image it drops down by one scan line and returns to the left hand side of the registered SAR images.

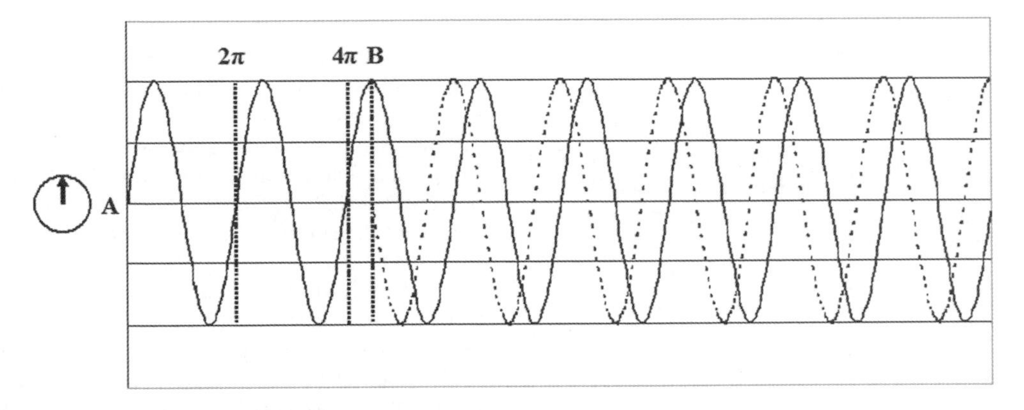

Figure 9.9 The need for phase unwrapping is shown by two signals, represented by solid and dashed lines. Imagine that these two curves are moving leftwards. The finger of the 'clock' at the origin, A, makes a complete revolution of 360° or 2π radians as each full sine wave passes. As point B passes the origin, the clock will show 2.25 revolutions, equivalent to an angular distance of 4.5π radians, which is the phase difference between the two curves. SAR interferometry measures only the phase difference in the range 0–2π and so the difference in the case shown here would be recorded as 0.5π. The process of phase unwrapping attempts to determine the number of integer multiples of 2π to add to the recorded phase difference. In this example, two complete cycles representing 4π radians must be added to produce the correct phase difference of 4.5π radians.

decorrelation has occurred (see above). Other factors causing loss of coherence are considered towards the end of this section.

A complication arises at this point because the phase differences shown in the raw interferogram are measured not in terms of the total number of full wavelength cycles but only in terms of an angular range of 2π radians (360°). Each full cycle of 0–360° or 0–2π radians represents one interferometric fringe. The phase differences must be 'unwrapped' by the addition of appropriate multiples of 2π before elevation can be extracted. This step is called phase unwrapping, and its implementation is difficult. A second difficulty also presents itself – an interferogram of a completely flat area has a fringe pattern that is parallel to the flight direction. This pattern is caused by Earth curvature. This 'flat Earth' fringe pattern must be removed before the fringes are calibrated in terms of elevation.

The 'flat Earth' correction can be accomplished by locating an area of flat terrain on the image and assuming that the fringe pattern shown in the corresponding region of the interferogram represents the desired 'flat Earth' pattern. This pattern is then removed from the interferogram as a whole. Recovering the unwrapped phase is a rather more difficult problem, and a number of algorithms have been developed to accomplish this task. None of them is completely satisfactory, and all are rather too complicated to be discussed here. Gens and van Genderen (1996 a) provide a useful summary of available methods. The problem that is faced is that the phase difference between the signals received at the two antennae can be measured only on a range of 2π radians (360°). An appropriate integer multiple of 2π must be added to the calculated phase difference in order to estimate the true phase difference (Figure 9.9). The unwrapped phase image may also contain empty areas or holes, representing pixels at which the phase coherence value is too low for a phase difference to be computed, or they may represent areas of radar shadow. The wrapped phase image may also be rather noisy, so smoothing may be performed by an adaptive median filter (Premelatha, 2001), or by wavelet denoising (section 9.3.2.2.3; Braunich, 2000) before phase unwrapping is begun. For the mathematically inclined reader, Gens (2003) and Ghiglia and Pritt (1998) provide reviews of phase unwrapping techniques.

At this point the interferogram represents the Earth surface elevation variations in the form of a DEM but is not calibrated in terms of height above a specific datum, nor does it fit a map projection. It is also expressed in 'slant range' form (the oblique view resulting from the fact that SAR is a sideways-looking instrument). The interferogram is first converted to 'ground range' form (meaning the vertical view from above). The next two steps are known as 'geocoding', which involves the warping of the interferometric DEM and interpolating or resampling the elevation values on to a regular grid. Both warping and resampling are discussed in section 4.3. Unless the platform position is known very accurately, ground control points (defined in section 4.3.2) will be needed in order to correct for global height offsets (i.e., the DEM may over- or under-estimate the surface elevation) and for other errors, described by Armour *et al.* (1998).

9.2.3 Problems in SAR interferometry

The quality of an InSAR DEM is affected by a number of factors (Premelatha, 2001). These are: system characteristics, baseline length, terrain characteristics, and processing parameters. System characteristics include wavelength and incidence angle. The property of height sensitivity is discussed in the preceding section, where it is related to baseline length. It is also inversely proportional to wavelength, incidence angle, and slant range distance. These properties of an imaging radar system are examined in section 2.4. Variations in slant range distance are not very significant for satellite-borne systems but may be significant for airborne InSAR. Height sensitivity increases as wavelength decreases, so it may be expected that X-band systems are capable of producing more detailed DEMs than are L-band systems. This is not a deterministic statement, but rather an indication of a tendency, for other factors such as temporal decorrelation may be more significant at shorter wavelengths. To some extent this is dependent on surface vegetation type, for longer wavelengths penetrate more deeply into the canopy than do short wavelengths, and so longer wavelengths are less influenced by short-term changes in the uppermost layers of the canopy. Also, the critical baseline for longer wavelengths is greater than for shorter wavelengths

The effect of vegetation on temporal decorrelation in repeat-pass interferometry has been mentioned already. Terrain relief is also an influential factor. In areas of rugged terrain, the effect of radar shadow is to create gaps in the spatial coverage of the radar, while foreshortening and overlay (section 2.4) can also provide additional problems. Where slopes are steep, spatial decorrelation occurs at lower baseline lengths than is the case on flat terrain, because the local incidence angle changes with surface slope (Figure 2.17). Finally, the choice of method of SLC image registration, phase unwrapping, and other processes such as filtering that are not mentioned here can produce results that differ substantially from each other. Refer to Gens and van Genderen (1996a and b) for more detailed discussion of the geometric factors that influence SAR interferometry.

9.2.4 Applications of SAR interferometry

The most common application of InSAR is to derive digital surface models of the Earth's terrain. This was the primary objective of the SRTM. Differential InSAR applications are generally found in accurate measurement of Earth surface movements, including land subsidence, ground movement associated with volcanic activity, movement of ice-streams and glaciers, and of ocean currents. A comprehensive survey of DInSAR applications is provided by Massonet and Feigl (1996) and Massonet (2000). Smith (2002) is a useful source of reference on InSAR applications in geomorphology and hydrology.

Elevation modelling, using repeat-pass interferometry or single-pass interferometry (represented by ERS, JERS-1 and RADARSAT for the former and SRTM and aircraft systems for the latter) is the most common InSAR application. Planimetric (x, y) accuracies of 10 m and elevation (z) accuracies of 10–15 m over swaths of up to 100 km are claimed for many applications. Papers reporting the use of InSAR in elevation modelling include Albertini and Pone (1996), Evans *et al.* (1992), Gabriel *et al.* (1989), Garvin *et al.* (1998), Madsen *et al.* (1998), and Zebker and Goldstein (1986). Sties *et al.* (2000) provide a comparative analysis of results achieved by InSAR and lidar (section 9.4) in elevation modelling. Hodgson *et al.* (2003) and Lin *et al.* (1994) compare elevation models generated from InSAR with DEMs generated from maps.

Studies of ocean currents using DInSAR techniques are described by Goldstein and Zebker (1986), and Shemer *et al.* (1993). Mouginis-Mark (1995) and Rosen *et al.* (1996) use differential interferometry to monitor ground movements related to volcanic activity, and Perski and Jura (1999) use similar methods to study land subsidence patterns. Bindschlander (1998), Goldstein *et al.* (1993), Joughin *et al.* (1995), Rabus and Fatland (2000) and Rignot *et al.* (1996) examine the potential uses of DInSAR in monitoring movements of ice-sheets and glaciers. The classic study of surface displacements following the Landers earthquake is Massonet *et al.* (1993), while Hooper *et al.* (2003) examine the potential of InSAR in mapping fault scarps for geomorphological purposes. Xu *et al.* (2001) use ERS SLC image pairs with a 105-day temporal separation to study subsidence in an oil field in southern California. Each fringe represents 30.4 mm vertical displacement. They were able to measure a vertical displacement of 25 cm over this time period.

Repeat-pass interferometric SAR has also been used in vegetation classification (Chapter 8) and in studies of canopy characteristics. The coherence image (described above) provides an indication of the rate of temporal change of surface conditions, as can differences in SAR intensity between the two dates of image acquisition. Some researchers create colour composite images (Chapter 3) by displaying coherence, average SAR intensity, and difference between SAR intensities in red, green and blue respectively. See Askne *et al.* (1997), Borgeaud and Wegmüller (1996), Dobson *et al.* (1995), Strozzi *et al.* (1999), Ulander (1995), Wegmüller and Werner (1997), Wegmüller *et al.* (1995) for examples and further details.

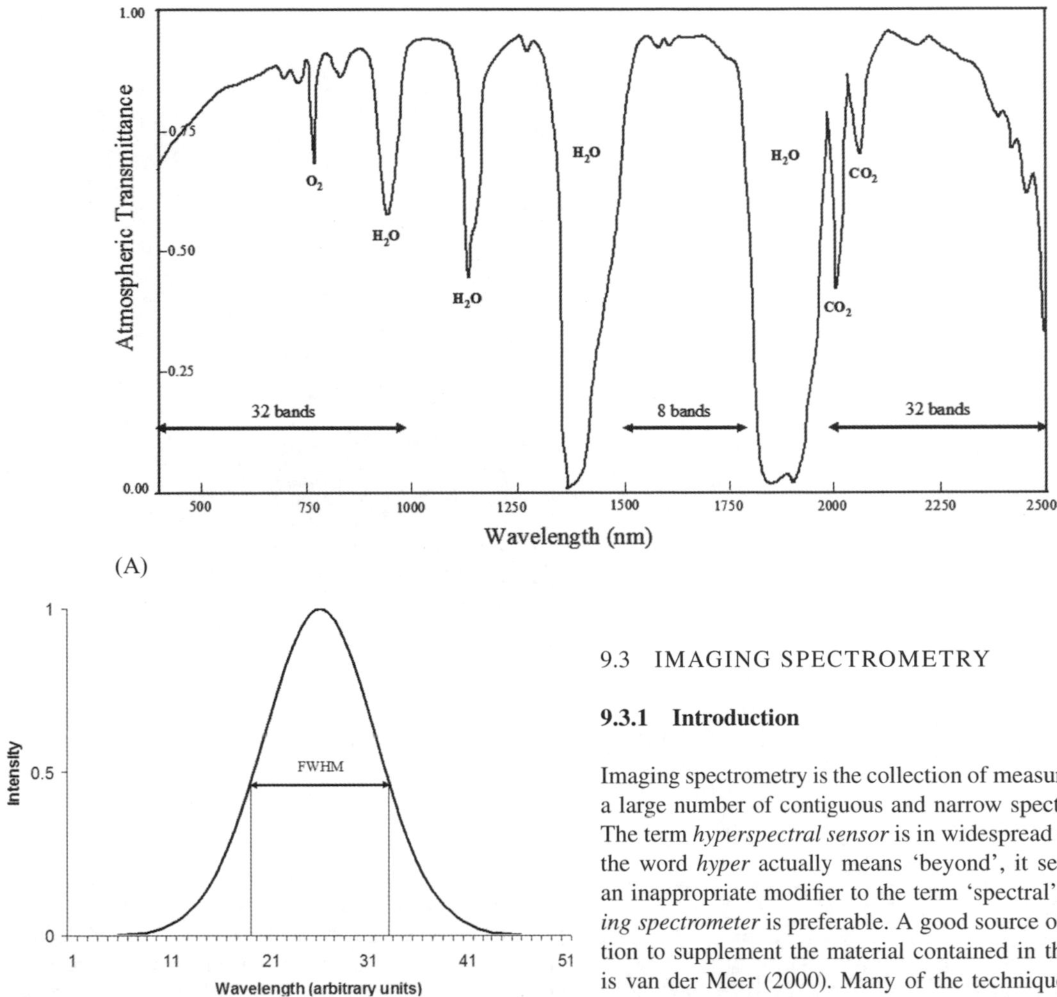

Figure 9.10 (A) Relationship between regions of high atmospheric transmittance and the location of DAIS wavebands. DAIS is an airborne spectrometer with 79 bands, 32 of which cover the range 400–1000 nm (the visible and near infrared), with another 32 bands located in the 2000–2500 nm region. Reflectance in the narrow 'window' around 1650 nm is detected by a further eight bands. Other bands not shown on this diagram are a single broad band covering the wavelengths between 3000 and 5000 nm, and six narrower bands in the thermal infrared (8000–12 600 nm). The width of the visible, near infrared, and short wave infrared bands varies from 15 to 45 nm. The bandwidth in the thermal infrared is 900 nm. (B) The Full Width Half Maximum is the wavelength range defined by the two points at which the intensity level is 50% of its peak value. In the example, the FWHM is $(33.5–19.5) = 14$ units. The bandpass profile for a remote sensing instrument tends to be Gaussian in shape.

9.3 IMAGING SPECTROMETRY

9.3.1 Introduction

Imaging spectrometry is the collection of measurements in a large number of contiguous and narrow spectral bands. The term *hyperspectral sensor* is in widespread use. Since the word *hyper* actually means 'beyond', it seems to be an inappropriate modifier to the term 'spectral', so *imaging spectrometer* is preferable. A good source of information to supplement the material contained in this section is van der Meer (2000). Many of the techniques used in the analysis of reflectance spectra were developed in the field of analytical chemistry. Huguenin and Jones (1986) provide an accessible account of developments in that subject.

It is noted in Chapter 1 that not all of the electromagnetic spectrum is available for remote sensing, due to the presence of absorption bands resulting from interactions between incoming radiation and molecules of gases such as water vapour, ozone and carbon dioxide. Regions known as *atmospheric windows* in which remote sensing is possible separate the absorption bands. The concept of the spectral reflectance curve that characterises the reflectance distribution of a specific material is introduced in section 1.3 and typical examples of such curves for materials such as the leaves of plants, rock surfaces and water are given in that section. The measurements made by an imaging spectrometer can be acquired only in regions of low atmospheric absorbance (atmospheric windows). The relationship between the positions of regions of atmospheric absorbance and the location of the wavebands used in a typical imaging spectrometer, the DAIS 7915, is illustrated in Figure 9.10 and Table 9.1. It is apparent that this instrument's spectral bands are located in regions of the spectrum with high transmittance.

The DAIS 7915 sensor is an airborne imaging spectrometer, built by the Geophysical Environmental Research Corporation and funded by the European Union and the German Space Agency, DLR. It has been used since spring 1995 for experimental remote sensing applications such as monitoring of land and marine ecosystems, vegetation status and stress studies, agriculture and forestry resource mapping, geological mapping, and mineral exploration. The instrument is mounted on an aircraft, and upwelling radiance is directed onto detectors using a four-faced mirror (each face producing four scan lines, a procedure that can result in banding or striping – see Example 9.1) and a beam splitter. Geometric, radiometric and atmospheric corrections (sections 4.3–4.7) are performed by DLR, and the data supplied to users are in the form of reflectance values.

EXAMPLE 9.1: DAIS 7915 IMAGING SPECTROMETER DATA

The three images shown in this example are extracts from a hyperspectral image set collected in June 2000 over a test site in the La Mancha region of central Spain, using the DAIS 7915 imaging spectrometer mounted on an aircraft. The data were collected by the German Space Agency on behalf of a research group led by Professor J. Gumuzzio of the Autonomous University of Madrid. The DAIS data were acquired, processed and distributed by the German Space Agency (DLR), Oberpfaffenhofen, Germany.

Example 9.1 Figure 1 measures reflectance in the green/red region of the electromagnetic spectrum (band 7, centre wavelength 602 nm). At this wavelength, the chlorophyll pigments in the leaves of photosynthically-active vegetation absorb incident energy, and so appear dark.

The centre waveband in Example 9.1 Figure 2 lies in the short wave infrared region, with a centre wavelength of 2378 nm (band 70). Solar irradiance in this region of the spectrum is very low relative to irradiance in the visible bands (Figure 1.17) so the signal-to-noise ratio (SNR) of the image is much lower than the SNR of the green/red image (top) or the thermal infrared image (bottom). The mechanism used by the scanner also results in banding of the image, which is prominent in the centre image.

Example 9.1 Figure 3 shows thermal emission in a waveband centred at 10.964 μm (band 76). This image shows emitted rather than reflected radiation. DAIS band 77 is close the Earth's emittence peak of approximately 10.5 μm. The relationship between land cover type and thermal emittence is evident from visual inspection of the patterns of light (higher thermal emission) and dark (lower thermal emission).

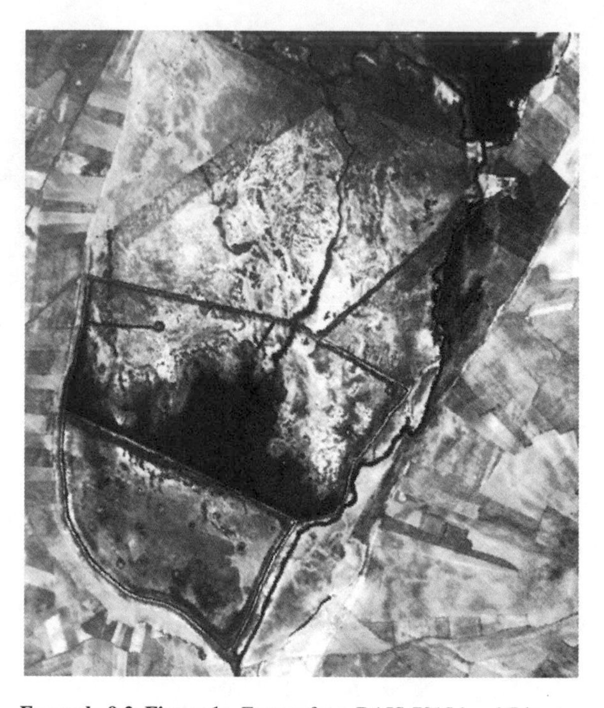

Example 9.2 Figure 1 Extract from DAIS 7915 band 7 image of part of the La Mancha area of central Spain. Band 7 is centred on a wavelength of 602 nm, in the red region of the electromagnetic spectrum. Vegetation absorbs electromagnetic energy in this region of the spectrum, and so photosynthetically active vegetation appears dark.

One of the differences between multispectral data sets, with fewer than ten bands of data, and imaging spectrometer data, collecting data in possibly more than 200 narrow spectral bands, is the fact that the focus of interest is not limited to the analysis of spatial patterns. With a Landsat ETM+ image, for example, we have $x = 6000$ pixels per line and $y = 6000$ scan lines, plus $z = 7$ bands. A DAIS image covers a smaller spatial area (up to $x = 512$ and $y = 3000$ pixels) but the z dimension is 79. Hence, we have two ways of looking at image spectrometer data. One is to consider spatial patterns in the $x - y$ plane. The other is to consider the properties of the z dimension at specific (x, y) pixel points (Figure 9.11). This distinction is similar to that made between the 'R-mode' and 'Q-mode' approaches in multivariate statistical analysis. R-mode analysis considers

Table 9.1 Bands 1–32 of the DAIS 7915 Imaging Spectrometer. The table shows the centre wavelength of each band together with the Full Width Half Maximum (FWHM) in nanometres (nm). The FWHM is related to the width of the band. See Figure 9.10.

Band number	Centre wavelength (nm)	Full width half maximum (nm)	Band number wavelength (nm)	Centre half maximum (nm)	Full width
1	502	23	17	783	29
2	517	21	18	802	27
3	535	20	19	819	30
4	554	18	20	837	27
5	571	22	21	854	28
6	589	20	22	873	28
7	607	20	23	890	27
8	625	22	24	906	31
9	641	22	25	923	28
10	659	24	26	939	30
11	678	25	27	955	28
12	695	25	28	972	28
13	711	28	29	990	38
14	729	27	30	1005	38
15	747	28	31	1020	36
16	766	29	32	1033	32

Example 9.1 Figure 2 This DAIS 7915 image extract covers the same geographical area as the image shown in Example 9.1 Figure 1. The centre waveband of this image is 2378 nm, on the edge of the optical region of the spectrum. The magnitude of incident radiation is low, and so the effects of striping are much more apparent than in Figure 1. The striping is diagonal as the image has been geometrically corrected to place north at the top.

Example 9.2 Figure 3 Whereas Example 9.2 Figures 1 and 2 show reflected Solar radiation at wavelengths of 602 and 2378 nm, respectively, this DAIS 7915 image was captured in the thermal infrared region of the spectrum. The centre wavelength is 10.964 μm, close to the Earth's emittence peak (see Figure 1.16).

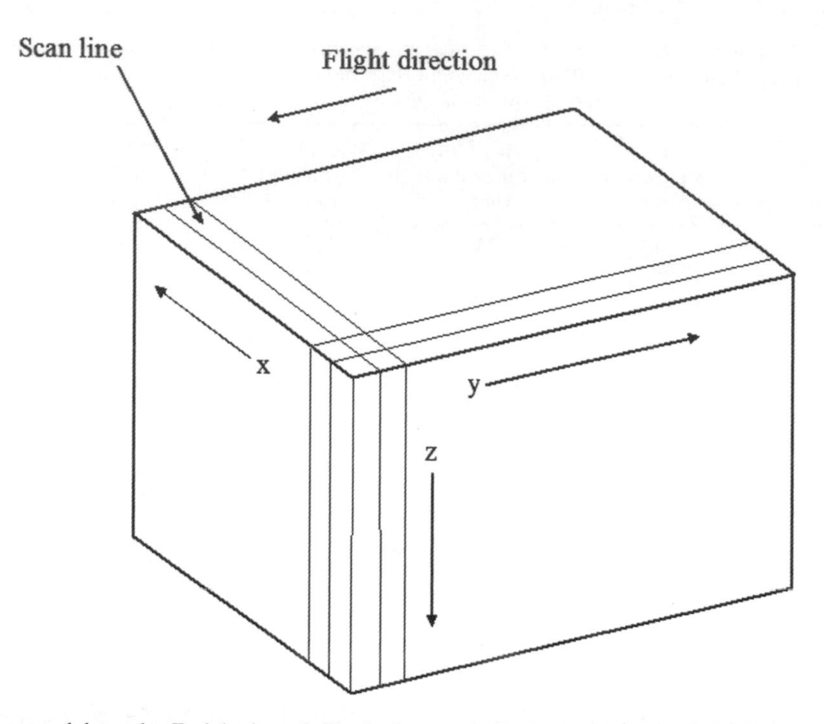

Figure 9.11 Hyperspectral data cube. Each horizontal slice (in the x–y plane) represents a spectral band. The columns (z direction) contain the spectra of individual pixels, as shown in Figure 9.13.

relationships such as correlations among the variables of interest that are measured on a sample of objects, while Q-mode analysis focuses on the relationships between the objects, each of which is characterised by a vector of measurements on a set of variables or features.

Thus, the analysis of imaging spectrometer data can take place in the 'spatial domain' (the x–y plane) as illustrated in Example 9.1. Because measurements are made in a large number of narrow and contiguous wavebands, it is also possible to analyse variability in the z direction (across wavebands) at one or more points. Figure 9.12 shows the spectral reflectance curves derived from two pixels selected from the DAIS image data set shown in Example 9.1. Methods of processing imaging spectrometer data are considered below (section 9.3.2). These methods have been derived from procedures used in analytical chemistry. In addition, some of the methods described elsewhere in this book in the context of multispectral data analysis can also be used to process imaging spectrometry data; for example, spatial and frequency domain filtering (Chapter 7), image classification (Chapter 8), and linear spectral unmixing (section 8.5.1) as well as procedures used in filtering and de-noising.

A second example of an airborne imaging spectrometer is Hymap ('Hyperspectral Mapper'), produced by Integrated Spectronics Pty. Ltd., an Australian company. Like the DAIS 7915 sensor, Hymap can collect data in wavebands ranging from the visible to the thermal infrared. The version operated by Hyvista Ltd. collects data in the optical region only, in 126 spectral bands (Table 9.2). The Hymap sensor uses the opto-mechanical principle described in Chapter 2 to direct upwelling radiance on to a beam splitter and thence to the detector elements. Note that Hymap has more spectral bands available in the 0.4–2.5 μm region than has DAIS 7915 (126 against 72) and that the bands are more closely spaced (spectral sampling interval 13–17 nm, compared with Full Width Half Maximum (FWHM) values of 18–40 nm for DAIS 7915; see Figure 9.12). Also note that Hymap bands are collected by four spectrometers (rows of Table 9.2) and that there is an overlap between the data collected by spectrometers 1 and 2. The bands in the overlap region are not presented in wavelength order, and this can lead to problems with some software. Specifically, band 31 has a longer wavelength than band 32.

The AVIRIS (Airborne Visible/Infrared Imaging Spectrometer) was first deployed by NASA in the late 1980s. Since then, the instrument has been continuously updated. AVIRIS has 224 spectral bands covering the region 0.4 μm to 2.45 μm. Each image is 614 pixels wide. Pixel size depends upon aircraft altitude; at a flying height of 20 km the

Table 9.2 Summary of Hymap imaging spectrometer wavebands, bandwidths and sampling intervals (Based on information from Cocks *et al*., 1998).

Spectral region	Wavelength range (nm)	Bandwidth (nm)	Average spectral sampling interval (nm)
Visible	450–890	15–16	15
Near infrared	890–1350	15–16	15
Short-wave infrared 1	1400–1800	15–16	13
Short-wave infrared 2	1950–2480	18–20	17

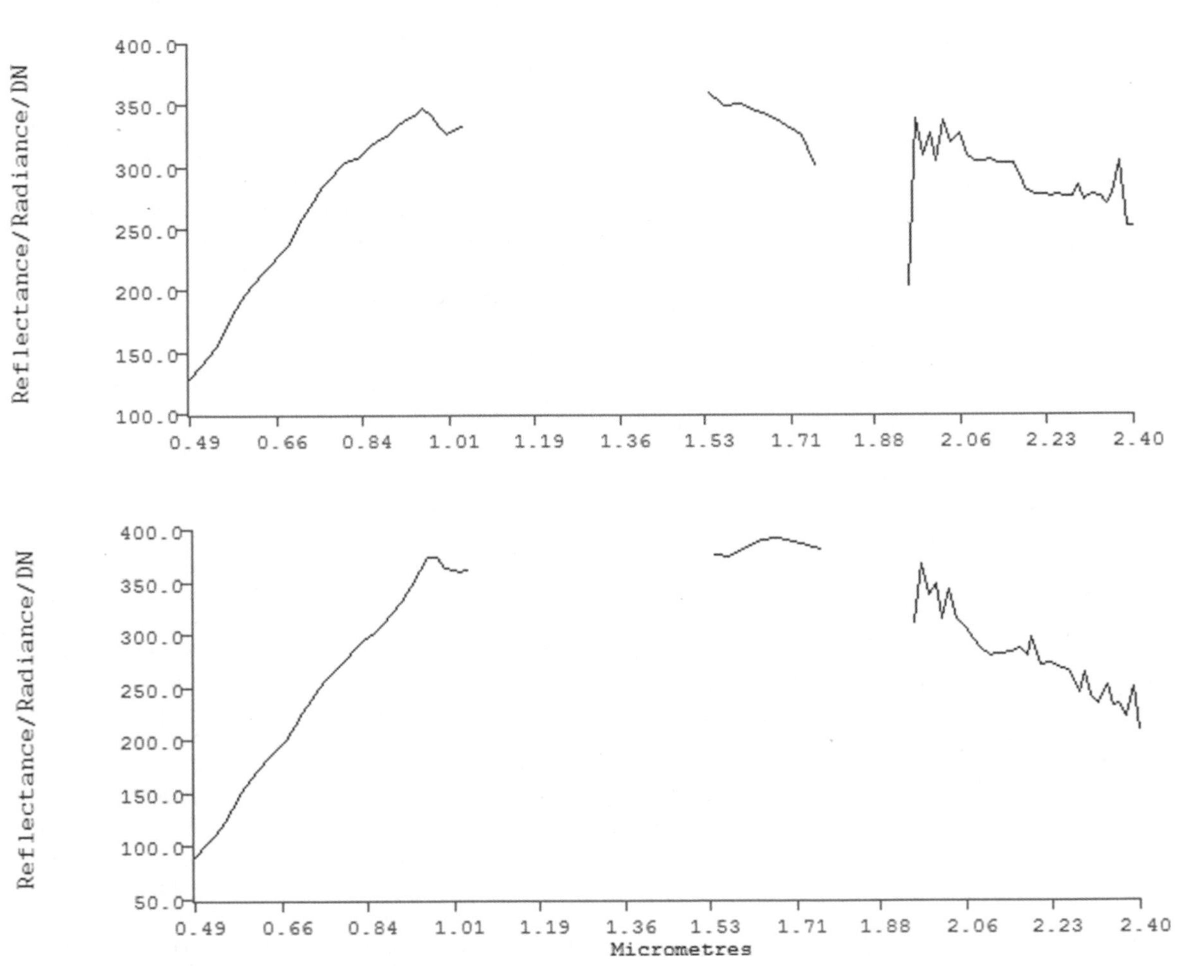

Figure 9.12 Pixel spectra measured by the DAIS imaging spectrometer in the range 0.4–2.4 μm (see Figure 9.10(a)).

pixel size is 20 × 20 m. AVIRIS is described in detail in Vane (1987).

NASA's experimental Earth Observer-1 (EO-1) satellite was launched on 21 November 2000. It is in the same orbit as Landsat-7 with an equatorial crossing time of one minute later than Landsat-7. It carries two Earth-observing sensors. The Advanced Land Imager (ALI) is a prototype for a Landsat-7 ETM+ replacement instrument, while the Hyperion Imaging Spectrometer is the first civilian high spatial resolution imaging spectrometer to be carried in orbit. The third instrument carried by EO-1 is a spectrometer that measures atmospheric water vapour content. The data produced by this instrument are used in the process of atmospheric correction (section 4.4). WWW links to sites describing the Hyperion instrument are included on the CD-ROM accompanying this book.

9.3.2 Processing imaging spectrometry data

9.3.2.1 Derivative analysis

A derivative measures the rate of change of the variable being differentiated (x) with respect to some other variable (y), and is written as $\delta x/\delta y$. For example, given a moving object, one could compute the derivative of its velocity with respect to distance. The result would show whether velocity was increasing as distance travelled increased (acceleration, positive derivative), decreasing with distance (deceleration, negative derivative) or remaining constant (zero derivative). Acceleration is independent of velocity, so that two objects having quite different velocities could have the same acceleration (section 7.3.2).

Derivatives can be computed only for continuous and single-valued functions. A function is simply an expression that returns a value when provided with an input. For instance, if x is an angle then the function $\cos(x)$ will return a value between $+1$ and -1 for any finite value of x, such as $-125.985°$ or $319\,276.241\,35°$. A continuous function returns a value for any permissible input value. In the case of the cos function, the only restriction on the argument x is that it is finite. A single-valued function returns only one answer, in contrast to a function such as sqrt, which returns two values, except when the argument is zero. For example, sqrt(4) returns two values ($+2$ and -2) for a single value of the argument, and so the square root function cannot be differentiated.

A digital image can be considered to be an example of a two-dimensional function that returns the pixel value (grey level) at a given point defined by the image row r and column c, so that we could write *pixel value* $= f(r, c)$. However, it is not possible to differentiate $f(r, c)$ because it is not a continuous function that is defined for every possible value of r and c. The row and column indices must be integer values. Instead of the method of differentiation being used to calculate the rate of change at a particular point, a procedure called the method of *differences* is used instead.

The spectral reflectance curve of a target, as collected by a field radiometer or an imaging spectrometer, is drawn by interpolating between measured, discrete points which are spaced apart at intervals such as 15 nm. The measurements on which the curve is based are discrete or separate, and so the derivatives are estimated using the method of differences. If y_i and y_j represent adjacent, discrete, reflectance values on a spectral reflectance curve at wavelengths x_i and x_j then the first difference value is given by the expression $\dfrac{\Delta y}{\Delta x} = \dfrac{y_i - y_j}{x_i - x_j}$. The terms Δx and Δy are pronounced 'delta-x' and 'delta-y', and the left-hand side of this equation is pronounced 'delta-x by delta-y'. The second difference (i.e., the difference of the first difference) is calculated in a similar way from the formula $\dfrac{\Delta^2 y}{\Delta^2 x} = \dfrac{\Delta y_i - \Delta y_j}{\Delta x_i - \Delta x_j}$. The first difference gives the rate of change of the function y with distance along the x-axis, which is the same as the slope of the graph representing the function. The second difference is the rate of change of slope with distance along the x-axis. If the curve is flat then both first and second derivatives are zero. If the curve slopes upwards to the right then the slope and the first differences are positive, and increase in magnitude as the curve becomes steeper until, if the curve becomes vertical, the first difference is infinite in magnitude. Conversely, as the slope decreases the first difference reduces in magnitude. When a turning point is reached, for example at a maximum or minimum of the curve, the first difference is zero. If the graph slopes down to the right, the slope is negative, and it decreases in value until the curve reaches a minimum. The second difference shows how rapidly the slope is changing. Where slope is constant, such as at a maximum or minimum, then the second difference is zero. Where the slope gets steeper, the second difference becomes larger and, conversely, where the slope gets less steep so the second difference becomes smaller in magnitude. These ideas are illustrated in Figure 9.13.

First and second differences calculated for 1-D spectra or 2-D images are often described as 'derivatives', though it is clear that they provide a means to approximate the derivatives of a discrete function that cannot be calculated. Nevertheless, the term 'derivative' is used in the remainder of this section in order to ensure compatibility with the literature. The first derivative measures a rate of change. It is not dependent on the magnitude of the function. For example, if $x_1 = 6$ and $x_2 = 12$ then the difference is 6. The difference is also 6 if $x_1 = 106$ and $x_2 = 112$. If $y_1 = 1$ and $y_2 = 3$ then value of the first difference is $6/2 = 3$ in both cases. This means spectral reflectance curves with the same shape will have the same first derivative curve, irrespective of their measured reflectance values. This property can be useful because it means that objects with similar spectral reflectance properties located in shadow and in direct sunlight, as shown in Figure 6.5, have the same derivative though their apparent (at-sensor) radiances are different.

The first and second derivatives of a single image band are used in Chapter 7 without particular reference to the concepts outlined above. For instance, the Roberts Gradient edge detection operator is an example of a first derivative, while the Laplacian operator is an example of a second derivative function (section 7.3.2). Other examples of the use of the derivative in image processing include the analysis of the position and magnitude of absorption bands in the pixel spectrum (Blackburn, 1998; Demitriades-Shah

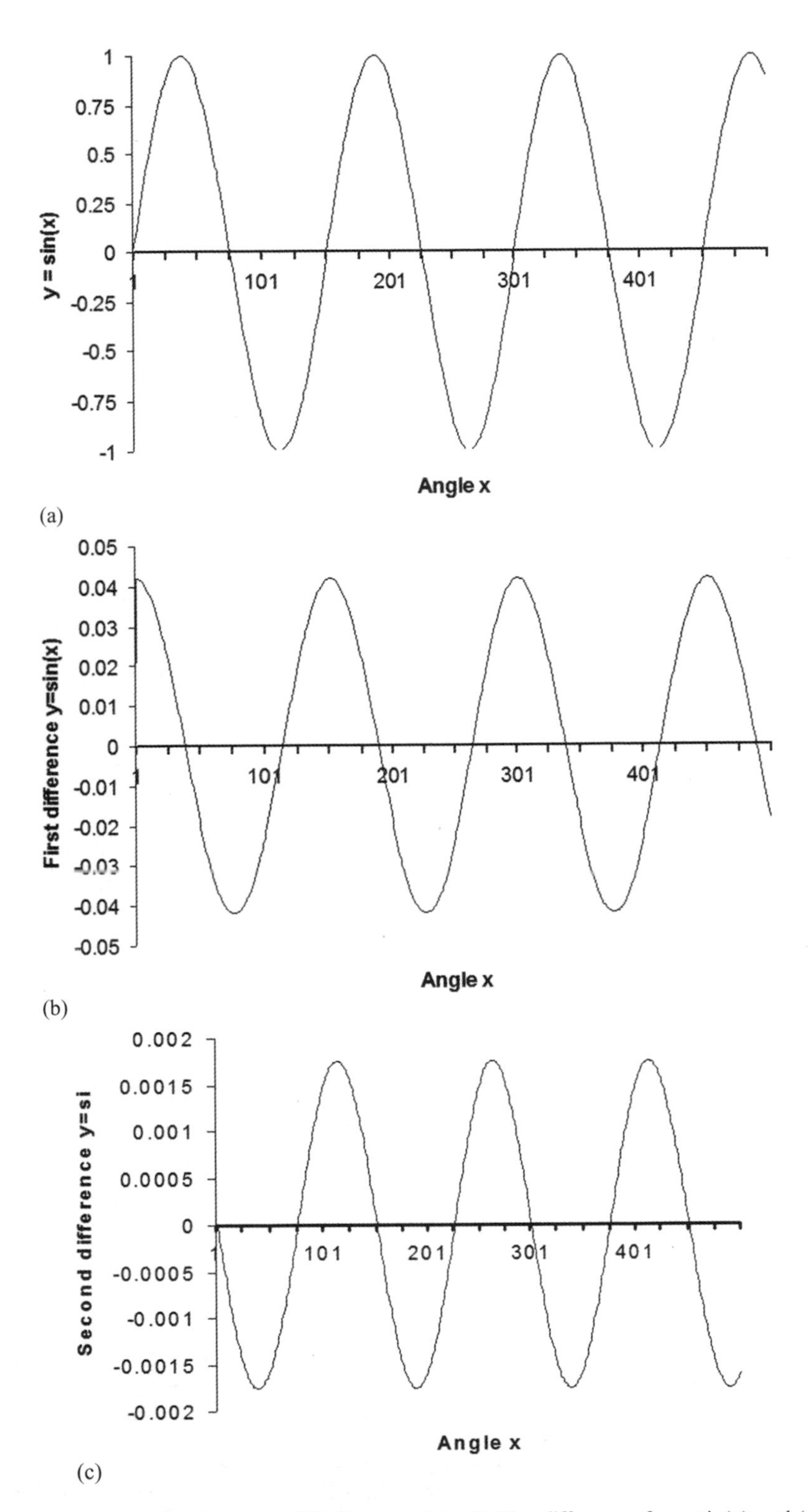

Figure 9.13 (a) Graph of $y = \sin(x)$ calculated at 500 discrete points. (b) First difference of $y = \sin(x)$, and (c) Graph of second differences of $y = \sin(x)$. Note differences in the scale of the y-axis. The slope of $y = \sin(x)$ is initially positive, then decreases to zero at a turning point at $x = 39$, $y = 1$ .The slope then becomes negative, increases towards $y = 0$ and reaches a minimum at $x = 114$, $y = -1$. The same pattern then repeats for each cycle of the sine wave. Figures 9.13(b) and (c) show the first and second differences, which you should attempt to interpret.

et al., 1990; Huguenin and Jones, 1986; Philpot, 1991; Tsai and Philpot, 1998). Gong *et al.* (1997) use first derivatives as inputs to a neural network for classifying coniferous species. Bruce and Li (2001) and Schmidt and Skidmore (2004) consider the role of wavelets in smoothing the pixel reflectance spectrum prior to derivative calculation. The topic of smoothing and de-noising is described in the next section.

9.3.2.2 Smoothing and de-noising the reflectance spectrum

One of the characteristics of derivative-based methods is that they amplify any noise that is present in the data. Even the presence of moderate noise can make the derivative spectrum unusable. Various methods of noise removal, ranging from simple filtering to more complex wavelet-based methods, have been applied to remote sensing data. In section 9.3.2.2.1 a method based on the fitting of local polynomials that have special properties is described. It was first described by Savitzky and Golay (1964) in the context of analytical chemistry. Not surprisingly, it is known as the Savitzky–Golay (S–G) method. An alternative procedure using the one-dimensional discrete wavelet transform (DWT, section 6.7) is described in section 9.3.2.2.2. The DWT decomposes a data series into a set of scale components. The lower-order scale components (levels 1, 2, ...) are referred to as 'detail coefficients', and the basic wavelet-based de-noising method involves the thresholding of these detail coefficients in order to separate noise and high frequency information. In contrast, simple filtering methods, such as the moving average, make no distinction between high frequency information and noise, and are best described as smoothing rather than de-noising functions.

9.3.2.2.1 Savitzky–Golay polynomial smoothing

Savitzky and Golay (1964) introduced a technique that combines smoothing (i.e., low-pass filtering, section 7.2) and calculation of derivatives in an elegant and computationally effective fashion. Smoothing is performed by approximating the data series by a low-order local polynomial, using a moving window technique. One might reasonably distinguish between smoothing and denoising. Smoothing is a purely mathematical operation that is designed to remove some or all of the high-frequency components of the data series, either to reduce the level of detail or to eliminate noise. Denoising has the aim of characterising the statistical nature of the noise and of using estimates of these characteristics to reduce or remove the effects of that noise.

The user of the S–G method must specify *a priori* (i) the order of the polynomial and (ii) the size of the moving window. The larger the window the greater the smoothing effect. Calculation of a least-squares polynomial at every moving window position in a data series might seem to be a lengthy task, but Savitzky and Golay's method is a clever one. Only one set of coefficients is calculated, and this is applied to the data in every window simply by multiplying the value at each data point in the window by the corresponding coefficient value (Figure 9.14). Usually, a polynomial of order two is selected for data smoothing, and an order of four is recommended for derivative calculation. The method is easy to program but more difficult to describe verbally in the abstract, so a simple example is used.

Assume that we specify a window size of five, as shown in Figure 9.14. The centre point (where the coefficient is labelled '0') has two coefficients to the left (N_L, labelled –1 and –2) and two to the right (N_R, labelled 1 and 2). A second-order polynomial ($M = 2$) is selected, as the example relates to data smoothing. Recall from section 4.3.2 that the classical least-squares equation is:

$$\mathbf{c} = (\mathbf{A}'\mathbf{A})^{-1}\mathbf{A}'\mathbf{g} \tag{9.1}$$

where $\mathbf{A}$ is the design matrix, $\mathbf{c}$ is the vector of least-squares coefficients, and $\mathbf{g}$ is the data vector. The elements a_{ij} of the Savitzky–Golay design matrix $\mathbf{A}$ are given by $a_{i0} = 1$

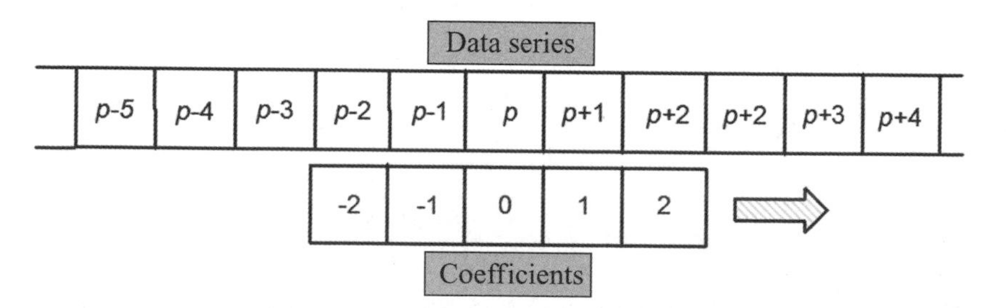

Figure 9.14 One-dimensional moving window. The window starts on the left of the data series and the filter coefficients are multiplied by the corresponding data value. The products are added to give the filter output. In this example, the output is {data value at $(p-2)$ × coefficient (-2)} + {data value at $(p-1)$ × coefficient (-1)} + { data value at (p)× coefficient (0)} {data value at $(p+1)$ × coefficient (1)} {data value at $(p+2)$ × coefficient(2)}.

Table 9.3 Two-dimensional moving window. The cell values are referenced by the x and y coordinates in the usual way.

1.000	−2.000	4.000
1.000	−1.000	1.000
1.000	−0.000	0.000
1.000	1.000	4.000
1.000	2.000	4.000

(a) Design matrix **A**

5.000	0.000	10.000
0.000	10.000	0.000
10.000	0.000	34.000

(b) Matrix product **A′A**

0.486	0.000	−0.143
0.000	0.100	0.000
0.143	0.000	0.071

(c) Inverse of (**A′A**)

−0.086	0.343	0.486	0.343	0.086
−0.200	−0.100	0.000	0.100	0.200
0.143	−0.071	−0.143	−0.071	0.143

(d) Matrix $(\mathbf{AvA})^{-1}\mathbf{A}'$

and $a_{ij} = i^j$ for $i = -N_L$ to N_R and j from 1 to M, the order of the polynomial. We are using a window size of 5 so $N_L = N_R = 2$ and we can define $N_{TOT} = N_R + N_L + 1 = 5$. For this example, **A** has N_{TOT} rows and $M + 1$ columns and its contents are shown in Table 9.3(a).

Equation 9.1 shows that we need to compute the matrix product **A′A** and then find its inverse $(\mathbf{A'A})^{-1}$. The final calculation involves the premultiplication of **A′** by $(\mathbf{A'A})^{-1}$. These matrices are shown in Tables 9.3(b)–(d). The first row of Table 9.3(d) contains the required coefficients for data smoothing. None of the real data is required in the computation of these coefficients – that is the advantage of the Savitzky–Golay method. All we need do to calculate the smoothed value at data point i is to perform an element-by-element multiplication of the raw spectrum values $\{r_{i-2}r_{i-1}r_ir_{i+1}r_{i+2}\}$ for $i = N_L + 1, N - N_R$ with the coefficients {–0.086 0.343 0.486 0.343 –0.086}, as shown in Figure 9.15. Figure 9.16 shows the result of this operation carried out on a small sample of arbitrary data. The smoothed curve was calculated using the coefficients vector shown above. Note that smoothed values for points 1 and 2 at the beginning and points 14 and 15 at the end of the data series cannot be calculated as $N_L = 2$ and $N_R = 2$. Remember too that one assumption of the method is that the data points are equally spaced.

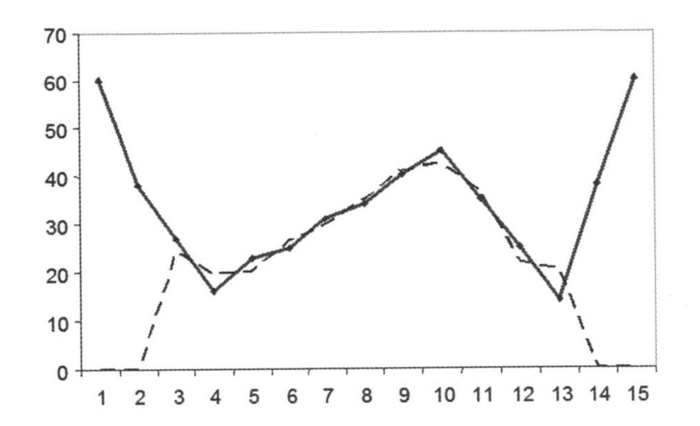

Figure 9.15 Application of Savitzky–Golay smoothing to an arbitrary set of set of 15 data points. The raw data points are indicated by the diamond symbols, which are joined by a solid line. The smoothed curve (values of which can only be computed for data points 3–13 inclusive) is shown by the dashed line.

The values making up the second row of Table 9.3(d) are, perhaps surprisingly, the coefficients required to calculate the first derivative of the Savitzky–Golay smoothed polynomial. To illustrate this, the result of applying the second row of coefficients (Table 9.3(d)) to the set of arbitrary data shown by the solid line in Figure 9.15 is given in Figure 9.16. The third row of coefficients is used to compute the second derivative, and so on. Usually, however, one would use a polynomial order of four rather than two for derivative calculation (Press *et al.*, 1992).

In summary, the one-dimensional Savitzky–Golay procedure provides a computationally efficient means of smoothing a one-dimensional data series such as a reflectance spectrum. It is, however, a smoothing procedure and, as such, differs from denoising techniques based on the Discrete Wavelet Transform, as noted earlier in this section. The following points should be remembered:

- The data points are assumed to be equally spaced. Press *et al.* (1992) suggest that moderate departures from this assumption are not likely to have any considerable effect on the result.

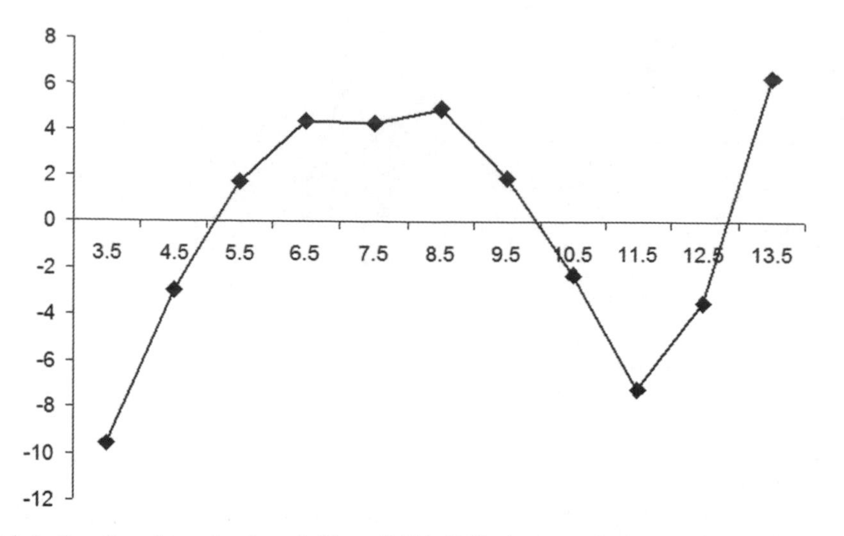

Figure 9.16 First derivative of raw data series shown in Figure 9.15 (solid line) computed using Savitzky–Golay smoothing polynomial procedure. Note that the data points are shifted right along the x-axis by half a unit because the derivative is estimated using differences between adjacent raw data points.

- The degree of smoothing depends on the order of the polynomial and on the number of points in the moving window.
- A polynomial order of two for smoothing and four for derivative calculation is generally used.
- The number of points to the left and to the right of the point of interest (i.e., N_L and N_R in the discussion above) should normally be equal for smoothing reflectance spectra.

The S–G method can also be applied to two-dimensional data such as images, in which case a local smoothing polynomial surface is fitted. The computational procedure is described by Krumm (2001). The moving window is now rectangular, and the design matrix is defined in terms of powers of x and y. Again, it easier to explain by example. The moving window is size $x = 5$ and $y = 5$, but the x and y values are counted from -2 to $+2$ rather than 1 to 5. The 25 cells in the 5×5 moving window are labelled 0–24, counting sequentially left to right along rows in a zigzag fashion starting from the top row (Table 9.4). The leftmost cell in row 1 is labelled '0'

Firstly, think of the 25 cells in Table 9.4 as being stored as a column vector, with P_0 at the top and P_{24} at the bottom. The design matrix **A** is formed by calculating, for each of these 25 rows, a vector of powers and cross-products of the (x, y) values associated with that cell. The arrangement of the terms in this vector is exactly the same as that used in section 4.3.1 in the context of geometric correction. For example, a first order polynomial has terms $\{1, x, y\}$. At

Table 9.4 Two-dimensional moving window. The cell values are referenced by the x and y coordinates in the usual way.

		X				
		−2	−1	0	1	2
	−2	P_0	P_1	P_2	P_3	P_4
	−1	P_5	P_6	P_7	P_8	P_9
Y	0	P_{10}	P_{11}	P_{12}	P_{13}	P_{14}
	1	P_{15}	P_{16}	P_{17}	P_{18}	P_{19}
	2	P_{20}	P_{21}	P_{22}	P_{23}	P_{24}

position P_0, the values $x = -2$ and $y = -2$ would be substituted into this expression, so that the vector of powers and cross-products at this position for a first-order polynomial is $\{1, -2, -2\}$. At position 17 (P_{17} in Table 9.4), $x = 0$ and $y = 1$, so the vector of powers and cross products for a first-order polynomial is $\{1, 0, 1\}$. The vectors of powers and cross products for low-order polynomials are as follows:

Order 1: $\{1, x, y\}$

Order 2: $\{1, x, y, x^2, xy, y^2\}$

Order 3: $\{1, x, y, x^2, xy, y^2, x^3, x^2y, xy^2, y^3\}$

Order 4: $\{1, x, y, x^2, xy, y^2, x^3, x^2y, xy^2, y^3, x^4, x^3y, x^2y^2, xy^3, y^4\}$

For example, the vector of powers and cross products at position P_{11} (which has coordinates $x = -1$ and $y = 0$) for a third-order polynomial would be $\{1, -1, 0, 1, 0, 0, 1,$

$0, 0, 0\}$, a total of 10 coefficients. This vector is calculated simply by substituting $x = -1$ and $y = 0$ into the definition of the order 3 powers and cross products given above, i.e. $\{1, x, y, x^2, xy, y^2, x^3, x^2y, xy^2, y^3\}$.

Thus, the S–G design matrix **A**, in the case of a third-order polynomial, has 25 rows corresponding to the 25 elements of the 5 × 5 moving window and 10 columns, corresponding to the 10 power and cross-product terms for a third-order polynomial. We now compute $\mathbf{C} = (\mathbf{A'A})^{-1}\mathbf{A'}$ as before and find that the 25 smoothing polynomial coefficients are contained in row 1 of **C**. Rows 2 and 3 of **C** contain the coefficients for the first partial derivatives with respect to x and y, respectively. Again, note that the values of the elements of matrix **C** do not depend on the real data values in the image. These coefficients are the same for all images. The 25 values underlying the 5 × 5 moving window are multiplied by the corresponding coefficient values in the first row of **C**, using the labelling scheme shown in Table 9.4. The 25 products are summed to give the smoothed value at point ($x = 0$, $y = 0$) in the output image. The moving window, like the Biblical finger, moves on and ($x = 0$, $y = 0$) now overlies the next pixel to the left. The concept of the moving window is illustrated in Figure 9.8.

'The various matrices involved in the 2D S–G technique are too large to be reproduced here, but the CD contains a program called *TeachmeSavitky 2D.exe* in the *\Programs* folder. Double-click on the program name in Windows Explorer to activate the program. The results will be placed in a file of your choice. You can read this file into Notepad or Wordpad in order to read it or print it out. Example 9.2 illustrates the use of the S–G technique in image smoothing.

EXAMPLE 9.2: SAVITZKY–GOLAY SMOOTHING.

This example shows the result of applying a Savitzky–Golay smoothing filter to a single band (band 30) of the DAIS 7915 imaging spectrometer data set discussed in Example 9.1. The moving average window size is $x = 5$ and $y = 5$. A second-order polynomial is used for image smoothing (Figure 9.15) and a fourth-order polynomial is used to calculate the first derivative images. Example 9.2 Figure 1 shows the original DAIS 7915 band 30 image. Next, Example 9.2 Figure 2 shows the image after smoothing using the Savitzky–Golay procedure. Three first derivative images are displayed in Example 9.2 Figure 3–5. The first, in Example 9.2 Figure 3, estimates the horizontal grey level gradient along the scan lines (rows) of the image. The second (Example 9.2 Figure 4) is the vertical grey level gradient, measured down the columns of the image, and the third (Example 9.2 Figure 5) is the spatial derivative (calculated with respect to both x and y). You can use the *Filter|Savitzky–Golay* item on the MIPS main menu to carry out a similar exercise. Compare

Example 9.2 Figure 1 Band 30 of the DAIS 7915 imaging spectrometry data set described in Example 9.1. The bright area in the lower centre is a dry salt lake, and the dark area in the upper left corner is a water-filled lake.

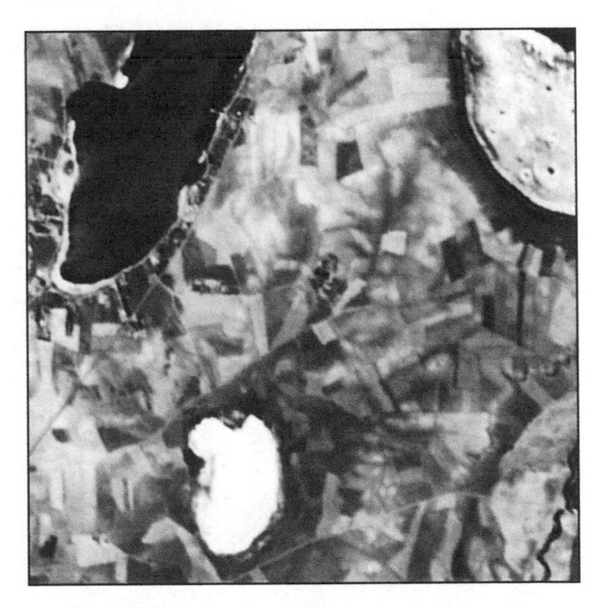

Example 9.2 Figure 2 Image shown in Example 9.2 Figure 1 after smoothing using a Savitzky–Golay polynomial filter (second-order polynomial, window size 5 × 5).

Example 9.2 Figure 3 First partial derivative (horizontal) of the image shown in Example 9.2 Figure 1 using a Savitzky–Golay polynomial filter.

Example 9.2 Figure 4 First partial derivative (vertical) of the image shown in Figure 1 using a Savitzky–Golay polynomial filter.

Example 9.2 Figures 3–5. Can you say that the filter separates the horizontal (3) and vertical (4) edges? Why are the edges shown in the spatial derivative image (5) more marked than those in Example 9.2 Figures 1 (3) and (4)?

Example 9.2 Figure 5 First spatial derivative (with respect to $x$and y) of the image shown in Example 9.2 Figure 1 using a Savitzky–Golay polynomial filter

9.3.2.3 De-noising using the Discrete Wavelet Transform

The Discrete Wavelet Transform (DWT) is introduced in section 6.7.1. It is shown there that a one-dimensional data series can be decomposed into a collection of sub-sets of detail coefficients, with $n/2$ first-level detail coefficients, $n/4$ second-level detail coefficients, and so on, as illustrated in Figures 6.22 and 6.23. Donoho and Johnstone (1995) show that the variance of Gaussian-distributed white noise with zero mean can be estimated from the higher-order wavelet detail coefficients, and that a threshold value, which they call the Universal Threshold, can be computed from this noise variance. Other studies, such as those reported by Cai and Harrington (1998), Zervakis *et al.* (2001) and Horgan (1999) discuss the use of the Donoho/Johnstone method, and suggest that the Universal Threshold may overestimate the noise level in the data. Other problems include the choice of wavelet function (the 'mother wavelet'), the selection of the number of levels of detail coefficients, and the fact that the procedure ideally requires equispaced data points that number a power of two. The assumption is also made that the data series is circular in the sense that we can use points from the end of the series in to precede the first data point. This may work reasonably well with functions such as sine waves measured over a range that is a multiple of 180°, in which the data values are similar in magnitude at both ends of the series. If there is a discrepancy in data magnitude at the start and end of the series then some instability is present in the wavelet coefficients. Shafri (2003) gives a detailed analysis of these problems in relation to both one- and two-dimensional data series, and Taswell (2000) gives a tutorial guide to wavelet shrinkage denoising.

The steps involved in the Donoho and Johnstone (1995) procedure are:

1. If the data series length, n, is not a power of two then add sufficient zero values to the series so that $n = 2^j$.
2. Select an appropriate mother wavelet (section 6.7) and decompose the data series into a set of j detail coefficients using the forward DWT.
3. Determine the noise variance from the detail coefficients at levels $1 - p$ $(p < j)$
4. Evaluate the Universal Threshold and determine an appropriate multiple of the threshold for this data (usually between 0.4 and 1.0).
5. Use either hard or soft thresholding (see below) to modify all wavelet coefficients (levels 1 to j).
6. Perform an inverse DWT to reconstruct the denoised data.

The results of each stage are shown graphically in Figure 9.17. Although it is difficult to see any difference between the graphs shown in Figure 9.17(a), there does appear to be a small amount of noise present when the two series are differenced (Figure 9.17(e)). Because the derivative is very sensitive to noise in the raw data, the use of the wavelet denoising procedure can be justified on the grounds that the data series is essentially unaffected if no noise is present, yet the procedure will remove noise when it is present.

The degree or severity of denoising is related to the choice of mother wavelet. The Daubechies wavelet (section 6.7), for example, is implemented in MIPS in three different forms, with four, 12 and 20 coefficients. The greater the number of coefficients, the greater the degree of smoothing. In the example illustrated in Figure 9.17, a Daubechies-20 wavelet was employed. The noise variance is computed (following Donoho and Johnstone, 1995) by firstly selecting the number of levels of detail coefficients to be used in noise estimation. In the example above, in which the series length is 512 (after zero padding), a total of five out of a possible nine levels of detail coefficients were selected ($512 = 2^9$; see also Figure 6.22). The median of the selected detail coefficients is computed first, then the absolute deviations of the detail coefficients from this median are calculated. The median of the absolute deviations (MAD) is multiplied by $\sqrt{2n}$, where n is the number of detail coefficients, and divided by the constant 0.6435 to give the value of the Donoho and Johnstone (1995) Universal Threshold (UT).

The UT can be multiplied or divided by a scaling factor to increase or decrease its value (Horgan, 1999 provides some examples using synthetic data). If hard thresholding is used, all wavelet detail coefficients in the chosen levels (five in the example above) that are smaller in magnitude than the scaled UT are set to zero. Other detail coefficients are left unaltered. If soft thresholding is used then hard thresholding is followed by the subtraction of the scaled UT from all the remaining detail coefficients. Again, Horgan (1999) provides some examples. The inverse DWT is then used to transform the thresholded detail coefficients back to the spatial domain; this operation generates the denoised data series.

Two-dimensional denoising follows a similar pattern. Figure 6.36 shows a three-level wavelet decomposition of an image. The detail coefficients at each level are divided into horizontal, vertical and diagonal components. Any or all of these can be selected at a chosen number of levels for the computation of the UT. Often the diagonal detail coefficients at level one are used. The computation of the noise variance and the UT follow the same steps as described above, and hard or soft thresholding is applied before the inverse wavelet transform is computed. See Example 9.3.

9.3.2.4 *Determination of 'red edge' characteristics of vegetation*

The 'red edge' in the reflectance spectrum of active vegetation has become more widely used as a diagnostic feature as the volume of data collected by imaging spectrometers has increased (Bochs and Kupfer, 1990; Curran *et al.*, 1991; Horler *et al.*, 1983; Ustin *et al.*, 1999). An example of a typical vegetation spectrum is shown in Figure 1.21. The biophysical factors that give the spectrum of active vegetation its typical shape can be summarised as follows: leaf chemistry is responsible for the absorption characteristics of the leaf spectrum in the visible wavebands, while the high reflectivity in the near-infrared wavebands is explained by the internal leaf structure. Different vegetation types generally have different spectral reflectance curves, and these differences can be sufficient to allow these different vegetation types to be discriminated and mapped using classification techniques (Chapter 8). Also, the spectrum of a given plant (or group of plants, depending size of the individual ground element that is 'seen' by the sensor) will change during the day and from day to day, depending on season, moisture availability, and other stress factors. A number of methods of characterising vegetation in terms of biomass, leaf area index, vigour (or response to stress) using features of its spectral reflectance curve have been developed, such as vegetation indices (section 6.2.4). Both the simple vegetation ratio (IR reflectance divided by red

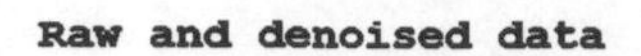

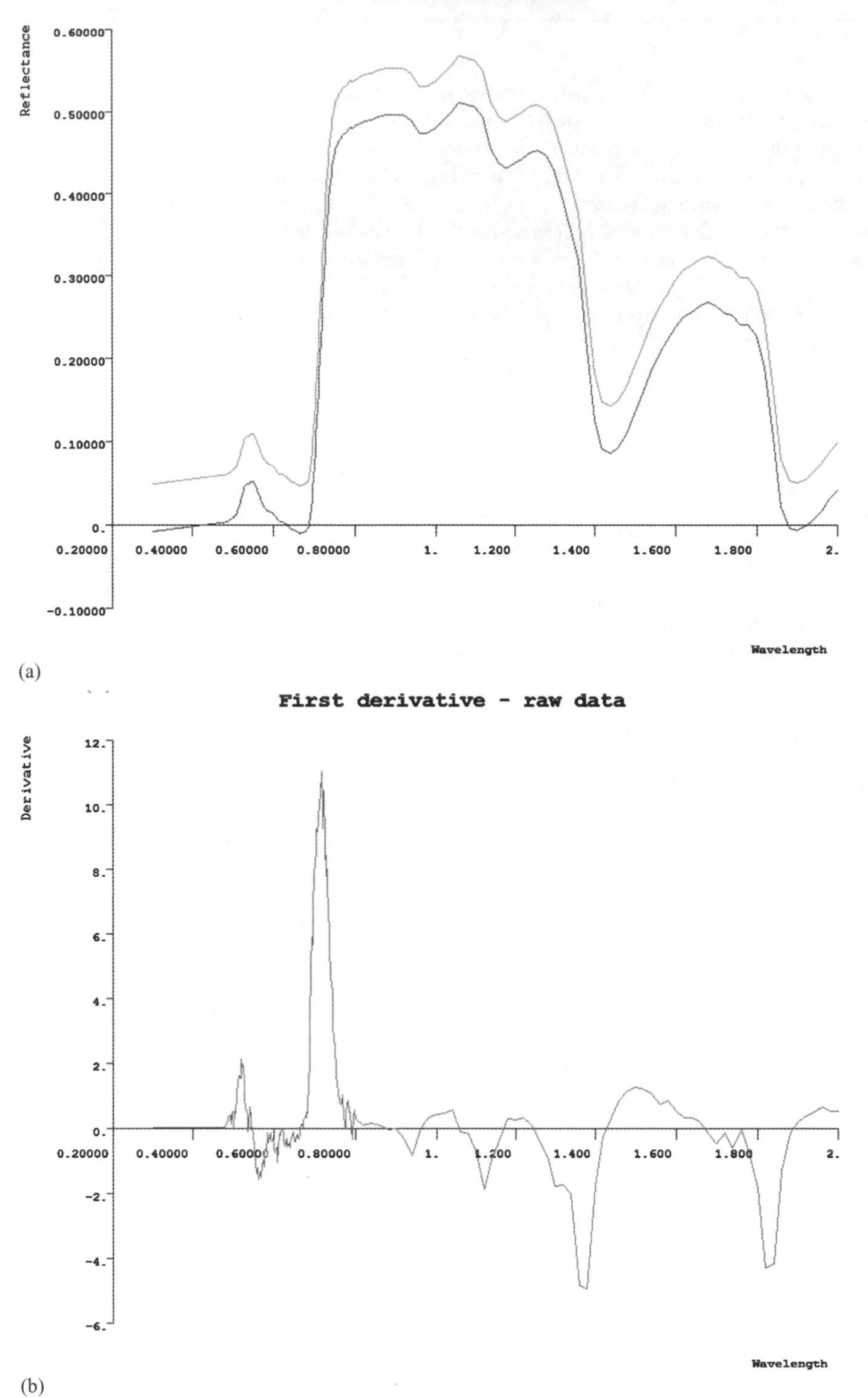

Figure 9.17 Stages in denoising a one-dimensional spectrum of a deciduous leaf. (a) Raw and denoised data. The denoised data is offset vertically downwards for clarity. (b) First derivative calculated from raw data. (c) Wavelet coefficients. (d) First derivative calculated from denoised data. (e) Difference between (b) and (d). Data from the ASTER Spectral Library through the courtesy of the Jet Propulsion Laboratory, California Institute of Technology, Pasadena, California. © 1999, California Institute of Technology. ALL RIGHTS RESERVED.

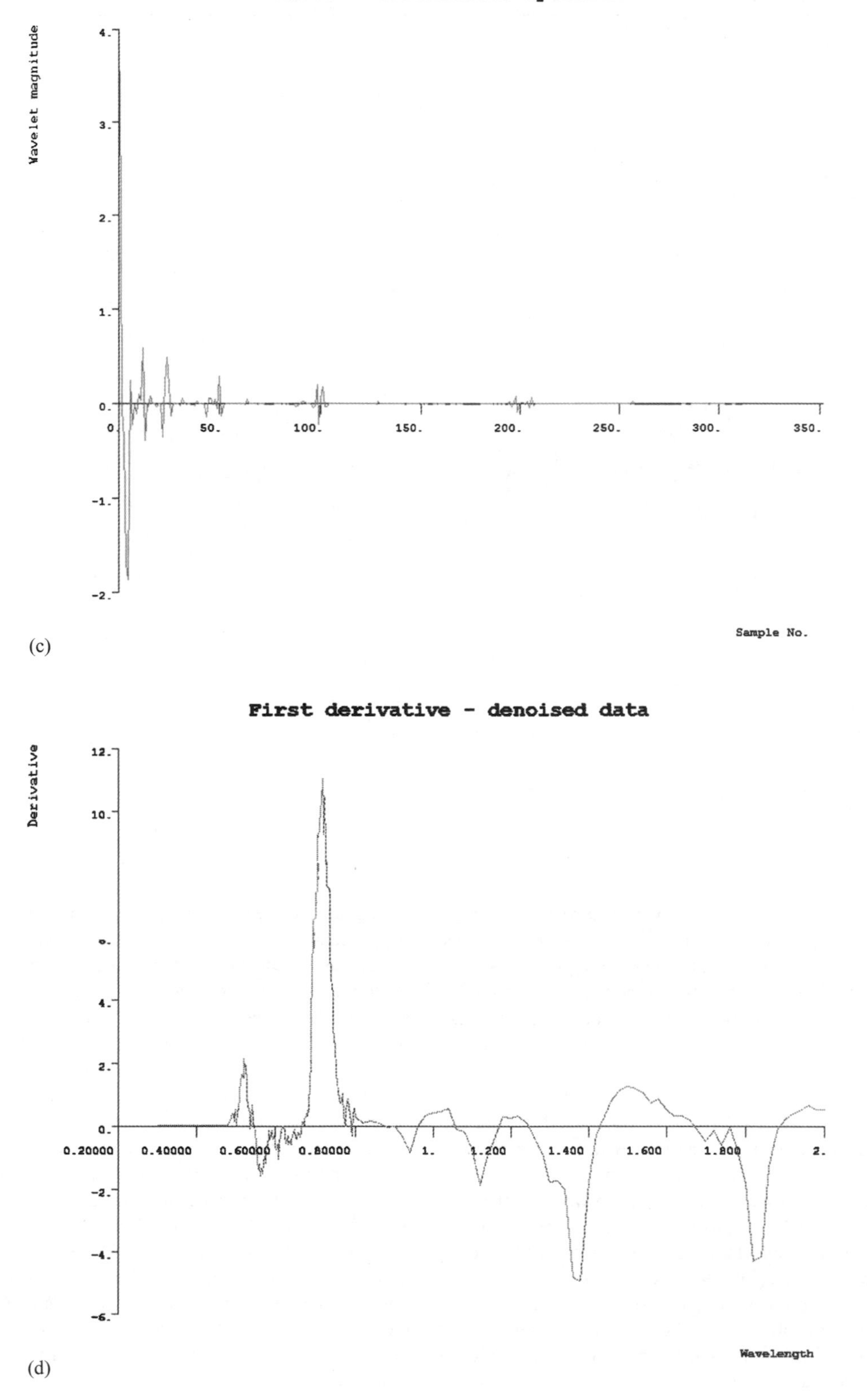

Figure 9.17 (*Cont.*)

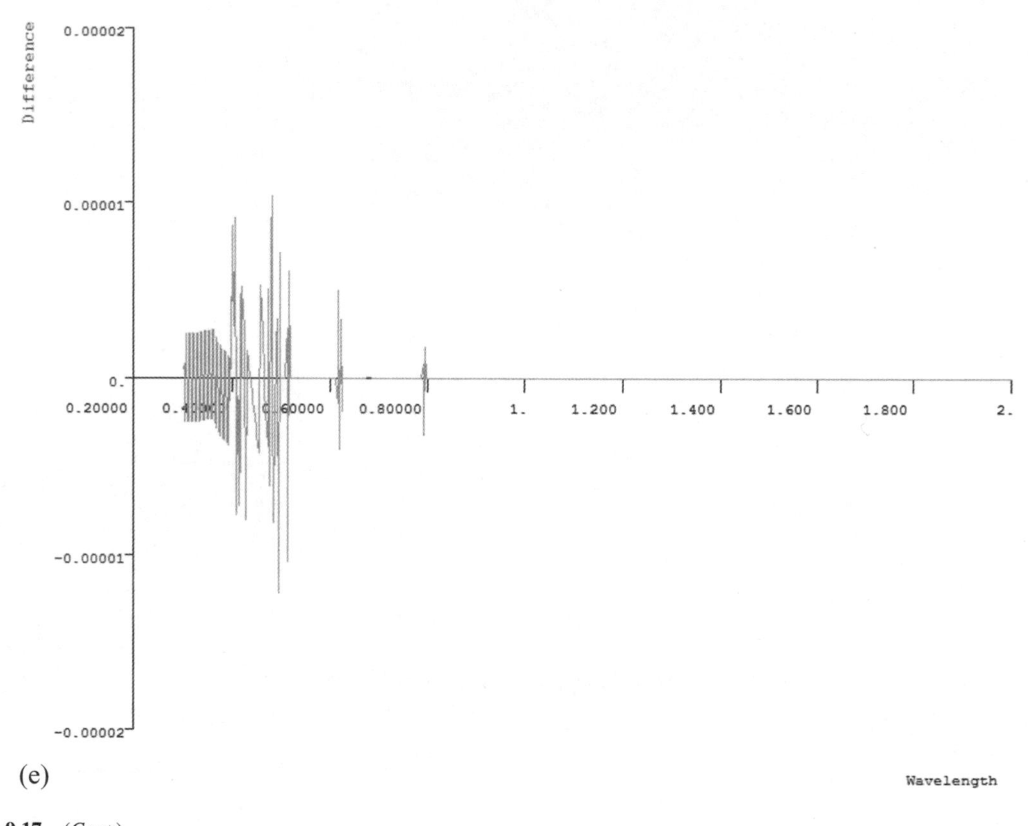

Figure 9.17 (*Cont.*)

reflectance) and the Normalised Difference Vegetation Index or NDVI attempt to characterise the spectral reflectance curve of vegetation by contrasting the high reflectance in the near-infrared wavebands and the low reflection in the visible red wavebands. In effect, such ratios are measuring the steepness of the red-infrared region of the spectral reflectance curve, but not the position in the spectrum of the steep section. Using imaging spectrometer data it is now possible to attempt to characterise this steep rise in the reflectance curve in terms of a single wavelength (though in practice the accuracy of such a determination depends on the width and the spacing of the wavebands in which data are acquired by the imaging spectrometer). The position of this steep rise in reflectance can be characterised by the *red edge wavelength* and the *red edge magnitude*.

The most common definition of the red edge position is the point of inflection of the spectral reflectance curve of vegetation in the red/near-infrared region. A point of inflection is that point on an upward-sloping curve at which the gradient (steepness) of the curve stops rising and starts falling, i.e. it is the point of maximum gradient and it is also the point at which the rate of change of gradient is zero.

This idea is illustrated in Figure 9.18, in which the uppermost curve represents an idealised vegetation reflectance spectrum for the wavelength range 650–800 nm. This curve is referred to as 'the function' in the following sentences. The central plot in Figure 9.18 shows the slope or first derivative of the function, defined as the rate of change of the function value (y-axis) per unit step along the x-axis. The first derivative increases in value, reaches a maximum, and then declines again. The rate of change of slope (bottom plot in Figure 9.18) per unit step along the x-axis is the second derivative. The second derivative curve crosses the x-axis at the same wavelength as the first derivative reaches a maximum. This is the point of inflection – defined as a point at which the first derivative reaches a maximum and the value of the second derivative changes from negative to positive (or vice versa). The point at which the value of a function changes from positive to negative (or vice versa) is known as a 'zero crossing'.

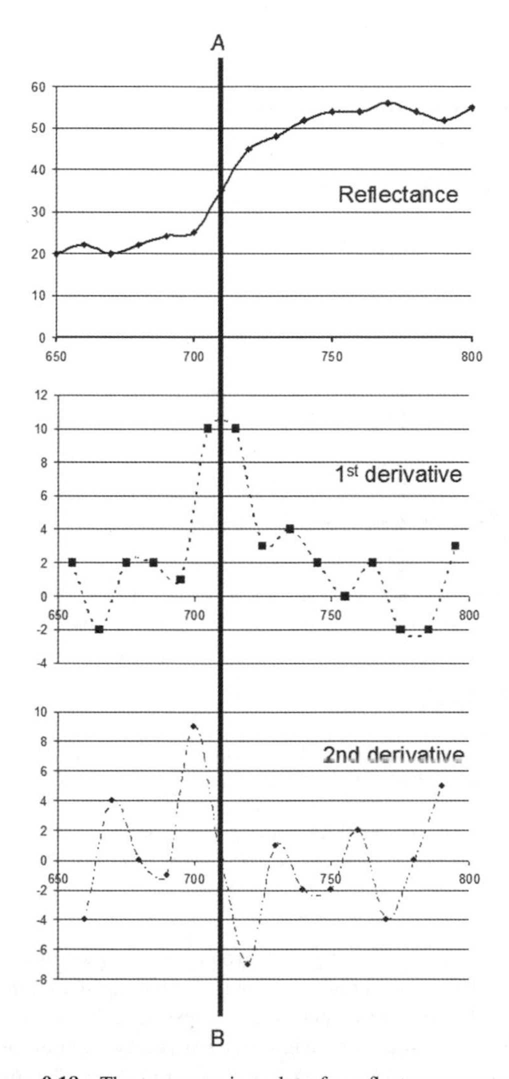

Figure 9.18 The top curve is a plot of a reflectance spectrum covering from wavelengths 600 nm to 800 nm. Note the sudden increase in the gradient of the curve between 680 and 750 nm. The middle graph shows the gradient of the reflectance spectrum, calculated by the method of first differences, which approximates the first derivative. The gradient increases from zero in the green-red wavelengths (600 nm), reaches a maximum, and declines back to zero in the near-infrared wavelengths (800 nm). The rate of change of this gradient (the second difference/derivative of the reflectance spectrum) is shown in the bottom graph. The point at which the second difference (derivative) curve crosses the x-axis (i.e., changes from positive to negative, or vice versa) is called a zero crossing. A point of inflection on a curve is indicated by the correspondence of (i) a zero crossing in the second derivative and (ii) a maximum in the first derivative. In this example, such a point is indicated by the vertical line AB. This point of inflection is often used as an estimate of the red edge wavelength (approximately 710 nm in this case).

If there is random noise in a data series then derivative analysis will amplify it. Some researchers, for example Clevers and Jongschaap (2001), whose ideas on red edge determination are summarised below, suggest that derivative-based methods are not robust. However, noise removal ('denoising') using methods such as the one-dimensional discrete wavelet transform (section 6.7) are effective in removing additive random noise from both one and two-dimensional data sets. Thus, it is sensible to use denoising procedures before carrying out derivative-based red edge determinations.

An alternative and simpler method of computing the red edge wavelength is given by Guyot and Baret (1988) (an accessible account is provided by Clevers and Jongschaap (2001)). Their method requires only four reflectance measurements in the red/near-infrared region of the spectrum. It is therefore suitable for use with image data that are measured in relatively broad wavebands, such as the data collected by the MODIS sensor (section 2.3.3). Call these four wavebands R_1, R_2, R_3 and R_4 and assume that they are measured at points 1, 2, 3 and 4 on the spectrum, with point 1 being close to 670 nm and point 4 being near 780 nm. The red edge radiance R_e is simply the average of R_1 and R_4. The red edge wavelength λ_e is found by linear interpolation: $\lambda_e = R_2 + \mathrm{WI}((R_e - R_2)/(R_3 - R_2))$. The term WI is the wavelength interval. In Figure 9.19 the wavelength interval from 700 to 780 nm is 40 nm. Points R_1, R_2, R_3 and R_4 are measured at wavelengths of 670, 700, 740 and 780 nm (shown in Figure 9.20 as $R_{670}R_{700}$, R_{740} and R_{780}). Clevers and Jongschaap (2001) suggest that the Guyot/Baret procedure is more robust than the derivative analysis described above, and that it produces results that are comparable with those achieved by more complicated methods. Bonham-Carter (1988) describes another method, based in fitting an inverted Gaussian model. He also provides a Fortran program to implement the method.

Two procedures are available in MIPS to compute the red edge position. The first uses the derivative-based approach as follows:

- Denoise the individual pixel spectra using a discrete wavelet transform (for example, based on the Daubechies-4 wavelet).
- For each pixel, compute the ratio between the reflectance values at the wavebands closest to 800 and 660 nm. If the magnitude of this ratio is less than a specified threshold (such as 2.0) then mark this pixel as 'non-vegetation' (MIPS uses a 'flag' of –999 to indicate 'non-vegetation'.
- For all 'vegetation' pixels, calculate the first and second derivatives of the spectrum.

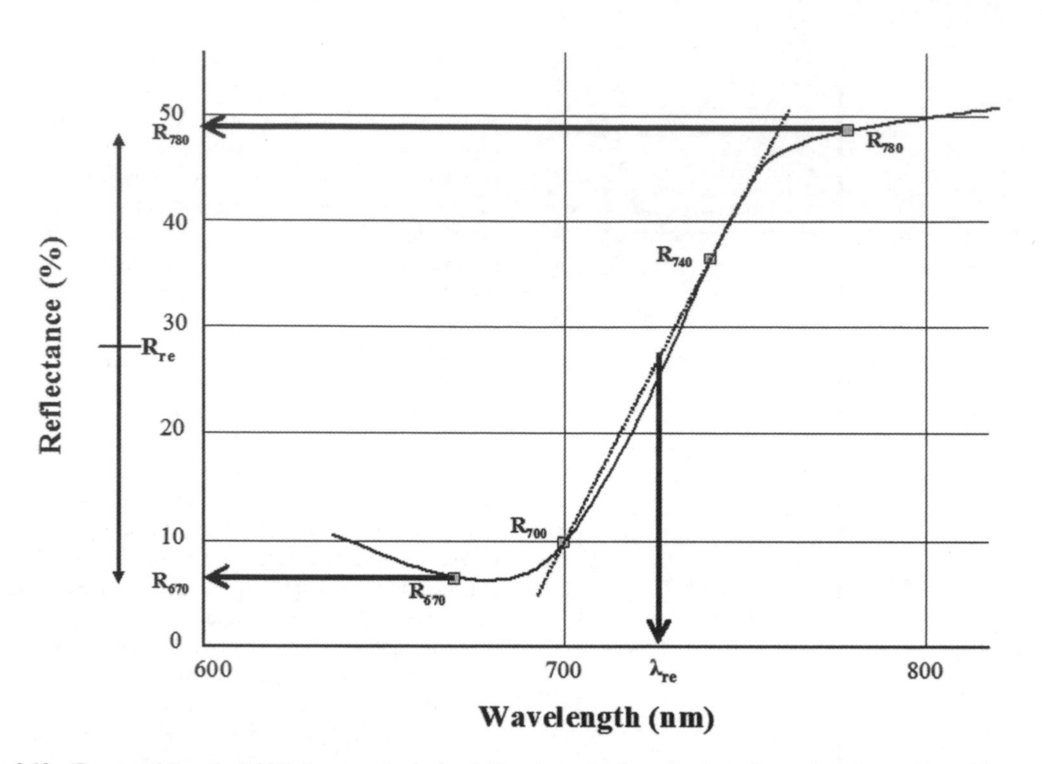

Figure 9.19 Guyot and Baret's (1988) linear method of red edge determination uses the reflectance at four points on the spectrum (the values 670, 700, 740 and 780 nm are used here). These points are marked R_{670}, R_{700}, R_{740} and R_{780}. The reflectance R_e at the red edge is the average of the reflectance at 780 and 670 nm. The red edge wavelength is determined by a linear interpolation between the 700 and 780 nm points. From Figure 2 of J.P.G.W. Clevers and R.E.E. Jongschaap (2001), Imaging Spectrometry for Agriculture. In F. van der Meer and S.M. de Jong, Imaging Spectrometry: Principles and Applications. Dordrecht: Kluwer Academic Publishers, pp. 157–199. © Kluwer Academic Publishers.

- Locate a zero crossing in the second derivative in the 660–820 nm spectral region that corresponds to a maximum of the first derivative.
- Use linear interpolation to estimate the wavelength of the zero crossing. For example, the zero crossing is indicated by a positive value in waveband i and a negative value in waveband $i + 1$. The magnitudes of the second derivative at points i and $i + 1$ are known, so the wavelength at which the value of the second derivative are zero can be interpolated.
- Output the red edge wavelength, the magnitude of the first derivative at the red edge, and the area under the first derivative curve between (red edge –30) and (red edge +30) nm.

The second approach to red edge wavelength determination in MIPS uses the Guyot-Baret method. The output from the derivative approach using data collected over Thetford Forest in eastern England by the Hymap sensor is shown in Figure 9.20(a). The same image data, showing spatial variations in the red edge wavelength, is shown in Figure 9.20(b) after the application of a 3 × 3 median filter. Longer wavelengths are displayed in lighter shades of grey. The black area is that which has been masked by the application of a vegetation index mask, as described above. The spatial variations in red edge wavelength and magnitude correlate well with information about tree species and age for each stand. Hansen and Schjoerring (2003) do not specifically use the red edge wavelength position but find, in a study of a variety of narrow-band vegetation ratios applied to imaging spectrometer data, that most of the ratios used bands that were located in the red edge region of the spectrum.

EXAMPLE 9.3: IMAGE DE-NOISING USING THE WAVELET TRANSFORM

In this example, a single-band greyscale image is decomposed using the DWT, then the noise variance is estimated

(a)

(b)

Figure 9.20 Red edge wavelength for a 512 × 512 area of a Hymap imaging spectrometer image of part of Thetford Forest, eastern England. (a) Raw output from the MIPS derivative-based red edge wavelength procedure, and (b) image shown in (a) after the application of a 3 × 3 median filter. Black areas are masked using a vegetation index threshold and represent bare soil and non-vegetated areas, plus left and right marginal areas resulting from geometric correction of the image. Based on data collected for the BNSC/NERC SHAC campaign, 2000.

Example 9.3 Figure 1 Landsat ETM+ panchromatic image of an agricultural area of eastern England.

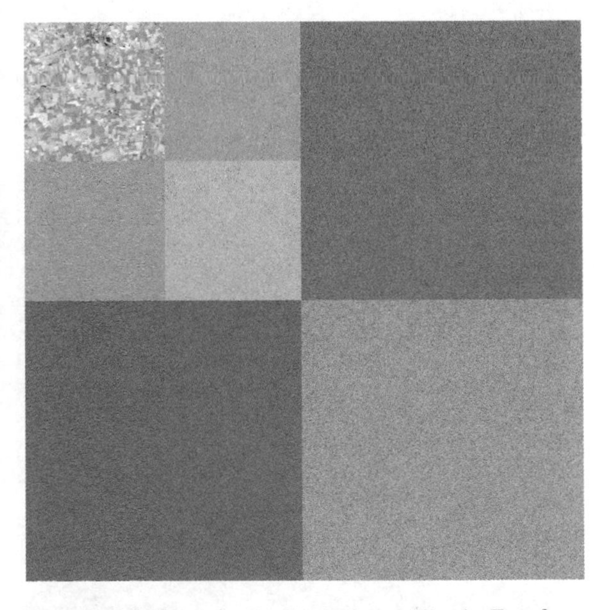

Example 9.3 Figure 2 Two-level Discrete Wavelet Transform of the image shown in figure 1, using the Daubechies-4 mother wavelet.

from the detail coefficients, and the de-noised image is reconstructed. The image used in this example is referenced by the dictionary file *etm_pan2.inf*. You should locate this file before starting the exercise.

Example 9.3 Figure 3 Noise removed from figure 1 using wavelet shrinkage. The noise was estimated from the level 1 (horizontal, vertical and diagonal) detail coefficients of figure 2 using hard thresholding.

Example 9.3 Figure 4 De-noised image

Begin by displaying the image *etm_pan2.img*, which is a 1024 × 1024 Landsat ETM panchromatic band image of an agricultural area in eastern England. Display the image using *View|Display Image* and enhance it using *Enhance|Stretch|Use Percentage Limits*, selecting 5% and 95% as the lower and upper bounds. This step is not strictly needed, but it will be useful later to have the original image on-screen for comparative purposes.

Now follow these steps:

1. Choose *Transform|Wavelet (2D) Transform*
2. Check the radio button for *Mode 1: Single Band, 3 output images*
3. Identify the INF file to be used by double clicking on the entry *etm_pan2.inf* when the File Selection dialogue box appears.
4. There is only a single band in this image set, so enter 1 in the next dialogue box.
5. Choose the default number of decomposition levels, *i.e.*, two.
6. Check all three radio buttons to compute the transformed, noise, and de-noised images.
7. Check all three radio buttons to select the horizontal, vertical, and diagonal detail coefficients at level 1 for noise estimation.
8. Do not select any of the three radio buttons for level 2 noise estimation.
9. After a short wait, select the Daubechies 4 mother wavelet.
10. Opt to save the specially stretched transformed output image (shown in Example 9.3 Figure 2) in which each quadrant is separately stretched (you can see why that is done at a later stage). Supply a data (BSQ), header (HDR) and dictionary (INF) file name for this special 8-bit image.
11. Provide the name of an output band sequential (BSQ) file and the corresponding header (HDR) file that will hold the three output images (in order: transformed, noise, de-noised).
12. Wait a while as the DWT is applied to each of the 1024 rows and 1024 columns of the image.
13. Use a threshold multiplier value of 1.0 (the default) plus hard thresholding.
14. The image is now de-noised and the inverse DWT is computed. Eventually, you will see the message *Finished.*
15. Select *File|INF File Operations|Create ENVI INF File* and follow the instructions to create an INF file that references the three output files created at step 6.
16. Finally, use *View|Display Image* to view each of the three output files separately.

Example 9.3 Figure 1 shows the original image. The DWT decomposition is shown in Example 9.3 Figure 2. We chose to do a two-level decomposition simply to discover what the resulting decomposed image would look like, and we see that the image at level 1 has been transformed into quadrants, as explained in the main text. The top left quadrant is decomposed into four sub-quadrants at level 2.

Noise is computed from the $3n/2$ detail coefficients in the top right, bottom left and bottom right quadrants. The noise image is shown in Example 9.3 Figure 3. The denoised image, after hard thresholding using the Universal Threshold, is shown in Example 9.3 Figure 4.

There is no noticeable (visible) noise in the original image. You might like to investigate the validity of this statement by repeating the experiment and using a different mother wavelet. Other questions that you could investigate are:

- Is there any apparent difference between the results of hard and soft thresholding?
- What happens if you increase or decrease the threshold for wavelet shrinkage by changing the value of the multiplier at step 13? (The default is 1.0.)
- Wavelet shrinkage is designed for the removal of *additive* noise. Could you use the method to remove speckle noise in SAR images? If so, how?
- What happens if you base the noise threshold on the level 2 detail coefficients (steps 7 and 8) rather than on level 1? Or if you base the noise estimate on a single quadrant (one of horizontal, vertical, and diagonal) at either level 1 or level 2? Or even level 3 (step 5)?

Note that the output BSQ file contains the transformed, noise and denoised images. However, you created a special 8-bit output image at step 10. Display this image, and compare it to the transformed image in the main output file (it is the first of three images, the others being the noise image and the denoised image). You will see that a single contrast stretch cannot accommodate the range of values present in the transformed image; that is why the special image is created. Each of the quadrants and sub-quadrants in the special image is stretched individually in order to chieve the optimum display.

9.3.2.5 Continuum removal

Workers in the field of analytical chemistry have found that removal of the local trend from a one-dimensional derivative spectrum can enhance its interpretability. Continuum removal emphasises absorption bands that depart from their local trend line. The local trend line is usually defined as the upper surface of a convex hull surrounding the data points. Consider a plot of reflectance (y-axis) against centre wavelength for a number of spectral bands at a given pixel (x, y) position. The convex hull is a line that surrounds the scatter of points representing measurements of reflectance at each waveband centre (Figure 9.21). For present purposes the upper surface of the convex hull is required. The upper convex hull is defined by a series of unequally spaced points. Values on the hull at each waveband centre position are interpolated, and the ratio of the reflectance value at waveband centre i to the corresponding interpolated hull value is computed to give the continuum removed spectrum. An example is shown in Figure 9.22.

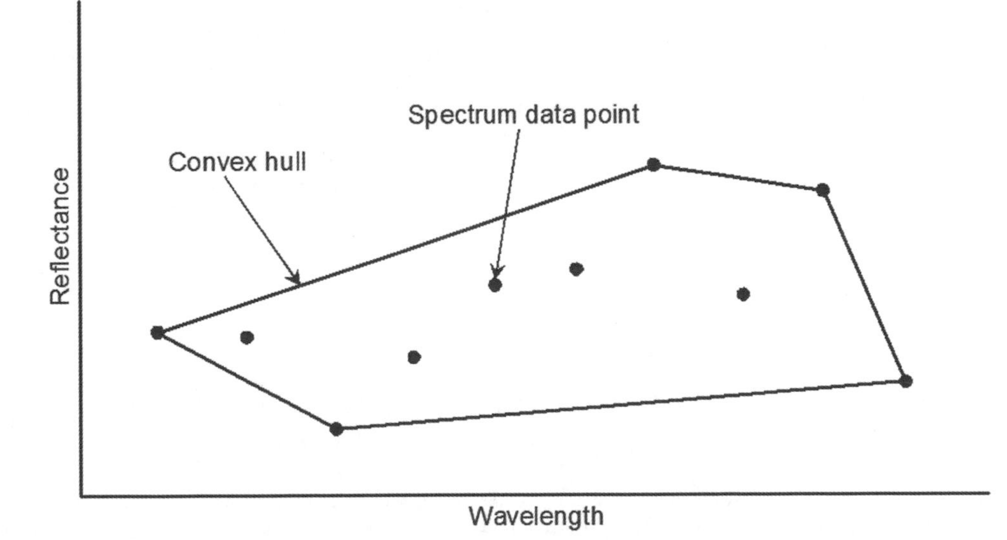

Figure 9.21 The black circles show the reflectance (y-axis) plotted against waveband centre (x-axis) for 10 spectral wavebands. The solid line joining the extreme points is the convex hull. Only the upper surface of the hull between the first and last data points is required.

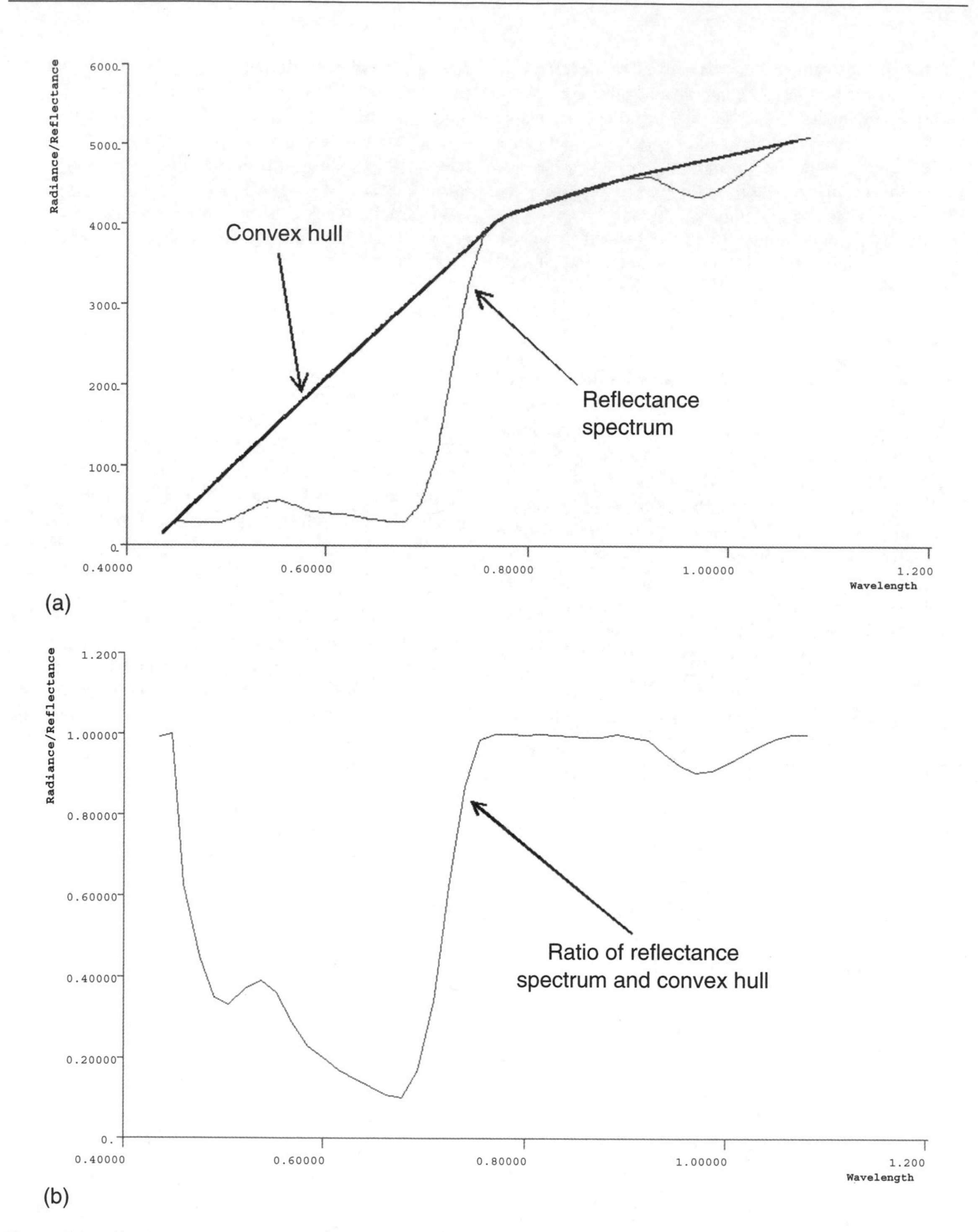

Figure 9.22 Continuum removal. (a) Reflectance spectrum of the selected pixel and the upper segment of the convex hull. (b) Continuum-removed spectrum derived from the ratio of the reflectance spectrum and the corresponding convex hull value at each waveband centre. The positions and features of the absorption bands are more clearly perceived.

The data used for this example were derived from a pixel representative of an area of deciduous woodland in the top right of the region shown in Figure 9.20 near the north-south river valley. Clark (1999) and Ustin *et al.* (1999) describe the use of continuum removal in the context of rock and mineral identification and geobotany, respectively.

9.4 LiDAR

9.4.1 Introduction

The word 'LiDAR' is an acronym derived from *Li*ght *D*etection *a*nd *R*anging. The same system is also known as LADAR (*LA*SER *D*etection *a*nd *R*anging), and also as LASER Altimetry. The word LASER is yet another acronym, of *L*ight *A*mplification by *S*timulated *E*mission of *R*adiation. The lower-case word 'lidar' will be used here to be consistent with our use of the word 'radar', which is also an acronym. Like radar, a lidar sensor is an active sensor, the differences between lidar and radar being: (i) lidar uses electromagnetic energy in the visible and near-infrared (VNIR) wavelengths, whereas a radar sensor uses microwave energy, (ii) lidar is a nadir-looking, but radar is a side-looking instrument, (iii) a lidar records information at discrete points across the swath, which is not therefore formed of contiguous pixels, as is the case with a radar sensor, and (iv) because lidar operates in the visible and near-infrared wavelengths, its signal is affected by atmospheric conditions, whereas at the wavelengths used in remote sensing, a radar sensor is weather-independent. Unlike imaging sensors operating in the VNIR wavebands, and which record upwelling electromagnetic energy that is emitted by or reflected from objects on the Earth's surface, a lidar instrument measures the time taken by an energy pulse to reach the ground, and for a part of the scattered radiation to return to the sensor. A lidar thus measures the distance from the sensor to the ground, because electromagnetic energy travels at the speed of light and so the time taken for the energy pulse to travel from the lidar instrument to the ground and back can easily be converted to a distance. If the position of the instrument is known to a sufficient level of accuracy then these distances can be converted to elevations above a specified geodetic datum, and a raster map of these elevations can be generated. The set of elevation values produced by a lidar does not necessarily define a digital elevation model (DEM), because the lidar pulse is reflected back by the first object of sufficient size and density that it meets as it travels downwards from the instrument. This object may be a branch, a tree crown, the ground surface, or the top of a building, depending on the area being viewed and the size (or footprint) of the lidar pulse. The set of point elevation values can be used to generate a digital surface model (DSM), which shows the elevation of the highest reflective object on the ground. Thus, data from lidar sensors can be used to map the highest point on a building, or of a tree. In order to generate a DEM from the DSM, the surface objects must be removed, or the height of the superimposed object must be estimated. The DSM may be useful in itself. For example, Figure 9.23 is a lidar image produced by NOAA showing the site of 'Ground Zero' in Manhattan, New York, in September 2001. Lidar images collected by aircraft can be used to generate 3D urban models, or they can show objects such as forests that project above the ground surface.

We saw in the preceding paragraph that the conversion of a lidar DSM to a DEM requires that we estimate the height of the highest reflective object on the ground at a given point. The lidar instrument itself can in fact, perform this task. The account given above of lidar operation describes what is known as a 'first return' or 'first bounce' system, which – as the name implies – records the time taken between the emission of the pulse of light energy and the reception at the sensor of the first backscattered return. More sophisticated instruments (with more sophisticated signal processing software) can generate two signals for each pulse. One of these is the first return, as described earlier, while the second is the position of the last indication of backscatter. This second event is called the last return, so these systems give 'first return – last return' data for each grid cell of the raster. There is no difference between the first and last return if the target object does not transmit light; if it does, then the time difference between the first and last bounce is proportional to the height of the object above the ground. A concrete surface does not transmit light, but a forest canopy does. Even more sophisticated systems can record the backscatter events between the first and last return, and so provide a profile of this backscattering between the first and last bounce points (Figure 9.24). As noted already, solid objects such as buildings are not penetrated by the electromagnetic energy emitted by a lidar. For these targets, and for the ground itself, only a single return is recorded. The 'first return – last return' and the profile data are returned by objects such as trees, forests, and other types of vegetation that are capable of transmitting light energy.

So far, the operation of a lidar system has been described in terms of the emission of energy pulses, and the timing of the returned (backscattered) energy. Not all lidars operate in this way. Some use the continuous wave (CW) principle, which is described in more detail in section 9.2 in the context of SAR interferometry. Instead of generating discrete pulses of energy, a CW lidar emits energy in the form of a sinusoidal wave of known wavelength. Recall that the phase of a wave (section 1.2) is the offset between the y-axis and

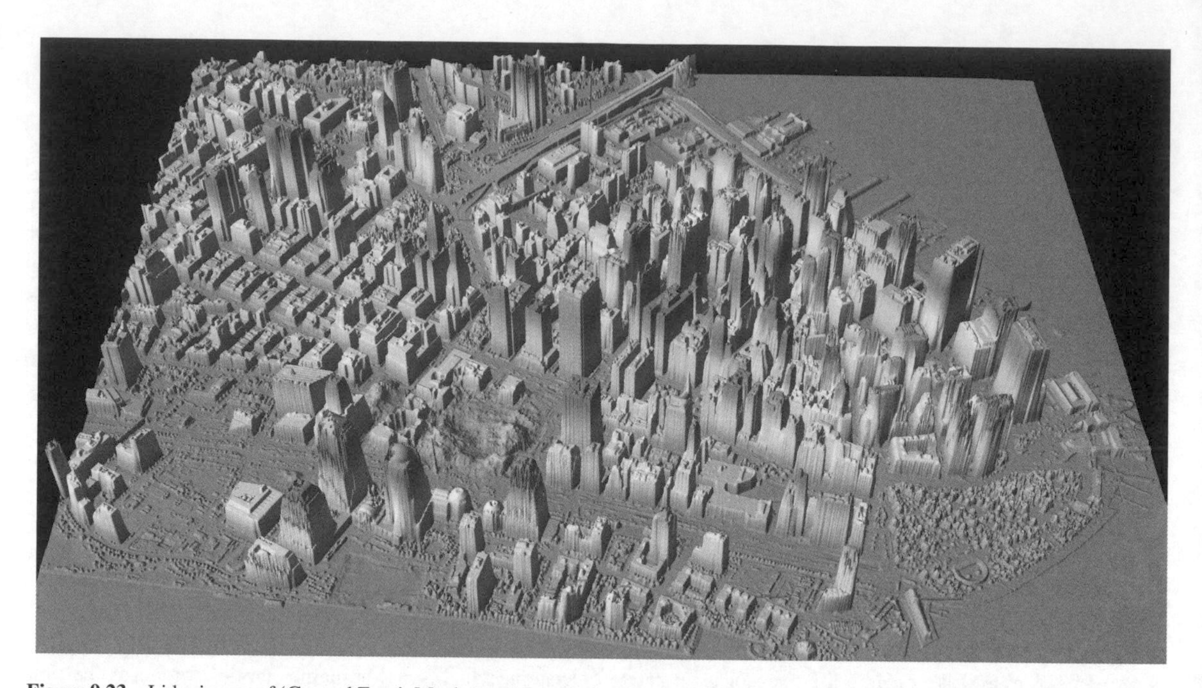

Figure 9.23 Lidar image of 'Ground Zero', Manhattan, New York, taken on 17 September 2001. The data collection programme was the result of a collaboration between the US National Atmospheric and Oceanic Administration (NOAA), the US Army Joint Precision Strike Demonstration (JSD), and the University of Florida, using an aircraft-mounted Optech lidar. The shades of grey represent elevations between 0 and 200 m. The 3-D model helped to locate original support structures, stairwells, elevator shafts, basements, etc. Credit: NOAA/U.S. Army JPSD. [Original in colour at *http://www.noaanews.noaa.gov/stories/s781.htm*]

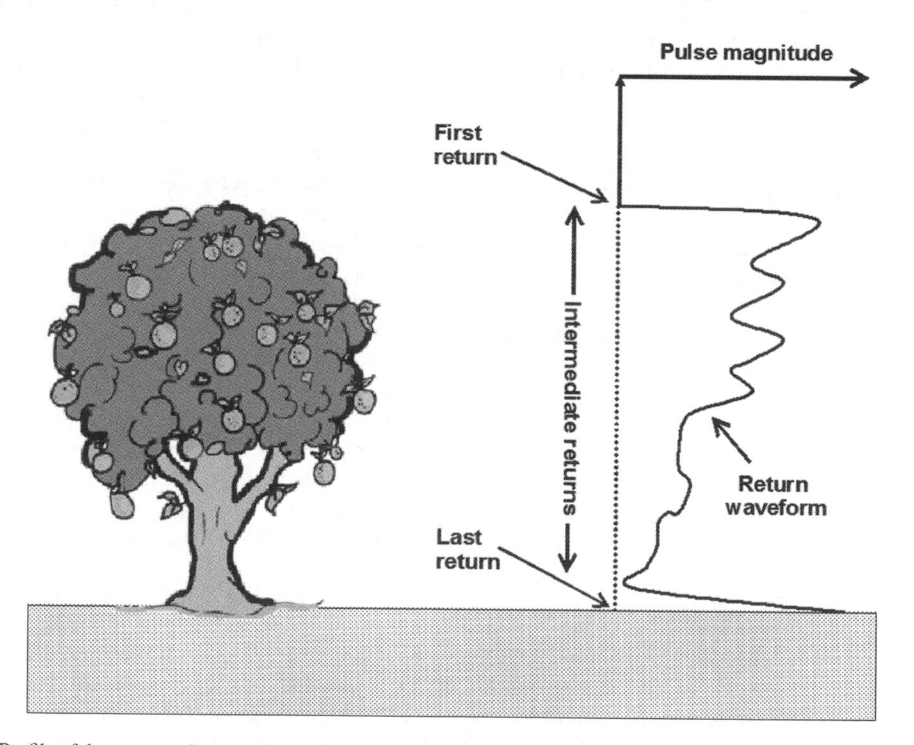

Figure 9.24 Profile of the return pulse magnitude of lidar interaction with a tree. The lidar sensor receives the returned (back-scattered) signal (solid line on right). Some systems record the time from pulse transmission to receipt of the first return, others record the time to the last return, while more sophisticated systems take a sample of the intermediate returns. The distance (range) from the sensor to the target is computed from these timings. Clipart tree from http://www.clipsahoy.com/webgraphics2/as3313.htm.

the waveform crest (Figure 9.1). In effect, the number of waveforms that are required to cover the distance between the lidar sensor and the ground is calculated, with the fractional part being estimated by the phase difference between the original and the received wave.

An important distinction can be made between 'small footprint' (5–30 cm) and 'large footprint' (10–25 m) lidar sensors. A large-footprint system has a greater swath width than a small-footprint system. Small-footprint systems are used for detailed local mapping of surface elevations, as might be required for floodplain mapping. However, the spacing between the points at which the lidar pulse hits the target in a small-footprint system may be such that several points on the surface of a vegetation canopy may be measured, giving a detailed representation of that canopy, whereas the large-footprint system will collect an average value for a greater area of the canopy surface. This latter measurement may be more useful for studies of forest canopy response.

Most lidar sensors are flown onboard aircraft. One experimental lidar sensor, the Lidar In-space Technology Experiment, or LITE, was flown in September 1994 as part of the STS-64 mission. LITE is a three-wavelength profiling lidar developed by NASA Langley Research Center, and is primarily designed for measuring atmospheric rather than terrestrial phenomena. It takes simultaneous measurements in three harmonically related wavelengths of 1064 nm (infrared), 532 nm (visible green), and 355 nm (ultraviolet) along a profile measuring approximately 300 m wide at the Earth's surface.

The SLA (Shuttle Laser Altimeter) was carried onboard two Space Shuttle missions, in January 1996 and August 1997, respectively. SLA incorporates a laser operating at a wavelength of 1068 nm, with a sampling rate of 10 Hz. It has a footprint radius of 100 m. The first SLA mission (SLA-1) was not as successful as expected, as the dynamic range of the backscattered echoes was greater than had been allowed for in the system design. SLA-2 was modified to provide more flexibility. Data from the SLA-2 mission can be downloaded from the WWW (search for 'SLA data'). Garvin *et al.* (1998) provide a review of the SLA program.

Two space-borne lidars are planned for the near future. The payload of the Vegetation Canopy Lidar (VCL) Mission, which is being planned and supervised by a group at the University of Maryland, led by Professor R. Dubayah. VCL is a large-footprint system, with a sensor footprint of 25 m radius, and an 8 km swath. The laser operates at a wavelength of 1064 nm, in the photographic infrared region. The aim is to measure canopy top and ground elevations to an accuracy of 1 m in order to provide measurements of the age and condition of forest ecosystems. The global scale measurements provided by VCL will provide an inventory of forest biomass as well as observations of the spatial texture of the Earth's land cover, which is important in climate modelling (Dubayah *et al.*, 1997). The VCL mission is currently scheduled for launch in 2003. Another satellite, ICESat (Ice, Cloud and land Elevation SATellite), was launched on 13 January 2003. It carries the Geoscience Laser Altimeter System (GLAS), which operates at two wavelengths – 1064 and 532 nm. The position of the ICESat platform is determined by GPS and by stellar navigation systems. The footprint of the GLAS is 75 m, and the spacing between points is 175 m. A comprehensive description of the GLAS instrument and its applications is provided by Zwally *et al.* (2002).

9.4.2 Lidar details

The material in this section summarises two review papers, by Baltsivias (1999) and Wehr and Lohr (1999), to which readers should refer for more detailed accounts.

The basic principle of operation of a lidar sensor is described briefly above. A typical lidar instrument incorporates (i) a laser ranging unit, (ii) an opto-mechanical scanner, and (iii) a control and processing unit. The laser ranging unit contains the laser transmitter and receiver. The transmitter is able to generate a narrow beam of electromagnetic energy, while the receiver 'looks' along the same path as the transmitter in order to capture the backscattered return. Most of the present generation of lidar sensors use the pulse principle, described above; Wehr and Lohr (1999) note that only one commercial airborne lidar employs the continuous wave (CW) principle to calculate range. The principle of the pulsed lidar is quite straightforward; if the time between transmission and reception of the pulse is t and if R is the distance between the lidar transmitter and the target then $R = ct/2$, where c is the speed of light. A CW lidar transmits a continuous signal on which a sinusoidal wave of known period is superimposed. If the phase difference between the transmitted and received signals is computed then the range R is related to the number of full waveforms plus the phase difference. Since the phase difference could be more than 360° some ambiguity could be introduced, analogous to the 'phase unwrapping' problem in SAR interferometry (section 9.2).

The maximum range of a pulsed lidar system depends on the maximum time interval that can be measured by the control unit and on the strength of the backscattered signal, which is to some extent dependent on the power of the transmitted pulse (it also depends on the reflectivity of the surface). An analogy can be made between a laser and a torch. The range of the torch depends on the battery power,

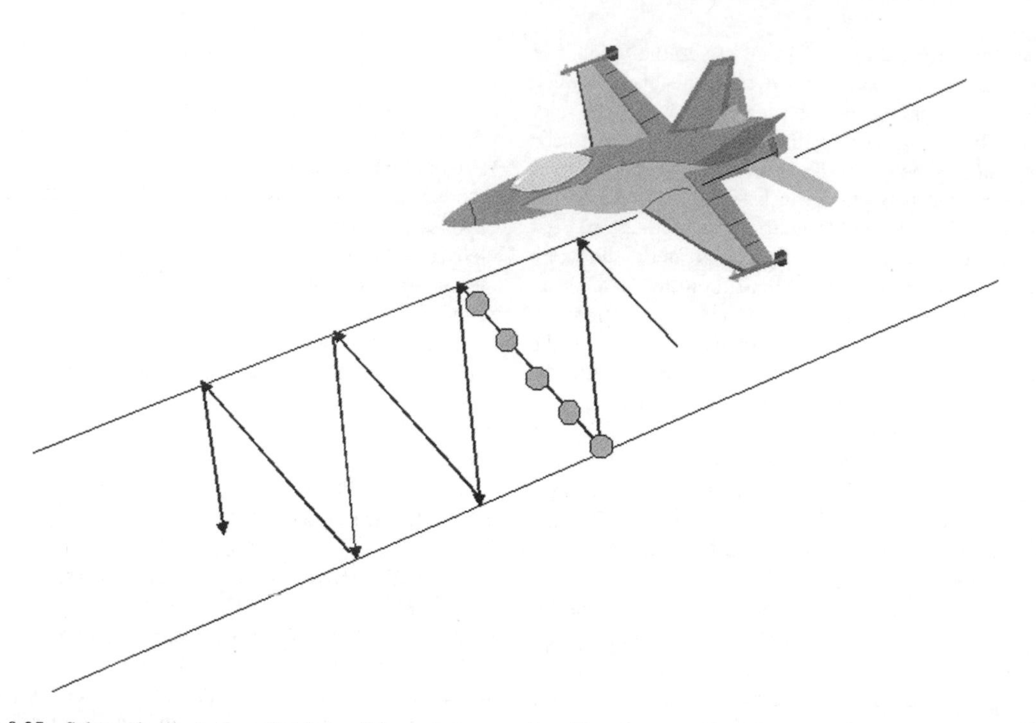

Figure 9.25 Schematic illustration of airborne lidar scanner operation. The mirror oscillates from side to side in the vertical plane, and the forward motion of the platform results in a scan line that is oblique to the flight direction. Further disturbances result from the pitch, roll and yaw of the platform. The measurement points (grey circles) are collected at equal-angle steps across the scan, so their ground spacing is not constant.

the properties of the bulb, and the focusing power of the lens. A 'high-power' torch can transmit a narrow beam of light over a considerable distance. If you take the torch to a dark and isolated location and direct the torchlight upwards, you will not see anything because there are no reflective objects within the maximum range of the torch. The accuracy of the measurements made by a lidar instrument depends upon the signal to noise ratio, which in turn is dependent on electronic noise in the components of the lidar sensor, as well as on the power of the transmitted signal.

Wehr and Lohr (1999) note that the most sensitive detectors for use in the receiver unit operate in the 800–1000 nm region (photographic infrared). In this wavelength range, eye safety is a consideration and so longer wavelengths (around 1500 nm) are employed, because higher power lasers can be used at these wavelengths without compromising safety. A further advantage that comes from the use of longer wavelengths is that the background level of solar radiation is lower than in the 800–1000 nm region (see Figure 1.7). Lidar systems that are used for measuring bathymetry, rather than the properties of terrestrial targets, use shorter wavelengths (of the order of 500–550 nm) because electromagnetic energy at longer wavelengths is absorbed by water bodies, rather than transmitted or reflected (Figure 1.23).

The laser ranging unit described above emits and receives a pulse of light energy, and the range (or distance to the target) is calculated from the time difference between transmission and reflection. A two-dimensional field of measurements is generated firstly by the forward movement of the platform and secondly by the employment of a side-to-side scanning system (Figure 9.25). The lidar footprint, that is, the size of the small area on the ground that is viewed by the lidar, depends on the instantaneous field of view of the instrument, on the altitude of the platform, and on the angle of view. The radius of the footprint is greater at the edge of the scan than at the centre. The footprint points are collected at equal angle intervals across the scan, so that their ground spacing is unequal. The number of points collected is related to the pulse rate of the lidar transmitter and to the height of the aircraft above the ground. Most lidar sensors employ the scanning mirror principle, as used by the Landsat ETM+ and NOAA AVHRR sensors. The distribution of the observed ground points when an oscillating mirror is used results in a zigzag pattern, shown in Figure 9.25. One problem experienced with some oscillating mirror scanners is that the mirror has to slow down, stop, and accelerate at the end of each scan. Other problems are caused by variations in the altitude and attitude (pitch, roll and yaw) of the aircraft. These variations result in displacements of the ground points from their theoretical positions.

The range or distance from an aircraft to a point on the ground is merely of local interest in that it must be more than zero at nadir, unless the aircraft has landed. To be of scientific use, range information must be placed in the context of a co-ordinate system, that is, it must be converted to a height above an accepted datum such as WGS84. This transformation can only be achieved if the position of the sensor relative to some reference point is known to an acceptable degree of accuracy. Information on the aircraft's attitude is also required if the positions of the points on the ground are to be calculated accurately. Data relating to position and accuracy are collected by the control and processing unit, which contains a GPS receiver and an inertial navigation system. The results of a lidar mission thus consist of two data sets. The first consists of the measurements made by the lidar sensor, and the second contains the positional data collected by the GPS and inertial navigation unit. Both data sets are used at the processing stage, in which the lidar measurements are converted to a regular raster format. The range and position of each measured ground point are computed, and the resulting irregular spatial pattern is re-sampled on a regular grid to produce the output data set. Further processing is necessary if a 'bare earth' DEM is required, as the lidar range is measured between the sensor and the first reflector, in the case of a 'first-return' lidar. Maas (2002) provides a useful survey of methods of analysing errors in lidar data.

The preceding description of the *modus operandi* of airborne lidar sensors makes only one reference to the instantaneous field of view (IFOV) of the instrument, and implicitly assumes that the purpose of any investigation using lidar is to generate a DEM or a digital surface model (DSM). This is an over-simplification. One of the main areas of research using lidar is ecology. Here, some interest lies in the measurement of canopy heights, especially of forests, but there is an equal if not greater interest in the measurement of the 3D characteristics of vegetation. Small-footprint lidars 'see' only a small area on the ground, and these small footprints are separated by 'unseen' areas (Figure 9.25). The spatial distribution of these small footprints over a forest canopy may be such that gaps between trees are missed, and it is also possible for the crown of a tree to be left unobserved by a small footprint system. The small footprint systems also tend to record either or both of the first return and the last return. Such systems have many valuable applications, such as floodplain mapping, in which a dense grid of points associated with accurate elevation measurements is required (Holmes and Atkinson, 2000). Some of these applications are described in the next section. Other applications need a larger footprint and a 'profile' of laser returns between the first and last returns. These large-footprint systems are most use in large area surveys of vegetation characteristics. The VCL mission, described in section 9.4.1, will employ a large footprint system. The applications of lidar systems are briefly described in the next section.

9.4.3 Lidar applications

Data from lidar sensors has been used in a range of applications. Those described here are hydrographic and coastal mapping, glacier monitoring, ecological studies, flood modelling, and DEM/DSM generation. The use of first/last return lidar data in forest studies is illustrated in Example 9.4.

EXAMPLE 9.4: LIDAR FIRST/LAST RETURN

Example 9.4 Figure 1 shows the difference between a first-return lidar dataset and a last-return lidar dataset for a

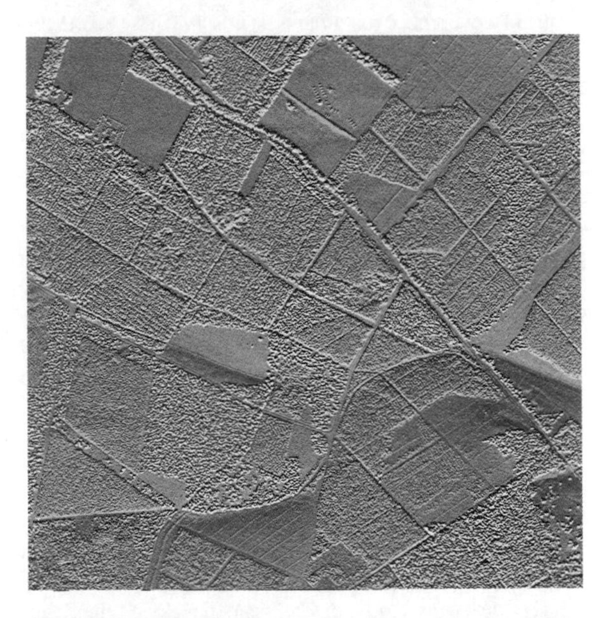

Example 9.4 Figure 1 The difference between the first and the last return lidar data for a 2 × 2 km area near Thetford, Norfolk. The difference between the two returns is related to canopy hight and to the vertical structure of the vegetation. The contrast between smooth and busy textured areas is clear. The visual interpretability (and impact) of this type of image can be enhanced via the use of a pseudocolour transform.

2 × 2 km area near Thetford, Norfolk, in eastern England. The lidar data were acquired for the Environment Agency by Infoterra Ltd, using an Optech ALTM (Airborne Laser Terrain Mapping) 2033 instrument. The data are processed by the provider, and are supplied in gridded form as first and last return measurements, with a nominal spatial resolution of 2 m. The accuracy of the surface elevation measurements is claimed to be 0.15 m at 1200 m altitude (one sigma, i.e. 66% of all points will be within 15 cm of the true elevation, and 95% will be within 30 cm). The horizontal accuracy is quoted as 0.002 x aircraft altitude.

Images such as this are used in studies of forest properties. Canopy height can be estimated from the (first – last return) image, within the accuracy limits noted above. Canopy height correlates with other biophysical variables such as above-ground biomass, basal area, and mean stem diameter.

Example 9.4 Figure 1 was produced using the MIPS software that is distributed with this book. The *Utilities|Subtract images on disc* function was used to compute the difference between the first and last return on a pixel-by-pixel basis, with the resulting difference image stored in 32-bit real format.

Data coverage from lidar systems flown onboard aircraft is of the order of 25 km^2 per hour. For small areas, data collection is very rapid, and the problems of ground or sea-based fieldwork are avoided. In the coastal zone, water depths are often too shallow for ship-based bathymetric surveys, and so wading techniques are often employed. Because the coastal environment is so dynamic, the collection of data by these methods is often less than adequate, as measurements are not collected simultaneously. Lidar remote sensing can provide coverage both of beach topography and bathymetry, without expensive field measurement programmes. Since the remote sensing mission can be carried out quickly and relatively cheaply, repeat surveys to provide more frequent coverage of the area of interest becomes feasible, and it then becomes possible to establish a dynamic model of the coastal zone system. The data provided by lidar remote sensing are used in developing sediment budgets, monitoring and predicting shoreline erosion, creating nautical charts and managing navigation projects (Irish and Lillycrop, 1999).

The US Army Corps of Engineers developed the Scanning Hydrographic Operational Airborne Lidar Survey (SHOALS) instrument in the early 1990s, and it became operational in 1994. SHOALS is a small-footprint lidar that is able to collect bathymetric and topographic data simultaneously from an aircraft or a helicopter. It uses two lasers. One, operating at a wavelength of 1064 nm in the near infrared, is used to determine the land-water interface. The wavelength used by the second laser is 532 nm, in the blue-green region of the visible spectrum. The discussion of the absorption, transmission and reflectance properties of water in chapter 1 explains why these particular wavelengths were selected. Water absorbs strongly in wavelengths longer than the visible green, and so the 1064 nm laser will ‘see’ the water surface but not the interior of the water body. Shorter wavelength visible radiation does penetrate water, with the penetration depth increasing as wavelength decreases. The first return will come from the water surface, and the second (last) return will be from the bed of the water body, providing that the depth is not too great and the water is clear. The 532 nm laser will thus penetrate to a depth that depends mainly on the organic and inorganic content of the water body. For optically clear waters, the depth of penetration is as much as 70 m. The 532 nm laser is also used by the SHOALS instrument to measure the range to points on the beach, and so can produce both bathymetric and elevation data for the coastal zone. The lasers have a pulse frequency of 400 kHz, which means that the system collects 400 range measurements per second. A swath 220 m wide is generated from an operating aircraft height of 400 m, with a point spacing of 8 m. The vertical accuracy is claimed to be ±15 cm, while horizontal accuracy is dependent on the GPS used to fix the aircraft position. When kinematic GPS is used then horizontal accuracy is about one metre.

Another example of a dynamic environment is the alpine glacier. It is physically difficult as well as time-consuming to make measurements on glaciers. The traditional method is to hammer stakes into the glacier surface and to survey them at regular intervals in order to calculate the rate of movement of the glacier surface. Photogrammetric methods have been used as an alternative, but it may be difficult to identify the glacier surface on the photograph and the collection of identifiable control points to link the photograph to a coordinate system is also a problem. Lidar may seem to be a potential solution, as the reflectivity of snow is high in the 800 nm region, and the use of GPS can provide accurate location in terms of an established map projection. Favey *et al.* (2000) describe a comparative study using airborne lidar and digital photogrammetry to map the Lauteraar and Untersar glaciers in the Swiss Alps. The purpose of the mapping exercise was to produce a series of DEMs of the glacier surfaces. Subtraction of successive DEMs provides an estimate of the volume of ice gained or lost.

The lidar used by Favey *et al.* (2000) incorporated a laser scanning system operating at a wavelength of 810 nm, as ice and snow have a high reflectivity at this wavelength. The system was mounted onboard an aircraft, which flew at altitudes between 600 and 1100 m, this latter height being the upper limit for successful use of the lidar sensor. Sub-metre accuracy was achieved for the higher parts of the glacier, but problems were experienced for parts of the glacier covered by debris having a much lower reflectance than snow and ice. These authors conclude that laser altimetry is a feasible tool for glacier monitoring when the flying height is below 750 m. Apart from the speed of data collection, an added advantage of the lidar method is that the survey can be extended to the upper surface of the glacier. Other papers addressing the use of lidar in glacier mass balance studies are Kennett and Eitken (1997) and Thomas *et al.* (1995).

The third example is the use of a small-footprint lidar in topographic mapping for DEM production. High-resolution DEM are required for a variety of purposes, including flood-plain mapping, urban models, and determining line of sight in telecommunications applications. The use of interferometric SAR (InSAR) to generate high-resolution DEM is described in section 9.2. Hodgson *et al.* (2003) provide a comparison between DEMs derived from conventional surveying, and DEMs generated from lidar and interferometric SAR (InSAR) data.

A lidar records the position of the first object encountered during the downward passage of the energy pulse, and so produces a DSM (digital surface model) rather than a 'bare-earth' DEM. This, of course, is an advantage in the production of urban models, and in modelling intervisibility, but is a disadvantage when a 'bare-earth' DEM is required. The distance to the first return will depend on the nature of the target. For example, the first return distance for a dense forest canopy will provide an over-estimate of ground surface elevation. Ground slope also has an effect on vertical accuracy, with higher errors occurring on steep slopes. The RMSE (root mean square error) for lidar-derived ground elevation measurements is of the order of 0.2 m (Lefsky *et al.*, 2002) though it should be recalled that the 95% confidence estimate is derived by multiplying the RMSE by the factor 1.96. The removal of unwanted surface detail in order to convert a DSM to the corresponding DEM is an active research topic. Some algorithms are proprietary and thus confidential; for example, Sties *et al.* (2000) note that

> '... a detailed functionality of the selection method [i.e., the method of identifying 'above ground objects'] is not known.'

Other methods are based on filtering. InSAR-generated DEMs are also affected by the presence of vegetation, but microwave energy penetrates the vegetation canopy to a degree that is dependent on the radar wavelength (section 2.4).

Highly accurate DEMs are required for floodplain modelling (Cobby *et al.*, 2001). The best-available conventional DEMs have a spatial resolution of 10 m and a vertical accuracy of ± 0.5 m, which is too great for the effective use of hydraulic models. DEMs derived from lidar data have, at least in theory, a horizontal spacing of less than 10 m and a vertical accuracy of ± 0.15–0.25 m. Cobby *et al.* (2001) discuss the factors that should be taken into account when using lidar data to generate high-resolution DEMs, using the Severn floodplain near Shrewsbury as a test site. They measured ground elevation at 25 points along a transect between two Ordnance Survey benchmarks, and compared these ground measurements with those derived from lidar. They find that the root mean square error (RMSE) is related to the nature of the vegetation, including its height. Thus, for short vegetation the RMSE is 24 cm, while for dense deciduous woodland the RMSE rises to 4 m, even though the lidar DSM had been filtered to remove surface objects such as vegetation and buildings. They attribute this disappointing performance to (i) the ground slope, as it is suggested that both height and planimetric accuracy decrease as slope increases, and (ii) incomplete penetration of dense vegetation canopies, i.e. the 'last return' does not come from the ground.

The final example is taken from the field of ecosystem research (Dubayah and Drake, 2000; Harding *et al.*, 2001; Lefsky *et al.*, 2002; Lim *et al.*, 2003). In this example, interest focuses not so much on the determination of surface elevations but in estimating the height and structure of forest canopies. Such information augments the spectral reflectance information provided by other remote sensing systems, which provide two-dimensional information about the spatial distribution of objects. Lidar remote sensing adds information about the third dimension, which is important in modelling the ecological properties of a forest area. It has been noted already that small-footprint lidars are less useful in forest studies than are large-footprint lidars, especially as it is more common for the latter type to produce a digital profile made up of the first, last, and intermediate returns (Figure 9.24). Of particular interest to ecological and forestry studies are measurements of canopy height (which correlates closely with other biophysical indices such as above-ground biomass, basal area, and mean stem diameter), and the vertical distribution of backscatter, which can also be used to estimate above-ground biomass and the successional state of forest vegetation (Dubayah *et al.*, 2000).

9.5 SUMMARY

SAR interferometers, imaging spectrometers, and lidar represent developments in remote sensing that are capable of

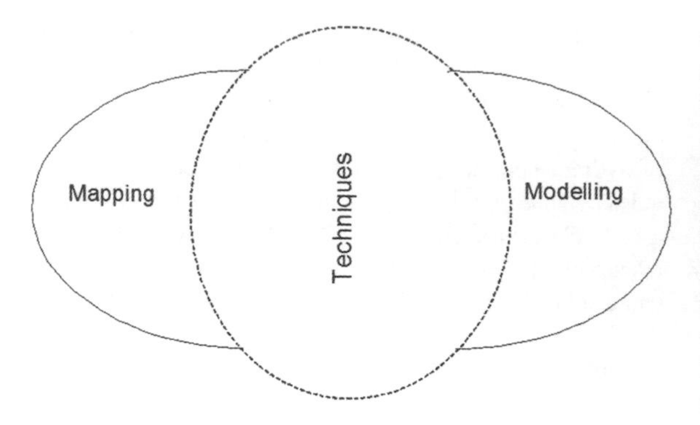

Figure 9.26 Remotely sensed data are used both for mapping and modelling purposes, using a common body of information processing methods.

providing more extensive and more detailed information about the Earth's surface than ever before. The success of the SRTM in producing global digital elevation models, plus developments in the analysis of high-dimensional optical and infrared data, as well as the increasing use of lidar to provide detailed local topographic information, has led to the introduction of increasingly sophisticated methods of processing remotely-sensed data.

Earth observation by remote sensing is expanding in several directions. One view of the organisation of the discipline is shown in Figure 9.26, in which two separate but interlinked 'user communities' (mapping and modelling) share a common interest in data processing and information extraction techniques. Mapping is the representation of the state of the Earth's surface in terms of the nature of the Earth surface cover (e.g., soil, vegetation and water). The modellers, on the other hand, use remotely-sensed data to provide quantitative estimates of properties of the materials making up the Earth's surface, as well as those of the atmosphere. This taxonomy of remote sensing is similar to that proposed by Verstraete *et al.* (1996), whose paper should be widely read.

The aims of these two groups are, in reality, inter-related in several ways. The mapping community requires quantitative estimates of atmospheric and topographic conditions in order to provide a standardised image product and permit the comparison of map-like outputs at different points in time. These estimates could be derived from the outputs of the modelling community, which in turn can develop the analytical procedures used by the 'mappers' so as to extend their applicability. For example, the red edge wavelength can be used both as a surrogate for biomass and as a parameter to be mapped in its own right, relating (for example) to tree species type and tree age. Another case in point is the use of land cover maps derived by remote sensing as inputs to climate models. The two communities are not as distinct and separate, as Figure 9.26 indicates. They are both reliant on each other, and on the development (or adaptation) of techniques of data analysis and information extraction that allow them to make optimal use of the growing amount of increasingly sophisticated data that are being collected by airborne and spaceborne remote sensing instruments.

Appendix A

Using the CD-ROM

The CD-ROM accompanying this book contains the MIPS installation files, a set of test images, four Examples[1], and the Useful Links WWW pages, all of which are described below.

A.1 SUMMARY OF CD-ROM CONTENTS

The CD-ROM should auto-start when it is placed in your CD drive. Instructions on what to do if it does not auto-start are provided below. The auto-start script runs a program, *mips_opening_page.exe*, which provides menu items for:

1. The MIPS installation program, *install_mips.exe*. This program optionally copies the test images to your hard disc.
2. The *useful links* folder, which contains a set of web pages, indexed according to subject, providing WWW links to sites of interest.
3. The *examples* folder, which contains four sub-folders, each containing an exercise of an advanced remote sensing application, plus the associated datasets. Read the material in this folder for more details.

To access the *useful links* and *examples* folders, simply click on the associated menu item. If the CD-ROM does not auto-start, try double-clicking on *mips_opening_page.exe*, which is located in the top-level folder of the CD-ROM. Alternatively, copy the two folders to your hard disc using Windows Explorer. You can then start the *useful_links* web pages by clicking on *useful_links\index.html*.

The *examples* folder contains four exercises, each in a separate subfolder. Each subfolder contains a PDF file (you need Acrobat Reader to access this file) and two subfolders. One subfolder contains the images used to illustrate the text of the exercise (in TIFF and JPEG formats). The other contains the data sets that are used in the exercises.

[1] *Contributed by Magaly Koch, Center for Remote Sensing, Boston University, Boston, MA.*

A.2 INSTALLING MIPS

A.2.1 CD-ROM auto-starts

If the CD_ROM auto-starts when placed in the drive, and *mips_opening_page.exe* runs automatically, start by clicking the item *install mips* on the top menu bar. If the CD-ROM does not auto-start, see section 2.2. You will then be prompted to provide the names of the source folder (i.e., drive letter of the CD-ROM, for example, *d:*) and the destination folder (i.e., the folder on your hard drive where MIPS will be placed, such as *c:\mips*). The installer program (*install_mips.exe*) will copy the program files to this destination folder. You then have the option to copy the *Help* files from the CD-ROM to a subfolder called *c:\ mips\html* and the example images to a second subfolder, *c:\mips\images*. You are recommended to do this. Your directory structure should be of this form:

c:\mips
c:\ mips\html
c:\mips\images

If you do not want to use the folder name *c:\mips*, it can be replaced by any other directory such as *c:\program files\mips*. The sub-folders will then become *c:\program files\html* and *c:\program files\images*.

A.2.2 CD-ROM does not auto-start

Start Windows Explorer, navigate to the top-level folder of the CD, and double-click on *mips_opening_page.exe*. This program should run, and the procedure described in section 2.1 can then be followed.

A.3 PROBLEMS WITH THE INSTALL PROGRAM

You may find that the *install_mips* program starts, but then fails with a message such as "Cannot copy file". Try running *install_mips* again before taking any other steps. If the

Computer Processing of Remotely-Sensed Images: An Introduction, Third Edition. Paul M. Mather.
 ISBNs: 0-470-84918-5 (HB); 0-470-84919-3 (PB)

```
[username]
username = paul mather                    <<< Put your name here
[filepath]
path = C:\mips          <<< Name of folder where MIPS.EXE is located
[date]
day = 18
month = 10
year = 2003

[compname]
compname = PLGDOM2                        <<< Not relevant
[help path]
help = C:\mips\html     <<< Name of sub-folder containing the Help files
[toolbar]
status = of                  <<< Toolbar status – either of (f) or on
[files]
blank line followed by a carriage return.
```

files cannot be copied successfully by *install_mips.exe*, use either Method 1 or Method 2, described below.

A.3.1 Method 1

1. Create a (new) temporary folder on your hard disc.
2. Using Windows Explorer, copy the entire contents of the CD-ROM to this temporary directory.
3. Double-click on the filename *mips_opening_page.exe* in this temporary directory
4. Click the menu item *install_mips* and proceed as described in section 2.1, remembering that the source folder is not the CD-ROM but the temporary directory created at step 1.
5. Delete the temporary directory and all its contents when installation is complete.

A.3.2 Method 2

1. Create a folder called *c:\mips*, with subfolders *c:\mips\html* and *c:\mips\images*.
2. Use Windows Explorer to copy the files in the top-level folder of the CD-ROM to the folder created at step 1.
3. Use Windows Explorer to copy the subfolder *html* from the CD-ROM to *c:\mips\html*.
4. Repeat step 3, this time copying the contents of the *images* subfolder on the CD to *c:\mips\images*.

There are two problems in using Method 2. The first is the fact that a MIPS INI file may not have been created. Check your Windows directory (usually *c:\Windows*) for a file called MyMIPS.ini. If this file exists, open it with a text editor such as Windows Notepad and check its structure, as described below.

Notes

1. The directory name following the bracketed term *[help path]|path =* should be the directory specified at step 3.
2. The directory name following the bracketed term
   [file path]
   path =
   should be the directory specified at step 4; MIPS will start in this directory
3. The two-letter character string following
   [toolbar]
   status =
   should be either *on* or *of*; putting *off* shouldn't matter, but the case does – OF will not work.
4. The name following the line *[compname]* is an historical anomaly; put *oemcomputer*.
5. The keyword *[files]* is used to keep track of the last 5 MIPS Inf files that you have accessed.
6. Ensure that all the headers are present, i.e. *[username]*, *[filepath]*, *[date]*, *[compname]*, *[help]* and *[files]*.
7. Check the entries for each header, editing them as appropriate (e.g. change *username = Paul Mather* to *username = Lionel Jospin*, if that is your name).
8. Save *myMIPS.ini* as a text file to folder *c:\Windows*.

The second problem is that all your INF files will have incorrect path information. For example, they may refer to *e:\mips\test\new\images* rather than to the path *c:\mips\images* or *e:\thisone\images* that you created earlier. You can change each INF file manually or start MIPS, choose *File|Inf File Operations*, and use the *Update Inf File* module to change all the INF file paths. You cannot use this option for ENVI or HYPERSPECTRAL *INF* files. For these, you can use the *File|INF File Operations|Edit Inf File* module. INF files are discussed in chapter 3.

A.4 UNINSTALLING MIPS

Delete all files in the MIPS folder and all subfolders (e.g., */html* and */images*). The installation does not affect the registry in any way. You can also delete the file *c:\windows\mymips.ini.*

References

Abramowitz, M. and Stegun, I.A., (eds.), 1972, *Handbook of Mathematical Functions*. New York: Dover Books.

Abrams, M., 2000, The Advanced Spaceborne Thermal Emission and Reflection Radiometer (ASTER): data products for the high spatial resolution imager on NASA's Terra platform. *International Journal of Remote Sensing*, **21**, 847–859.

Ackermann, F., 1984, Digital image correlation: performance and potential application in photogrammetry. *Photogrammetric Record*, **64**, 429–439.

Adams, J.B., Smith, M.O. and Gillespie, A.R., 1989, Simple models for complex natural surfaces: a strategy for the hyperspectral era of remote sensing. *Proceedings of the IEEE International Geoscience and Remote Sensing Symposium (IGARSS'89), 10–14 July 1989, Vancouver, British Columbia, Canada*. New York: IEEE Press, volume 1, 16–21.

Adams, J.B., Smith, M.O. and Gillespie, A.R., 1993, Imaging spectroscopy: interpretation based on spectral mixture analysis. In: Pieters, C.M. and Englert, P.A.J. (eds.), 1993, *Remote Geochemical Analysis: Elemental and Mineralogical Composition*. Cambridge: Cambridge University Press, 145–166.

Addison, P.S., 2002, *The Illustrated Wavelet Transform Handbook: Introductory Theory and Applications in Science, Engineering, Medicine and Finance*. Bristol: Institute of Physics Publishing.

Aha, D.W. and Bankert, R.L., 1996, A comparative evaluation of sequential feature selection algorithms. In: Fisher, D. and Lenz, J.-H. (eds.), *Learning from Data: Artifical Intelligence and Statistics V*. New York: Springer-Verlag, 199–206.

Albertini, G. and Ponte, S., 1996, Three dimensional digital elevation model of Mt Vesuvius from NASA/JPL TOPSAR. *International Journal of Remote Sensing*, **17**, 1797–1801.

Aleksander, I. and Morton, J., 1990, *An Introduction to Neural Computing*. London: Chapman and Hall.

Alfoldi, T.T., 1982, Remote sensing for water quality monitoring. In: Johanssen, C.J. and Sanders, J.L. (eds.), *Remote Sensing for Resource Management*. Ankeny, Iowa: Soil Conservation Society of America, 317–328.

Al-Hinai, K.G., Khan, M.A. and Canaas, A.A., 1991, Enhancement of sand dune texture from Landsat imagery using difference of Gaussian filter. *International Journal of Remote Sensing*, **12**, 1063–1069.

Allan, J.A., 1984, The role and future of remote sensing. *Proceedings of the Tenth Anniversary International Conference of the Remote Sensing Society*, Remote Sensing Society, Nottingham, 23–30.

Alley, R.E., 1995, *Algorithm Theoretical Basis Document, Version 2.0, March 1, 1995*. Jet Propulsion Laboratory, Pasadena, CA.

Alparone, L., Baronti, S., Carla, R. and Pugilisi, C., 1996, An adaptive order-statistics filter for SAR images. *International Journal of Remote Sensing*, **17**, 1357–1365.

Al-Rousan, N. and Petrie, G., 1998, System calibration, geometric accuracy testing and validation of DEM and orthoimage data extracted from SPOT stereopairs using commercially available image processing systems. *International Archives of Photogrammetry and Remote Sensing*, **32**, 8–15.

Al-Rousan, N., Cheng, P., Petrie, G., Toutin, T. and Valadan Zoej, M.J., 1997, Automated DEM extraction and orthoimage generation from SPOT Level 1B imagery. *Photogrammetric Engineering and Remote Sensing*, **63**, 965–974.

Alsberg, B., Woodward, A.M. and Kell, D.B., 1997, An introduction to wavelet transforms for chemometricians: a time-frequency approach. *Chemometrics and Intelligent Laboratory Systems*, **37**, 215–239.

Anderberg, M.R., 1973, *Cluster Analysis for Applications*. New York: Academic Press.

Andreadis, I., Glavas, E. and Tsalides, Ph., 1995, Image enhancement using colour information. *International Journal of Remote Sensing*, **16**, 2285–2289.

Anuta, P.E., 1970, Spatial registration of multispectral and multitemporal digital imagery using Fast Fourier Transform techniques. *IEEE Transactions on Geoscience Electronics*, **8**, 353–368.

Ardö, J., Pilesjö, P. and Skidmore, A., 1997, Neural networks, multitemporal Landsat Thematic Mapper data and topographic data to classify forest damage in the Czech Republic. *Canadian Journal of Remote Sensing*, **23**, 217–219.

Armour, B., Tanaka, A., Ohkura, H. and Saito, G., 1998, Radar interferometry for environmental change detection. In: Lunetta, R.S. and Elvidge, C.D. (eds.), *Remote Sensing Change Detection: Environmental Monitoring Methods and Applications*. Ann Arbor, MI: Sleeping Bear Press.

Computer Processing of Remotely-Sensed Images: An Introduction, Third Edition. Paul M. Mather.

ISBNs: 0-470-84918-5 (HB); 0-470-84919-3 (PB)

Arnaud, M., 1994, The SPOT programme. In: Mather, P.M. (ed.) (1994), 29–39.

Askne, J. (ed.), 1995, *Sensors and Environmental Applications of Remote Sensing (Proceedings of the 14th EARSeL Symposium, Gõteborg, Sweden, 6–8 June, 1994)*. Rotterdam: A.A. Balkema.

Askne, J.I.H., Dammert, P.B.G., Ulander, L.M.H. and Smith, G., 1997, C-band repeat-pass interferometric SAR observations of the forest. *IEEE Transactions on Geoscience and Remote Sensing*, **35**, 25–35.

Aspinall, R.J., Marcus, W.A. and Boardman, J.W., 2002, Considerations in collecting, processing, and analysing high spectral resolution hyperspectral data for environmental investigations. *Journal of Geographical Systems*, **4**, 15–29.

Asrar, G., (ed.), 1989, *Theory and Applications of Optical Remote Sensing*. New York: Wiley-Interscience.

Atkinson, P.M. and Curran, P.J., 1997, Choosing an appropriate spatial resolution. *Photogrammetric Engineering and Remote Sensing*, **63**, 1345–1351.

Austin, J., Harding, S., Kanellopoulos, I., Lees, K., McNaughton, H., Roli, F., Vernazza, G. and Wilkinson, G., 1997, *Connectionist Computation in Earth Observation*. Joint Research Centre, European Commission, Report EUR 17314 EN, Brussels, Belgium.

Babey, S.K. and Anger, C.D., 1989, A Compact Airborne Spectrographic Imager (CASI). *Proceedings of the IEEE International Geoscience and Remote Sensing Symposium (IGARSS'89), 10–14 July 1989, Vancouver, British Columbia, Canada*. New York: IEEE Press, 1028–1031.

Bakker, W.H., 2000, Satellite and sensor systems for environmental monitoring. In: Meyers, R.A. (ed.),*Encyclopaedia of Analytical Chemistry*. Chichester: John Wiley and Sons Ltd. 86938746.

Baltsavias, E.P., 1999, Airborne laser scanning: basic relations and formulas. *ISPRS Journal of Photogrammetry and Remote Sensing*, **54**, 199–214.

Bamler, R. and Hartl, P., 1998, Synthetic aperture radar interferometry. *Inverse Problems*, **14**, R1–R54.

Bannari, A., Morin, D., Bénié, G.B. and Bonn, F.J., 1995a, A theoretical review of different mathematical models of geometric corrections applied to remote sensing images. *Remote Sensing Reviews*, **13**, 27–47.

Bannari, A., Morin, D., Bonn, F. and Huete, A., 1995b, A review of vegetation indices. *Remote Sensing Reviews*, **13**, 95–120.

Baret, F. and Guyot, G., 1991, Potential and limits of vegetation indices for LAI and PAR assessment. *Remote Sensing of Environment*, **35**, 161–173.

Baret, F., Jacquemond, S. and Hanocq, J.F., 1993, The soil line concept in remote sensing. *Remote Sensing Reviews*, **7**, 65–82.

Barnea, D.I. and Silverman, H.F., 1972, A class of algorithm for fast digital image registration. *IEEE Transactions on Computers*, **21**, 179–186.

Barnsley, M.J., 1983, The implications of view angle effects on the use of multispectral data for vegetation studies. *Proceedings of the International Conference on Remote Sensing for Rangeland Monitoring and Management, Silsoe, Bedfordshire, England*. Remote Sensing Society, Nottingham, 173–177.

Barnsley, M.J., 1994, Environmental monitoring using multiple-view-angle (MVA) remotely-sensed data. In: Foody, G. and Curran, P. (eds.), *Environmental Remote Sensing from Regional to Global Scales*. Chichester: John Wiley and Sons, 181–201.

Barnsley, M.J. and Kay, S.A.W., 1990, The relationship between sensor geometry, vegetation-canopy geometry and image variance. *International Journal of Remote Sensing*, **11**, 1075–1083.

Bastin, L., 1997, Comparison of fuzzy c-mean classification, linear mixture modelling and MLC probabilities as tools for unmixing coarse pixels. *International Journal of Remote Sensing*, **18**, 3629–3648.

Basu, J.P. and Odell, P.L., 1974, Effects of intraclass correlation among training samples on the misclassification probabilities of Bayes' procedure. *Pattern Recognition*, **6**, 13–16.

Bateson, A. and Curtiss, B., 1996, A method for manual end-member selection and spectral unmixing. *Remote Sensing of Environment*, **55**, 229–243.

Bauer, E. and Kohavi, R., 1999, An empirical comparison of voting classification algorithms: bagging, boosting and variants. *Machine Learning*, **36**, 105–142.

Baumgartner, M.F., Silva, L.F., Biel, L.L. and Stoner, E.R., 1985, Reflectance properties of soils. *Advances in Agronomy*, **38**, 1–44.

Beale, R. and Jackson, T., 1990, *Neural Computing: An Introduction*. Bristol: Adam Hilger.

Beauchemin, M., Thomson, K.B.P. and Edwards, G., 1996, Edge detection and speckle adaptive filtering based on a second-order textural measure. *International Journal of Remote Sensing*, **17**, 1751–1759.

Begni, G., 1988, Absolute calibration of SPOT data. *SPOT Newsletter*, **10**, 2–3.

Begni, G., Dinguirard, M.C., Jackson, R.D. and Slater, P.N., 1988, Absolute calibration of the SPOT-1 HRV cameras. *SPIE*, **660**, 66–76.

Belward, A., 1991, Spectral characteristics of vegetation, soil and water in the visible, near-infrared and middle-infrared wavelengths. In: Belward, A. and Valenzuela, C.R. (eds.), 1991, 31–53.

Belward, A. and Valenzuela, C.R. (eds.), 1991, *Remote Sensing and Geographical Information Systems for Resource Management in Developing Countries*. EuroCourses: Remote Sensing, Volume 1. Dordrecht: Kluwer Academic Publishers.

Ben-Dor, E. and Kruse, F.A., 1995, Surface mineral mapping of the Makhtesh Ramon Negev, Israel, using GER 63 channel scanner data. *International Journal of Remote Sensing*, **16**, 3529–3553.

Benediktsson, J.A. and Sveinsson, J.R., 1997, Feature extraction for neural network classifiers. In Kanellopoulos, I. *et al* (eds.) (1997), 97–104.

Benediktsson, J.A., Swain, P.H. and Ersoy, O.K., 1990, Neural network approaches versus statistical methods in the classification of multisource remote sensing data. *IEEE Transactions of Geoscience and Remote Sensing*, **28**, 540–552.

Benny, A.H., 1981, Automatic relocation of ground control points in Landsat imagery. *Proceedings of the International Conference Matching Remote Sensing Technologies and their Applications*, Remote Sensing Society, Nottingham, 307–315.

Bergland, G.D., 1969, A guided tour of the Fast Fourier Transform. *IEEE Spectrum*, **6**, 41–45.

Bergland, G.D. and Dolan, M.T., 1979, Fast Fourier transform algorithms. New York: IEEE Acoustics, Speech and Signal Processing Society, IEEE Press/Wiley, *Programs for Digital Signal Processing*. Section 1.2-1.

Berk, A.L., Anderson, G.P., Bernstein, L.S., Acharya, P.K., Dothe, H., Matthew, M.W., Adler-Golden, S.M., Chetwynd, J.H.J., Richtsmeier, S.C., Pukall, B., Allred, C.L., Jeong, L.S., and Hoke, M.L., 1999, MODTRAN4: Radiative transfer modelling for atmospheric correction. *AVIRIS Proceedings 1999* (JPL Publication 99-17). Pasadena, CA.: NASA Jet Propulsion Laboratory. Available from http://popo.jpl.nasa.gov/docs/workshops/99_docs/toc.html. Accessed 9 September 2003.

Bernard, A.C., Kanellopoulos, I. and Wilkinson, G.G., 1996, Neural net classification of mixtures. In: Binaghi, E., Brivio, P.A. and Rampini, A. (eds.) 53–58.

Bernstein, R., Lotspiech, J.B., Myers, J., Kolsky, H.G., and Lees, R.D., 1984, Analysis and processing of Landsat-4 sensor data using advanced image processing techniques and technologies. *IEEE Transactions on Geoscience and Remote Sensing*, **22**, 192–221.

Bezdek, J.C., 1993, Editorial: Fuzzy models – what are they, and why? *IEEE Transactions on Fuzzy Systems*, 1, 1–6.

Bezdek, J.C., 1994, The thirsty traveler visits Gamont: A rejoinder to 'Comments on fuzzy sets – what are they and why?' *IEEE Transactions on Fuzzy Systems*, **2**, 43–45.

Billingsley, F.C., (ed.), 1983, Data processing and reprocessing. In Colwell, R.N., (1983) (ed.), 719–792.

Binaghi, E., Brivio, P.A. and Rampini, A. (eds.), 1996, *Proceedings of the 14th International Workshop on Soft Computing in Remote Sensing Data Analysis, Milan, 4-5 December 1995*. Singapore: World Scientific.

Bindschlander, R., 1998, Monitoring ice sheet behaviour from space. *Reviews of Geophysics*, **36**, 79–104.

Bird, A.C., 1991a, Principles of remote sensing: electromagnetic radiation, reflectance and emissivity. In: Belward, A. and Valenzuela, C.R. (eds.), 1991, 1–15.

Bird, A.C., 1991b, Principles of remote sensing: interaction of electromagnetic radiation with the atmosphere and the Earth. In: Belward, A. and Valenzuela, C.R. (eds.), 1991, 17–30.

Bishof, H., Schneider, W. and Pinz, A.J., 1992, Multispectral classification of Landsat images using neural networks. *IEEE Transactions on Geoscience and Remote Sensing,* **30**, 482–490.

Bishop, C.M., 1995, *Neural Networks for Pattern Recognition*. Oxford: Clarendon Press.

Bishop, M.P. and Colby, J.D., 2002, Anisotropic reflectance correction of SPOT-3 HRV imagery. *International Journal of Remote Sensing*, **23**, 2125–2131.

Bishop, Y.M., Fienberg, S.E. and Holland, P.W., 1975, *Discrete Multivariate Analysis: Theory and Practice*. Cambridge, MA: MIT Press.

Blackburn, G.A., 1998, Quantifying chlorophyll and carotenoids at leaf and canopy scales: an evaluation of some hyperspectral approaches. *Remote Sensing of Environment*, **66**, 273–285.

Blamire, P., 1996, The influence of relative sample size in training artificial neural networks. *International Journal of Remote Sensing*, **17**, 223–230.

Blaser, T.J. and Caloz, R., 1991, Digital ortho-image registration from a SPOT panchromatic image using a digital elevation model. *IEEE Transactions on Geoscience and Remote Sensing*, **29**, 2431–2434.

Blom, R.G., 1988, Effects of variations in look angle and wavelength in radar images of volcanic and aeolian terrains, or now you see it, now you don't. *International Journal of Remote Sensing*, **9**, 945–965.

Blom, R.G. and Daily, M., 1982, Radar image processing for rock-type discrimination. *IEEE Transactions on Geoscience Electronics*, **20**, 343–351.

Blonda, P.N. and Pasquariello, G., 1991, An experiment for the interpretation of multitemporal remotely sensed images based on a fuzzy logic approach. *International Journal of Remote Sensing*, **12**, 463–476.

Boardman, J., 1989, Inversion of imaging spectrometer data using singular value decomposition. *Proceedings of the IEEE International Geoscience and Remote Sensing Symposium (IGARSS'89), 10–14 July 1989, Vancouver, British Columbia, Canada*. New York: IEEE Press, **4**, 2069–2072.

Bodechtel, J. and Zilger, J., 1996, MOMS – History, concepts, goals. *Proceedings of the MOMS-02 Symposium, Cologne, Germany, 5–7 July, 1995*. Paris: European Association of Remote Sensing Laboratories (EARSeL), 12–25.

Bolstad, P.V. and Lillesand, T.M., 1992, Semi-automated training approaches for spectral class definition. *International Journal of Remote Sensing*, **13**, 3157–3166.

Bolstad, P.V. and Stowe, T., 1994, An evaluation of DEM accuracy: elevation, slope and aspect. *Photogrammetric Engineering and Remote Sensing*, **60**, 1327–1332.

Bolstad, P.V., Gessler, P. and Thomas, M.L., 1990, Positional uncertainty in manually-digitised map data. *International Journal of Geographical Information Systems*, **4**, 39–42.

Bonham-Carter, G.F., 1988. Numerical procedures and computer program for fitting an inverted Gaussian model to vegetation reflectance data. *Computers and Geosciences*, **14**, 339–356.

Bonham-Carter, G.F., 1994, *Geographic Information Systems for Geoscientists*. Kidlington, Oxford: Pergamon/ Elsevier Science Publications.

Bonhomme, R., The solar radiation: characteristics and distribution in the canopy. In Varlet-Grancher, C., Bonhomme, R. and Sinoquet, H. (eds.), *Crop Structures and Light Microclimate, Characteristics and Applications*. Paris, France: INRA Editions, 17–28.

Boochs, F. and Kupfer, G., 1990, Shape of the red edge as vitality indicator for plants. *International Journal of Remote Sensing*, **11**, 1741–1753.

Borgeaud, M. and Wegmüller, U., 1996, On the use of ERS SAR interferometry for the retrieval of geo- and bio-physical information. *Proceedings of FRINGE'96 – ESA Workshop on Applications of ERS SAR Interferometry*, 30 September – 2 October, 1996, Remote Sensing Laboratories, University of Zurich, Switzerland, URL: http://www.geo.unizh.ch/rsl/fringe96/papers/borgeaud-wegmuller/. Date accessed: 14 January 2003.

Borgeson, W.T., Batson, R.M. and Kieffer, H.H., 1985. Geometric accuracy of Landsat-4 and Landsat-5 Thematic Mapper images. *Photogrammetric Engineering and Remote Sensing*, **51**, 1893–1898.

Bouman, B.A.M., 1992, Accuracy of estimating the Leaf Area Index from vegetation indices derived from crop reflectance characteristics: a simulation study. *International Journal of Remote Sensing*, **13**, 3069–3084.

Bow, S.-T., 1992, *Pattern Recognition and Image Preprocessing*. New York: Marcel Dekker.

Box, E.O., Holben, B.N. and Kalb, V., 1989, Accuracy of the AVHRR Vegetation Index as a predictor of biomass, primary productivity and net CO_2 flux. *Vegetatio*, **80**, 71–89.

Brady, M., 1982, Computational approaches to image understanding. *Association of Computer Manufacturers' (ACM) Computing Surveys*, **14**, 3–71.

Braunich, H., Wu, B.-I. and Kong, J.A., 2000, Phase unwrapping of SAR interferograms after wavelet denoising. *Proceedings of the IEEE International Geoscience and Remote Sensing Symposium (IGARSS) 2000*, Honolulu, Hawaii. New York: IEEE. 752–754.

Breiman, L., 1996, Bagging predictors. *Machine Learning*, **24**, 123–140.

Brown, S.R. and Scholz, C.H., 1985, Broad bandwidth study of the topography of natural rock surfaces. *Journal of Geophysical Research*, **90**, 12575–12582.

Brownrigg, D.R.K., 1984, The weighted median filter. *Communications of the Association of Computer Manufacturers (ACM)*, **27**, 807–818.

Bruce, L.M. and Li, J., 2001, Wavelets for computationally efficient hyperspectral derivative analysis. *IEEE Transactions on Geoscience and Remote Sensing*, **39**, 1540–1546.

Brush, R.J.H., 1985, A method for real-time navigation of AVHRR imagery. *IEEE Transactions on Geoscience and Remote Sensing*, **23**, 876–887.

Bruzzone, L., Conese, C., Maselli, F. and Roli, F., 1997, Multisource classification of complex rural areas using statistical and neural-network approaches. *Photogrammetric Engineering and Remote Sensing*, **63**, 523–533.

Buckingham, W.F. and Sommer, S.E., 1983, Mineralogical characterization of rock surfaces formed by hydrothermal alteration and weathering: application to remote sensing. *Economic Geology*, **78**, 664–674.

Buckley, J.J. and Hayashi, Y., 1994, Fuzzy neural networks: A survey. *Fuzzy Sets and Systems*, **66**, 1–13.

Budkewitsch, P., Newton, G. and Hynes, A.J., 1994, Characterisation and extraction of linear features from digital images. *Canadian Journal of Remote Sensing*, **20**, 268–279.

Burrough, P.A. and Frank, A., 1996, *Geographic Objects with Indeterminate Boundaries*. London: Taylor and Francis.

Cai, C. and Harrington, P.D.B., 1998, Different discrete wavelet transforms applied to denoising analytical data. *Journal of Chemical Information and Computer Sciences*, **38**, 1161–1170.

Calder, N., 1991, *Spaceship Earth*. London: Viking Books/Channel Four Television.

Campbell, J.B., 1981, Spatial correlation effects upon accuracy of supervised classification of land cover. *Photogrammetric Engineering and Remote Sensing*, **47**, 355–357.

Campbell, J.B. and Esaias, W.E., 1983, Basis for spectral curvature analysis in remote sensing of chlorophyll. *Applied Optics*, **22**, 1084–1093.

Campbell, N.A., 1980, Robust procedure in multivariate analysis. I: Robust covariance estimation. *Applied Statistics*, **29**, 231–237.

Campbell, N.A., 1996, The decorrelation stretch transform. *International Journal of Remote Sensing*, **17**, 1939–1949.

Cannon, R.J., Dave, J.A., Bezdek, J.C. and Trivedi, M.M., 1986, Segmentation of a Thematic Mapper image using the fuzzy c-means clustering algorithm. *IEEE Transactions on Geoscience and Remote Sensing*, **24**, 400–408.

Canters, F., 1997, Evaluating the uncertainty of area estimates derived from fuzzy land cover classification. *Photogrammetric Engineering and Remote Sensing*, **63**, 403–414.

Cappellini, V., Chiuderi, A. and Fini, S, 1995, Neural networks in remote sensing multisensor data processing. In: Askne, J. (ed.) (1995), 457–462.

Carara, A., Bitelli, G. and Carla, T., 1997, Comparison of techniques for generating digital terrain models from contour lines. *International Journal of Geographical Information Science*, **11**, 451–473.

Carlson, G.E. and Ebell, W.J., 1995, Co-occurrence matrices for small region texture measurements and comparison. *International Journal of Remote Sensing*, **16**, 1417–1423.

Carpenter, G.A., Gjaja, M.N., Gopal, S. and Woodcock, C.E., 1997, ART neural networks for remote sensing:

vegetation classification from Landsat TM and terrain data. *IEEE Transactions on Geoscience and Remote Sensing*, **35**, 308–325.

Carpenter, G.A., Grossberg, S., Markuzon, N., Reynolds, J.H. and Rosen, D.B., 1992, Fuzzy ARTMAP: A neural network architecture for incremental supervised learning for analog multidimensional maps. *IEEE Transactions on Neural Networks*, **3**, 698–712.

Carter, G.A., 1994, Ratios of leaf reflectances in narrow wavebands as indicators of plant stress. *International Journal of Remote Sensing*, **15**, 697–703.

Carter, G.A. and Miller, R.L., 1994, Early detection of plant stress by digital imaging within narrow stress-sensitive wavebands. *Remote Sensing of Environment*, **50**, 295–302.

Chalmers, A.I. and Harris, R., 1979, Band ratios in multispectral analysis of Landsat digital data. *Proceedings of the 8th Annual Conference of the Remote Sensing Society*. Nottingham: The Remote Sensing Society, 139–146.

Chan, J.C.-W., Huang, C. and DeFries, R., 2001, Enhanced algorithm performance for land cover classification from remotely sensed data using bagging and boosting. *IEEE Transcations on Geoscience and Remote Sensing*, **39**, 693–695.

Chander, G. and Markham, B., undated, Revised Landsat 5 TM radiometric calibration procedures and post-calibration dynamic ranges. *http://landsat7.usgs.gov/documents/L5TMCal2003.pdf*. Accessed 9 January 2004.

Chang, C.G., Yu, B. and Vetterli, M., 2000a, Adaptive wavelet thresholding for image denoising and compression. *IEEE Transactions on Image Processing*, **9**, 1532–1546.

Chang, C.G., Yu, B. and Vetterli, M., 2000b, Spatially adaptive wavelet thresholding with context modelling for image denoising. *IEEE Transactions on Image Processing*, **9**, 1522–1531.

Chang, W.C., 1983, On using principal components before separating a mixture of two multivariate normal distributions. *Applied Statistics*, **32**, 267–275.

Chapman, R.E., 1995, *Physics for Geologists*. London: UCL Press.

Chavez, P.S., Jr., 1988, An improved dark-object subtraction technique for atmospheric scattering correction of multispectral data. *Remote Sensing of Environment*, **24**, 459–479

Chavez, P.S., Jr., 1996, Image-based atmospheric corrections- revisited and improved. *Photogrammetric Engineering and Remote Sensing*, **62**, 1025–1036.

Chavez, P.S., Jr., 1986, Processing techniques for digital sonar images from GLORIA. *Photogrammetric Engineering and Remote Sensing*, **52**, 1133–1145.

Chavez, P.S., Jr. and Bauer, B., 1982, An automatic kernel-size selection technique for edge enhancement. *Remote Sensing of Environment*, **12**, 23–38.

Che, N. and Price, J.C., 1992, Survey of radiometric calibration results and methods for visible and near-infrared channels of NOAA-7, -9, and -11 AVHRRs. *Remote Sensing of Environment*, **41**, 19–27.

Chilar, J., St.-Laurent, L. and Dyer, J.A., 1991, Relation between the normalised vegetation index and ecological variables. *Remote Sensing of Environment*, **35**, 279–298.

Chittenini, C.B., 1983, Edge and line detection in multidimensional noisy imagery. *IEEE Transactions on Geoscience and Remote Sensing*, **21**, 163–174.

Chiuderi, A. and Cappellini, V., 1996, A Kohonen's self organising map for land cover classification. In: Parlow, E. (ed.) (1996), 107–112.

Chopping, M.J., Rango, A., Havstad, K.M., Scheibe, F.R., Richie, J.C., Schmugge, T.J., French, A.N., Su, L., McKee, L. and Davis, M.R., 2003, Canopy attributes of desert grassland and transition communities derived from multiangular airborne imagery. *Remote Sensing of Environment*, **85**, 339–354.

Christianini, N. and Shawe-Taylor, J., 2000, *Support Vector Machines and Other Kernel-based Learning Methods*. Cambridge: Cambridge University Press.

Clark, C. and Cañas, A., 1995, Spectral identification by artificial neural network and genetic algorithm. *International Journal of Remote Sensing*, **16**, 2255–2275.

Clark, R.N., 1999, Spectroscopy of rocks and minerals and principles of spectroscopy. In: Rencz, A.N. (ed.) (1999), 3–58.

Clavet, D.M., Lassere, M. and Pouliot, J., 1993, GPS control for 1:50,000 scale topographic mapping from satellite images. *Photogrammetric Engineering and Remote Sensing*, **59**, 107–111.

Clevers, J.P.G.W. and Jongschamp, R., 2001, Imaging spectrometry for agricultural applications. In: van der Meer and de Jong (eds.) (2001), 157–199.

Clevers, J.P.G.W., de Jong, S.M., Epema, G.F., van der Meer, F.D., Bakker, W.H., Skidmore, A.K. and Scholte, K.H., 2002, Derivation of the red edge index using MERIS standard band setting. *International Journal of Remote Sensing*, **23**, 3169–3184.

Cliff, A.D. and Ord, J.K., 1973, *Spatial Autocorrelation*. London: Pion Press.

Cloude, S.R. and Papathanassiou, K.P., 1998, Polarimetric SAR interferometry. *IEEE Transactions on Geoscience and Remote Sensing*, **36**, 1551–1565.

Cobby, D.M., Mason, D.C. and Davenport, I.J., 2001, Image processing of airborne scanning laser altimetry data for improved river flood modelling. *ISPRS Journal of Photogrammetry and Remote Sensing*, **56**, 121–138.

Cocks, T., Jenssen, R., Stewart, A., Wilson, I. and Shields, T., 1998, The Hymap airborne hyperspectral sensor: the system, calibration and performance. *First EARSeL Workshop on Imaging Spectroscopy*. Zurich, Switzerland. Paris: European Association of Remote Sensing Laboratories (EARSeL).

Coley, D.A., 1999, *An Introduction to Genetic Algorithms for Scientists and Engineers*. Singapore: World Scientific.

Collins, J.B. and Woodcock, C.E., 1996, Assessment of several linear change detection techniques for mapping forest mortality using multitemporal Landsat TM data. *Remote Sensing of Environment*, **56**, 66–77.

Colwell, R.N., (ed.), 1983, *Manual of Remote Sensing*. Two volumes. Falls Church, VA: American Society of Photogrammetry.

Combal, B. and Isaka, H., 2002, The effect of small topographic variations on reflectance. *IEEE Transactions on Geoscience and Remote Sensing*, **40**, 663–670.

Conese, C., Gilabert, M.A., Maselli, F. and Bottai, L., 1993, Topographic normalisation of TM scenes through the use of an atmospheric correction method and digital terrain models. *Photogrammetric Engineering and Remote Sensing*, **59**, 1745–1753.

Congalton, R.G., 1991, A review of assessing the accuracy of classifications of remotely sensed data. *Remote Sensing of Environment*, **37**, 35–46.

Congalton, R.G., Oderwald, R. and Mead, R., 1983, Landsat classification accuracy using discrete multivariate analysis statistical techniques. *Photogrammetric Engineering and Remote Sensing*, **49**, 1671–1678.

Cook, A.E. and Pinder, J.E. III, 1996, Relative accuracy of rectifications of coordinates determined from maps and the Global Positioning System. *Photogrammetric Engineering and Remote Sensing*, **62**, 73–77.

Cormen, T.H., Leiserson, C.E., Rivest, R.L. and Clifford Stein, C., 2001, *Introduction to Algorithms, Second Edition*. Cambridge, MA: MIT Press.

Cracknell, A.P., 1997, *The Advanced Very High Resolution Radiometer*. London: Taylor and Francis.

Cracknell, A.P., 1998, Synergy in remote sensing – what's in a pixel? *International Journal of Remote Sensing*, **19**, 2025–2047.

Craig, R.G., 1979, Autocorrelation in Landsat data. *Proceedings of the 13th International Symposium on Remote Sensing of Environment*, Environmental Research Institute of Michigan, Ann Arbor, Michigan, USA, 1517–1524.

Crawford, P.S., Brooks, A.R. and Brush, R.J.H., 1996, Fast navigation of AVHRR images using complex orbital models. *International Journal of Remote Sensing*, **17**, 197–212.

Crippen, R.W., 1989, A simple spatial filtering routine for the cosmetic removal of scan line noise from Landsat TM P-tape imagery. *Photogrammetric Engineering and Remote Sensing*, **55**, 327–331.

Crist, E.P., 1983, The TM tasseled cap – a preliminary formulation. *Proceedings of the Symposium on Machine Processing of Remotely- Sensed Data 1983, Purdue University, West Lafayette, Indiana*. 357– 364.

Crist, E.P. and Cicone, R.C., 1984a, A physically-based transformation of thematic mapper data – the TM Tasseled Cap. *IEEE Transactions on Geoscience and Remote Sensing*, **22**, 256–263.

Crist, E.P. and Cicone, R.C., 1984b, Comparison of the dimensionality and features of simulated Landsat-4 MSS and TM data. *Remote Sensing of Environment*, **14**, 235–246.

Crist, E.P. and Kauth, R.J., 1986, The Tasselled Cap demystified. *Photogrammetric Engineering and Remote Sensing*, **52**, 81–86.

Crosetto, M., 2002, Calibration and validation of SAR interferometry for DEM generation. *ISPRS Journal of Photogrammetry and Remote Sensing*, **57**, 213–227.

Crosetto, M. and Crippa, B., 2000, Quality assessment of interferometric SAR DEMs. *International Archives of Photogrammetry and Remote Sensing*, **33**, 46–53.

Cross, A.M., Settle, J.J., Drake, N.A. and Paivinen, R.T.M., 1991, Subpixel measurement of tropical forest cover using AVHRR data. *International Journal of Remote* Sensing, **12**, 1119–1129.

Curran, P.J., 1980, Multispectral remote sensing of vegetation amount. *Progress in Physical Geography*, **4**, 315–341.

Curran, P.J., 1983, Multispectral remote sensing for estimation of green leaf area index. *Philosophical Transactions of the Royal Society of London*, **309A**, 257–270.

Curran, P.J., 1994, Imaging spectroscopy. *Progress in Physical Geography*, **18**, 247–266.

Curran, P.J. and Atkinson, P.M., 1998, Geostatistics and remote sensing. *Progress in Physical Geography*, **22**, 61–78.

Curran, P.J., Dungan, J.L., Maesler, B.A. and Plummer, S.E., 1991, The effect of a red leaf pigment on the relationship between red edge and chlorophyll concentration. *Remote Sensing of Environment*, **35**, 69–76.

Czaplewski, R.L., 1992, Misclassification bias in areal estimates. *Photogrammetric Engineering and Remote Sensing*, **58**, 189–192.

Daily, M., 1983, Hue-saturation-intensity split-spectrum processing of Seasat radar images. *Photogrammetric Engineering and Remote Sensing*, **49**, 349–355.

Dale, P.E.R., Chandica, A.L. and Evans, M., 1996, Using image subtraction and classification to evaluate change in sub-tropical inter-tidal wetlands. *International Journal of Remote Sensing*, **17**, 703–719.

Danaher, T.J., 2002, An empirical BRDF correction for Landsat TM and ETM+ imagery. *Proceedings of the 11th Australasian Remote Sensing and Photogrammetry Conference*. Brisbane. 966–977.

Danaher, T.J., Xiolaing, W. and Campbell, N.A., 2001, Bi-directional reflectance distribution function approaches to radiometric calibration of Landsat ETM+ imagery. *Proceedings of the IEEE International Geoscience and Remote Sensing Symposium (IGARSS) 2001*. University of New South Wales, Sydney, Australia. IEEE. 2654–2657.

Danielsson, P.E., 1981, Getting the median faster. *Computer Graphics and Image Processing*, **15**, 71–78.

Danson, F.M. and Plummer, S.E., 1995, Red-edge response to forest leaf area index. *International Journal of Remote Sensing*, **16**, 183–188.

Dawson, T.P. and Curran, P.J., 1998, A new technique for interpolating the reflectance red edge position. *International Journal of Remote Sensing*, **19**, 2133–2139.

Dayhoff, J.E., 1990, *Neural Network Architectures: An Introduction*. New York: Van Nostrand Reinhold.

de Cola, L., 1994, Simulating and mapping spatial complexity using multi-scale techniques. *International Journal of Geographical Information Systems*, **8**, 411–427.

de Jong, S.M., 1998, Imaging spectrometry for monitoring tree damage caused by volcanic activity in the Long Valley caldera, California. *ITC Journal*, **1**, 1–10.

de Jong, S.M. and Burrough, P.A., 1995, A fractal approach to the classification of Mediterranean vegetation types using remotely sensed data. *Photogrammetric Engineering and Remote Sensing*, **61**, 1041–1063.

de Souza Filho, C.R., Drury, S.A., Denniss, A.M., Carlton, R.W.T. and Rothery, D.A., 1996, Restoration of corrupted optical Fuyo-1 (JERS-1) data using frequency domain techniques. *Photogrammetric Engineering and Remote Sensing*, **62**, 1037–1047.

Decell, H.P. and Guseman, L.F., 1979, Linear feature selection with applications.*Pattern Recognition*, **11**, 55–63.

Dehaan, R.L. and Taylor, G.R., 2002, Field-derived spectra of salinized soils and vegetation as indicators of irrigation-induced soil salinity. *Remote Sensing of Environment*, **80**, 406–417.

Demetriades-Shah, T.H., Steven, M.D. and Clark, A.C., 1990. High resolution derivative spectra in remote sensing. *Remote Sensing of Environment*, **33**, 55–64.

Depczynski, U., Jetter, K., Molt, K. and Niemöller, A., 1999, The fast wavelet transform on compact intervals as a tool in chemometrics. II: Boundary effects, denoising and compression. *Chemometrics and Intelligent Laboratory Systems*, **49**, 151–161.

Deschamps, P.Y., Herman, M. and Tanré, D., 1983, Definitions of atmospheric radiance and transmittances in remote sensing. *Remote Sensing of Environment*, **13**, 89–92.

Desnos, Y.-L. and Matteini, V., 1993, Review on structural detection and speckle filtering on ERS-1 images. *EARSeL Advances in Remote Sensing*, **2**, 52–65. Paris: European Association of Remote Sensing Laboratories.

Devereux, B.J., Fuller, R.M., Carter, L. and Parsell, R.J., 1990, Geometric correction of airborne scanner data by matching Delaunay triangles. *International Journal of Remote Sensing*, **11**, 2237–2251.

Dikshit, O., 1996, Textural classification for ecological research using ATM images. *International Journal of Remote Sensing*, **17**, 887–915.

Diner, D.J., Beckert, J.C., Reilly, T.H., Bruegge, C.J., Conel, J.E., Kahn, R.A., Martonchick, J.V., Ackerman, T.P., Davies, R., Gerstl, S.A.W., Gordon, H.R., Muller, J.-P., Myeni, R.B., Sellers, P.J., Pinty, B. and Verstraete, M.M., 1998, Multi-angle imaging spectroradiometer (MISR) instrument description and overview. *IEEE Transactions on Geoscience and Remote Sensing*, **36**, 1072–1087.

Diner, D.J., Bruegge, C.J., Martonchik, J.V., Bothwell, G.W., Danielson, E.D., Floyd, E.L., Ford, V.G., Hovland, L.E., Jones, K.L. and White, M.L., 1991, A multi-angle imaging spectroradiometer for terrestrial remote sensing from the Earth Observing System. *International Journal of Imaging Systems and Technology*, **3**, 92–107.

Dixon, T.H., 1994, *SAR Interferometry and Surface Change Detection*. Jet Propulsion Laboratory, California Institute of Technology, URL: http://southport.jpl.nasa.gov/scienceapps/dixon/. Date accessed: 15 January 2003.

Dobbertin, M. and Biging, G.S., 1996, A simulation study of the effect of scene autocorrelation, training size and sampling method on classification accuracy. *Canadian Journal of Remote Sensing*, **22**, 360–367.

Dobson, M.C., Ulaby, F.T., Pierce, L.E., Sharik, T.L., Bergen, K.M., Kellndorfer, J., Kendra, J.R., Li, E., Lin, Y.C., Nashashibi, A., Sarabandi, K. and Siqueira, P., 1995, Estimation of forest biophysical characteristics in Northern Michigan with SIR-C/X-SAR. *IEEE Transactions on Geoscience and Remote Sensing*, **33**, 877–895.

Donoho, D.L., 1995, Denoising by soft thresholding. *IEEE Transactions on Information Theory*, **41**, 613–627.

Donoho, D.L. and Johnstone, L.M., 1994. Ideal spatial adaption by wavelet shrinkage. *Biometrika*, **81**(3), 425–455.

Donoho, D.L. and Johnstone, L.M., 1995, Adapting to unknown smoothness via wavelet shrinkage. *Journal of the American Statistical Association*, **90**, 1200–1224.

Dowman, I., 1992, The geometry of SAR images for geocoding and stereo applications. *International Journal of Remote Sensing*, **13**, 1609–1617.

Dowman, I. and Dolloff, J., 2000, An evaluation of rational functions for photogrammetric restitution. *International Archives of Photogrammetry and Remote Sensing*, **33** (B3), 254–266.

Dowman, I. and Neto, F., 1994, The accuracy of along track stereoscopic data for mapping: Results from simulations and JERS OPS. *International Archives of Photogrammetry and Remote Sensing*, **30**, 216–221.

Dowman, I., Laycock, J. and Whalley, J., 1993, Geocoding in the UK. In Schreier, G., (1993 a) (ed.), 373–387.

Drewniok, C., 1994, Multi-spectral edge detection. Some experiments on data from Landsat-TM. *International Journal of Remote Sensing*, **15**, 3743–3765.

Drucker, H., Schapire, R. and Simard, P.Y., 1993, Boosting performance in neural networks. *International Journal of Pattern Recognition and Artificial Intelligence*, **7**, 705–719.

Drury, S.A., 1993, *Image Interpretation in Geology*, Second edition. London: Allen and Unwin.

Du, L. and Lee, J.S., 1996, Fuzzy classification of earth terrain covers using complex polarimetric SAR data. *International Journal of Remote Sensing*, **17**, 809–826.

Du, Y., Guindon, B. and Cihlar, J., 2002, Haze detection and removal in high resolution satellite image with wavelet analysis. *IEEE Transactions on Geoscience and Remote Sensing*, **40**, 210–217.

Dubayah, R., Blair, J.B., Bufton, J., Clark, D., JaJa, J., Knox, R., Luthcke, S., Prince S. and Weishampel, J., 1997, The vegetation canopy Lidar mission, *Land Satellite Information in the Next Decade. II: Sources and Applications*. Bethesda, MD: American Society for Photogrammetry and Remote Sensing, 100–112.

Dubayah, R. and Drake, J., 2000, Lidar remote sensing for forestry applications. *Journal of Forestry*, **98**, 44–46.

Dubayah, R., Knox, R.G., Hofton, M.A., Blair, J.B. and Drake, J.B., 2000, Land surface characteristics using lidar remote sensing. In: Hill, M. and Aspinall, R. (eds.), *Spatial Information for Land Use Management*. Singapore: International Publishers Direct.

Duda, O. and Hart, P.E., 1973, *Pattern Classification and Scene Analysis*. New York: Wiley.

Duggin, M.J., 1985, Factors influencing the discrimination and quantification of terrestrial features using remotely-sensed radiance. *International Journal of Remote Sensing*, **6**, 3–28.

Duguay, C.R. and Peddle, D.R., 1996, Comparison of evidential reasoning and neural network approaches in a multi-source classification of alpine tundra vegetation. *Canadian Journal of Remote Sensing*, 22, 433–440.

Dymond, J.R., 1992, How accurately do image classifiers estimate area? *International Journal of Remote Sensing*, **13**, 1735–1742.

Eberlein, R.B. and Wezka, J.S., 1975, Mixtures of derivative operators as edge detectors. *Computer Graphics and Image Processing*, **4**, 180–185.

Egbert, D.D. and Ulaby, F.T., 1972, Effect of angles on reflectivity. *Photogrammetric Engineering*, **29**, 556–564.

Eghbali, H.J., 1979, A K.-S. test for detecting changes from Landsat imagery data. *IEEE Transactions on Systems, Man and Cybernetics*, **9**, 17–23.

Ehleringer, J.R. and Field, C.B. (eds.), 1991, *Scaling Physiological Processes – Leaf to Globe*. San Diego: Academic Press.

Elachi, C., 1987, *Introduction to the Physics and Techniques of Remote Sensing*. New York: Wiley.

Elachi, C., 1988, *Spaceborne Remote Sensing: Applications and Techniques*. New York: IEEE Press.

Eliason, E.M. and McEwen, A.S., 1990, Adaptive box filters for the removal of random noise from digital images. *Photogrammetric Engineering and Remote Sensing*, **56**, 453–458.

Elvidge, C.D. and Lyon, R.J.P., 1985, Influence of rock-soil spectral variation on the assessment of green biomass. *Remote Sensing of Environment*, **17**, 265–279.

Elvidge, C.D., Yuan, D., Weerackoon, R.D. and Lunetta, R.S., 1995, Relative radiometric normalisation of Landsat Multispectral Scanner (MSS) data using an automatic scattergram-controlled regression. *Photogrammetric Engineering and Remote Sensing*, **61**, 1255–1260.

Esaias, W.E., Abbot, M.R., Barton, I., Brown, O.B., Campbell, J.W., Carder, K.L., Clark, D.K., Evans, R.H., Hoge, F.E., Gordon, H.R., Balch, W.M., Letelier, R. and Minnett, P.J., 1998, An overview of MODIS capabilities for ocean science observations. *IEEE Transactions on Geoscience and Remote Sensing*, **36**, 1250–1265.

Esch, T., Roth, A., Strunz, G. and Dech, S., 2003, Object-oriented classification of Landsat-7 data for regional planning purposes. *The International Archives of the Photogrammetry, Remote Sensing and Spatial Information Sciences (CD-ROM)*, vol. XXXIV-7/W9, Regensburg, Germany, 27–29 June 2003. Links to this and other documents relating to the E-Cognition software can be found at URL: http://www.definiens-imaging.com/documents/reference.htm.

Evans, D.L., Farr, T.G., van Zyl, J.J. and Zebker, H.A., 1988. Radar polarimetry: Analysis tools and applications. *IEEE Transactions on Geoscience and Remote Sensing*, **26**, 774–789.

Evans, D.L., Farr, T.G., Zebker, H.A., van Zyl, J.J. and Mouginis-Mark, P.J., 1992. Radar interferometry studies of the Earth's topography. *Eos: Transactions of the American Geophysical Union*, **73**, 553–558.

Evans, D.L., Plant, J.J. and Stofan, E.R., 1997, Overview of the Spaceborne Imaging Radar-C/X-Band Synthetic Aperture Radar (SIR-C/X-SAR) missions. *Remote Sensing of Environment*, **59**, 135–140.

Farag, A.A., 1992, Edge-based image segmentation. *Remote Sensing Reviews*, **6**, 95–122.

Farr, T.G. and M. Kobrick, 2000, Shuttle radar topography mission produces a wealth of data. *Eos: Transactions of the American Geophysical Union*, **81**, 583–585.

Favey, E., Pateraki, M., Baltsavias, E.P., Bauder, A. and Bösch, H., 2000, Surface modelling for Alpine glacier monitoring by airborne laser scanning and digital photogrammetry. *XIXth Congress of the ISPRS*. Amsterdam, 269–277.

Feng, J., Rivard, B. and Sánchez-Azofeifa, A., 2003, The topographic normalisation of hyperspectral data: Implications for the selection of spectral end members and lithologic mapping. *Remote Sensing of Environment*, **85**, 221–231.

Ferrari, M.C., 1992, Improved decorrelation stretching of TM data for geological applications: first results in Northern Somalia. *International Journal of Remote Sensing*, **13**, 841–851.

Ferrier, G., 1995, Evaluation of apparent surface reflectance methodologies. *International Journal of Remote Sensing*, **16**, 2291–2297.

Feynman, R., 1985. *QED: The Strange Theory of Light and Matter*. Princeton: Princeton University Press. Also published by Penguin Books, London, 1990.

Filella, I. and Penuelas, J., 1994, The red edge position and shape as indicators of plant chlorophyll content, biomass and hydric status. *International Journal of Remote Sensing*, **15**, 1459–1470.

Fioravanti, S., 1994, Multifractals: theory and application to image texture recognition. In: Wilkinson, G.G., Kanellopoulos, I. and Mégier, J. (eds.), *Fractals in Geoscience and Remote Sensing: Proceedings of a Joint JRC/*

EARSeL Expert Meeting, Ispra, Italy, 14–15 April 1994. Report EUR 16092. Luxembourg: Office for Official Publications of the European Communities, 152–175.

Fisher, P., 1997, The pixel: a snare and a delusion. *International Journal of Remote Sensing*, **18**, 679–685.

Fitzgerald, R.W. and Lees, B.G., 1994, Assessing the classification accuracy of multisource remote sensing data. *Remote Sensing of Environment*, **47**, 362–368.

Flygare, A.-M., 1997, A comparison of contextual classification methods using Landsat TM. *International Journal of Remote Sensing*, **18**, 3835–3842.

Fogel, D.N. and Tinney, L.R., 1996, *Image Registration using Multiquadric Functions, the Finite Element Method, Bivariate Mapping Polynomials and Thin Plate Spline*. Technical Report 96-1, National Center for Geographic Information and Analysis (NCGIA), Simonett Center for Spatial Analysis (University of California), the State University of New York, and the University of Maine.

Foley, J.D., van Dam, A., Feiner, S.K. and Hughes, J.F., 1990, *Computer Graphics – Principles and Practice, Second Edition in C*. Reading, MA: Addison-Wesley.

Foley, J.D., van Dam, A., Feiner, S.K., Hughes, J.F. and Phillips, R.L., 1994, *Introduction to Computer Graphics*. Reading, MA: Addison-Wesley.

Foody, G.M., 1996, Approaches for the production and evaluation of fuzzy land cover classification from remotely-sensed data. *International Journal of Remote Sensing*, **17**, 1317–1340.

Foody, G.M., 2002, Status of land cover classification accuracy assessment. *Remote Sensing of Environment*, **80**, 185–201.

Foody, G.M. and Arora, M.K., 1966, Fuzzy thematic mapping: incorporating mixed pixels in the training, allocation and testing stages of supervised image classification. In: Binaghi *et al.* (eds.) (1996), 43–52.

Foody, G.M. and Atkinson, P.M. (eds.), 2002, *Uncertainty in Remote Sensing and GIS*. Chichester: John Wiley and Sons.

Foody, G.M., Lucas, R.M., Curran, P.J. and Honzak, M., 1996, Estimation of the areal extent of land-cover classes that only occur at the sub-pixel level. *Canadian Journal of Remote Sensing*, **22**, 428–432.

Foody, G.M, Lucas, R.M., Curran, P.J. and Honzak, M., 1997, Non-linear mixture modelling without end members using an artificial neural network. *International Journal of Remote Sensing*, **18**, 937–953.

Foody, G.M., McCullagh, M.B. and Yates, W.B., 1995, The effect of training set size and composition on artificial neural net classification. *International Journal of Remote Sensing*, **16**, 1707–1723.

Forshaw, M.R.B., Haskell, A., Miller, P.F., Stanley, D.J. and Townshend, J.R.G., 1983, Spatial resolution of remotely-sensed imagery: a review paper. *International Journal of Remote Sensing*, **4**, 497–520.

Frank, T.D., 1985, Differentiating semiarid environments using Landsat reflectance data. *Professional Geographer*, **37**, 36–46.

Franklin, S.E. and Giles, P.T., 1995, Radiometric processing of aerial and satellite remote-sensing imagery. *Computers and Geosciences*, **21**, 413–425.

Franklin, S.E. and Peddle, D.R., 1987, Texture analysis of digital image data using spatial co-occurrence. *Computers and Geosciences*, **13**, 293–311.

Frankot, R.T. and Chellappa, R., 1987, Lognormal random-field models and their applications to radar image synthesis. *IEEE Transactions on Geoscience and Remote Sensing*, **25**, 195–207.

Fraser, C., Hanley, H. and Yarnakawa, T., 2002, 3D geopositioning accuracy of Ikonos imagery. *The Photogrammetric Record*, **17**, 465–479.

Freeman, A., Villasenor, J., Klein, J.D., Hoogeboom, P. and Groot, J., 1994, On the use of multi-frequency and polarimetric radar backscatter features for classification of agricultural crops. *International Journal of Remote Sensing*, **15**, 1799–1812.

Frei, U., Graf, K. Chr. and Meier, E., 1993, Cartographic reference systems. In: Schreier, G. (1993a) (ed.), 213–234.

Friedl, M.A., Brodley, C.E. and Strahler, A.H., 1999, Maximising land cover classification accuracies produced by decision trees at continental to global scales. *IEEE Transactions on Geoscience and Remote Sensing*, **37**, 969–977.

Freund, Y. and Schapire, R.E., 1996, Experiments with a new boosting algorithm. *Proceedings of the 13th International Conference on Machine Learning*, Bari, Italy, 3–6 July 1996, 148–156.

Frulla, L.A., Milovich, J.A. and Gagliardini, D.A., 1995, Illumination and observational geometry for NOAA-AVHRR imagery. *International Journal of Remote Sensing*, **16**, 2233–2253.

Fusco, L. and Trevese, D., 1985, On the reconstruction of lost data in images of more than one band. *International Journal of Remote Sensing*, **6**, 1535–1544.

Gabriel, A., Goldstein, R. and Zebker, H.A., 1989. Mapping small elevation changes over large areas: Differential radar interferometry. *Journal of Geophysical Research*, **94**, 9183– 9191.

Gahegan, M. and West, G., 1998, The classification of complex datasets: an operational comparison of artificial neural networks and decision tree classifiers. *Proceedings of the 3rd International Conference on GeoComputation*. University of Bristol, UK. URL: http://divcom.otago.ac.nz/SIRC/GeoComp/GeoComp98/61/gc_61.htm. Accessed 15 July 2003.

Gamon, J.A., Field, C.B., Roberts, D.A., Ustin, S.L. and Valentini, R., 1993, Functional patterns in an annual grassland during and AVIRIS overflight. *Remote Sensing of Environment*, **44**, 239–253.

Gao, B.-C., Heidebrecht, K.B. and Goetz, A.F.H., 1993, Derivation of scaled surface reflectances from AVIRIS data. *Remote Sensing of Environment*, **44**, 145–163.

Garcia-Haro, F.J., Gilabert, M.A. and Melia, J., 1996, Linear spectral mixture modelling to estimate vegetation amount from optical spectral data. *International Journal of Remote Sensing*, **17**, 3373–3400.

Gardner, B.P., Blad, B.L., Thompson, D.R. and Henderson, K., 1985, Evaluation and interpretation of Thematic Mapper ratios for estimating corn growth parameters. *Remote Sensing of Environment*, **18**, 225–234.

Garguet-Duport, B., Girel, J., Chassery, J. and Pautou, G., 1996, The use of multiresolution analysis and wavelets transform for merging SPOT panchromatic and multispectral image data. *Photogrammetric Engineering and Remote Sensing*, **62**, 1057–1066.

Garvin, J., Bufton, J., Blair, J., Harding, D., Lutcke, S., Frawley, J. and Rowlands, D., 1998, Observations of the Earth's topography from the Shuttle laser altimeter (SLA): Laser-pulse echo-recovery measurements of terrestrial surfaces. *Physics and Chemistry of the Earth*, **23**, 1053–1068.

Gens, R., 2003, Two-dimensional phase unwrapping for radar interferometry: developments and new challenges. *International Journal of Remote Sensing*, **24**, 703–710.

Gens, R. and van Genderen, J.L., 1996, SAR interferometry – issues, techniques, applications. *International Journal of Remote Sensing*, **17**, 1803–1835.

Gens, R. and van Genderen, J.L., 1996, Analysis of the geometric parameters of SAR interferometry for spaceborne systems. *International Archives of Photogrammetry and Remote Sensing*, **31**(B2), 107–110.

German, G.W.H. and Gahegan, M.N., 1996, Neural network architectures for the classification of temporal image sequences. *Computers and Geosciences*, **22**, 969–979.

Ghiglia, D.C. and Pritt, M.D., 1998, *Two-Dimensional Phase Unwrapping: Theory, Algorithms and Software*. New York: John Wiley and Sons.

Gilabert, M.A., Gonzales-Piqueras, J., Garcia-Haro, F.J. and Melia, J., 2002, A generalised soil-adjusted vegetation index. *Remote Sensing of Environment*, **82**, 303–310.

Giles, P.T. and Franklin, S.E., 1996, Comparison of derivative topographic surfaces of a DEM generated from stereoscopic SPOT images with field measurements. *Photogrammetric Engineering and Remote Sensing*, **62**, 1165–1171.

Gillespie, A.R., Kahle, A.B. and Walker, R.E., 1986, Color enhancement of highly correlated images. I: Decorrelation and HSI contrast stretches. *Remote Sensing of Environment*, **20**, 209–235.

Goetz, A.F.H., 1984, High spectral resolution remote sensing of the land. *Proceedings of the Society of Photo-Optical Instrumentation Engineers (SPIE)*, **268**, 56–68.

Goetz, A.F.H., 1991, Spectral remote sensing in geology. In: Asrar, G. (ed.) (1991), 491–526.

Goetz, A.F.H., Rock B.N. and Rowan, L.C., 1983, Remote sensing for exploration: an overview. *Economic Geology*, **78**, 573–589.

Goetz, A.F.H., Vane, G., Solomon, J.E. and Rock, B.N., 1985, Imaging spectroscopy for earth remote sensing. *Science*, **228**, 1147–1153.

Goldstein, R.M., Engelhardt, H., Kamb, B. and Frolich, R.M., 1993, Satellite radar interferometry for monitoring ice-sheet motion: Application to Antarctic ice stream. *Science*, **262**, 1525.

Goldstein, R.M. and Zebker, H.A., 1987, Interferometric radar measurement of ocean surface currents. *Nature*, **328**, 707–709.

Gong, P. and Howarth, P.J., 1990, An assessment of some factors influencing multispectral land-cover classification. *Photogrammetric Engineering and Remote Sensing*, **56**, 597–603.

Gong, P., Ledrew, E.F. and Miller, J.R., 1992, Registration noise reduction in difference images for change detection. *International Journal of Remote Sensing*, **13**, 773–739.

Gong, P., Pu, R. and Chen, J., 1996, Mapping ecological land systems and classification uncertainties from digital elevation and forest-cover data using neural networks. *Photogrammetric Engineering and Remote Sensing*, **62**, 1249–1260.

Gong, P., Pu, R. and Yu, B., 1997, Conifer species recognition: An exploratory analysis of *in situ* hyperspectral data. *Remote Sensing of Environment*, **62**, 189–200.

Gonzales, R.C. and Woods, R.E., 1992, *Digital Image Processing*. Reading, MA: Addison-Wesley.

Goodenough, D.G., Narendra, P.M. and O'Neill, K., 1978, Feature subset selection in remote sensing. *Canadian Journal of Remote Sensing*, **4**, 143–148.

Gopal, S. and Woodcock, C., 1994, Theory and methods for accuracy assessment of thematic maps using fuzzy sets. *Photogrammetric Engineering and Remote Sensing*, **60**, 181–188.

Gopal, S. and Woodcock, C., 1996, Remote sensing of forest change using artificial neural networks. *IEEE Transactions on Geoscience and Remote Sensing*, **34**, 398–403.

Govaerts, Y.M., Verstraete, M.M., Pinty, B. and Gobron, N., 1999, Designing optimal spectral indices: a feasibilty and proof of concept study. *International Journal of Remote Sensing*, **20**, 1853–1873.

Green, A.A., Berman, M., Switzer, P. and Craig, M.D., 1988, A transformation for ordering multispectral data in terms of image quality with implications for noise removal. *IEEE Transactions on Geoscience and Remote Sensing*, **26**, 65–64.

Greenfield, S., 1997, *The Human Brain: A Guided Tour*. London: Weidenfeld and Nicholson.

Gregory, R.L., 1998, *Eye and Brain: The Psychology of Seeing* 5th edition. Oxford: Oxford University Press.

Gribben, J., 1984. *In Search of Schrödinger's Cat*. London: Wildwood House.

Grossman, Y.L., Ustin, S.L., Jacquemond, S., Sanderson, E.W., Schmuck, G. and Verdebout, J., 1996, Critique of stepwise multiple linear regression for the extraction of leaf biochemistry information from leaf reflectance data. *Remote Sensing of Environment*, **56**, 182–193.

Gu, D., Gillespie, A.R., Adams, J.B. and Weeks, R., 1999, A statistical approach for topographic correction of satellite images using spatial context information. *IEEE Transactions on Geoscience and Remote Sensing*, **37**, 236–246.

Guenther, G.C., Brooks, M.W. and LaRoque, P.E., 2000, New capabilities of the 'SHOALS' airborne Lidar bathymeter. *Remote Sensing of Environment*, **73**, 247–255.

Guo, L.J. and Moore, J.McM., 1996, Direct decorrelation stretch technique for RGB colour composition. *International Journal of Remote Sensing*, **17**, 1005–1018.

Gupta, R.P., 1991, *Remote Sensing in Geology*. Berlin: Springer-Verlag.

Gurney, C.M., 1980, Threshold selection for line detection algorithms. *IEEE Transactions on Geoscience and Remote Sensing*, **18**, 204–211.

Gurney, C.M. and Townshend, J.R.G., 1983, The use of contextual information in the classification of remotely sensed data. *Photogrammetric Engineering and Remote Sensing*, **49**, 55–64.

Gutman, G. and Ignatov, A., 1995, Global land monitoring from AVHRR: potential and limitations. *International Journal of Remote Sensing*, **16**, 2301–2309.

Guyot, G. and Baret, F., 1988, Utilisation de la haute resolution spectrale pour suivre l'etat des couverts vegetaux. *Fourth International Colloquium on Spectral Signatures of Objects in Remote Sensing*. Aussois, France. Paris: European Space Agency. 279–286.

Guyot, G. and Gu, X.-F., 1994, Effect of radiometric corrections on NDVI determined from SPOT-HRV and Landsat-TM data. *Remote Sensing of Environment*, **49**, 169–180.

Hall, F.G., Strebel, D.E., Nickeson, J.R. and Goetz, S.J., 1991, Radiometric rectification: toward a common radiometric response among multidate, multisensor images. *Remote Sensing of Environment*, **35**, 11–27.

Hansen, P.M. and Schjoerring, J.K., 2003, Reflectance measurement of canopy biomass and nitrogen status in wheat crops using normalised difference vegetation indices and partial least squares regression. *Remote Sensing of Environment*, **86**, 542–533.

Hanssen, R., 2001, *Radar Interferometry*. Dordrecht, The Netherlands: Kluwer Academic Publishing.

Haralick, R.M., 1980, Edge and region analysis of digital image data. *Computer Graphics and Image Processing*, **12**, 50–73.

Haralick, R.M. and Fu, K.-S., 1983, Pattern recognition and classification. In: Colwell, R.N. (1983) (ed.), 793–805.

Haralick, R.M. and Shanmugam, K.S., 1974, Combined spectral and spatial processing of ERTS imagery data. *Remote Sensing of Environment*, **3**, 3–13.

Haralick, R.M., Shanmugam, K.S. and Dinstein, I., 1973, Textural features for image classification. *IEEE Transactions on Systems, Man and Cybernetics*, **3**, 610–622.

Harding, D.J., Lefsky, M.A., Parker, G.G. and Blair, J.B., B., 2001, Laser altimetry waveform measurement of vegetation canopy structures: Measurement and validation for closed-canopy, broadleaved forests. *Remote Sensing of Environment*, **76**, 283–297.

Hardy, R.L., 1971, Multiquadric equations of topography and other irregular surfaces. *Journal of Geophysical Research*, **76**, 1904–1915.

Hardy, R.L., 1990, Theory and applications of the multiquadric-biharmonic method. *Computers and Mathematical Applications*, **19**, 163–208.

Harries, J., Llewellyn-Jones, D.T., Mutlow, C., Murray, M.J., Barton, I.J. and Prata, A.J., 1994, The ATSR programme: instruments, data and science. In: Mather, P.M. (ed.) (1994), 20–28.

Harris, J.R., Murray, R. and Hirose, T., 1996, IHS transform for the integration of radar imagery and other remotely sensed data. *Photogrammetric Engineering and Remote Sensing*, **56**, 1631–1641.

Harris, R., 1981, An experiment in probabilistic relaxation for terrain cover classification of Kuwait from Landsat imagery. *Proceedings of the 15th International Conference on Remote Sensing of Environment*, Environmental Research Institute of Michigan, Ann Arbor, MI, 1245–1252.

Harris, R., 1985, Contextual classification post-processing of Landsat data using a probabilistic relaxation model. *International Journal of Remote Sensing*, **6**, 847–866.

Hartigan, J.A., 1975, *Clustering Algorithms*. New York: Wiley.

Hawwar, Y. and Reza, A., 2002, Spatially adaptive multiplicative noise image denoising technique. *IEEE Transactions on Image Processing*, **11**, 1397–1404.

Hay, J.C. and Mackay, D.C., 1985, Estimating solar irradiances on inclined surfaces: a review and assessment of methodologies. *International Journal of Solar Energy*, **3**, 203–240.

Healy, M.J.R., 1968, Multivariate normal plotting. *Applied Statistics*, **17**, 157–161.

Hearn, D. and Baker, M.P., 1994, *Computer Graphics. Second Edition*. London: Prentice-Hall International.

Hellwich, O., 1999a, Basic principles and current issues of SAR interferometry. ISPRS Joint Workshop *Sensors and Mapping from Space 1999*. Hannover. Institute for Photogrammetry and Engineering Surveying, University of Hannover, Germany.

Hellwich, O., 1999b, SAR interferometry: Principles, processing and perspectives. In: Heipke, C. and Mayer, H. (eds.), *Festschrift für Prof. Dr.-Ing. Heinrich Ebner*. Munich: Lehrstuhl für Photogrammetrie und Fernerkundung, Technische Universität München.109–120.

Henebry, G.M., 1997, Advantages of principal components analysis for land cover segmentation from SAR image series. *Proceedings of the 3rd ERS Symposium, 18–21 March, Florence, Italy*. ESA: Noordwijk, The Netherlands, 175–178. URL: http://earth.esa.int/symposia/papers/henebry3/index.html. Accessed 15 July 2003.

Hepner, G.F., Logan, T., Ritter, N. and Bryant, N., 1990, Artificial neural network classification using a minimal training set: comparison to conventional supervised classification. *Photogrammetric Engineering and Remote Sensing*, **56**, 469–473.

Hill, I.D., 1973, The normal integral. Algorithm AS66. *Applied Statistics*, **22**, 424.

Hill, J., 1991, A quantitative approach to remote sensing: sensor calibration and comparison. In: Belward, A.S. and Valenzuela C.R. (eds.), 1991, 97–110.

Hill, J., 1993–1994, Monitoring land degradation and soil erosion in Mediterranean environments. *ITC Journal*, **1993–1994**, 323–331.

Hill, J. and Aifadopoulou, D. 1990, Comparative analysis of Landsat-5 TM and SPOT HRV-1 data for use in multiple sensor approaches. *Remote Sensing of Environment*, **34**, 55–70.

Hill, J. and Horstert, P.O., 1996, Monitoring the growth of a Mediterranean metropolis based on the analysis of spectral mixtures – a case study on Athens. In: Parlow, E. (ed.), 1996, 21–31.

Hill, J., Mehl, W. and Radeloff, V., 1995, Improved terrain mapping by combining correction of atmospheric and topographic effects in Landsat TM imagery. In: Askne, J. (ed.), 143–151.

Hirano, A., Welch, R. and Lang, H., 2003, Mapping from ASTER stereo image data: DEM validation and accuracy assessment. *ISPRS Journal of Photogrammetry and Remote Sensing*, **57**, 356–370.

Hobbs, R.J. and Mooney, H.A. (eds.), 1990, *Remote Sensing of Biospheric Functioning*. New York: Springer-Verlag.

Hochberg, E.J. and Atkinson, M.J., 2000, Spectral discrimination of coral reef benthic communities. *Coral Reefs*, **19**, 164–171.

Hodgson, M.E., Jensen, J.R., Schmidt, L., Schill, S. and Davis, B., 2003, An evaluation of LiDAR- and IFSAR-derived digital elevation models in leaf-on conditions with USGS Level 1 and Level 2 DEMs. *Remote Sensing of Environment*, **84**, 295–308.

Hoffer, R.M., 1978, Biological and physical considerations in applying computer-aided analysis techniques to remote sensor data. In: Swain, P.H. and Davis, S.M. (1978) (eds.), *Remote Sensing – the Quantitative Approach*. New York: McGraw-Hill, chapter 5.

Holben, B.N. and Fraser, R.S., 1984, Red and near-infrared sensor response to off-nadir viewing. *International Journal of Remote Sensing*, **5**, 145–160.

Holben, B.N. and Kimes, D., 1986, Directional reflectance response in AVHRR red and near-infrared bands for three cover types and varying atmospheric conditions. *Remote Sensing of Environment*, **19**, 213–226.

Holland, J.H., 1992, Genetic algorithms. *Scientific American*, **267**, 44–50.

Holm, R.G., Moran, M.S., Jackson, R.D., Slater, P.N., Yuan, B. and Biggar, S.F., 1989, Surface reflectance factor retrieval from Thematic Mapper data. *Remote Sensing of Environment*, **27**, 47–57.

Hooper, D.M., Bursik, M.I. and Webb, F.E., 2003, Application of high-resolution, interferometric DEMs to geomorphic studies of fault scarps, Fish Lake Valley, Nevada-California, USA. *Remote Sensing of Environment*, **84**, 255–267.

Hoque, E. and Hutzler, P.J.S., 1992, Spectral blue-shift of red edge monitors damage class of beech trees. *Remote Sensing of Environment*, **39**, 81–84.

Horgan, G.W., 1998, Wavelets for SAR image smoothing. *Photogrammetric Engineering and Remote Sensing*, **64**, 1171–1177.

Horgan, G.W., 1999, Using wavelets for data smoothing: A simulation study. *Journal of Applied Statistics*, **26**, 923–932.

Horler, D.N.H., Dockray, M. and Barber, J., 1983. The red edge of plant leaf reflectance. *International Journal of Remote Sensing*, **4**, 273–288.

Horn, B.K.P. and Woodham, R.J., 1979, Destriping Landsat MSS images by histogram modification. *Computer Graphics and Image Processing*, **10**, 69–83.

Huang, C., Townshend, J.R.G., Liang, S., Kalluri, S.N.V. and DeFries, R.S., 2002, Impact of a sensor's point spread function on land cover characterization: assessment and deconvolution. *Remote Sensing of Environment*, **80**, 203–212.

Huang, C., Wylie, B., Yang, L., Homer, C. and Zylstra, G., 2002, Derivation of a tasseled cap transformation based on Landsat 7 at-satellite reflectance. *International Journal of Remote Sensing*, **23**, 1741–1748.

Huang, T.S., Yang, G.J. and Tang, G.Y., 1979, A fast two-dimensional median filtering algorithm. *IEEE Transactions on Acoustics, Speech and Signal Processing*, **27**, 13–18.

Hudak, A.T., Lefsky, M.A., Cohen, W.A. and Berterretche, M., 2002, Integration of lidar and Landsat ETM+ data for estimating and mapping forest canopy height. *Remote Sensing of Environment*, **82**, 397–416.

Huete, A., 1989, Soil influences in remotely sensed vegetation-canopy spectra. In: Asrar, G. (ed.) (1929), 107–141.

Huguenin, R.L. and Jones, J.L., 1986, Intelligent information extraction from reflectance spectra: Absorption band positions. *Journal of Geophysical Research*, **19**(B9), 9585–9598.

Hung, C.C., 1993, Competitive learning networks for unsupervised training. *International Journal of Remote Sensing*, **14**, 2411–2415.

Hunt, G.R., 1977, Spectral signatures of particulate minerals in the visible and near-infrared. *Geophysics*, **42**, 501–513.

Hunt, G.R., 1979, Near-infrared (1.3–2.4 μm) spectra of alteration minerals: potential for use in remote sensing. *Geophysics*, **44**, 1974–1986.

Hunt, G.R. and Ashley, R.P., 1979, Spectra of altered rocks in the visible and near infrared. *Economic Geology*, **74**, 1613–1629.

Hunt, G.R. and Salisbury, J.W., 1970, Visible and near-infrared spectra of minerals and rocks: I silicate minerals. *Modern Geology*, **1**, 283–300.

Hunt, G.R. and Salisbury, J.W., 1971, Visible and near-infrared spectra of minerals and rocks: II carbonates. *Modern Geology*, **2**, 23–30.

Hunt, G.R., Salisbury, J.W. and Lenhoff, D.J., 1971, Visible and near-infrared spectra of minerals and rocks: III oxides and hydroxides. *Modern Geology*, **2**, 195–205.

Hunter, G.J. and Goodchild, M.F., 1997, Modeling the uncertainty of slope and aspect estimates derived from spatial databases. *Geographical Analysis*, **29**, 35–49.

Huseby, R.B. and Solberg, R., 1998, *A Model-based Approach for Geometrical Correction of Optical Satellite Images*. Report No. 932, Norwegian Computing Centre, Oslo, Norway.

Hussein, H.A., Rabie, S.I. and Abdel Nabie, S., 1996, Structural interpretation of aeromagnetic data, Gabel Zubeir area, North Eastern Desert, Egypt. *International Journal of Remote Sensing*, **17**, 1997–2012.

Hutchinson, C.F., 1982, Techniques for combining Landsat and ancillary data for digital classification improvement. *Photogrammetric Engineering and Remote Sensing*, **48**, 123–130.

Hyman, A.H. and Barnsley, M.J., 1997, On the potential for land cover mapping from multiple-view-angle (MVA) remotely sensed images. *Proceedings of the 23rd Annual Conference and Exhibition of the Remote Sensing Society, University of Reading, 2–4 September 1997*. Nottingham: The Remote Sensing Society, 135–140.

Hyppänen, H., 1996, Spatial autocorrelation and optimum spatial resolution of optical remote sensing data in boreal forest environment. *International Journal of Remote Sensing*, **17**, 3441–3452.

Hyyppä, J., Pyysalo, U., Hyyppä, H. and Samberg, A., 2000, Elevation accuracy of laser scanning-derived digital terrain and target modules in forest environment. *EARSeL SIG-Workshop Lidar*. Dresden, Germany. Paris: European Association of Remote Sensing Laboratories (EARSeL). 139–147.

Ichoku, C. and Karnieli, A., 1996, A review of mixture modeling techniques for sub-pixel land cover estimation. *Remote Sensing Reviews*, **13**, 161–186.

Ichoku, C., Karnieli, A., Meisels, A. and Chorowicz, J., 1996, Detection of drainage channel networks on satellite imagery. *International Journal of Remote Sensing*, **17**, 1659–1678.

Imbrie, J., 1963, Factor and vector analysis programs for analyzing geologic data. *Technical Report*, **6**, Office of Naval Research Geography Branch, Task No. 389-135, Contract Nonr 1228(26), Northwestern University, Evanston, Illinois, October 1963.

Irish, J.L. and Lillycrop, W.J., 1999, Scanning laser mapping of the coastal zone: the SHOALS system. *ISPRS Journal of Photogrammetry and Remote Sensing*, **54**, 123–129.

Irish, R., 2002, *Landsat 7 Science Data Users Handbook*. NASA Goddard Spaceflight Centre, MD, URL: http://ltpwww.gsfc.nasa.gov/IAS/handbook/handbook_toc.html. Date accessed: 13 May 2003.

Irons, J.R. and Petersen, G.W., 1981, Texture transforms of remotely sensed data. *Remote Sensing of Environment*, **11**, 359–370.

Irons, J.R., Weismiller, R.A. and Petersen, G.W., 1989, Soil reflectance. In: Asrar, G. (ed.) (1991), 66–106.

Itten, K.I. and Meyer, P., 1993, Geometric and radiometric correction of TM data of mountainous forested areas. *IEEE Transactions on Geoscience and Remote Sensing*, **31**, 764–770.

Jackson, R.D., 1983, Spectral indices in n-space. *Remote Sensing of Environment*, **13**, 409–421.

Jackson, R.D., Slater, P.N. and Pinter, P.J., Jr., 1983, Discrimination of growth and water stress in wheat by various vegetation indices through clear and turbid atmospheres. *Remote Sensing of Environment*, **13**, 187–208.

Jansen, M., 2000, *Wavelet Thresholding and Noise Reduction*, Catholic University of Leuven, Leuven.

Jarvis, C.H. and Stuart, N., 1996, The sensitivity of a neural net for classifying remotely sensed imagery. *Computers and Geosciences*, **22**, 959–967.

Jensen, J.R., 1986, *Introductory Digital Image Processing – A Remote Sensing Perspective*. Englewood Cliffs, NJ: Prentice-Hall.

Jensen, J.R. and Toll, D.L., 1982, Detecting residential land-use development at the urban fringe. *Photogrammetric Engineering and Remote Sensing*, **48**, 629–643.

Johnsen, H., Lauknes, L. and Guneriussen, T., 1995, Geocoding of fast-delivery SAR image mode products using DEM data. *International Journal of Remote Sensing*, **16**, 1957–1968.

Johnson, R.D. and Kasischke, E.S., 1998, Change vector analysis: a technique for the multispectral monitoring of land cover and condition. *International Journal of Remote Sensing*, **19**, 411–426.

Jones, A.R., Settle, J.J. and Wyatt, B.K., 1988, Use of digital terrain data in the interpretation of SPOT-1 HRV data. *International Journal of Remote Sensing*, **9**, 669–682.

Joseph, G., 1996, Imaging sensors for remote sensing. *Remote Sensing Reviews*, **13**, 257–342.

Joseph, G., 2000, How well do we understand Earth observation electro-optical sensor parameters? *Remote Sensing of Environment*, **55**, 9–12.

Joughin, I.R., Winebrenner, D.P. and Fahnestock, M.A., 1995, Observations of ice-sheet motion in Greenland using satellite radar interferometry. *Geophysics Research Letters*, **22**, 571–574.

Jupp, D.L.B. and Mayo, K.K., 1982, The use of residual images in Landsat image analysis. *Photogrammetric Engineering and Remote Sensing*, **48**, 595–604.

Jupp, D.L.B., Strahler, A.H. and Woodcock, C.E., 1988, Autocorrelation and regularisation in digital images. I: Basic theory. *IEEE Transactions on Geoscience and Remote Sensing*, **26**, 463–473.

Jupp, D.L.B., Strahler, A.H. and Woodcock, C.E., 1989, Autocorrelation and regularisation in digital images. II: Simple image models. *IEEE Transactions on Geoscience and Remote Sensing*, **27**, 247–258.

Justice, C.O., Townshend, J.R.G., Holben, B.N. and Tucker, C.J., 1985, Analysis of the phenology of global vegetation using meteorological satellites. *International Journal of Remote Sensing*, **6**, 1271–1318.

Jutz, S.L. and Chorowicz, J., 1993, Geological mapping and detection of oblique extensional structures in the Kenyan Rift Valley with a SPOT/Landsat data merge. *International Journal of Remote Sensing*, **14**, 1677–1688.

Kahle, A.B. and Rowan, L.C., 1980, Evaluation of multispectral infrared aircraft images for lithological mapping in the East Tintic Mountains, Utah. *Geology*, **8**, 234–239.

Kailath, T., 1967, The divergence and Bhattacharyya distance measures in feature selection. *IEEE Transactions on Communication Theory*, **15**, 52–60.

Kalkhan, M.A., Reich, R.M. and Czaplewski, R.L., 1997, Variance estimates and confidence intervals for the kappa measure of classification accuracy. *Canadian Journal of Remote Sensing*, **23**, 210–216.

Kanal, L., 1974, Patterns in pattern recognition: 1968–1974. *IEEE Transactions on Information Theory*, **20**, 697–722.

Kaneko, T., 1976, Evaluation of Landsat image registration accuracy. *Photogrammetric Engineering and Remote Sensing*, **42**, 1285–1299.

Kanellopoulos, I., Varas, A., Wilkinson, G.G. and Mégier, J., 1992, Land-cover discrimination in SPOT HRV imagery using an artificial neural network – a 20-class experiment. *International Journal of Remote Sensing*, **13**, 917–924.

Kanellopoulos, I., Wilkinson, G.G., Roli, F. and Austin, J. (eds.), 1997, *Neurocomputation in Remote Sensing Data Analysis*. Heidelberg: Springer.

Kardoulas, N.G., Bird, A.C. and Lawan, A.I., 1996, Geometric correction of SPOT and Landsat imagery: a comparison of map and GPS-derived control points. *Photogrammetric Engineering and Remote Sensing*, **62**, 1173–1177.

Karpouzli, E. and Malthus, T., 2003, The empirical line method for atmospheric correction of IKONOS imagery. *International Journal of Remote Sensing*, **24**, 1143–1150.

Kasischke, E., Melack, J. and Dobson, M.C., 1997, The use of imaging radars for ecological applications – A review. *Remote Sensing of Environment*, **59**, 141–156.

Katawa, Y., Ueno, S. and Kusaka, T., 1986, Radiometric correction for atmospheric and topographic effects on Landsat MSS images. *International Journal of Remote Sensing*, **9**, 729–748.

Kaufmann, H. Berger, M., Meissner, M. and Süssenguth, G., 1996, MOMS-02/STS-55 design and validation of spectral and panchromatic modules. *Proceedings of the MOMS Symposium, Cologne, Germany, 5–7 July, 1995*. Paris: European Association of Remote Sensing Laboratories (EARSeL). 26-38.

Kaufman, Y.J., Herring, D.D., Ranson, K.R. and Collatz, G.J., 1998, Earth observing system AM1 mission to Earth. *IEEE Transactions on Geoscience and Remote Sensing*, **36**, 1045–1055.

Kauth, R.J. and Thomas, G., 1976, The tasselled cap – a graphic description of the spectral-temporal development of agricultural crops as seen by Landsat. *Proceedings of the Symposium on Machine Processing of Remotely- Sensed Data 1976, Purdue University, West Lafayette, IN*, volume 4B, 41–51.

Kavzoglu, T. and Mather, P.M., 1999, Pruning artificial neural networks: an example using land cover classification of multi-sensor images. *International Journal of Remote Sensing*, **20**, 2787–2803.

Kavzoglu, T. and Mather, P.M., 2002, The role of feature selection in artificial neural network applications. *International Journal of Remote Sensing*, **23**, 2919–2937

Keller, J.M., Chen, S. and Crownover, R.M., 1989, Texture description and segmentation through fractal geometry. *Computer Graphics, Vision and Image Processing*, **45**, 10–166.

Kennet, M. and Eiken, T., 1997, Airborne measurements of glacier surface elevation by scanning laser altimetry. *Annals of Glaciology*, **24**, 293–296.

Kerdiles, H. and Grondona, M.O., 1995, NOAA-AVHRR NDVI decomposition and sub-pixel classification using linear mixing in the Argentine Pampa. *International Journal of Remote Sensing*, **16**, 1303–1325.

Kess, B.L., Steinwand, D.R. and Reichenbach, S.E., 1996, Compression of Global Land 1-km AVHRR dataset. *International Journal of Remote Sensing*, **17**, 2955–2969.

Key, J.R., Maslanik, J.A. and Barry, R.G., 1989, Cloud classification from satellite data using a fuzzy sets algorithm: a polar example. *International Journal of Remote Sensing*, **10**, 1823–1842.

Khan, B., Hayes, L.W.B. and Cracknell, A.P., 1995, The effects of higher-order resampling on AVHRR data. *International Journal of Remote Sensing*, **16**, 147–163.

Kim, H. and Swain, P.H., 1995, Evidential reasoning approach to multisource data classification in remote sensing. *Proceedings of the International Geoscience and Remote Sensing Symposium (IGARSS'89), 10–14 July 1989, Vancouver, British Columbia, Canada*. New York: IEEE Press, volume 2, 829–832.

Kimble, G., 1951, The inadequacy of the regional concept. In: Dudley Stamp, L. and Wooldridge, S.W (eds.), *London Essays in Geography: Rodwell Jones Memorial Volume*. London: Longmans Green and Co., 151–174.

Kingsley, S. and Quegan, S., 1992, *Understanding Radar Systems*. London: McGraw-Hill.

Kittler, J., 1983, Image processing for remote sensing. *Philosophical Transactions of the Royal Society of London*, Series A, **309**, 323–335.

Kittler, J. and Föglein, J., 1984, Contextual classification of multispectral pixel data. *Image and Vision Computing*, **2**, 13–29.

Kneizys, F.X., Shettle, E.P., Abreu, L.W., Chetwynd, J.H., Anderson, G.P., Gallery, W.O., Selby, J.E.A., and Clough, S.A., 1988, *Users Guide to LOWTRAN*

7. Hanscomb Air Force Base, MA: U.S. Air Force Geophysics Laboratory, 137 pp.

Koch, M. and Mather, P.M., 1997, Lineament mapping for groundwater resource assessment: A comparison of digital synthetic aperture radar imagery and stereoscopic large format camera photographs in the Red Sea Hills, Sudan. *International Journal of Remote Sensing*, **18**, 1465–1482.

Kohl, H.G. and Hill, J., 1988, Geometric registration of multi-temporal TM data over mountainous areas by use of a low resolution digital elevation model. *Proceedings of the 8th EARSeL Symposium on Alpine and Mediterranean Areas: A Challenge for Remote Sensing*. Capri, Italy, 17–20 May, 1998, 323–355. Paris: European Association of Remote Sensing Laboratories (EARSeL).

Konecny, G., 2003, *Geoinformation: Remote Sensing, Photogrammetry and Geographic Information Systems*. London: Taylor and Francis.

Kontoes, C.C. and Rokos, D., 1996, The integration of spatial context information in an experimental knowledge-based system and the supervised relaxation algorithm – two successful approaches to improving SPOT-XS classification. *International Journal of Remote Sensing*, **17**, 3093–3106.

Kowalik, W.S., Lyon, R.J.P. and Switzer, P., 1983, The effects of additive radiance terms on ratios of Landsat data. *Photogrammetric Engineering and Remote Sensing*, **49**, 659–669.

Kramer, H.J., 1994, *Observation of the Earth and its Environment – Survey of Missions and Sensors*. Berlin: Springer-Verlag.

Krishnamurthy, J., Manalavan, P. and Saivasan, V., 1992, Application of digital enhancement techniques for groundwater exploration in hard rock terrains. *International Journal of Remote Sensing*, **13**, 2925–2942.

Kropatsch, W.G. and Strobl, D., 1990, The generation of SAR layover and shadow maps from digital elevation models. *IEEE Transactions on Geoscience and Remote Sensing*, **28**, 90–107.

Krumm, J., 2001, *Savitzky-Golay filters for 2D images*. URL: http://www.research.microsoft.com/users/ jckrumm/savgol/savgol.htm. Date accessed: 27 February 2003.

Kruse, F.A., Lefkoff, A.B., Boardman, J.W., Heidebrecht, K.B., Shapiro, A.T., Barloon, P.J. and Goetz, A.F.H., 1993, The Spectral Image Processing System (SIPS) – Interactive visualization and analysis of imaging spectrometer data. *Remote Sensing of Environment*, **44**, 145–163.

Kumar, R., 1979, Comparison of feature selection techniques for Earth resources data. In: W.E. Gardner (ed.) *Proceedings of an International Conference on Applications of Machine-Aided Image Analysis, Oxford, 1979*. Bristol: Institute of Physics, Conference Series 44, 238–243.

Kwok, R., Curlander, J.C. and Pang, S.S., 1987, Rectification of terrain-induced distortions in radar images. *Photogrammetric Engineering and Remote Sensing*, **53**, 507–513.

Labovitz, M.L. and Marvin, J.W., 1986. Precision in geodetic correction of TM data as a function of the number, spatial distribution, and success in matching control points: A simulation. *Remote Sensing of Environment*, **20**, 237–252.

Labovitz, M.L. and Matsuoko, E.J., 1984, The influence of autocorrelation on signature extraction – an example from a geobotanical investigation of Cotter Basin, Montana. *International Journal of Remote Sensing*, **5**, 315–332.

Labovitz, M.L., Toll, D.L. and Kennard, R.E., 1982, Preliminary evidence for the influence of physiography and scale upon the autocorrelation function of remotely sensed data. *International Journal of Remote Sensing*, **3**, 13–30.

Lambin, E.F., 1996, Change detection at multiple temporal scales: seasonal and annual variations in landscape variables. *Photogrammetric Engineering and Remote Sensing*, **62**, 931–938.

Lambin, E.F. and Strahler, A.H., 1994, Indicators for land cover change for change-vector analysis in multitemporal space at coarse spatial scales. *International Journal of Remote Sensing*, **15**, 2099–2119.

Land, E.H., 1977, The retinex theory of color vision. *Scientific American*, December 1977, 108–128.

Landgrebe, D.A., 1980, The development of a spectral-spatial classifier for Earth observation data. *Pattern Recognition*, **12**, 165–175.

Landgrebe, D.A. and staff, 1974, *A study of the utilisation of ERTS-1 data from the Wabash River Basin*. Laboratory for the Applications of Remote Sensing, Purdue University, West Lafayette, IN, LARS Information Note 052375.

Lang, H. and Welch, R., 1996, *Algorithm Theoretical Basis Document for ASTER Digital Elevation Models (Standard Product AST14), Version 3*. ATBD-AST-08, Jet Propulsion Laboratory, Pasadena, CA.

Lark, R.M., 1995, A reappraisal of unsupervised classification, I: Correspondence between spectral and conceptual classes. *International Journal of Remote Sensing*, **16**, 1425–1443.

Lark, R.M., 1996, Geostatistical description of texture on an aerial photograph for discriminating classes of land cover. *International Journal of Remote Sensing*, **17**, 2115–2133.

Lark, R.M., 1998, Forming spatially-coherent regions by classification of multi-variate data: an example from the analysis of maps of crop yield. *International Journal of Remote Sensing*, **19**, 83–98.

Laur, H., Bally, P., Meadows, P., Sanchez, J., Schaettler, B., Lopinto, E. and Esteban, D., 2002, *ERS SAR Calibration: Derivation of the Backscattering Coefficient Sigma-0 in ESA ERS SAR PRI Products*. Document

No. ES-TN-RS-PM-HL09, Issue 2, Rev. 5d. Paris: European Space Agency.

Lawrence, R.L. and Ripple, W.J., 1996, Determining patch perimeters in raster image processing and geographic information systems. *International Journal of Remote Sensing*, **17**, 1255–1259.

Lawson, C.L. and Hansen, R.J. (eds.), 1995, *Solving Least Squares Problems*. Englewood Cliffs, NJ: Prentice-Hall.

Le Cun, Y., Denker, J.S. and Solla, S.A., 1990, Optimal brain damage. In: Touretsky, D.S. (ed.) 1990, *Advances in Neural Information*. San Mateo: Morgan Kaufmann, 598–605.

Leachtenauer, J.C., 1977, Optical power spectrum analysis: scale and resolution effects. *Photogrammetric Engineering and Remote Sensing*, **43**, 1117–1125.

Leberl, F.W., 1990, *Radargrammetric Image Processing*, Dedham, MA: Artech House.

Lee, C. and Landgrebe, P.A., 1993, Decision boundary feature extraction for non-parametric classifiers. *IEEE Transactions on Systems, Man, and Cybernetics* , **23**, 433–444.

Lee, J.S., 1983a, Digital image smoothing and the sigma filter. *Computer Graphics, Vision and Image Processing*, **17**, 24–32.

Lee, J.S., 1983b, A simple speckle smoothing algorithm for synthetic aperture radar images. *IEEE Transactions on Systems, Man and Cybernetics*, **13**, 85–89.

Lee, J.S., Jurkevich, I., Dewaele, P., Wambacq, P. and Oosterlinck, A., 1994. Speckle filtering of synthetic aperture radar images: A review. *Remote Sensing Reviews*, **8**, 313–340.

Lee, T. and Richards, J.A., 1985, A low-cost classifier for multitemporal applications. *International Journal of Remote Sensing*, **6**, 1405–1418.

Lee, T., Richards, J.A. and Swain, P.H., 1987, Probabilistic and evidential approaches for multisource data analysis. *IEEE Transactions on Geoscience and Remote Sensing*, **25**, 283–293.

Lefsky, M.A., Cohen, W.B., Parker, G.G. and Harding, D.J., 2002, Lidar remote sensing for ecological studies. *BioScience*, **52**, 19–30.

Lei, Q., Henkel, J., Frei, M., Mehl, H., Lõrchner, G. and Bodechtel, J., 1996, Radiometric noise correction of panchromatic high resolution data of MOMS-02. *Proceedings of the MOMS Symposium, Cologne, Germany, 5–7 July, 1995*. Paris: European Association of Remote Sensing Laboratories (EARSeL), 303–313.

Leprieur, C., Kerr, Y.H. and Pichon, J.M., 1996, Critical assessment of vegetation indices from AVHRR in a semi-arid environment. *International Journal of Remote Sensing*, **17**, 2549–2563.

Leung, A.K., Chau, F. and Gao, J., 1998, A review on applications of wavelet transform techniques in chemical analysis: 1989–1997. *Chemometrics and Intelligent Laboratory Systems*, **43**, 165–184.

Levin, S.A., Concepts of scale at the local level. In: Ehleringer, J.R. and Field, C.B. (eds.), 1991, 7–19.

Li, M., Daels, L. and Antrop, M., 1996, Lambertian and Minnaert relation simulation for topographic normalization. *Proceedings of the 11th Thematic Conference and Workshops on Applied Geologic Remote Sensing*, Las Vegas, Nevada, 27–29 February 1996, **2**, 133–141. Ann Arbor, MI: Environmental Research Institute of Michigan (ERIM).

Li, Z., 1993, Theoretical models of the accuracy of digital terrain models: An evaluation and some observations. *Photogrammetric Record*, **14**, 651–660.

Lillesand, T.M. and Keifer, R.W., 1994, *Remote Sensing and Image Interpretation* Third edition. New York: Wiley.

Lim, K., Treitz, P., Wulder, M., St-Onge, B. and Flood, M., 2003, LiDAR remote sensing of forest structure. *Progress in Physical Geography*, **27**, 88–106.

Lin, Q., Vesesky, J.F. and Zebker, H.A., 1994. Comparison of elevation derived from InSAR data with DEM over large relief terrain. *International Journal of Remote Sensing*, **15**, 1775–1790.

Lloyd, C.D. and Atkinson, P.M., 2002, Deriving DSMs from LiDAR data with kriging. *International Journal of Remote Sensing*, **23**, 2519–2524.

Lobo, A., Chic, O., and Casterad, A., 1996, Classification of Mediterranean crops with multi-sensor data: per-pixel versus per-object statistics and image segmentation. *International Journal of Remote Sensing*, **17**, 2385–2400

Logan, T.L. and Strahler, A.H., 1983, Optimal Landsat transforms for forest applications. *Proceedings of the Symposium on Machine Processing of Remotely Sensed Data 1983*. West Lafayette, IN: Purdue University, 146–153.

Loizides, M., 1996, *Wavelets Theory and Applications in Signal and Image Processing*. M.Sc. Thesis, University of Manchester Institute of Science and Technology, Manchester.

Lopes, A., Nezry, E., Touzi, R. and Laur, H., 1993, Structure detection and statistical adaptive speckle filtering in SAR images. *International Journal of Remote Sensing*, **14**, 1735–1758.

Lowitz, G.E., 1978, Stability and dimensionality of Karhunen-Loeve multispectral image expansions. *Pattern Recognition*, **10**, 359–363.

Lowman, P.D., Jr., 1994, Radar geology of the Canadian Shield: a 10-year review. *Canadian Journal of Remote Sensing*, **20**, 198–209.

Lunetta, R.S. and Elvidge, C.D. (eds.), 1998, *Remote Sensing Change Detection: Environmental Monitoring, Methods and Applications*. Chelsea, MI: Ann Arbor Press.

Lynn, P.A., 1982, *An Introduction to the Analysis and Processing of Signals. Second edition*. London: Macmillan.

Maas, H.-G., 2002, Methods for measuring height and planimetry discrepancies in airborne laserscanner data. *Photogrammetric Engineering and Remote Sensing*, **68**, 933–940.

Madsen, S.N. and Zebker, H.A., 1998, Synthetic aperture radar interferometry: Principles and applications. In: Henderson, F.M. and Lewis, A.J. (eds.), *Manual of Remote Sensing, Volume 2, Principles and Applications of Imaging Radar*. New York: John Wiley and Sons.

Madsen, S.N., Zebker, H.A. and Martin, J., 1993, Topographic mapping using radar interferometry: processing techniques. *IEEE Transactions on Geoscience and Remote Sensing*, **31**, 303–306.

Maguire, D.J., Goodchild, M.F. and Rhind, D.W. (eds.), 1991, *Geographic Information Systems – Principles and Applications*. Harlow, Essex: Longmans Scientific and Technical.

Mallat, S., 1998, *A Wavelet Tour of Signal Processing*. San Diego: Academic Press.

Man, K.F., Tang, K.S. and Kwong, 1999, *Genetic Algorithms: Concepts and Designs*. London: Springer-Verlag.

Mannan, B., Roy, J. and Ray, A.K., 1998, Fuzzy ARTMAP supervised classification of multi-spectral remotely-sensed images. *International Journal of Remote Sensing*, **19**, 767–774.

Marceau, J., Howarth, P.J., Dubois, J.M. and Gratton, D.J., 1990, Evaluation of the grey-level co-occurrence matrix method for land cover classification using SPOT imagery. *IEEE Transactions on Geoscience and Remote Sensing*, **28**, 513–519.

Markham, B.L. and Barker, J.L., 1987, Thematic Mapper bandpass solar exoatmospheric irradiances. *International Journal of Remote Sensing*, **8**, 517–523.

Markham, B.L., Halthore, R.N. and Goetz, S.J., 1992, Surface reflectance retrieval from satellite and aircraft sensors: result of sensor and algorithm comparisons during FIFE. *Journal of Geophysical Research*, **97**, 18785–18795.

Marr, D. and Hildreth, 1980, Theory of edge detection. *Proceedings of the Royal Society of London*, Series B **209**, 187–217.

Martin, F.J. and Turner, R.W., 1993, SAR speckle reduction by weighted filtering. *International Journal of Remote Sensing*, **14**, 1759–1774.

Masek, J.G., Honzak, M., Goward, S.N., Liu, P. and Pak, E., 2001, Landsat-7 ETM+ as an observatory for land cover: Initial radiometric and geometric comparisons with Landsat-5 Thematic Mapper. *Remote Sensing of Environment*, **78**, 118–130.

Maselli, F., Conese, C., de Filippis, T. and Norcini, S., 1995, Estimation of forest parameters through fuzzy classification. *IEEE Transactions on Geoscience and Remote Sensing*, **33**, 77–84.

Maselli, F., Conese, C., de Filippis, T. and Romani, M., 1995, Integration of ancillary data into a maximum likelihood classification with nonparametric priors. *ISPRS Journal of Photogrammetry and Remote Sensing*, **50**, 2–11.

Maselli, F., Rodolfi, A. and Conese, C., 1996, Fuzzy classification of spatially degraded TM data for the estimation of sub-pixel components. *International Journal of Remote Sensing*, **17**, 537–551.

Mason, D.C., Corr, D.G., Cross, A., Hoggs, D.C., Lawrence, D.H., Petrou, M. and Tailor, A.M., 1988, The use of digital map data in the segmentation and classification of remotely-sensed images. *International Journal of Remote Sensing*, **9**, 195–205.

Massonet, D., 1993, Geoscientific applications at CNES. In: Schreier, G. (1993a) (ed.), 397–415.

Massonet, D., 2000, Elevation modelling and displacement mapping using radar interferometry. In: Meyers, R.A. (ed.), *Encyclopaedia of Analytical Chemistry*. Chichester: John Wiley and Sons, 8533–8543.

Massonet, D. and Feigl, K., 1998, Radar interferometry and its application to changes in the Earth's surface. *Reviews of Geophysics*, **36**, 441–500.

Massonnet, D., Rossi, M., Carmona, C., Adragna, F., Peltzer, G., Feigl, K. and Rabaute, T., 1993. The displacement field of the Landers earthquake mapped by radar interferometry. *Nature*, **364**, 138–142.

Masuoka, E., Fleig, A., Wolfe, R.E. and Patt, F., 1998, Key characteristics of MODIS data products. *IEEE Transactions on Geoscience and Remote Sensing*, **36**, 1313–1323.

Mather, P.M., 1976, *Computational Methods of Multivariate Analysis in Physical Geography*. Chichester: Wiley.

Mather, P.M., 1985. A computationally-efficient maximum likelihood classifier employing prior probabilities for remotely-sensed data. *International Journal of Remote Sensing*, **6**, 369–376.

Mather, P.M., 1991, *Computer Applications in Geography*, Chichester, John Wiley and Sons.

Mather, P.M. (ed.), 1994, *TERRA 2: Understanding the Terrestrial Environment – Remote Sensing Data Systems and Networks*. Chichester: John Wiley and Sons.

Mather, P.M., 1995, Map-image registration using least-squares polynomials. *International Journal of Geographical Information Systems*, **9**, 543–554.

Mather, P.M., 1999, Land cover classification revisited. In P.M. Atkinson and B.J. Tate (eds.) *Advances in Remote Sensing and GIS Analysis*. Chichester: John Wiley and Sons, 7–16.

Mather, P.M., Tso, B. and Koch, M., 1998, An evaluation of Landsat TM spectral data and SAR-derived textural information for lithological discrimination in the Red Sea Hills, Sudan. *International Journal of Remote Sensing*, **19**, 587–604.

Maxwell, E.L., 1976, Multivariate system analysis of multispectral images. *Photogrammetric Engineering and Remote Sensing*, **42**, 1173–1186.

McDonnell, M.J., 1981, Box filtering techniques. *Computer Graphics and Image Processing*, **17**, 65–70.

Meadows, P., 1995, The calibration of ERS-1 synthetic aperture radar images using UK-PAF imagery. In: Askne, J. (ed.) (1995) 423–430.

Menenti, M., Azzali, S., Verhoef, W. and van Swol, R., 1993, Mapping agroecological zones and time-lag in vegetation growth by means of Fourier analysis of time series of NDVI images. *Advances in Space Research*, **13**, 233–237.

Milovich, J.A., Frulla, L.A. and Gagliardini, D.A., 1995, Environmental contribution to the atmospheric correction for Landsat-MSS images. *International Journal of Remote Sensing*, 16, 2515–2537.

Milton, E.J., 1986, Principles of field spectroscopy and its role in remote sensing. In: Bradbury, P.A. and Rollin, E.A. (1986) (eds.), *Ground Truth for Remote Sensing. Proceedings of a Remote Sensing Workshop*, Department of Geography, The University of Nottingham, May 1986. Remote Sensing Society, Nottingham, 14–39.

Misra, P.N. and Wheeler, S.G., 1978, Crop classification with Landsat multispectral Scanner data. *Pattern Recognition*, **10**, 1–13.

Mitchell, M., 1996, *An Introduction to Genetic Algorithms*. Cambridge, MA: MIT Press.

Miura, T., Huete, A.R. and Yoshioka, H., 2000, Evaluation of sensor calibration uncertainties on vegetation indices for MODIS. *IEEE Transactions on Geoscience and Remote Sensing*, **38**, 1399–1409.

Mohr, J.J. and Madsen, S.N., 2001, Geometric calibration of ERS satellite SAR images. *IEEE Transactions on Geoscience and Remote Sensing*, **39**, 842–850.

Moik, J., 1980, *Digital Processing of Remotely Sensed Images*. NASA Special Publication 431. Washington, D.C.: NASA:.

Mojsilovic, A., Popovic, M.V. and Rackov, D.M., 2000, On the selection of an optimal wavelet basis for texture characterization. *IEEE Transactions on Image Processing*, **9**, 2043–2050.

Monte Guarnieri, A., Prati, C., Rocca, F. and Desnos, Y.-L., 1998, Wide baseline interferometry with very low resolution SAR systems. *EUSAR'98: European Conference on Synthetic Aperture Radar*. Friedrichshafen, Germany. Berlin: VDE-Verlag GmbH. 361–364.

Montserud, R.A. and Leamans, R., 1992, Comparing global vegetation maps with the kappa statistic. *Ecological Modelling*, **62**, 275–293.

Moon, W.M., Li, B., Singhroy, V., So, C.S. and Yamaguchi, Y., 1994, Notes on JERS-1 SAR data characteristics for geological applications. *Canadian Journal of Remote Sensing*, **20**, 329–332.

Moore, G.K. and Waltz, F.A., 1983, Objective procedures for lineament enhancement and extraction. *Photogrammetric Engineering and Remote Sensing*, **49**, 641–647.

Morad, M., Chalmers, A.I. and O'Regan, P.R., 1996, The role of root-mean-square error in the geo-transformation of images in GIS. *International Journal of Geographical Information Systems*, **10**, 347–353.

Moran, M.S., Bryant, R., Thome, K., Ni, W., Nouvellon, Y., Gonzales-Dugo, M.P., Qi, J. and Clarke, T.R., 2001, A refined empirical line approach for reflectance factor retrieval from Landsat-5 TM and Landsat-7 ETM+. *Remote Sensing of Environment*, **78**, 71–82.

Moran, M.S., Jackson, R.D., Clarke, T.R., Qi, J., Cabot, F., Thome, K.J. and Markham, B.L., 1995, Reflectance factor retrieval from Landsat TM and SPOT HRV data for bright and dark targets. *Remote Sensing of Environment*, **52**, 218–230.

Moran, M.S., Jackson, R.D., Hart, G.F., Slater, P.N., Bartell, R.J., Biggar, S.F., Gellman, D.I. and Santer, R.P., 1990, Obtaining surface reflectance factors from atmospheric and view angle corrected SPOT-1 HRV data. *Remote Sensing of Environment*, **32**, 203–214.

Moran, M.S., Jackson, R.D., Slater, P.N. and Teillet, P.M., 1992, Evaluation of simplified procedures for retrieval of land surface reflectance factors from satellite sensor output. *Remote Sensing of Environment*, **41**, 169–184.

Moreno, J.E. and Melia, J., 1993, A method for accurate geometric correction of NOAA AVHRR HRPT data. *IEEE Transactions on Geoscience and Remote Sensing*, **31**, 204–226.

Motrena, P. and Rebordão, J.M., 1998, Invariant models for ground control points in high-resolution images. *International Journal of Remote Sensing*, **19**, 1359–1375.

Mouginis-Mark, P.J., 1995, Analysis of volcanic hazards using radar interferometry. *Earth Observation Quarterly*, **47**, 6–10.

Muasher, M.J. and Landgrebe, D.A., 1984, A binary tree feature selection technique for limited training sample size. *Remote Sensing of Environment*, **16**, 183–194.

Muchoney, D., Borak, J., Chi, H., Friedl, M.A., Gopal, S., Hodges, J., Morrow, N. and Strahler, A.H., 2000, Application of MODIS global supervised classification model to vegetation and land cover mapping in Central America. *International Journal of Remote Sensing*, **21**, 1115–1138.

Muchoney, D. and Strahler, A.H., 2002, Pixel- and site-based calibration and validation methods for evaluating supervised classification of remotely sensed data. *Remote Sensing of Environment*, **81**, 290–299.

Mulder, N.J., 1980, A view on digital image processing. *ITC Journal*, **1980–1983**, 452–476.

Muller, E., 1993, Evaluation and correction of angular anisotropic effects in multidate SPOT and Thematic Mapper data. *Remote Sensing of Environment*, **45**, 295–309.

Mustard, J.F. and Sunshine, J.M., 1999, Spectral analysis for Earth science: Investigations using remote sensing data. In: Rencz, R.N. (ed.) (1999), 251–306.

Myneni, R.B., Hall, F.G., Sellers, P.J. and Marshak, A.L., 1995, The interpretation of spectral vegetation indices. *IEEE Transactions on Geoscience and Remote Sensing*, **33**, 481–486.

Nagao, M. and Matsuyama, T., 1979, Edge preserving smoothing. *Computer Graphics and Image Processing*, **9**, 394–407.

Nalbant, S.S. and Alptekin, Õ., 1995, The use of Landsat Thematic Mapper imagery for analysing lithology and structure of Korucu-Duìla area in Western Turkey. *International Journal of Remote Sensing*, **16**, 2357–2374.

Narendra, P.M. and Fukunaga, K., 1977, A branch and bound algorithm for feature subset selection. *IEEE Transactions on Computers*, **26**, 917–921.

NASA, 1988, *Earth Observing System Instrument Panel Report, Volume IIf, Synthetic Aperture Radar*. Washington, D.C.: NASA.

Nasrabadi, N.M., Ibikunle, J.O. and King, R.A., 1984, A new line detection technique in noisy images. *Proceedings of the Tenth Anniversary International Conference of the Remote Sensing Society*. Nottingham: The Remote Sensing Society, 237–246.

Neilsen, A.A., 1994. *Analysis of Regularly and Irregularly Sampled Spatial, Multivariate and Multitemporal Data*. Ph.D. Thesis, Technical University of Denmark, Lyngsby, Denmark.

Novak, K., 1992, Rectification of digital imagery. *Photogrammetric Engineering and Remote Sensing*, **58**, 339–344.

O'Leary, D.W., Friedmann, J.D. and Pohn, H.A., 1976, Lineament, linear, lineation: some proposed new standards for old terms. *Bulletin of the Geological Society of America*, **87**, 1463–1469.

Olsson, H., 1993, Regression functions for multitemporal relative calibrations of Thematic Mapper data over boreal forests. *Remote Sensing of Environment*, **46**, 89–102.

Olsson, H., 1995, Radiometric calibration of thematic mapper data for forest change detection. *International Journal of Remote Sensing*, **16**, 81–96.

Olsson, L. and Ekhlund, L., 1994, Fourier series for analysis of temporal sequences of satellite sensor imagery. *International Journal of Remote Sensing*, **15**, 3735–3741.

O'Neill, M.A. and Dowman, I.J., 1993, A simulation study of the ASTER sensor using a versatile general purpose rigid sensor modelling system. *International Journal of Remote Sensing*, **14**, 565–585.

Ormsby, J., 1992, Evaluation of natural and man-made features using Landsat TM data. *International Journal of Remote Sensing*, **13**, 303–318.

Overheim, R.D. and Wagner, D.L., 1982, *Light and Color*. New York: Wiley.

Pal, M. and Mather, P.M., 2003, An assessment of the effectiveness of decision tree methods for land cover classification. *Remote Sensing of Environment*, **86**, 554–565.

Palà, V. and Pons, X., 1995, Incorporation of relief in polynomial-based geometric corrections. *Photogrammetric Engineering and Remote Sensing*, **61**, 935–944.

Pan, J.-J. and Chang, C.-I., 1992, Destriping of Landsat MSS images by filtering techniques. *Photogrammetric Engineering and Remote Sensing*, **58**, 1417–1423.

Paola, J. and Schowengerdt, R.A., 1995a, A review and analysis of backpropagation neural networks for classification of remotely-sensed multi-spectral imagery. *International Journal of Remote Sensing*, **16**, 3033–3058.

Paola, J. and Schowengerdt, R.A., 1995b, A detailed comparison of backpropagation neural networks and maximum-likelihood classifiers for urban land use classification. *IEEE Transactions on Geoscience and Remote Sensing*, **33**, 981–996.

Paola, J. and Schowengerdt, R.A., 1997, The effect of neural network structure on multispectral land-use/land-cover classification. *Photogrammetric Engineering and Remote Sensing*, **63**, 535–544.

Parlow, E. (ed.), 1996a, Progress in environmental remote sensing research and applications. *Proceedings of the 15th EARSeL Symposium, Basle, Switzerland, 4–6 September 1996*. Rotterdam: A.A. Balkema.

Parlow, E., 1996b, Correction of terrain controlled illumination effects in satellite data. In: Parlow, E. (ed.) (1996a) 139–145.

Pavlidis, T., 1982, *Algorithms for Graphics and Image Processing*. Berlin: Springer-Verlag.

Peddle, D.R., 1993, An empirical comparison of evidential reasoning, linear discriminant analysis, and maximum likelihood algorithms for land cover classification. *Canadian Journal of Remote Sensing*, **19**, 31–44.

Peddle, D.R., 1995a, Knowledge foundation for supervised evidential classification. *Photogrammetric Engineering and Remote Sensing*, **61**, 409–417.

Peddle, D.R., 1995b, Mercury⊕: an evidential reasoning classifier. *Computers and Geosciences*, **21**, 1163–1176.

Peddle, D.R., Foody, G.M., Zhang, A., Franklin, S.E. and LeDrew, E.F., 1994, Multi-source image classification II: an empirical comparison of evidential reasoning and neural network approaches. *Canadian Journal of Remote Sensing*, **20**, 396–407.

Peleg, S., 1980, A new probabilistic relaxation scheme. *IEEE Transactions on Pattern Analysis and Machine Intelligence*, **2**, 362–369

Peli, T. and Malah, D., 1982, A study of edge-detection algorithms. *Computer Graphics and Image Processing*, **20**, 1–21.

Perry, C.R. and Lautenschlager, L.F., 1984, Functional equivalence of spectral vegetation indices. *Remote Sensing of Environment*, **14**, 169–182.

Perski, Z. and Jura, D., 1999, ERS SAR interferometry for land subsidence detection in coal mining areas (amended version). *ESA Earth Observation Quarterly*, **64**, Available at URL: http://esapub.esrin.esa.it/eoq/eoq64/reprint63.pdf.

Philpott, W.D., 1991, The derivative ratio algorithm: Avoiding atmospheric effects in remote sensing. *IEEE Transactions on Geoscience and Remote Sensing*, **29**, 350–357.

Pinter, P.J., Jackson, R.D., Idso, S.B. and Reginato, R.J., 1983, Diurnal patterns of wheat spectral reflectance. *IEEE Transactions on Geoscience and Remote Sensing*, **21**, 156–163.

Pinty, B., Leprieur, C. and Verstraete, M.M., 1993, Towards a quantitative interpretation of vegetation indices. Part I: Biophysical canopy properties and classical indices. *Remote Sensing Reviews*, **7**, 127–150.

Pitas, I., 1993, *Digital Image Processing Algorithms*. New York: Prentice Hall.

Pohl, C., and van Genderen, J.L., 1998, Multisensor image fusion in remote sensing: concepts, methods and application. *International Journal of Remote Sensing*, **19**, 823–854.

Popp, T., 1995, Correcting atmospheric masking to retrieve the spectral albedo of land surfaces from satellite measurements. *International Journal of Remote Sensing*, **16**, 3843–3508.

Prasad, L. and Iyengar, S.S., 1997, *Wavelet Analysis with Applications to Image Processing*. Boca Raton: CRC Press.

Pratt, W.K., 1978, *Digital Image Processing*. New York: Wiley.

Premelatha, M., 2001, *Quality Assessment of Interferometrically Derived Digital Elevation Models*. Ph.D. Thesis, University of Nottingham, Nottingham.

Press, W.A., Teukolsky, S.A., Vetterling, W.T. and Flannery, B.P., 1992. *Numerical Recipes in Fortran,* Second edition. Cambridge: Cambridge University Press.

Price, J.C., 1987, Calibration of satellite radiometers and the comparison of vegetation indices. *Remote Sensing of Environment*, **21**, 15–27.

Price, J.C., 1988, An update on visible and near infrared calibration of satellite instruments. *Remote Sensing of Environment*, **24**, 419–422.

Price, J.C., 1994, How unique are spectral signatures? *Remote Sensing of Environment*, **49**, 181–186.

Proy, C., Tanré, D. and Deschamps, P.Y., 1989, Evaluation of topographic effects in remotely sensed data. *Remote Sensing of Environment*, **30**, 21–32.

Quegan, S. and Rhodes, I., 1994, Relating polarimetric SAR data to surface properties – the MAC-Europe experiment. In: Mather, P.M. (ed.) (1994), 159–174.

Quegan, S. and Wright, A., 1984, Automatic segmentation techniques for satellite-borne synthetic aperture radar (SAR) images. *Proceedings of the Tenth Anniversary Conference of the Remote Sensing Society*. Nottingham: The Remote Sensing Society, 161–167.

Quinlan, J.R., 1993, *C4.5: Algorithm for Machine Learning*. San Mateo: Morgan Kaufmann.

Rabus, B., Eineder, M., Roth, A. and Bamler, R., 2003, The Shuttle Radar Topographic Mission – a new class of digital elevation models acquired by spaceborne radar. *ISPRS Journal of Photogrammetry and Remote Sensing,* **57**, 241–262.

Rabus, B. and Fatland, D.R., 2000, Comparison of SAR-interferometric and surveyed velocities on a mountain glacier: Black Rapids Glacier. *Journal of Glaciology*, **46**, 119–128.

Ramirez, R.W., 1985, *The Fast Fourier Transform: Fundamentals and Concepts*. Englewood Cliffs, NJ: Prentice-Hall.

Ranchin, T. and Wald, L., 1993, The wavelet transform for the analysis of remotely sensed images. *International Journal of Remote Sensing*, **14**, 615–619.

Ranson, K.J., Biehl, L.L. and Bauer, M.E., 1985, Variation in spectral response of soybeans with respect to illumination, view and canopy geometry. *International Journal of Remote Sensing*, **6**, 1827–1842.

Rao, C.R.N. and Chen, J., 1996, Post-launch calibration of the visible and near-infrared channels on the advanced very high resolution radiometer on the NOAA-14 spacecraft. *International Journal of Remote Sensing,* **17**, 2743–2747.

Rao, K.S. and Rao, Y.S., 1995, Frequency dependence of polarisation phase difference. *International Journal of Remote Sensing*, **16**, 3605–3617.

Rast, M., 1999, Special issue: ESA medium resolution imaging spectrometer (MERIS). *International Journal of Remote Sensing*, **20**, 1677–1927.

Ray, T.W. and Murray, B.C., 1996, Nonlinear spectral mixing in desert vegetation. *Remote Sensing of Environment*, **55,** 59–74.

Rayner, J.N., 1971, *An Introduction to Spectral Analysis*. London: Pion Press.

Reddy, B.S. and Chatterji, B.N., 1996, An FFT-based technique for translation, rotation and scale-invariant image registration. *IEEE Transactions on Image Processing*, **5**, 1266–1271.

Rees, W.G. and Satchell, M.J.F., 1997, The effect of median filtering on synthetic aperture radar images. *International Journal of Remote Sensing*, **18**, 2887–2893.

Rees, W.G., 1990, *Physical Principles of Remote Sensing*. Cambridge: Cambridge University Press.

Reigber, C., Xia, Y., Kaufmann, H., Massman, F.H., Timmen, L., Bodechtel, J. and Frei, M., 1996, Impact of precise orbits on SAR interferometry. *FRINGE '96: ESA Workshop on Applications of ERS SAR Interferometry.* Zurich, Switzerland. ESA SP-406. Paris: European Space Agency. Available at: http://www.geo.unizh.ch/rsl/fringe96/papers/reigber-et-al/. Date visited: 17 January 2003.

Rencz, A.N. (ed.), 1999, *Manual of Remote Sensing, Volume 3, Remote Sensing for the Earth Sciences*, Third Edition. New York: John Wiley and Sons.

Riaño, D., Chuvieco, E., Salas, J. and Aguado, I., 2003, Assessment of different topographic corrections on Landsat-TM data for mapping vegetation types. *IEEE Transactions on Geoscience and Remote Sensing*, **41**, 1056–1061.

Riazanoff, S., Cervelle, B. and Chorowicz, J., 1990, Parametrizable skeletonisation of binary and multi-level images. *Pattern Recognition Letters*, **11**, 25–33.

Ribed, P.S. and Lopez, A.M., 1995, Monitoring burnt areas by principal components analysis of multitemporal TM data. *International Journal of Remote Sensing*, **16**, 1577–1587.

Richards, J.A., 1993, *Remote Sensing Digital Image Analysis: An Introduction*. Second Edition. Berlin: Springer.

Richardson, A.J. and Wiegand, C.L., 1977, Distinguishing vegetation from soil background information.

Photogrammetric Engineering and Remote Sensing, **43**, 1541–1552.

Richter, R., 1996, A spatially adaptive fast atmospheric correction algorithm. *International Journal of Remote Sensing*, **17**, 1201–1214.

Rignot, E.J.M., Forster, R.R. and Isacks, B.L., 1996, Mapping of glacial motion and surface topography of Hielo Patagonico Norte, Chile, using satellite SAR L-band interferometry data. *Annals of Glaciology*, **23**, 209–216.

Roberts, D., Adams, J.B. and Smith, M.O., 1993, Discriminating green vegetation, non-photosynthetic vegetation and soils in AVIRIS data. *Remote Sensing of Environment*, **44**, 1–25.

Roger, R.E., 1996, Principal components transform with simple, automatic noise adjustment. *International Journal of Remote Sensing*, **17**, 2719–2727.

Rondeaux, G., 1995, Vegetation monitoring by remote sensing: a review of biophysical indices. *Photo-Interpretation*, No. 1995/3, 197–216.

Rondeaux, G., Steven, M.D. and Baret, F., 1996, Optimisation of Soil-Adjusted Vegetation Indices. *Remote Sensing of Environment*, **55**, 95–107.

Rosen, P.A., Hensley, S., Zebker, H.A., Webb, F.H. and Fielding, E.J., 1996, Surface deformation and coherence measurements of Kilauea Volcano, Hawaii, from SIR-C radar interferometry. *Journal of Geophysical Research*, **101**, 101–123.

Rosen, P.A., S. Hensley, Joughin, I., Li, F., Madsen, S., Rodriguez, E. and Goldstein, R.M., 2000, Synthetic aperture radar interferometry. *Proceedings of the IEEE*, **88**, 333–382.

Rosenfeld, A., 1976, Iterative methods in image analysis. *Pattern Recognition*, **10**, 181–187.

Rosenfeld, A. and Kak, A.C., 1982, *Digital Picture Processing*. Second Edition, Volume 1. New York: Academic Press.

Rothery, D.A. and Hunt, G.A., 1990, A simple way to perform decorrelation stretching and related techniques on menu-driven image processing systems. *International Journal of Remote Sensing*, **11**, 133–137.

Rowan, L.C., Wetlaufer, P., Goetz, A.F.H., Billingsley, F. and Stewart, J., 1974, Discrimination of rock types and detection of hydrothermally-altered areas in south-central Nevada by the use of computer-enhanced ERTS images. *U.S. Geological Survey, Professional Paper 883*, Washington, D.C.: U.S. Government Printing Office.

Running, S.W., Justice, C.O., Salomonson, V.V., Strahler, A.H., Huete, A.R., Muller, J.-P., Vanderbilt, V., Wan, Z.M., Teillet, P.M. and Carneggie, D., 1994, Terrestrial remote sensing science and algorithms planned for EOS/MODIS. *International Journal of Remote Sensing*, **15**, 3587–3620.

Samet, H., 1990, *The Design and Analysis of Spatial Data Structures*. Reading, MA: Addison-Wesley.

Sammon, J.W., Jr., 1969, A nonlinear mapping algorithm for data structure analysis. *IEEE Transactions on Computers*, **18**, 401–409.

Sansosti, E., Lanari, R., Fornaro, G., Franceschetti, G., Tesauro, M., Pugulisi, G. and Coltelli, M., 1999, Digital elevation model generation using ascending and descending ERS-1/ERS-2 tandem data. *International Journal of Remote Sensing*, **20**, 1527–1547.

Sardy, S., Percival, D.B., Bruce, A.G., Gao, H.-Y. and Stuetzle, W., 1999, Wavelet shrinkage for unequally spaced data. *Statistics and Computing*, **9**, 65–75.

Savitsky, A. and Golay, M.J.E., 1964. Smoothing and differentiation of data by simplified least squares procedures. *Analytical Chemistry*, **36**, 1627–1639.

Sawter, R., Deuze, J.L., Devaux, G., Vermote, E., Guyot, G., Tu, X.F., Verbrugge, M. and Leroy, M., 1991, SPOT calibration on the test site at La Crau, France. *Cinquième Colloque International Mesures Physiques et Signatures on Telédétection, Courcheval, 14–18 January 1991*. ESA-SP-319. Paris: European Space Agency, 77–80.

Schaale, M. and Furrer, R., 1995, Land surface classification by neural networks. *International Journal of Remote Sensing*, **16**, 3003–3031.

Schetselaar, E.M., 1998, Fusion by the IHS transform: should we use cylindrical or spherical coordinates? *International Journal of Remote Sensing*, **19**, 759–765.

Schistad, A.H. and Jain, A.K., 1992, Texture analysis in the presence of speckle noise. *Proceedings of the International Symposium on Geoscience and Remote Sensing (IGARSS'92), May 1992, Houston, Texas*. New York: IEEE Press, 884–886.

Schmidt, K.S. and Skidmore, A.K., 2004, Smoothing vegetation spectra with wavelets. *International Journal of Remote Sensing*, **25**, 1167–1184.

Schott, J.R. and Volchock, W.J., 1985, Thematic mapper thermal infrared calibration. *Photogrammetric Engineering and Remote Sensing*, **51**, 1351–1357.

Schotten, C.G.J., van Rooy, W.W.L. and Janssen, L.L.F., 1995, Assessment of the capabilities of multi-temporal ERS-1 SAR data to discriminate between agricultural crops. *International Journal of Remote Sensing*, **16**, 2619–2637.

Schowengerdt, R.A., 1997, *Remote Sensing: Models and Methods for Image Processing*. Second Edition. San Diego: Academic Press.

Schreier, G. (ed.), 1993a, *SAR Geocoding: Data and Systems*. Karlsruhe, Germany: Herbert Wichmann.

Schreier, G., 1993b, Geometrical properties of SAR images. In: Schreier, G., (ed.) (1993a), 103–134.

Seftor, J.L. and Larch, D., 1995, The use of the genetic algorithm to optimise rule-based classifiers for land cover categorization. *Canadian Journal of Remote Sensing*, **21**, 412–420.

Sellers, P., 1989, Vegetation-canopy reflectance and biophysical properties. In: Asrar, G. (ed.) (1989), 297–335.

Settle, J.J. and Campbell, N., 1998, On the errors of two estimators of sub-pixel fractional cover when mixing is linear. *IEEE Transactions on Geoscience and Remote Sensing*, **36**, 163–170.

Settle, J.J. and Drake, N.A., 1993, Linear mixing and the estimation of ground cover proportions. *International Journal of Remote Sensing*, **14**, 1159–1177.

Shandley, J., Franklin, J. and White, T., 1996, Testing the Woodcock-Harward image segmentation algorithm in an area of southern California chaparral and woodland vegetation. *International Journal of Remote Sensing*, **17**, 983–1004.

Shafri, H., 2003, *An Assessment of the Potential of Wavelet-based De-noising in the Analysis of Remotely Sensed Data*. Ph.D. Thesis, School of Geography, The University of Nottingham.

Shaw, R., Sowers, L. and Sanchez, E., 1982, A comparative study of linear and nonlinear edge finding techniques for Landsat multispectral data. In Richason, B.F., Jr., (ed.), *Proceedings of the Pecora VII Symposium, Sioux Falls, South Dakota* Falls Church, VA: American Society of Photogrammetry, 529–542.

Shemer, L., Marom, M. and Markman, D., 1993, Estimates of currents in the nearshore ocean region using interferometric synthetic aperture radar. *Journal of Geophysical Research*, **98**, 7001–7010.

Shih, T.-Y., 1995, The reversibility of six geometric color spaces. *Photogrammetric Engineering and Remote Sensing*, **61**, 1223–1232.

Shimabukuro, Y.E. and Smith, J.A., 1991, The least-squares mixing models to generate fraction images derived from remote sensing multispectral data. *IEEE Transactions on Geoscience and Remote Sensing*, **29**, 16–21.

Shipman, H. and Adams, J.B., 1987, Detectability of minerals on desert alluvial fans using reflectance spectra. *Journal of Geophysical Research*, **92**, 10931–10402.

Shlien, S., 1979, Geometric correction, registration and resampling of Landsat imagery. *Canadian Journal of Remote Sensing*, **5**, 74–87.

Sietsma, J. and Dow, R.J.F., 1988, Neural net pruning – why and how. *IEEE International Conference on Neural Networks*, New York: IEEE, I-325 to I-333.

Siljeströ̃m, P.A. and Moreno, A., 1995, Monitoring burnt areas by principal components analysis of multi-temporal TM data. *International Journal of Remote Sensing*, **16**, 1577–1587.

Siljeströ̃m, P.A., Moreno, A., Vikgren, G. and Cáceres, L.M., 1997, The application of selective principal components analysis (SPCA) to a Thematic Mapper (TM) image for the recognition of geomorphologic features configuration. *International Journal of Remote Sensing*, **18**, 3843–3852.

Simonett, D.S., 1983, (ed.) The development and principles of remote sensing. In Colwell, R.N. (1983) (ed.), 1–36.

Singh, A., 1984, Some clarifications about the pairwise divergence method in remote sensing. *International Journal of Remote Sensing*, **5**, 623–627.

Singleton, R.C., 1979a, Mixed radix fast Fourier transform. *Programs for Digital Signal Processing*. New York: IEEE Acoustics, Speech and Signal Processing Society, IEEE Press /Wiley. Section 1.4-1.

Singleton, R.C., 1979b, Two-dimensional mixed radix mass storage Fourier transform. *Programs for Digital Signal Processing*. New York: IEEE Acoustics, Speech and Signal Processing Society, IEEE Press /Wiley. Section 1.9-1.

Slater, P.N., 1980, *Remote Sensing: Optics and Optical Systems*. Reading, MA: Addison-Wesley.

Slater, P.N., Biggar, S.F., Holm, R.G., Jackson, R.D., Mao, Y., Moran, M.S., Palmer, M. and Yuan, B., 1987, Reflectance- and radiance-based methods for the in-flight calibration of multi-spectral sensors. *Remote Sensing of Environment*, **22**, 11–37.

Smith, A.R., 1978, Color gamut transform pairs. *Computer Graphics*, **12**, 12–19.

Smith, D.M., 1996, Speckle reduction and segmentation of synthetic aperture radar images. *International Journal of Remote Sensing*, **17**, 2043–2057.

Smith, G.M. and Curran, P., 1996, The signal-to-noise ratio (SNR) required for the estimation of foliar biochemical concentrations. *International Journal of Remote Sensing*, **17**, 1031–1058.

Smith, G.M. and Milton, E.J., 1999, The use of the empirical line method to calibrate remotely sensed data to reflectance. *International Journal of Remote Sensing*, **20**, 2653–2662.

Smith, J.A., Lin, T.L., and Ranson, K.J., 1980, The Lambertian assumption and Landsat data. *Photogrammetric Engineering and Remote Sensing*, **46**, 1183–1189.

Smith, L.C., 2002, Emerging applications of Interferometric Synthetic Aperture Radar (InSAR) in geomorphology and hydrology. *Annals of the Association of American Geographers*, **93**, 385–398.

Smith, M.O., Ustin, S.L., Adams, J.B. and Gillespie, A.R., 1990, Vegetation in deserts: I. Regional measure of abundance from multispectral images. *Remote Sensing of Environment*, **31**, 1–26.

Smits, P.C., Dellepiane, S.G. and Schowengerdt, R.A., 1999, Quality assessment of image classification algorithms for land-cover mapping: a review and a proposal for a cost-based approach. *International Journal of Remote Sensing*, **20**, 1461–1486.

Snyder, J.P., 1982, *Map Projections used by the U. S. Geological Survey*. United States Geological Survey Bulletin 1532. Washington, D.C.: U.S. Government Printing Office.

Soares, J.V., Rennó, C.D., Formaggio, A.R., Yanasse, C.C.F. and Frery, A.C., 1997, An investigation into the selection of texture features for crop discrimination using SAR imagery. *Remote Sensing of Environment*, **59**, 234–247.

Sohn, Y. and McCoy, R.M., 1997, Mapping desert shrub rangeland using spectral unmixing and modelling spectral mixtures with TM data. *Photogrammetric Engineering and Remote Sensing*, **63**, 707–716.

Solaas, G.A., 1994. *ERS-1 Interferometric Baseline Algorithm Verification*. Report ES-TN-DPE-OM-GS02. Paris: European Space Agency.

Song, C., Woodcock, C.E., Seto, K.C., Lenney, M.P. and Macomber, S.A., 2001. Classification and change detection using Landsat TM data: When and how to correct atmospheric effects? *Remote Sensing of Environment*, **75**, 230–244.

Spanner, M.A., Brass, J.A. and Peterson, D.L., 1984, Feature selection and the information content of Thematic Mapper simulator data for forest structural assessment. *IEEE Transactions on Geoscience and Remote Sensing*, **22**, 482–289.

Sparks, D.N., 1985, Half-normal plotting. In: Griffiths, P. and Hill, I.D. (1985) (eds.), *Applied Statistical Algorithms*, Chichester: Ellis-Horwood, 65–69.

Srinavasan, R., Cannon, M. and White, J., 1988, Landsat data destriping using power spectral filtering. *Optical Engineering*, **27**, 939–943.

Srinivasana, A. and Richards, J.A., 1990, Knowledge-based techniques for multisource classification. *International Journal of Remote Sensing*, **11**, 505–525.

Srokosz, M.A., 2000, Biological oceanography by remote sensing. In: Meyers, R.A. (Ed.), *Encyclopaedia of Analytical Chemistry*. Chichester: John Wiley and Sons, Ltd., 8506–8533.

Starck, J.-L., Murtagh, F. and Bijaou, A., 1998, *Image Processing and Data Analysis: The Multiscale Approach*. Cambridge: Cambridge University Press.

Steers, J.A., 1962, *An Introduction to the Study of Map Projections (13th Edition)*. London: University of London Press.

Stehman, S.V., 1997, Selecting and interpreting measures of thematic classification accuracy. *Remote Sensing of Environment*, **62**, 77–89.

Stehman, S.V., 1999, Basic probability sampling designs for thematic map accuracy assessment. *International Journal of Remote Sensing*, **20**, 2423–2441.

Stehman, S.V., 1999, Comparing thematic maps based on map value. *International Journal of Remote Sensing*, **20**, 2347–2366.

Steigler, S.E., 1978, *Dictionary of Earth Sciences*. London: Pan Books.

Steven, M.D., 1998, The sensitivity of the OSAVI vegetation index to observational parameters. *Remote Sensing of Environment*, **63**, 49–60.

Sties, M., Kruger, S., Mercer, J.B. and Schnick, S., 2000, Comparison of digital elevation data from airborne laser and interferometric SAR systems. *XIXth Congress of the ISPRS*. Amsterdam. 866–873.

Stimson, A., 1974, *Photometry and Radiometry for Engineers*. New York: Wiley.

Stimson, G.W., 1998, *Introduction to Airborne Radar*. Mendham, NJ: Scitech.

Stolz, R. and Mauser, W., 1996, A fuzzy approach for improving landcover classification by integrating remote sensing and GIS. In: Parlow, E. (ed.) (1996a), 33–41.

Story, M. and Congalton, R.G., 1986, Accuracy assessment: a user's perspective. *Photogrammetric Engineering and Remote Sensing*, **52**, 397–399.

Strahler, A.H., 1980, The use of prior probabilities in maximum likelihood classification of remotely sensed data. *Remote Sensing of Environment*, **10**, 135–163.

Strahler, A.H., Logan, T.L. and Bryant, N.A., 1978, Improving forest classification from Landsat by incorporating topographic information. *Proceedings of the 12th International Conference on Remote Sensing of Environment*, Environmental Research Institute of Michigan, Ann Arbor, Michigan, USA, 927–942.

Strang, G., 1994, Wavelets. *American Scientist*, **82**, 250–255.

Strang, G. and Truong, N., 1996, *Wavelets and Filter Banks*. Wellesley, MA: Wellesley-Cambridge Press.

Stromberg, W.D. and Farr, T.G., 1986, A Fourier-based textural feature extraction procedure. *IEEE Transactions on Geoscience and Remote Sensing*, **24**, 722–731.

Strozzi, T., Dammert, P., Wegmüller, U., Martinez, J.-M., Beaudoin, A., Askne, J. and Hallikainen, M., 1999, Forest mapping with SAR interferometry. *ESA Earth Observation Quarterly*, **62**, 17–20.

Suits, G.H., 1983, The nature of electromagnetic radiation. In: Colwell, R.N. (1983) (ed.), volume I, 37-60.

Swain, P.H. and King, R.C., 1973, Two effective feature selection criteria for multispectral remote sensing. *Proceedings of the 1st International Joint Conference on Pattern Recognition*, New York: IEEE 73 CHO821-9C, 536–540.

Switzer, P., Kowalik, W.S. and Lyon, R.J.P., 1981, Estimation of atmospheric path radiance by the covariance matrix method. *Photogrammetric Engineering and Remote Sensing*, **47**, 1469–1476.

Tanré, D., Deroo, C., Duhaut, P., Herman, M., Morcrette, J.J., Perbos, J. and Deschamps, P.Y., 1986, *Simulation of the Satellite Signal in the Solar Spectrum*. Laboratoire d'Optique Atmospherique, Universite des Sciences et Techniques de Lille, 59655 Villeneuve D'Ascq Cedex, France/Centre Spatiale de Toulouse, 31055 Toulouse Cedex, France.

Tao, C.V. and Hu, Y., 2001, A comprehensive study on the rational function model for photogrammetric processing. *Photogrammetric Engineering and Remote Sensing*, **67**, 1347–1357.

Taswell, C., 2000, The what, how and why of wavelet shrinkage denoising. *Computing in Science and Engineering*, **2**, 12–19.

Teillet, P.M., Barker, J.L., Markham, B.L., Irish, R.R., Fedosejevs, G.J.C. and Storey, J.C., 2001, Radiometric cross-calibration of the Landsat-7 ETM+ and Landsat-5 TM sensors based on tandem data sets.*Remote Sensing of Environment*, **78**, 39–54.

Teillet, P.M. and Fedosejevs, G, 1995, On the dark target approach to atmospheric correction of remotely-sensed data. *Canadian Journal of Remote Sensing*, **21**, 374–387.

Teillet, P.M., Guindon, B. and Goodenough, D.G., 1982, On the slope-aspect correction of multispectral data. *Canadian Journal of Remote Sensing*, **8**, 84–106.

Terhalle, U. and Bodechtel, J., 1986, Landsat TM data enhancement technique for mapping arid geomorphic features. *Proceedings of the ISPRS/Remote Sensing Society Symposium, Mapping from Modern Imagery, Edinburgh, September 1986*. Nottingham: The Remote Sensing Society, 725–729.

Theodossiou, E.I. and Dowman, I.J., 1990, Heighting accuracy of SPOT. *Photogrammetric Engineering and Remote Sensing*, **56**, 1643–1649.

Thomas, G., Hobbs, S.E. and Dufour, M., 1996, Woodland area estimation by spectral mixing: applying a goodness of fit solution method. *International Journal of Remote Sensing*, **17**, 291–301.

Thomas, I.L., 1980, Spatial postprocessing of spectrally-classified Landsat data. *Photogrammetric Engineering and Remote Sensing*, **46**, 1201–1206.

Thomas, R., Krabill, W., Frederick, E. and Jezek, K., 1995, Thickening of Jacobshavns Isbrae, West Greenland, measured by airborne laser altimetry. *Annals of Glaciology*, **21**, 259–262.

Thome, K., Markham, B., Barker, J., Slater, P. and Biggar, S., 1997, Radiometric calibration of Landsat. *Photogrammetric Engineering and Remote Sensing*, **63**, 853–858.

Thome, K.J., 2001, Absolute radiometric calibration of Landsat 7 ETM+ using the reflectance-based method. *Remote Sensing of Environment*, **78**, 27–38.

Thome, K.J., Gellman, D.I., Parada, R.J., Biggar, S.F., Slater, P.N. and Moran, M.S., 1993, Absolute radiometric calibration of Thematic Mapper. *SPIE Proceedings*, **600**, 2–8.

Thome, K., Markham, B., Barker, J., Slater, P. and Biggar, S., 1997, Radiometric Calibration of Landsat. *Photogrammetric Engineering and Remote Sensing*, **63**, 853–858.

Tidemann, J. and Nielsen, A.A., 1997, A simple neural network contextual classifier. In: Kanellopoulos, I. *et al* (eds.) (1997) 186–193.

Todd, S.W., Hoffer, R.M. and Milchunas, D.G., 1998, Biomass estimation on grazed and ungrazed rangelands using spectral indices. *International Journal of Remote Sensing*, **19**, 427–438.

Tokunaga, M. and Hara, S., 1996, DEM accuracy derived from ASTER data. *Proceedings of the 17th Asian Conference on Remote Sensing*, Colombo, Sri Lanka, 4–8 November 1996, J-7-1-J-7-5.

Tompkins, S., Mustard, J.F., Pieters, C.M. and Forsythe, D.W., 1997, Optimization of end members for spectral mixture analysis. *Remote Sensing of Environment*, **59**, 472–489.

Torres, J. and Infante, S.O., 2001, Wavelet analysis for the elimination of striping noise in satellite images. *Optical Engineering*, **40**, 1309–1314.

Toselli, F. and Bodechtel, J., (eds.) 1992, *Imaging Spectroscopy: Fundamentals and Prospective Applications*. Dordrecht: Kluwer.

Tou, J. and Gonzales, R., 1974, *Pattern Recognition Principles*. Reading, MA: Addison-Wesley.

Toutin, T., 1995, Multisource data integration with integrated and unified geometric modelling. In: Askne, J., 163–174.

Toutin, T., 2002, Three-dimensional topographic mapping with ASTER stereo data in rugged topography. *IEEE Transactions on Geoscience and Remote Sensing*, **40**, 2241–2247.

Toutin, T. and Cheng, P., 2002, QuickBird – A milestone for high-resolution mapping. *Earth Observation Magazine*, **11**(4). Available from URL: http://www.eomonline.com/Common/currentissues/Apr02/cheng.htm. Accessed 2 September 2003.

Townshend, J.G.H. and Skole, D.L., 1994, The global 1 km data set from the Advanced Very High Resolution Radiometer. In: Mather, P.M. (ed.) (1994), 75–82.

Townshend, J.R.G., 1980, The spatial resolving power of Earth resources satellites: a review. *NASA Technical Memorandum 82020*, Goddard Spaceflight Center, Greenbelt, Maryland. See also: *Progress in Physical Geography*, **5**, 33–35.

Townshend, J.R.G. and Harrison, A., 1984, Estimation of the spatial resolving power of the Thematic Mapper of Landsat-4. *Proceedings of the Tenth Anniversary International Conference of the Remote Sensing Society*, Remote Sensing Society, Nottingham, 67–72.

Townshend, J.R.G., Justice, C., Gurney, C. and McManus, J., 1992, The impact of misregistration on change detection. *IEEE Transactions on Geoscience and Remote Sensing*, **30**, 1054–1060.

Tozawa, Y., 1983, Fast geometric correction of NOAA AVHRR. *Proceedings of the Symposium on Machine Processing of Remotely-Sensed Data 1983*. Purdue University, West Lafayette, IN, 1983, 46–53.

Tsai, F. and Philpott, W.D., 1998, Derivative analysis of hyperspectral data. *Remote Sensing of Environment*, **66**, 41–51.

Tso, B., 1997, *The Investigation of Alternative Strategies for incorporating Spectral, Textural and Contextual Information in Remote Sensing Image Classification*. Ph.D. Thesis, Department of Geography, The University of Nottingham.

Tso, B. and Mather, P.M., 2001, *Classification Methods for Remotely Sensed Data*. London: Taylor and Francis.

Tucker, C.J., 1979, Red and photographic infrared linear combinations for monitoring vegetation. *Remote Sensing of Environment*, **10**, 127–150.

Tukey, J.W., 1977, *Exploratory Data Analysis*. Reading, MA: Addison-Wesley

Tyo, J.S., Konsolakis, A., Diersen, D.I. and Olsen, R.C., 2003, Principal-Components based display strategy for spectral imagery. *IEEE Transactions on Geoscience and Remote Sensing*, **41**, 708–718.

Ulaby, F.T. and Elachi, C., 1990, *Radar Polarimetry for Geoscience Applications*. Dedham, MA: Artech House.

Ulaby, F.T., Moore, R.K. and Fung, A.K., 1981–1986, *Microwave Remote Sensing: Active and Passive*. 3 volumes. Dedham, MA: Artech House.

Ulander, L.M.H., Dammert, P.B.G. and Hagberg, J.O., 1995, Measuring tree height using ERS-1 SAR interferometry. *International Geoscience and Remote Sensing Symposium* (IGARSS'95). Florence, Italy. New York: IEEE, 544–546.

Unwin, D.J. and Mather, P.M., 1998, Selecting and using ground control points in image rectification and registration. *Geographical Systems*, **5**, 239–260.

Unwin, D.J. and Wrigley, N., 1987, Towards a general theory of control point distribution effects in trend surface models. *Computers and Geosciences*, **13**, 351-355.

US Geological Survey (U.S.G.S.), 2003, *Earth Observing-1 Extended Mission Fact Sheet 032-03 (March 2003)*, URL: http://erg.usgs.gov/isb/pubs/factsheets/fs03203.html. Date accessed: 22 May 2003.

Ustin, S.L., Hart, Q.J., Duan, L. and Scheer, G, 1996, Vegetation mapping on hardwood rangelands in California. *International Journal of Remote Sensing*, **17**, 3015–3036.

Ustin, S.L., Smith, M.O. and Adams, J.B., 1993, Remote sensing of ecological processes – a strategy for developing and testing ecological models through spectral mixture analysis. In: Ehleringer, J.R. and Field, C.B. (eds.) (1993), 339–357.

Ustin, S.L., Smith, M.O., Jacquemoud, S., Verstraete, M. and Govaerts, Y., 1999, Geobotany: Vegetation mapping for earth sciences. In: Rencz, A.N. (ed.) (1999), 189–248.

van der Meer, F., 1994, Extraction of mineral absorption features from high spectral resolution data using nonparametric geostatistical techniques. *International Journal of Remote Sensing*, **15**, 2193–2214.

van der Meer, F., 1996a, Classification of remotely-sensed imagery using an indicator kriging approach: application to the problem of calcite-dolomite mineral mapping. *International Journal of Remote Sensing*, **17**, 1233–1249.

van der Meer, F., 1996b, Performance characteristics of the indicator classifier on simulated data. *International Journal of Remote Sensing*, **17**, 621–627.

van der Meer, F., 1996c, Metamorphic facies zonation in the Ronda peridotites: spectroscopic results from field and GER imaging spectrometer data. *International Journal of Remote Sensing*, **17**, 1633–1657.

van der Meer, F., van Dijk, P.M. and Westerhof, A.B., 1995, Digital classification of the contact metamorphic aureole along the Los Pedroches batholith, south-central Spain, using Landsat Thematic Mapper data. *International Journal of Remote Sensing*, **16**, 1043–1062.

van der Meer, F.D., 2000, Imaging spectrometry for geological applications. In: Meyers, R.A. (ed.), *Encyclopedia of Analytical Chemistry*. Chichester: John Wiley and Sons, 8601–8638.

van der Meer, F.D. and de Jong, S.M. (eds.), 2001, *Imaging Spectrometry: Basic Principles and Applications*. Dordrecht, The Netherlands.: Kluwer Academic Publishers.

van Gardingen, P.R., Foody, G.M. and Curran, P.J. (eds.), 1997, *Scaling-Up*. Cambridge: Cambridge University Press.

van Wie, P. and Stein, M., 1977, A Landsat digital image rectification system. *IEEE Transactions on Geoscience Electronics*, **15**, 130–137.

Vanderbrugg, G.J., 1976, Line detection in satellite imagery. *IEEE Transactions on Geoscience Electronics*, **14**, 37–44.

Vane, G., (ed.) 1987, *Airborne Visible/Infrared Imaging Spectrometer (AVIRIS): A Description of the Sensor, Ground Data Processing Facility, Laboratory Calibration and First Results*. Pasadena, CA: NASA Jet Propulsion Laboratory. JPL Publication, 87–38

Vane, G., Green, R.O., Chrien, T.G., Enmark, H.T., Hansen, E.G. and Porter, W.M., 1993, The airborne visible/infrared imaging spectrometer (AVIRIS). *Remote Sensing of Environment*, **44**, 127–143.

Varjo, J., 1996, Controlling continuously updated forest data by satellite remote sensing. *International Journal of Remote Sensing*, **17**, 43–67.

Vermote, E. and Kaufman, Y.J., 1995, Absolute calibration of AVHRR visible and near-infrared channels using ocean and cloud views. *International Journal of Remote Sensing*, **16**, 2317–2340.

Vermote, E.F., Tanré, D., Deuze, J.L., Herman, M. and Morcrett, J.J., 1997, Second simulation of the satellite signal in the solar spectrum, 6*S*: an overview. *IEEE Transactions on Geoscience and Remote Sensing*, **35**, 675–686.

Verstraete, M.M. and Pinty, B., 1992, Extracting surface properties from satellite data in the visible and near-infrared wavelengths. In: Mather, P.M. (ed.), *TERRA-1: Understanding the Terrestrial Environment – The Role of Earth Observations from Space*. London: Taylor and Francis, 203–209.

Verstraete, M.M., Pinty, B. and Myeni, R.B., 1996, Potential and limitations of information extraction on the terrestrial biosphere from satellite remote sensing. *Remote Sensing of Environment*, **58**, 201–214.

Vidal-Pantaleone, A., Martí, D. and Ferraldo, M., 2000, An adaptive soft thresholding method for speckle noise reduction in Synthetic Aperture Radar images. In: Casanova, J.L. (ed.), *Remote Sensing in the 21st Century: Economic and Environmental Applications*. Rotterdam: A.A. Balkema, 267–274.

Vieira, C.A.O. and Mather, P.M., 2000, Visualisation of measures of classification reliability and error in remote sensing. In: Heuvelink, G.B.M. and Lemmens, M.J.P.M. (eds.) *Proceedings of the Fourth International Symposium on Spatial Accuracy Assessment in Natural Resources and Environmental Science*. Delft: Delft University Press, 701–708.

Vogelmann, J.E., Helder, D., Morfitt, R., Choate, M.J., Merchant, J.W. and Bulley, H., 2001, Effects of Landsat 5 Thematic Mapper and Landsat 7 Enhanced Thematic

Mapper Plus radiometric and geometric calibrations and corrections on landscape characterisation. *Remote Sensing of Environment*, **78**, 55–70.

Wakabayeshi, H. and Arai, K., 1996, A new method for SAR speckle noise reduction (CST filter). *Canadian Journal of Remote Sensing*, **22**, 190–197.

Walker, J.S., 1999, *A Primer on Wavelets and their Scientific Applications*. Studies in Advanced Mathematics. Boca Raton, FL: Chapman and Hall/CRC Press.

Wang, F., 1990, Fuzzy supervised classification of remote sensing images. *IEEE Transactions on Geoscience and Remote Sensing*, **28**, 194–201.

Wang, J., 1993, LINDA – A system for automated linear feature detection. *Canadian Journal of Remote Sensing*, **19**, 9–21.

Wang, R-Y., 1986a, An approach to tree-classifier design based on hierarchical clustering. *International Journal of Remote Sensing*, **7**, 75–88.

Wang, R-Y., 1986b, An approach to tree-classifier design based on a splitting algorithm. *International Journal of Remote Sensing*, **7**, 89–104.

Wardley, N.W., 1984, Vegetation index variability as a function of viewing geometry. *International Journal of Remote Sensing*, **5**, 861–870.

Warrender, C.E. and Augusteijn, M.F., 1999, Fusion of image classifications using Bayesian techniques with Markov random fields. *International Journal of Remote Sensing*, **20**, 1987–2002.

Wasserman, G., 1978, *Color Vision: An Historical Introduction*. New York: Wiley.

Wecksung, G.W. and Breedlove, J.R., Jr., 1977, Some techniques for digital preprocessing, display and interpretation of ratio images in multispectral remote sensing. *Proceedings of the Society of Photo-Optical Instrumentation Engineers*, **119**, 47–54.

Wegener, M., 1990, Destriping multiple sensor imagery by improved histogram matching. *International Journal of Remote Sensing*, **11**, 859–975.

Wegmüller, U. and Werner, C.L., 1997, Retrieval of vegetation parameters with SAR interferometry. *IEEE Transactions on Geoscience and Remote Sensing*, **35**, 18–24.

Wegmüller, U., Werner, C.L., Nüesch, D. and Borgeaud, M., 1995, Forest mapping using ERS repeat-pass SAR interferometry. *ESA Earth Observation Quarterly*, **49**, 4–7.

Wehr, A. and Lohr, U., 1999, Airborne laser scanning – an introduction and overview. *ISPRS Journal of Photogrammetry and Remote Sensing*, **54**, 68–82.

Weinand, H.C., 1974, Cosine theta in components analysis. *Annals of the Association of American Geographers*, **64**, 353.

Welch, R., Jordan, T., Lang, H. and Murakami, H., 1998, ASTER as a source of topographic data for the late 1990's. *IEEE Transactions on Geoscience and Remote Sensing*, **36**, 1282–1289.

Welch, R., Jordan, T.R. and Ehlers, N., 1985. Comparative evaluations of the geodetic accuracy and cartographic potential of Landsat-4 and Landsat-5 Thematic Mapper image data. *Photogrammetric Engineering and Remote Sensing*, **51**, 1249–1262.

Westin, T., 1990, precision rectification of SPOT imagery. *Photogrammetric Engineering and Remote Sensing*, **56**, 247–253.

Wezka, J.S., Dyer, C.R. and Rosenfield, A., 1976, A comparative study of texture measures for terrain classification. *IEEE Transactions on Systems, Man and Cybernetics*, **6**, 269–285.

Wharton, S.W., 1980, A contextual classification method for recognising land use patterns in high-resolution remotely-sensed data. *Pattern Recognition*, **15**, 317–324.

White, K., 1993, Image processing of thematic mapper data for discriminating piedmont surficial materials in the Tunisian Southern Atlas. *International Journal of Remote Sensing*, **14**, 961–977.

White, R.G., 1994, Cross-section estimation by simulated annealing. *Proceedings of the International Geoscience and Remote Sensing Symposium (IGARSS'94), 9–12 August 1994, Pasadena, California*. New York: IEEE Press, 2188–2190.

Wilkinson, J.J. and Mégier, J., 1990, Evidential reasoning in a pixel classification hierarchy – a potential method for integrating image classifiers and expert system rules based on geographic context. *International Journal of Remote Sensing*, **11**, 1963–1968.

Williams, C.S. and Becklund, O.A., 1972, *Optics: A Short Course for Scientists and Engineers*. New York: Wiley-Interscience.

Williams, J., 1995, *Geographic Information from Space*. Chichester: Wiley/Praxis.

Wilson, J.D., 1992, A comparison of procedures for classifying remotely-sensed data using simulated data sets incorporating autocorrelations between spectral responses. *International Journal of Remote Sensing*, **13**, 2701–2725.

Wolberg, G., 1990, *Digital Image Warping*. Los Alamitos, California: IEEE Computer Society Press.

Wolfe, P.R. and DeWitt, B.A., 2000, *Elements of Photogrammetry with Applications to GIS (third edition)*. New York: McGraw-Hill.

Wong, F., Orth, R. and Friedmann, D., 1981, The use of digital terrain models in the rectification of satellite-borne imagery. *Proceedings of the 15th International Symposium on Remote Sensing of Environment*, Ann Arbor, Michigan: Environmental Research Institute of Michigan (ERIM), 653–662.

Woodcock, C. and Harward, V.J., 1992, Nested-hierarchical scene models and image segmentation. *International Journal of Remote Sensing*, **16**, 3167–3187.

Woodcock, C.E. and Strahler, A.H., 1987, The factor of scale in remote sensing. *Remote Sensing of Environment*, **21**, 311–322.

Woodcock, C.E., Strahler, A.H. and Jupp, D.L.B., 1988a, The use of variograms in remote sensing. I: Scene models and simulated images. *Remote Sensing of Environment*, **25**, 323–348.

Woodcock, C.E., Strahler, A.H. and Jupp, D.L.B., 1988b, The use of variograms in remote sensing. II: Real digital images. *Remote Sensing of Environment*, **25**, 349–379.

Woodham, R.J., 1989, Determining intrinsic surface reflectance in rugged terrain and changing illumination. *Proceedings of the International Geoscience and Remote Sensing Symposium (IGARSS'89), 10–14 July 1989, Vancouver, British Columbia, Canada*. New York: IEEE Press, volume 1, 1–5.

Wooding, M.G., Zmuda, A.D. and Griffiths, G.H., 1993, Crop discrimination using multi-temporal ERS-1 SAR data. *Proceedings of the Second ERS-1 Symposium on Space at the Service of our Environment, Hamburg, Germany*. ESA SP-361. Paris: European Space Agency, 51–56.

Wu, H.-H.P and Schowengerdt, R.A., 1993, Improved fraction image estimation using image restoration. *IEEE Transactions on Geoscience and Remote Sensing*, **31**, 771–778.

Xie, H., Pierce, L.E. and Ulaby, F.T., 2002, SAR speckle reduction using wavelet denoising and Markov Random Fields. *IEEE Transactions on Geoscience and Remote Sensing*, **40**, 2196–2212.

Xu, H., Dvorkin, J. and Nur, A., 2001, Linking oil subsidence to surface subsidence from satellite SAR interferometry. *Geophysical Research Letters*, **28**, 1307–1310.

Yamaguchi, Y., Kahle, A., Tsu, H., Kawakami, T. and Pniel, M., 1998, An overview of Advanced Spaceborne Thermal Emission and Reflection Radiometer (ASTER). *IEEE Transactions on Geoscience and Remote Sensing*, **36**, 1062–1071.

Yang, H., van der Meer, F.D., Bakker, W.H. and Tan, Z.J., 1999, A back-propagation neural network for mineralogical mapping from AVIRIS data. *International Journal of Remote Sensing*, **20**, 97–110.

Yocky, D.A., 1996, Multiresolution wavelet decomposition image merger of Landsat Thematic Mapper and SPOT panchromatic data. *Photogrammetric Engineering and Remote Sensing*, **62**, 1067–1084.

Yool, S.R., Star, J.L., Estes, J.E., Botkin, D.B., Eckhardt, D.W. and Davis, F.W., 1986, Performance analysis of image processing algorithms for classification of natural vegetation in the mountains of Southern California. *International Journal of Remote Sensing*, **7**, 683–702.

Young, T.L. and Kaufman, Y.J., 1986, Non-Lambertian effects on remote sensing of surface reflectance and vegetation index. *IEEE Transactions on Geoscience and Remote Sensing*, **GE-24**, 699–707.

Yuan, D. and Elvidge, C.D., 1996, Comparison of radiometric normalization techniques. *ISPRS Journal of Photogrammetry and Remote Sensing*, **51**, 117–126.

Zebker, H.A. and Goldstein, R.M., 1986. Topographic mapping from interferometric synthetic aperture radar observations. *Journal of Geophysical Research*, **91**, 4993–4999.

Zebker, H.A., Werner, C.L., Rosen, P.A. and Hensley, S., 1994. Accuracy of topographic maps derived from ERS-1 interferometric radar. *IEEE Transactions on Geoscience and Remote Sensing*, **32**, 823–836.

Zervakis, M.E., Sundararajan, V. and Parhi, K.K., 2001, Vector processing of wavelet coefficients for robust image denoising. *Image and Vision Computing*, **19**, 435–450.

Zhang, Y., 1999, A new merging method and its spectral an spatial effects. *International Journal of Remote Sensing*, **20**, 2003–2014.

Zhou, J. and Civco, D.L., 1996, Using genetic learning neural networks for spatial decision making in GIS. *Photogrammetric Engineering and Remote Sensing*, **62**, 1287–1295.

Zhou, J., Civco, D.L. and Silander, J.A., 1999, A wavelet transform method to merge Landsat TM and SPOT data. *International Journal of Remote Sensing*, **19**, 743–757.

Zwally, H.J., Schutz, B., Abdalati, W., Abshire, J., Bentley, C., Brenner, A., Bufton, J., Dezio, J., Hancock, D., Harding, D., Herring, T., Minster, B., Quinn, K., Palm, S., Spinhirne, J. and Thomas, R., 2002, ICESat's laser measurements of polar ice, atmosphere, ocean, and land. *Journal of Geodynamics*, **34**, 405–445.

Index

Computer Processing of Remotely-Sensed Images: An Introduction, Third Edition. Paul M. Mather.
 ISBNs: 0-470-84918-5 (HB); 0-470-84919-3 (PB)